Simulation-Driven Design Optimization and Modeling for Microwave Engineering

Simulation-Driven Design Optimization and Modeling for Microwave Engineering

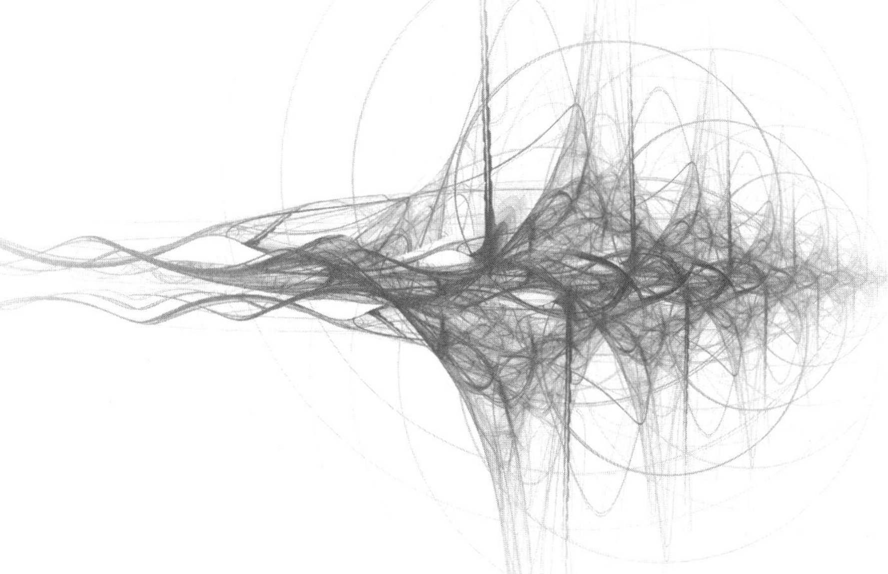

Edited by

Slawomir Koziel
Reykjavik University, Iceland

Xin-She Yang
Middlesex University, UK

Qi-Jun Zhang
Carleton University, Canada

Imperial College Press

Published by

Imperial College Press
57 Shelton Street
Covent Garden
London WC2H 9HE

Distributed by

World Scientific Publishing Co. Pte. Ltd.

5 Toh Tuck Link, Singapore 596224

USA office: 27 Warren Street, Suite 401-402, Hackensack, NJ 07601

UK office: 57 Shelton Street, Covent Garden, London WC2H 9HE

British Library Cataloguing-in-Publication Data
A catalogue record for this book is available from the British Library.

**SIMULATION-DRIVEN DESIGN OPTIMIZATION AND MODELING
FOR MICROWAVE ENGINEERING**

ISBN 978-1-84816-916-6

Typeset by Stallion Press
Email: enquiries@stallionpress.com

Printed in Singapore by World Scientific Printers.

List of Contributors

Editors

Slawomir Koziel

Engineering Optimization & Modeling Center, School of Science and Engineering, Reykjavik University, Menntavegur 1, 101 Reykjavik, Iceland. (koziel@ru.is)

Xin-She Yang

School of Science and Technology, Middlesex University, Hendon Campus, London NW4 4BT, United Kingdom. (x.yang@mdx.ac.uk)

Qi-Jun Zhang

Department of Electronics, Carleton University, 1125 Colonel By Drive, Ottawa, Canada. (qjz@doe.carleton.ca)

Contributors

Osman S. Ahmed

Department of Electrical and Computer Engineering, McMaster University, Hamilton, Ontario L8S 4K1, Canada. (mohammos@mcmaster.ca)

Mohamed H. Bakr

Department of Electrical and Computer Engineering, McMaster University, Hamilton, Ontario L8S 4K1, Canada. (mbakr@mail.ece.mcmaster.ca)

John W. Bandler

Simulation Optimization Systems Research Laboratory, Department of Electrical and Computer Engineering, McMaster University, L8S 4K1, Hamilton, ON, Canada. (bandler@mcmaster.ca)

Vicente Enrique Bôria Esbert

Instituto de Telecomunicaciones y Aplicaciones Multimedia Universitat Politècnica de València, Valencia, Spain. (vboria@dcom.upv.es)

Elena Díaz Caballero
Instituto de Telecomunicaciones y Aplicaciones Multimedia Universitat Politècnica de València, Valencia, Spain. (eldiaca@iteam.upv.es)

Yazi Cao
Department of Electronics, Carleton University, Ottawa, Ontario, Canada. (yacao@doe.carleton.ca)

Krishnan Chemmangat
Department of Information Technology, Ghent University — IBBT, Gaston Crommenlaan 8, Bus 201, B-9050, Gent, Belgium. (krishnan.cmc@intec. ugent.be)

Qingsha S. Cheng
Simulation Optimization Systems Research Laboratory, Department of Electrical and Computer Engineering, McMaster University, L8S 4K1, Hamilton, ON, Canada. (chengq@mcmaster.ca)

Tom Dhaene
Department of Information Technology, Ghent University — IBBT, Gaston Crommenlaan 8, Bus 201, B-9050, Gent, Belgium. (tom.dhaene@intec. ugent.be)

Francesco Ferranti
Department of Information Technology, Ghent University — IBBT, Gaston Crommenlaan 8, Bus 201, B-9050, Gent, Belgium. (francesco.ferranti@intec. ugent.be)

Venu-Madhav-Reddy Gongal-Reddy
Department of Electronics, Carleton University, Ottawa, Ontario, Canada. (vmgongal@doe.carleton.ca)

Héctor Esteban González
Instituto de Telecomunicaciones y Aplicaciones Multimedia Universitat Politècnica de València, Valencia, Spain. (hesteban@dcom.upv.es)

Humayun Kabir
COMDEV Ltd, Cambridge, Ontario, Canada. (humayun.kabir@comdev.ca)

Luc Knockaert
Department of Information Technology, Ghent University — IBBT, Gaston Crommenlaan 8, Bus 201, B-9050, Gent, Belgium. (luc.knockaert@intec.ugent.be)

Slawomir Koziel
Engineering Optimization & Modeling Center, School of Science and Engineering, Reykjavik University, Menntavegur 1, 101 Reykjavik, Iceland. (koziel@ru.is)

Leifur Leifsson
Engineering Optimization & Modeling Center, School of Science and Engineering, Reykjavik University, Menntavegur 1, 101 Reykjavik, Iceland. (leifurth@ru.is)

Tian-Hong Loh
National Physical Laboratory, Hampton Road, Teddington, Middlesex TW11 0LW, United Kingdom. (tian.loh@npl.co.uk)

Carmen Bachiller Martín
Instituto de Telecomunicaciones y Aplicaciones Multimedia Universitat Politècnica de València, Valencia, Spain. (mabacmar@dcom.upv.es)

Ángel Belenguer Martínez
Escuela Universitaria Politécnica de Cuenca Universidad de Castilla-La Mancha, Cuenca, Spain. (angel.belenguer@uclm.es)

Christos Mias
School of Engineering, University of Warwick, Gibber Hill Road, Coventry CV4 7AL, United Kingdom. (christos.mias@warwick.ac.uk)

José Vicente Morro Ros
Instituto de Telecomunicaciones y Aplicaciones Multimedia Universitat Politècnica de València, Valencia, Spain. (jomorros@dcom.upv.es)

Stanislav Ogurtsov
Engineering Optimization & Modeling Center, School of Science and Engineering, Reykjavik University, Menntavegur 1, 101 Reykjavik, Iceland. (stanislav@ru.is)

José E. Rayas-Sánchez
Research Group on Computer-Aided Engineering of Circuits and Systems (CAECAS), Department of Electronics, Systems and Informatics, Instituto Tecnológico y de Estudios Superiores de Occidente (ITESO – The Jesuit University of Guadalajara), Guadalajara, Mexico, 45604. (erayas@iteso.mx)

Ying Wang
Faculty of Engineering and Applied Sciences, University of Ontario Institute of Technology, Oshawa, Ontario, Canada. (ying.wang@uoit.ca)

Xin-She Yang
School of Science and Technology, Middlesex University, Hendon Campus, London NW4 4BT, United Kingdom. (x.yang@mdx.ac.uk)

Ming Yu
COMDEV Ltd, Cambridge, Ontario, Canada. (ming.yu@comdev.ca)

Chuan Zhang
School of Electronic Information Engineering, Tianjin University, Tianjin, China. (czhang-ann@sina.com)

Lei Zhang
RF Division, Freescale Semiconductor Inc., Tempe, Arizona, USA. (lei.zhang@freescale.com)

Qi-Jun Zhang
School of Electronic Information Engineering, Tianjin University, Tianjin, China Department of Electronics, Carleton University, Ottawa, Ontario, Canada. (qjz@doe.carleton.ca)

Preface

Computer-aided full-wave electromagnetic (EM) analysis has been used in microwave engineering for a few decades. Initially, its main application area was design verification. Today, EM-simulation-driven optimization and design closure become increasingly important due to the complexity of microwave structures and increasing demands for accuracy. In many situations, theoretical models of microwave structures can only be used to yield the initial designs that need to be further fine-tuned to meet given performance requirements. In addition, EM-based design is a must for a growing number of microwave devices such as ultrawideband (UWB) antennas, dielectric resonator antennas, and substrate-integrated circuits. For circuits like these, reliable and design-ready theoretical models are hardly available, so that design improvement can be only obtained through geometry adjustments based on repetitive, time-consuming simulations. On the other hand, various interactions between microwave devices and their environment such as feeding structures and housing must be often taken into account, which is only possible through full-wave EM analysis.

Electromagnetic simulations can be highly accurate, but, at the same time, they tend to be computationally expensive. Therefore, practical design optimization methods have to be computationally efficient, so that the number of CPU-intensive high-fidelity EM simulations is reduced as much as possible during the design process. For the same reasons, techniques for creating fast and yet accurate models of microwave structures become crucially important. In this edited book, we strive to review the state of the art of simulation-driven microwave design optimization and modeling. A group of international experts specializing in various aspects of microwave computer-aided design, will summarize and review a wide range of latest developments and real-world applications. Topics include conventional and surrogate-based design optimization techniques, methods exploiting adjoint sensitivity, simulation-based tuning, space mapping, as well as various modeling methodologies, among others, artificial neural networks and kriging. Applications and case studies include microwave filters, antennas, substrate integrated structures, as well as various active components and circuits. The book also contains a few introductory chapters highlighting fundamentals of optimization and modelling, among others, gradient-based and derivative-free algorithms, metaheuristics, and surrogate-based optimization techniques, as well as finite difference and finite-element methods.

Slawomir Koziel, Xin-She Yang, and Qi-Jun Zhang
March 2012

Acknowledgments

We would like to thank all the contributing authors for their contributions and their help during the peer-review process.

The Editors would like to thank the editors: Dr Kellye Curtis, Jackie Downs, Tasha D'Cruz, and staff at World Scientific and Imperial College Press for their help and professionalism.

SK, XSY, and QJZ

Contents

3. Surrogate-Based Optimization   41

Slawomir Koziel, Leifur Leifsson, and Xin-She Yang

Chapter 1

Introduction to Optimization
and Gradient-Based Methods

Xin-She Yang and Slawomir Koziel

Optimization is everywhere, and this is especially true for simulation-driven optimization and modeling in microwave engineering. Optimization is an important paradigm itself with a wide range of applications. In almost all applications in engineering and industry, we are always trying to optimize something — whether to minimize the cost and energy consumption, or to maximize the profit, output, performance, and efficiency. In reality, resources, time, and money are always limited; consequently, optimization is far more important in practice (Gill *et al.*, 1981; Arora, 1989; Yang, 2010; Yang and Koziel, 2010). The optimal use of available resources of any sort requires a paradigm shift in scientific thinking; this is because most real-world applications have far more complicated factors and parameters which affect how the system behaves.

1.1. Introduction

Contemporary engineering design is heavily based on computer simulations. This introduces additional difficulties to optimization. Growing demand for accuracy and ever-increasing complexity of structures and systems results in the simulation process being more and more time-consuming. In many engineering fields, the evaluation of a single design can take as long as several days or even weeks. On the other hand, simulation-based objective functions are inherently noisy, which makes the optimization process even more difficult. Still, simulation-driven design becomes a must for a growing number of areas, which creates a need for robust and efficient optimization

1

methodologies that can yield satisfactory designs even given the presence of analytically intractable objectives and limited computational resources.

For any optimization problem, the integrated components of the optimization process are optimization algorithm, an efficient numerical simulator, and a realistic representation of the physical processes we wish to model and optimize. This is often a time-consuming process, and in many cases, the computational costs are usually very high. Once we have a good model, the overall computation costs are determined by the optimization algorithms used for search and the numerical solver used for simulation.

Search algorithms are the tools and techniques of achieving optimality of the problem of interest. This search for optimality is complicated further by the fact that uncertainty is almost always presents in real-world systems. Therefore, we seek not only the optimal design but also robust design in engineering and industry. Optimal solutions, which are not robust enough, are not practical in reality. Suboptimal solutions or good robust solutions are often the choice in such cases.

Simulations are often the most time-consuming part. In many applications, an optimization process often involves the evaluation of objective function many times, often thousands and even millions of configurations. Such evaluations often involve the use of extensive computational tools such as a computational fluid dynamics simulator or a finite-element solver. This is the step that is most time-consuming, often taking 50% to 90% of the overall computing time. Therefore, we have to balance the accuracy (high-fidelity) and allowable computational time. In many cases, some fast approximations (often with reduced fidelity) can be used for most parts of the search iterations and the high-fidelity model used for double-checking the good designs. This combination of approximations and variable fidelity helps to reduce the overall computational time significantly.

Optimization problems can be formulated in many ways. For example, the commonly used method of least-squares is a special case of maximum-likelihood formulations. By far the most widely used formulation is to write a nonlinear optimization problem as

$$\text{minimize } f_i(x), \quad (i = 1, 2, \ldots, M), \tag{1.1}$$

subject to the constraints

$$h_j(x), \qquad (j = 1, 2, \ldots, J), \tag{1.2}$$

$$g_k(x) \le 0, \quad (k = 1, 2, \ldots, K), \tag{1.3}$$

where f_i, h_j and g_k are in general nonlinear functions. Here the design vector $\boldsymbol{x} = (x_1, x_2, \ldots, x_n)$ can be continuous, discrete, or mixed in n-dimensional space. The functions f_i are called objective or cost functions, and when $M > 1$, the optimization is multiobjective or multicriteria (Sawaragi *et al.*, 1985). It is possible to combine different objectives into a single objective, and we will focus on the single-objective optimization problems in most parts of this book. It is worth pointing out here that we write the problem as a minimization problem, but it can also be written as a maximization by simply replacing $f_i(\boldsymbol{x})$ by $-f_i(\boldsymbol{x})$.

When all functions are nonlinear, we are dealing with nonlinear constrained problems. In some special cases when f_i, h_j, g_k are linear, the problem becomes linear, and we can use the widely linear programming techniques such as the simplex method. When some design variables can only take discrete values (often integers), while other variables are real continuous, the problem is of mixed type, which is often difficult to solve, especially for large-scale optimization problems.

A very special class of optimization is convex optimization (Boyd and Vandenberghe, 2004), which has guaranteed global optimality. Any optimal solution is also the global optimum, and most importantly, there are efficient algorithms of polynomial time to solve such problems (Conn *et al.*, 2000). These efficient algorithms such as the interior-point methods (Karmarkar, 1984) are widely used and have been implemented in many software packages.

1.2. Main Challenges in Optimization

There are three main issues in simulation-driven optimization and modeling, and they are: the efficiency of an algorithm, and the efficiency and accuracy of a numerical simulator, and assigning the right algorithms to the right problem. Despite their importance, there is no satisfactory rule or guidelines to handle such issues. Obviously, we try to use the most efficient algorithms available, but the actual efficiency of an algorithm may depend on many factors such as the inner working of an algorithm, the information need (such as objective functions and their derivatives), and implementation details. The efficiency of a solver is even more complicated, depending on the actual numerical methods used and the complexity of the problem of interest.

In microwave engineering, however, it is reasonable to assume that electromagnetic (EM) simulation of sufficient fidelity (in particular, with a structure of interest discretized with sufficient density) gives accuracy that adequately represents the actual performance of the corresponding physical device (filter, antenna, etc.). To reach such accuracy, it is necessary, in many cases, to also include in the simulation the environment of the device such as housing, connectors, and others.

As for choosing the right algorithms for the right problems, there are many empirical observations, but there are no agreed guidelines. In fact, there are no universally efficient algorithms for all types of problems. Therefore, the choice may depend on many factors and is sometimes subjective to the personal preferences of researchers and decision-makers.

1.2.1. *Efficiency of an Algorithm*

An efficient optimizer is very important to ensure the optimal solutions are reachable. The essence of an optimizer is a search or optimization algorithm implemented correctly so as to carry out the desired search (though not necessarily efficiently). It can be integrated and linked with other modeling components. There are many optimization algorithms in the literature and no single algorithm is suitable for all problems, as dictated by the no free lunch theorems (Wolpert and Macready, 1997).

Optimization algorithms can be classified in many ways, depending on the focus or the characteristics we are trying to compare. Algorithms can be classified as gradient-based (or derivative-based methods) and gradient-free (or derivative-free methods). The classic method of steepest-descent and Gauss–Newton methods are gradient-based, as they use the derivative information in the algorithm, while the Nelder–Mead downhill simplex method (Nelder and Mead, 1965) is a derivative-free method because it only uses the values of the objective, not any derivatives.

Algorithms can also be classified as deterministic or stochastic. If an algorithm works in a mechanically deterministic manner without any random nature, it is called deterministic. For such an algorithm, it will reach the same final solution if we start with the same initial point. Hill-climbing and downhill simplex are good examples of deterministic algorithms. On the other hand, if there is some randomness in the algorithm, the algorithm will usually reach a different point every time we run the algorithm, even though we start with the same initial point. Genetic algorithms and hill-climbing with a random restart are good examples of stochastic algorithms.

Analyzing the stochastic algorithms in more detail, we can single out the type of randomness that a particular algorithm is employing.

For example, the simplest and yet often very efficient method is to introduce a random starting point for a deterministic algorithm. The well-known hill-climbing with random restart is a good example. This simple strategy is both efficient in most cases and easy to implement in practice. A more elaborate way to introduce randomness to an algorithm is to use randomness inside different components of an algorithm, and in this case, we often call such an algorithm heuristic or more often metaheuristic (Yang, 2008b; Talbi, 2009; Yang, 2010). A very good example is the popular genetic algorithms which use randomness for crossover and mutation in terms of a crossover probability and a mutation rate. Here, heuristic means to search by trial and error, while metaheuristic is a higher level of heuristics. However, modern literature tends to refer all new stochastic algorithms as metaheuristic. In this book, we will use metaheuristic to mean either. It is worth pointing out that metaheuristic algorithms form a hot research topic and new algorithms appear almost yearly (Yang, 2008b; Yang, 2010).

From the mobility point of view, algorithms can be classified as local or global. Local search algorithms typically converge towards a local optimum, not necessarily (often not) the global optimum, and such algorithms are often deterministic and have no ability of escaping local optima. Simple hill-climbing is an example. On the other hand, we always try to find the global optimum for a given problem, and if this global optimality is robust, it is often the best, though it is not always possible to find such global optimality. For global optimization, local search algorithms are not suitable. We have to use a global search algorithm. Modern metaheuristic algorithms in most cases are intended for global optimization, though not always successful or efficiently. A simple strategy such as hill-climbing with random restart may change a local search algorithm into a global search. In essence, randomization is an efficient component for global search algorithms. A detailed review of derivative-free methods including metaheuristics will be provided by Yang and Koziel in Chapter 2.

Straightforward optimization of a given objective function is not always practical. Particularly, if the objective function comes from a computer simulation, it may be computationally expensive, noisy, or non-differentiable. In such cases, so-called surrogate-based optimization algorithms may be useful where the direct optimization of the function of interest is replaced by iterative updating and re-optimization of its model surrogate (Forrester and Keane, 2009). The surrogate model is typically constructed from the sampled data of the original objective function, however, it is supposed to be cheap, smooth, easy to optimize, and yet reasonably accurate so that it can produce a good prediction of the function's optimum. Multi-fidelity or variable-fidelity optimization is a special case of the surrogate-based optimization where the surrogate is constructed from the low-fidelity model (or models) of the system of interest (Koziel *et al.*, 2008). Using variable-fidelity optimization is particularly useful if the reduction of the computational cost of the optimization process is of primary importance.

Whatever the classification of an algorithm is, we have to make the right choice to use an algorithm correctly and sometimes a proper combination of algorithms may achieve better results.

1.2.2. *The Right Algorithms?*

From the optimization point of view, the choice of the right optimizer or algorithm for a given problem is crucially important. The algorithm chosen for an optimization task will largely depend on the type of the problem, the nature of an algorithm, the desired quality of solutions, the available computing resource, time limit, availability of the algorithm implementation, and the expertise of the decision-makers (Yang, 2010).

The nature of an algorithm often determines if it is suitable for a particular type of problem. For example, gradient-based algorithms such as hill-climbing are not suitable for an optimization problem whose objective is discontinuous. Conversely, the type of problem we are trying to solve also determines the algorithms we possibly choose. If the objective function of an optimization problem at hand is highly nonlinear and multimodal, classic algorithms such as hill-climbing and downhill simplex are not suitable, as they are local search algorithms. In this case, global optimizers such as particle swarm optimization and cuckoo search are most suitable (Yang, 2010; Yang and Deb, 2010).

Obviously, the choice is also affected by the desired solution quality and computing resources. As in most applications, where computing resources are limited, we have to obtain good solutions (not necessary the best) in a reasonable and practical time. Therefore, we have to balance the resource and solution quality. We cannot achieve solutions with guaranteed quality, though we strive to obtain the quality solutions as best as we possibly can. If time is the main constraint, we can use some greedy methods, or hill-climbing with a few random restarts.

Sometimes, even with the best possible intention, the availability of an algorithm and the expertise of the decision-makers are the ultimate defining factors for choosing an algorithm. Even if some algorithms are better, we may not have that algorithm implemented in our system or we may not have such access to it, which limits our choice. For example, Newton's method, hill-climbing, Nelder–Mead downhill simplex, trust-region methods (Conn *et al.*, 2000), and interior-point methods (Nesterov and Nemirovskii, 1994) are implemented in many software packages, which may also increase their popularity in applications. In practice, even with the best possible algorithms and well-crafted implementation, we may still not get the desired solutions. This is the nature of nonlinear global optimization, as most of such problems are non-deterministic polynomial-time hard, or NP-hard, and no efficient algorithms (in the polynomial sense) exist for a given problem. Thus the challenge of research in computational optimization and applications is to find the algorithms most suitable for a given problem so as to obtain good solutions, which are hopefully also the best global solutions, in a reasonable timescale with a limited amount of resources. We aim to do it efficiently in an optimal way.

1.2.3. *Efficiency of a Numerical Solver*

To solve an optimization problem, the most computationally extensive part is probably the evaluation of the design objective to see if a proposed solution is feasible and/or if it is optimal. Typically, we have to carry out these evaluations many times, often thousands or millions of times (Yang, 2008a; Yang, 2010). Things become even more challenging computationally, when each evaluation task takes a long time via some black-box simulators. If this simulator is a finite-element or CFD solver, the running time of each evaluation can take from a few minutes to a few hours or even weeks. For example, microwave structures and devices are typically evaluated using electromagnetic (EM) solvers and typical simulation times range from a few minutes to many hours, depending on the system complexity as well as the required accuracy. In extreme cases (e.g., evaluation of antenna placed on a vehicle or a ship), simulation may take several days or even weeks. Therefore, any approach to save computational time either by reducing the number of evaluations or by increasing the simulator's efficiency will save time and money.

In general, a simulator can be simple function subroutines, a multi-physics solver, or external black-box evaluators.

The main way to reduce the number of objective evaluations is to use an efficient algorithm, so that only a small number of such evaluations are needed. In most cases, this is not possible. We have to use some approximation techniques to estimate the objectives, or to construct an approximation model to predict the solver's outputs without actually using the solver. Another approach is to replace the original objective function by its lower-fidelity model, e.g., obtained from a computer simulation based on the coarsely discretized structure of interest. The low-fidelity model is faster but not as accurate as the original one, and therefore it has to be corrected. Special techniques have to be applied to use an approximation or corrected low-fidelity model in the optimization process so that the optimal design can be obtained at a low computational cost. All of this falls into the category of surrogate-based optimization (Queipo *et al.*, 2005; Koziel *et al.*, 2008; Koziel *et al.*, 2009; Koziel and Yang, 2011).

Surrogate models are approximate techniques to construct response surface models, or metamodels (Simpson *et al.*, 2001). The main idea is to approximate or mimic the system behavior so as to carry out evaluations cheaply and efficiently, but still with accuracy comparable to the actual system. Widely used techniques include polynomial response surface or regression, radial basis functions, ordinary kriging, artificial neural networks, support vector machines, response correction, space mapping, and others. The data used to create the models comes from the sampling of the design space and evaluating the system at selected locations. Surrogate models can be used as predictive tools in the search for the optimal design of the system of interest. This can be realized by iterative re-optimization of the surrogate (exploitation), filling the gaps between sample points to improve glocal accuracy of the model (exploration of the design space), or a mixture of both. The new data is used to update the surrogate. A detailed review of surrogate-modeling techniques and surrogate-based optimization methods will be given by Koziel *et al.* in Chapter 3 of this book.

1.3. Gradient-Based Methods

1.3.1. *Newton's Method*

Newton's method is a root-finding algorithm, but it can be modified for solving optimization problems. This is because optimization is equivalent to finding the root of the first derivative $f'(x)$ based on the stationary conditions once the objective function $f(x)$ is given. For a continuously differentiable function $f(x)$, we have the Taylor expansion in terms of $\Delta x = x - x_n$ about a fixed point x_n:

$$f(x) = f(x_n) + (\nabla f(x_n))^T \Delta x + \frac{1}{2} \Delta x^T \nabla^2 f(x_n) \Delta x + \dots,$$

whose third term is a quadratic form. Hence $f(\boldsymbol{x})$ is minimized if $\Delta\boldsymbol{x}$ is the solution of the following linear equation:

$$\nabla f(\boldsymbol{x}_n) + \nabla^2 f(\boldsymbol{x}_n)\Delta\boldsymbol{x} = 0. \tag{1.4}$$

This leads to

$$\boldsymbol{x} = \boldsymbol{x}_n - \boldsymbol{H}^{-1}\nabla f(\boldsymbol{x}_n), \tag{1.5}$$

where $\boldsymbol{H}^{-1}(\boldsymbol{x}^{(n)})$ is the inverse of the Hessian matrix $\boldsymbol{H} = \nabla^2 f(\boldsymbol{x}_n)$, which is defined as

$$\boldsymbol{H}(\boldsymbol{x}) \equiv \nabla^2 f(\boldsymbol{x}) \equiv \begin{pmatrix} \frac{\partial^2 f}{\partial x_1^2} & \cdots & \frac{\partial^2 f}{\partial x_1 \partial x_n} \\ \vdots & & \vdots \\ \frac{\partial^2 f}{\partial x_n \partial x_1} & \cdots & \frac{\partial^2 f}{\partial x_n^2} \end{pmatrix}. \tag{1.6}$$

This matrix is symmetric due to the fact that

$$\frac{\partial^2 f}{\partial x_i \partial x_j} = \frac{\partial^2 f}{\partial x_j \partial x_i}. \tag{1.7}$$

If the iteration procedure starts from the initial vector $\boldsymbol{x}^{(0)}$, usually a guessed point in the feasible region, then Newton's formula for the nth iteration becomes

$$\boldsymbol{x}^{(n+1)} = \boldsymbol{x}^{(n)} - \boldsymbol{H}^{-1}(\boldsymbol{x}^{(n)})f(\boldsymbol{x}^{(n)}). \tag{1.8}$$

It is worth pointing out that if $f(\boldsymbol{x})$ is quadratic, then the solution can be found exactly in a single step. However, this method is not efficient for non-quadratic functions.

In order to speed up the convergence, we can use a smaller step size $\alpha \in (0, 1]$ and we have the modified Newton's method

$$\boldsymbol{x}^{(n+1)} = \boldsymbol{x}^{(n)} - \alpha\boldsymbol{H}^{-1}(\boldsymbol{x}^{(n)})f(\boldsymbol{x}^{(n)}). \tag{1.9}$$

Sometimes, it might be time-consuming to calculate the Hessian matrix for second derivatives. A good alternative is to use an identity matrix $\boldsymbol{I}$ to approximate $\boldsymbol{H}$ so that $\boldsymbol{H}^{-1} = \boldsymbol{I}$, and we have the quasi-Newton method

$$\boldsymbol{x}^{(n+1)} = \boldsymbol{x}^{(n)} - \alpha\boldsymbol{I}\nabla f(\boldsymbol{x}^{(n)}), \tag{1.10}$$

which is essentially the steepest-descent method.

1.3.2. Steepest-Descent Method

The essence of the steepest descent method is to find the lowest possible value of the objective function $f(\boldsymbol{x})$ from the current point $\boldsymbol{x}^{(n)}$. From the Taylor expansion of $f(\boldsymbol{x})$ about $\boldsymbol{x}^{(n)}$, we have

$$f(\boldsymbol{x}^{(n+1)}) = f(\boldsymbol{x}^{(n)} + \Delta \boldsymbol{s}) \approx f(\boldsymbol{x}^{(n)}) + (\nabla f(\boldsymbol{x}^{(n)}))^T \Delta \boldsymbol{s}, \qquad (1.11)$$

where $\Delta \boldsymbol{s} = \boldsymbol{x}^{(n+1)} - \boldsymbol{x}^{(n)}$ is the increment vector. Since we are trying to find a better approximation to the objective function, it requires that the second term on the right hand is negative. So

$$f(\boldsymbol{x}^{(n)} + \Delta \boldsymbol{s}) - f(\boldsymbol{x}^{(n)}) = (\nabla f)^T \Delta \boldsymbol{s} < 0. \qquad (1.12)$$

From vector analysis, we know that the inner product $\boldsymbol{u}^T \boldsymbol{v}$ of two vectors $\boldsymbol{u}$ and $\boldsymbol{v}$ is the largest when they are parallel but in opposite directions. Therefore, we have

$$\Delta \boldsymbol{s} = -\alpha \nabla f(\boldsymbol{x}^{(n)}), \qquad (1.13)$$

where $\alpha > 0$ is the step size. This is the case when the direction $\Delta \boldsymbol{s}$ is along the steepest descent in the negative gradient direction. In the case of finding maxima, this method is often referred to as hill-climbing.

The choice of the step size α is very important. A very small step size means slow movement towards the local minimum, while a large step may overshoot and subsequently makes it move far away from the local minimum. Therefore, the step size $\alpha = \alpha^{(n)}$ should be different at each iteration and should be chosen so that it minimizes the objective function $f(\boldsymbol{x}^{(n+1)}) = f(\boldsymbol{x}^{(n)}, \alpha^{(n)})$. Therefore, the steepest descent method can be written as

$$f(\boldsymbol{x}^{(n+1)}) = f(\boldsymbol{x}^{(n)}) - \alpha^{(n)} (\nabla f(\boldsymbol{x}^{(n)}))^T \nabla f(\boldsymbol{x}^{(n)}). \qquad (1.14)$$

In each iteration, the gradient and step size will be calculated. Again, a good initial guess of both the starting point and the step size is useful.

1.3.3. Line Search

In the steepest-descent method, there are two important parts: the descent direction and the step size (or how far to descent). The calculations of the exact step size may be very time-consuming. In reality, we intend to find the right descent direction. Then a reasonable amount of descent, not necessarily the exact amount, during each iteration will usually be sufficient. For this, we essentially use a line search method.

To find the local minimum of the objective function $f(\boldsymbol{x})$, we try to search along a descent direction $\boldsymbol{s}_k$ with an adjustable step size α_k so that

$$\psi(\alpha_k) = f(\boldsymbol{x}_k + \alpha_k \boldsymbol{s}_k), \tag{1.15}$$

decreases as much as possible, depending on the value of α_k. Loosely speaking, the reasonably right step size should satisfy the Wolfe's conditions:

$$f(\boldsymbol{x}_k + \alpha_k \boldsymbol{s}_k) \leq f(\boldsymbol{x}_k) + \gamma_1 \alpha_k \boldsymbol{s}_k^T \nabla f(\boldsymbol{x}_k), \tag{1.16}$$

and

$$\boldsymbol{s}_k^T \nabla f(\boldsymbol{x}_k + \alpha_k \boldsymbol{s}_k) \geq \gamma_2 \boldsymbol{s}_k^T \nabla f(\boldsymbol{x}_k), \tag{1.17}$$

where $0 < \gamma_1 < \gamma_2 < 1$ are algorithm-dependent parameters. The first condition is a sufficient decrease condition for α_k, often called the Armijo condition or rule, while the second inequality is often referred to as the curvature condition. For most functions, we can use $\gamma_1 = 10^{-4}$ to 10^{-2}, and $\gamma_2 = 0.1$ to 0.9. These conditions are usually sufficient to ensure the algorithm converge in most cases; however, stronger conditions may be needed for some tough functions.

1.3.4. *Conjugate Gradient Method*

The method of conjugate gradient belongs to a wider class of the so-called Krylov subspace iteration methods. The conjugate gradient method was pioneered by Magnus Hestenes, Eduard Stiefel, and Cornelius Lanczos in the 1950s. It was named as one of the top ten algorithms of the twentieth century.

The conjugate gradient method can be used to solve the following linear system

$$\boldsymbol{A}\boldsymbol{u} = \boldsymbol{b}, \tag{1.18}$$

where $\boldsymbol{A}$ is often a symmetric positive definite matrix. The above system is equivalent to minimizing the following function $f(\boldsymbol{u})$:

$$f(\boldsymbol{u}) = \frac{1}{2}\boldsymbol{u}^T \boldsymbol{A}\boldsymbol{u} - \boldsymbol{b}^T \boldsymbol{u} + \boldsymbol{v}, \tag{1.19}$$

where $\boldsymbol{v}$ is a vector constant and can be taken to be zero. We can easily see that $\nabla f(\boldsymbol{u}) = 0$ leads to $\boldsymbol{A}\boldsymbol{u} = \boldsymbol{b}$.

In general, the size of $\boldsymbol{A}$ can be very large and sparse with $n > 100{,}000$, but it is not required that $\boldsymbol{A}$ is strictly symmetric positive definite. In fact, the main condition is that $\boldsymbol{A}$ should be a normal matrix. A square matrix $\boldsymbol{A}$ is called normal if $\boldsymbol{A}^T\boldsymbol{A} = \boldsymbol{A}\boldsymbol{A}^T$. Therefore, a symmetric matrix is a normal matrix, and so is an orthogonal matrix because an orthogonal matrix $\boldsymbol{Q}$ satisfying $\boldsymbol{Q}\boldsymbol{Q}^T = \boldsymbol{Q}^T\boldsymbol{Q} = \boldsymbol{I}$.

The theory behind these iterative methods is closely related to the Krylov subspace $\mathcal{K}_n$ spanned by $\boldsymbol{A}$ and $\boldsymbol{b}$ as defined by

$$\mathcal{K}_n(\boldsymbol{A}, \boldsymbol{b}) = \{\boldsymbol{Ib}, \boldsymbol{Ab}, \boldsymbol{A}^2\boldsymbol{b}, \ldots, \boldsymbol{A}^{n-1}\boldsymbol{b}\}, \tag{1.20}$$

where $\boldsymbol{A}^0 = \boldsymbol{I}$.

If we use an iterative procedure to obtain the approximate solution $\boldsymbol{u}_n$ to $\boldsymbol{Au} = \boldsymbol{b}$ at nth iteration, the residual is given by

$$\boldsymbol{r}_n = \boldsymbol{b} - \boldsymbol{Au}_n, \tag{1.21}$$

which is essentially the negative gradient $\nabla f(\boldsymbol{u}_n)$. The search direction vector in the conjugate gradient method is subsequently determined by

$$\boldsymbol{d}_{n+1} = \boldsymbol{r}_n - \frac{\boldsymbol{d}_n^T \boldsymbol{A}\boldsymbol{r}_n}{\boldsymbol{d}_n^T \boldsymbol{A}\boldsymbol{d}_n}\boldsymbol{d}_n. \tag{1.22}$$

The solution often starts with an initial guess $\boldsymbol{u}_0$ at $n = 0$, and proceeds iteratively. The above steps can compactly be written as

$$\boldsymbol{u}_{n+1} = \boldsymbol{u}_n + \alpha_n\boldsymbol{d}_n, \quad \boldsymbol{r}_{n+1} = \boldsymbol{r}_n - \alpha_n\boldsymbol{A}\boldsymbol{d}_n, \tag{1.23}$$

and

$$\boldsymbol{d}_{n+1} = \boldsymbol{r}_{n+1} + \beta_n\boldsymbol{d}_n, \tag{1.24}$$

where

$$\alpha_n = \frac{\boldsymbol{r}_n^T\boldsymbol{r}_n}{\boldsymbol{d}_n^T\boldsymbol{A}\boldsymbol{d}_n}, \quad \beta_n = \frac{\boldsymbol{r}_{n+1}^T\boldsymbol{r}_{n+1}}{\boldsymbol{r}_n^T\boldsymbol{r}_n}. \tag{1.25}$$

Iterations stop when a prescribed accuracy is reached. This can easily be programmed in any programming language, especially Matlab.

It is worth pointing out that the initial guess $\boldsymbol{r}_0$ can be any educated guess; however, $\boldsymbol{d}_0$ should be taken as $\boldsymbol{d}_0 = \boldsymbol{r}_0$, otherwise, the algorithm may not converge. In the case when $\boldsymbol{A}$ is not symmetric, we can use the generalized minimal residual (GMRES) algorithm developed by Y. Saad and M.H. Schultz in 1986.

1.3.5. *BFGS Method*

The widely used BFGS method is an abbreviation of the Broydon–Fletcher–Goldfarb–Shanno method, and it is a quasi-Newton method for solving unconstrained nonlinear optimization. It is based on the basic idea of replacing the full Hessian matrix $\boldsymbol{H}$ by an approximate matrix $\boldsymbol{B}$ in terms of an iterative updating formula with rank-one matrices as its increment. Briefly speaking, a rank-one matrix is a matrix which can be written as $\boldsymbol{r} = \boldsymbol{a}\boldsymbol{b}^T$ where $\boldsymbol{a}$ and $\boldsymbol{b}$ are vectors, which has at most one nonzero eigenvalue, and this eigenvalue can be calculated by $\boldsymbol{b}^T\boldsymbol{a}$.

To minimize a function $f(\boldsymbol{x})$ with no constraint, the search direction $\boldsymbol{s}_k$ at each iteration is determined by

$$\boldsymbol{B}_k\boldsymbol{s}_k = -\nabla f(\boldsymbol{x}_k), \tag{1.26}$$

where $\boldsymbol{B}_k$ is the approximation to the Hessian matrix at kth iteration. Then, a line search is performed to find the optimal step size β_k so that the new trial solution is determined by

$$\boldsymbol{x}_{n+1} = \boldsymbol{x}_k + \beta_k\boldsymbol{s}_k. \tag{1.27}$$

Introducing two new variables

$$\boldsymbol{u}_k = \boldsymbol{x}_{k+1} - \boldsymbol{x}_k = \beta_k\boldsymbol{s}_k, \quad \boldsymbol{v}_k = \nabla f(\boldsymbol{x}_{k+1}) - \nabla f(\boldsymbol{x}_k), \tag{1.28}$$

we can update the new estimate as

$$\boldsymbol{B}_{k+1} = \boldsymbol{B}_k + \frac{\boldsymbol{v}_k\boldsymbol{v}_k^T}{\boldsymbol{v}_k^T\boldsymbol{u}_k} - \frac{(\boldsymbol{B}_k\boldsymbol{u}_k)(\boldsymbol{B}_k\boldsymbol{u}_k)^T}{\boldsymbol{u}_k^T\boldsymbol{B}_k\boldsymbol{u}_k}. \tag{1.29}$$

1.3.6. *Trust-Region Method*

The fundamental ideas of the trust-region have developed over many years with many seminal papers by a dozen pioneers. A good history review of the trust-region methods can be found in the book by Conn, Gould, and Toint (2000). Briefly speaking, the first important work was due to Levenberg in 1944, which proposed the usage of addition of a multiple of the identity matrix to the Hessian matrix as a damping measure to stabilize the solution procedure for nonlinear least-squares problems. Later, Marquardt in 1963 independently pointed out the link between such damping in the Hessian and the reduction of the step size in a restricted region. Slightly later in 1966, Goldfelt, Quandt, and Trotter essentially set the stage for trust-region methods by introducing an explicit updating formula for the maximum step size. Then, in 1970, Powell proved the global convergence for the trust-region method, though it is believed that the term 'trust region' was coined by Dennis in 1978, as earlier literature used various terminologies such as region of validity, confidence region, and restricted step method.

In the trust-region algorithm, a fundamental step is to approximate the nonlinear objective function by using truncated Taylor expansions, often the quadratic form

$$\phi_k(\boldsymbol{x}) \approx f(\boldsymbol{x}_k) + \nabla f(\boldsymbol{x}_k)^T \boldsymbol{u} + \frac{1}{2}\boldsymbol{u}^T \boldsymbol{H}_k \boldsymbol{u}, \tag{1.30}$$

in a so-called trust region Ω_k defined by

$$\Omega_k = \left\{ \boldsymbol{x} \in \Re^d \big| \|\boldsymbol{\Gamma}(\boldsymbol{x} - \boldsymbol{x}_k)\| \le \Delta_k \right\}, \tag{1.31}$$

where Δ_k is the trust-region radius. Here $\boldsymbol{H}_k$ is the local Hessian matrix. $\boldsymbol{\Gamma}$ is a diagonal scaling matrix that is related to the scalings of the optimization problem. Thus, the shape of the trust region is a hyperellipsoid, and an elliptical region in 2D centered at $\boldsymbol{x}_k$. If the parameters are equally scaled, then $\boldsymbol{\Gamma} = \boldsymbol{I}$ can be used.

The approximation to the objective function in the trust region will make it simpler to find the next trial solution $\boldsymbol{x}_{k+1}$ from the current solution $\boldsymbol{x}_k$. Then, we intend to find $\boldsymbol{x}_{k+1}$ with a sufficient decrease in the objective function. How good the approximation ϕ_k is to the actual objective $f(\boldsymbol{x})$ can be measured by the ratio of the achieved decrease to the predicted decrease

$$\gamma_k = \frac{f(\boldsymbol{x}_k) - f(\boldsymbol{x}_{k+1})}{\phi_k(\boldsymbol{x}_k) - \phi_k(\boldsymbol{x}_{k+1})}. \tag{1.32}$$

If this ratio is close to unity, we have a good approximation and then should move the trust region to $\boldsymbol{x}_{k+1}$.

Now the question is what radius should we use for the newly updated trust region centered at $\boldsymbol{x}_{k+1}$? Since the move is successful, and the decrease is significant, we should be bold enough to expand the trust region a little. A standard measure of such significance in decrease is to use a parameter $\alpha_1 \approx 0.01$. If $\gamma_k > \alpha_1$, the achieved decrease is noticeable, so we should accept the move (i.e., $\boldsymbol{x}_{k+1} \leftarrow \boldsymbol{x}_k$). What radius should we now use? Conventionally, we use another parameter $\alpha_2 > \alpha_1$ as an additional criterion. If γ_k is about $O(1)$ or $\gamma_k \ge \alpha_2 \approx 0.9$, we say that decrease is significant, and we can boldly increase the trust-region radius. Typically, we choose a value $\Delta_{k+1} \in [\Delta_k, \infty)$. The actual choice may depend on the problem, though typically $\Delta_{k+1} \approx 2\Delta_k$. If the decrease is noticeable but not so significant, that is $\alpha_1 < \gamma_k \leq \alpha_2$, we should shrink the trust region so that $\Delta_{k+1} \in [\beta_2 \Delta_k, \Delta_k]$ or

$$\beta_2 \Delta_k < \Delta_{k+1} < \Delta_k, \quad (0 < \beta_2 < 1). \tag{1.33}$$

Obviously, if the decrease is too small or $\gamma_k < \alpha_1$, we should abandon the move as the approximation is not good enough over this larger region. We should seek a better approximation on a smaller region by reducing the trust-region radius

$$\Delta_{k+1} \in [\beta_1 \Delta_k, \beta_2 \Delta_k], \tag{1.34}$$

where $0 < \beta_1 \leq \beta_2 < 1$, and typically $\beta_1 = \beta_2 = 1/2$, which means half the original size is used first.

To summarize, the typical values of the parameters are:

$$\alpha_1 = 0.01, \quad \alpha_2 = 0.9, \quad \beta_1 = \beta_2 = \frac{1}{2}. \tag{1.35}$$

1.4. Quadratic Programming

It is worth pointing out that the powerful linear programming or Dantzig's simplex method (Dantzig, 1963) requires detailed discussions, however, readers can easily find relevant textbooks for this topic. Therefore, we only briefly introduce quadratic programming in this section.

1.4.1. *Quadratic Programming*

A special type of nonlinear programming is quadratic programming whose objective function is a quadratic form

$$f(x) = \frac{1}{2} x^T Q x + b^T x + c, \tag{1.36}$$

where b and c are constant vectors. Q is a symmetric square matrix. The constraints can be incorporated using Lagrange multipliers and Karush–Kuhn–Tucker (KKT) formulation.

1.4.2. *Sequential Quadratic Programming*

Sequential (or successive) quadratic programming (SQP) represents one of the state-of-the-art and most popular methods for nonlinear constrained optimization. It is also one of the most robust methods. For a general nonlinear optimization problem

$$\text{minimize } f(\boldsymbol{x}), \tag{1.37}$$

$$\text{subject to } h_i(\boldsymbol{x}) = 0, \quad (i = 1, \ldots, p), \tag{1.38}$$

$$g_j(\boldsymbol{x}) \leq 0, \quad (j = 1, \ldots, q). \tag{1.39}$$

The fundamental idea of sequential quadratic programming is to approximate the computationally extensive full Hessian matrix using a quasi-Newton updating method. Subsequently, this generates a subproblem of quadratic programming (called QP subproblem) at each iteration, and the solution to this subproblem can be used to determine the search direction and the next trial solution.

Using the Taylor expansions, the above problem can be approximated, at each iteration, as follows

$$\text{minimize } \frac{1}{2}\boldsymbol{s}^T \nabla^2 L(\boldsymbol{x}_k)\boldsymbol{s} + \nabla f(\boldsymbol{x}_k)^T \boldsymbol{s} + f(\boldsymbol{x}_k), \tag{1.40}$$

$$\text{subject to } \nabla h_i(\boldsymbol{x}_k)^T \boldsymbol{s} + h_i(\boldsymbol{x}_k) = 0, \quad (i = 1, \ldots, p), \tag{1.41}$$

$$\nabla g_j(\boldsymbol{x}_k)^T \boldsymbol{s} + g_j(\boldsymbol{x}_k) \leq 0, \quad (j = 1, \ldots, q), \tag{1.42}$$

where the Lagrange function, also called merit function, is defined by

$$L(\boldsymbol{x}) = f(\boldsymbol{x}) + \sum_{i=1}^{p} \lambda_i h_i(\boldsymbol{x}) + \sum_{j=1}^{q} \mu_j g_j(\boldsymbol{x})$$

$$= f(\boldsymbol{x}) + \boldsymbol{\lambda}^T \boldsymbol{h}(\boldsymbol{x}) + \boldsymbol{\mu}^T \boldsymbol{g}(\boldsymbol{x}), \tag{1.43}$$

where $\boldsymbol{\lambda} = (\lambda_1, \ldots, \lambda_p)^T$ is the vector of Lagrange multipliers, and $\boldsymbol{\mu} = (\mu_1, \ldots, \mu_q)^T$ is the vector of KKT multipliers. Here we have used the notation $\boldsymbol{h} = (h_1(\boldsymbol{x}), \ldots, h_p(\boldsymbol{x}))^T$ and $\boldsymbol{g} = (g_1(\boldsymbol{x}), \ldots, g_q(\boldsymbol{x}))^T$.

To approximate the Hessian $\nabla^2 L(\boldsymbol{x}_k)$ by a positive definite symmetric matrix $\boldsymbol{H}_k$, the standard Broydon–Fletcher–Goldbarb–Shanno (BFGS) updating formula can be used

$$\boldsymbol{H}_{k+1} = \boldsymbol{H}_k + \frac{\boldsymbol{v}_k \boldsymbol{v}_k^T}{\boldsymbol{v}_k^T \boldsymbol{u}_k} - \frac{\boldsymbol{H}_k \boldsymbol{u}_k \boldsymbol{u}_k^T \boldsymbol{H}_k^T}{\boldsymbol{u}_k^T \boldsymbol{H}_k \boldsymbol{u}_k}, \tag{1.44}$$

where

$$\boldsymbol{u}_k = \boldsymbol{x}_{k+1} - \boldsymbol{x}_k, \tag{1.45}$$

and

$$v_k = \nabla L(x_{k+1}) - \nabla L(x_k). \tag{1.46}$$

The QP subproblem is solved to obtain the search direction

$$x_{k+1} = x_k + \alpha s_k, \tag{1.47}$$

using a line search method by minimizing a penalty function, also commonly called merit function,

$$\Phi(x) = f(x) + \rho \left[\sum_{i=1}^{p} |h_i(x)| + \sum_{j=1}^{q} \max\{0, g_j(x)\} \right], \tag{1.48}$$

where ρ is the penalty parameter.

It is worth pointing out that any SQP method requires a good choice of H_k as the approximate Hessian of the Lagrangian L. Obviously, if H_k is exactly calculated as $\nabla^2 L$, SQP essentially becomes Newton's method for solving the optimality condition. A popular way to approximate the Lagrangian Hessian is to use a quasi-Newton scheme as we used the BFGS formula described earlier.

There are other variants of gradient-based methods used in microwave engineering, and we will discuss them in the relevant context when necessary. Deterministic methods such as Newton's methods are often efficient for local search. For global search, especially for multimodal cases, nature-inspired metaheuristics are more reliable in finding the global optimality. Therefore, we will introduce some popular metaheuristics in the next chapter.

References

Arora, J. (1989). *Introduction to Optimum Design*, McGraw-Hill, New York: NY.

Boyd, S.P. and Vandenberghe, L. (2004). *Convex Optimization*, Cambridge University Press, Cambridge.

Conn, A.R., Gould, N.I.M. and Toint, P.L. (2000). *Trust-Region Methods*, Society for Industrial and Applied Mathematics, Philadelphia: PA.

Dantzig, G.B. (1963). *Linear Programming and Extensions*, Princeton University Press, Princeton: NJ.

Forrester, A.I.J. and Keane, A.J. (2009). Recent advances in surrogate-based optimization, *Prog. Aerosp. Sci.*, **45**(1–3), 50–79.

Gill, P.E., Murray, W. and Wright, M.H. (1981). *Practical Optimization*, Academic Press, Orlando: FL.

Karmarkar, N. (1984). A new polynomial-time algorithm for linear programming, *Combinatorica*, **4**(4), 373–395.

Koziel, S., Bandler, J.W. and Madsen, K. (2008). Quality assessment of coarse models and surrogates for space mapping optimization, *Optim. Eng.*, **9**(4), 375–391.

Koziel, S., Bandler, J.W. and Madsen, K. (2009). Space mapping with adaptive response correction for microwave design optimization, *IEEE T. Microw. Theory*, **57**(2), 478–486.

Koziel, S. and Yang, X.S. (2011). *Computational Optimization, Methods and Algorithms*, Springer-Verlag, Berlin/Heidelberg.

Nelder, J.A. and Mead, R. (1965). A simplex method for function optimization, *Comput. J.*, **7**, 308–313.

Nesterov, Y. and Nemirovskii, A. (1994). *Interior-Point Polynomial Methods in Convex Programming*, Society for Industrial and Applied Mathematics, Philadelphia: PA.

Queipo, N.V., Haftka, R.T., Shyy, W., Goel T., Vaidynathan, R. and Tucker, P.K. (2005). Surrogate-based analysis and optimization, *Prog. Aerosp. Sci.*, **41**(1), 1–28.

Sawaragi, Y., Nakayama, H. and Tanino, T. (1985). *Theory of Multiobjective Optimisation*, Academic Press, Orlando: FL.

Simpson, T.W., Peplinski, J. and Allen, J.K. (2001). Metamodels for computer-based engienering design: survey and recommendations, *Eng. Comput.*, **17**, 129–150.

Talbi, E.G. (2009). *Metaheuristics: From Design to Implementation*, Wiley, New York: NY.

Wolpert, D.H. and Macready, W.G. (1997). No free lunch theorems for optimization, *IEEE T. Evolut. Comput.*, **1**, 67–82.

Yang, X.S. (2008a). *Introduction to Computational Mathematics*, World Scientific Publishing, Singapore.

Yang, X.S. (2008b). *Nature-Inspired Metaheuristic Algoirthms*, Luniver Press, Frome.

Yang, X.S. (2010). *Engineering Optimization: An Introduction with Metaheuristic Applications*, Wiley, New York: NY.

Yang, X.S. and Deb, S., (2010). Engineering optimization by cuckoo search, *Int. J. Math. Modelling Num. Optimisation*, **1**(4), 330–343.

Yang, X.S. and Koziel, S. (2010). Computational optimization, modelling and simulation: a paradigm shift, *Procedia Computer Science*, **1**(1), 1291–1294.

Chapter 2

Derivative-Free Methods and Metaheuristics

Xin-She Yang and Slawomir Koziel

2.1. Derivative-Free Methods

Many search algorithms such as the steepest-descent method experience slow convergence near the local minimum. They are also memoryless because the pass information is not used to produce accelerated move. The only information they use is the current location $x^{(n)}$, gradient, and value of the objective itself at step n. However, in many applications, computation of derivatives may be very computationally expensive, even with some form of approximations, while in other cases, the objective function may have discontinuities and non-smoothness, and consequently, derivatives may not exist at all. In such cases, it is far better to use derivative-free methods.

Furthermore, if past information, such as the steps at $n-1$ and n, is properly used to generate a new move at step $n+1$, it may speed up the convergence. The Hooke–Jeeves pattern search method is one of such methods that incorporates the past history of iterations in producing a new search direction.

2.1.1. *Pattern Search*

In fact, pattern search is a class of derivative-free search methods for multidimensional nonlinear optimization. Pattern search consists of two moves: exploratory move and pattern move. The exploratory moves explore the local behavior and information of the objective function so as to identify any potential sloping valleys if they exist. For any given step size (each coordinate direction can have a different increment) $\Delta_i (i = 1, 2, \ldots, p)$, exploration movement performs from an initial starting point along each coordinate direction by increasing or decreasing $\pm\Delta_i$; if the new value of the objective function does not increase (for a minimization problem), that is $f(x_i^{(n)}) \leq f(x_i^{(n-1)})$, the exploratory move is considered successful. If it is not successful, then try a step in the opposite direction, and the result is updated only if it is successful. When all the p coordinates have been explored, the resulting point forms a base point $x^{(n)}$.

The pattern move intends to move the current base $x^{(n)}$ along the base line $(x^{(n)} - x^{(n-1)})$ from the previous (historical) base point to the current base point. The move is carried out by the following formula:

$$x^{(n+1)} = x^{(n)} + [x^{(n)} - x^{(n-1)}]. \tag{2.1}$$

Then $x^{(n+1)}$ forms a new temporary base point for further new exploratory moves. If the pattern move produces improvement (lower value of $f(x)$), the new base point $x^{(n+1)}$ is successfully updated. If the pattern move does not lead to improvement or a lower value of the objective function, then the pattern move is discarded and a new search starts from $x^{(n)}$. New search moves should use a smaller step size by reducing increments Δ_i/γ where $\gamma > 1$ is the step reduction factor. Iterations continue until the prescribed tolerance is met.

There are many variants of pattern search, including directional/direct search, generalized pattern search, mesh adaptive direct search (MADS), and generating set search. Interested readers can refer to more advanced literature (Conn *et al.*, 2009; Koziel and Yang, 2011; Yang and Koziel, 2011).

2.1.2. *Nelder–Mead's Simplex Method*

In the n-dimensional space, a simplex, which is a generalization of a triangle on a plane, is a convex hull with $n + 1$ distinct points. For simplicity, a simplex in the n-dimension space is referred to as n-simplex. Therefore, a 1-simplex is a line segment, a 2-simplex is a triangle, a 3-simplex is a tetrahedron, and so on.

The Nelder–Mead method is a downhill simplex algorithm for unconstrained optimization without using derivatives, and it was first developed by J.A. Nelder and R. Mead (1965). This is one of the most widely used methods since its computational effort is relatively small and is a way to get a quick grasp of the optimization problem. The basic idea of this method is to use the flexibility of the constructed simplex via amoeba-style manipulations by reflection, expansion, contraction, and reduction. In some books, such as the best-known *Numerical Recipes*, it is also called the 'Amoeba Algorithm'. It is worth pointing out that this downhill simplex method has nothing to do with the simplex method for linear programming.

There are a few variants of the algorithm, which use slightly different ways of constructing initial simplex and convergence criteria. However, the fundamental procedure is the same.

The first step is to construct an initial n-simplex with $n + 1$ vertices and to evaluate the objective function at the vertices. Then, by ranking the objective values and re-ordering the vertices, we have an ordered set so that

$$f(\boldsymbol{x}_1) \le f(\boldsymbol{x}_2) \le \ldots \le f(\boldsymbol{x}_{n+1}), \tag{2.2}$$

at $\boldsymbol{x}_1, \boldsymbol{x}_2, \ldots, \boldsymbol{x}_{n+1}$, respectively. As the downhill simplex method is for minimization, we use the convention that $\boldsymbol{x}_{n+1}$ is the worse point (solution), and $\boldsymbol{x}_1$ is the best solution. Then, at each iteration, similar ranking manipulations are carried out.

Then, we have to calculate the centroid $\boldsymbol{x}$ of simplex excluding the worst vertex $\boldsymbol{x}_{n+1}$:

$$\bar{\boldsymbol{x}} = \frac{1}{n} \sum_{i=1}^{n} \boldsymbol{x}_i. \tag{2.3}$$

Using the centroid as the basis point, we try to find the reflection of the worse point $\boldsymbol{x}_{n+1}$ by

$$\boldsymbol{x}_r = \bar{\boldsymbol{x}} + \alpha(\bar{\boldsymbol{x}} - \boldsymbol{x}_{n+1}), \quad (\alpha > 0), \tag{2.4}$$

though the typical value of $\alpha = 1$ is often used.

Whether the new trial solution is accepted or not and how to update the new vertex, depends on the objective function at $\boldsymbol{x}_r$. There are three possibilities:

- If $f(\boldsymbol{x}_1) \le f(\boldsymbol{x}_r) < f(\boldsymbol{x}_n)$, then replace the worst vertex $\boldsymbol{x}_{n+1}$ by $\boldsymbol{x}_r$, that is $\boldsymbol{x}_{n+1} \leftarrow \boldsymbol{x}_r$.
- If $f(\boldsymbol{x}_r) < f(\boldsymbol{x}_1)$ which means the objective improves, we then seek a more bold move to see if we can improve the objective even further by moving/expanding the vertex further along the line of reflection to a new trial solution

$$\boldsymbol{x}_e = \boldsymbol{x}_r + \beta(\boldsymbol{x}_r - \bar{\boldsymbol{x}}), \tag{2.5}$$

where $\beta = 2$. Now we have to test if $f(\boldsymbol{x}_e)$ improves even more. If $f(\boldsymbol{x}_e) < f(\boldsymbol{x}_r)$, we accept it and update $\boldsymbol{x}_{n+1} \leftarrow \boldsymbol{x}_e$; otherwise, we can use the result of the reflection. That is $\boldsymbol{x}_{n+1} \leftarrow \boldsymbol{x}_r$.

- If there is no improvement or $f(\boldsymbol{x}_r) > f(\boldsymbol{x}_n)$, we have to reduce the size of the simplex while maintaining the best sides. This is the contraction

$$\boldsymbol{x}_c = \boldsymbol{x}_{n+1} + \gamma(\bar{\boldsymbol{x}} - \boldsymbol{x}_{n+1}), \tag{2.6}$$

where $0 < \gamma < 1$, though $\gamma = 1/2$ is usually used. If $f(\boldsymbol{x}_c) < f(\boldsymbol{x}_{n+1})$ is true, we then update $\boldsymbol{x}_{n+1} \leftarrow \boldsymbol{x}_c$.

Other improved variants of Nelder–Mead methods such as modified Nelder–Mead methods can have global convergence. Other methods such as trust-region methods can also be modified and combined with derivative-free methods and surrogate functions.

2.1.3. *Surrogate-Based Methods*

In many cases, the objective function of interest is difficult to handle. Possible reasons include high computational cost of its evaluation, the presence of numerical noise, non-differentiability, or even discontinuity. In these cases, it is often advantageous to replace, in the optimization process, the original function by its suitable representation, the so-called surrogate model. Typically, the surrogate is constructed to be computationally cheap, analytically tractable, and yet sufficiently accurate so that it can be reliably used to make predictions about the original function's optimum. There are many ways of constructing the surrogate, either by approximating sampled data from the original function, or by correcting a properly chosen low-fidelity model. In some cases, it is sufficient to construct and optimize the surrogate model once; however, most approaches are based on iterative updating and re-optimization of the surrogate using appropriate strategies for allocating new samples to get the original function's data (so-called infill criteria). More detailed exposition of surrogate modeling and optimization techniques is contained in Chapter 3. It should be emphasized that most of the optimization problems in microwave engineering are challenging because of the high computational cost of evaluating the objective function which usually comes from high-fidelity electromagnetic (EM) simulation. Also, EM analysis results are inherently noisy (particularly due to adaptive meshing techniques). Therefore, using surrogate models for microwave design optimization has become quite popular. Certain surrogate-based techniques, such as space mapping (SM) (Bandler *et al.*, 1994) have been developed specifically to address design problems in this area. Chapter 4 of this book is entirely devoted to SM and related methods.

2.1.4. *Contemporary Derivative-Free Methods*

For derivative-free methods, classic approaches tend to use directions and search history to construct search moves to generate new solutions. However, contemporary approaches, especially those from the 1990s, tend to use swarm-intelligence-based methods to generated moves which somehow employ a historical best and some individual best solutions, in combination with randomization techniques, to guide search moves so as to generate promising new solutions. This forms a class of stochastic, nature-inspired metaheuristic algorithms. In fact, almost all metaheuristic algorithms are derivative-free methods, and we will introduce some of the most recent and widely used metaheuristics.

2.2. Metaheuristics

Metaheuristic algorithms are often nature-inspired, and they are now among the most widely used algorithms for optimization. They have many advantages over conventional algorithms, as discussed in Chapter 1 as part of the introduction and overview. There are a few recent books which are solely dedicated to metaheuristic algorithms (Yang, 2008; Talbi, 2009; Yang, 2010a; Yang, 2010b). Metaheuristic algorithms are very diverse, including genetic algorithms, simulated annealing, differential evolution, ant and bee algorithms, particle swarm optimization, harmony search, firefly algorithm, cuckoo search, and others. Here we will introduce some of these algorithms briefly.

2.2.1. *Ant Algorithms*

Ant algorithms, especially the ant colony optimization (Dorigo and Stütle, 2004), mimic the foraging behavior of social ants. Primarily, it uses pheromone as a chemical messenger and the pheromone concentration as the indicator of quality solutions to a problem of interest. As the solution is often linked with the pheromone concentration, the search algorithms often produce routes and paths marked by the higher pheromone concentrations, and therefore, ant-based algorithms are particularly suitable for discrete optimization problems.

The movement of an ant is controlled by pheromone, which will evaporate over time. Without such time-dependent evaporation, the algorithms will lead to premature convergence to the (often wrong) solutions. With proper pheromone evaporation, they usually behave very well.

There are two important issues here: the probability of choosing a route, and the evaporation rate of pheromone. There are a few ways of solving these problems, although it is still an area of active research. Here we introduce the current best method.

For a network routing problem, the probability of ants at a particular node i to choose the route from node i to node j is given by

$$p_{ij} = \frac{\phi_{ij}^{\alpha} d_{ij}^{\beta}}{\sum_{i,j=1}^{n} \phi_{ij}^{\alpha} d_{ij}^{\beta}}, \tag{2.7}$$

where $\alpha > 0$ and $\beta > 0$ are the influence parameters, and their typical values are $\alpha \approx \beta \approx 2$. ϕ_{ij} is the pheromone concentration on the route between i and j, and d_{ij} the desirability of the same route. Some *a priori* knowledge about the route such as the distance s_{ij} is often used so that $d_{ij} \propto 1/s_{ij}$, which implies that shorter routes will be selected due to their shorter traveling time, and thus the pheromone concentrations on these routes are higher. This is because the traveling time is shorter, and thus less of the pheromone has evaporated during this period.

2.2.2. *Bee Algorithms*

Bees-inspired algorithms are more diverse, and some use pheromone, however most do not. Almost all bee algorithms are inspired by the foraging behavior of honey bees in nature. Interesting characteristics such as waggle dance, polarization, and nectar maximization are often used to simulate the allocation of the foraging bees along flower patches and thus different search regions in the search space. For a more comprehensive review, please refer to Yang (2010a) and Parpinelli and Lopes (2011).

In the honeybee-based algorithms, forager bees are allocated to different food sources (or flower patches) so as to maximize the total nectar intake. The colony has to 'optimize' the overall efficiency of nectar collection; the allocation of the bees is thus dependant on many factors such as the nectar richness and the proximity to the hive (Nakrani and Tovey, 2004; Yang, 2005; Karaboga, 2005; Pham *et al.*, 2006).

The virtual bee algorithm (VBA), developed by Xin-She Yang in 2005, is an optimization algorithm specially formulated for solving both discrete and continuous problems (Yang, 2005). On the other hand, the artificial bee colony (ABC) optimization algorithm was first developed by D. Karaboga in 2005. In the ABC algorithm, the bees in a colony are divided into three groups: employed bees (forager bees), onlooker bees (observer bees), and scouts. For each food source, there is only one employed bee. That is to say, the number of employed bees is equal to the number of food sources. The employed bee of a discarded food site is forced to become a scout, searching new food sources randomly. Employed bees share information with the onlooker bees in a hive so that onlooker bees can choose a food source to forage. Unlike the honey bee algorithm which has two groups of the bees (forager bees and observer bees), bees in the ABC algorithm are more specialized (Afshar *et al.*, 2007; Karaboga, 2005).

Similar to the ant-based algorithms, bee algorithms are also very flexible in dealing with discrete optimization problems. Combinatorial optimization such as routing and optimal paths has been successfully solved by ant and bee algorithms. In principle, they can solve both continuous optimization and discrete optimization problems; however, they should not be the first choice for continuous problems.

2.2.3. *Bat Algorithm*

The bat algorithm (BA) is a relatively new metaheuristic, developed by Xin-She Yang in 2010 (Yang, 2010c). It was inspired by the echolocation behavior of microbats. Microbats use a type of sonar, called echolocation, to detect prey, avoid obstacles, and locate their roosting crevices in the dark. These bats emit a very loud sound pulse and listen for the echo that bounces back from the surrounding objects. Their pulses vary in properties and can be correlated with their hunting strategies, depending on the species. Most bats use short, frequency-modulated signals to sweep through about an octave, while others more often use constant-frequency signals for echolocation. Their signal bandwidth varies depends on the species, and is often increased by using more harmonics.

Inside the bat algorithm, there are three idealized rules:

1. All bats use echolocation to sense distance, and they also 'know' the difference between food/prey and background barriers in some magical way.
2. Bats fly randomly with velocity v_i at position x_i with a fixed frequency $f_{\min}$, varying wavelength λ, and loudness A_0 to search for prey. They can automatically adjust the wavelength (or frequency) of their emitted pulses and adjust the rate of pulse emission $r \in [0, 1]$, depending on the proximity of their target.
3. Although the loudness can vary in many ways, we assume that the loudness varies from a large (positive) A_0 to a minimum constant value $A_{\min}$.

BA has been extended to multiobjective bat algorithm (MOBA) by Yang (2011a), and preliminary results suggested that it is very efficient.

2.2.4. *Simulated Annealling*

Simulated annealing, developed by Kirkpatrick *et al.* (1983), is among the first metaheuristic algorithms, and it has been applied in almost every area of optimization. Unlike the gradient-based methods and other deterministic search methods, the main advantage of simulated annealing is its ability to avoid being trapped in local minima. The basic idea of the simulated annealing algorithm is to use random search in terms of a Markov chain, which not only accepts changes that improve the objective function, but also keeps some changes that are not ideal.

In a minimization problem, for example, any better moves or changes, that decrease the value of the objective function f, will be accepted; however, some changes that increase f will also be accepted with a probability p. This probability p, also called the transition probability, is determined by

$$p = \exp\left[-\frac{\Delta E}{k_B T}\right], \tag{2.8}$$

where k_B is the Boltzmann's constant, and T is the temperature for controlling the annealing process. ΔE is the change of the energy level. This transition probability is based on the Boltzmann distribution in statistical mechanics.

The simplest way to link ΔE with the change of the objective function Δf is to use

$$\Delta E = \gamma \Delta f, \tag{2.9}$$

where γ is a real constant. For simplicity without losing generality, we can use $k_B = 1$ and $\gamma = 1$. Thus, the probability p simply becomes

$$p(\Delta f, T) = e^{-\Delta f/T}. \tag{2.10}$$

Whether or not a change is accepted, a random number r is often used as a threshold. Thus, if $p > r$, or

$$p = e^{-\Delta f/T} > r, \tag{2.11}$$

the move is accepted.

Here the choice of the right initial temperature is crucially important. For a given change Δf, if T is too high ($T \to \infty$), then $p \to 1$, which means almost all the changes will be accepted. If T is too low ($T \to 0$), then any $\Delta f > 0$ (worse solution) will rarely be accepted as $p \to 0$, and thus the diversity of the solution is limited, but any improvement Δf will almost always be accepted. In fact, the special case $T \to 0$ corresponds to classical hill-climbing because only better solutions are accepted, and the system is essentially climbing up or descending down a hill. Therefore, if T is too high, the system is at a high energy state on the topological landscape, and the minima are not easily reached. If T is too low, the system may be trapped in a local minimum (not necessarily the global minimum), and there is not enough energy for the system to jump out of the local minimum to explore other minima including the global minimum. So a proper initial temperature should be calculated.

Another important issue is how to control the annealing or cooling process so that the system cools down gradually from a higher temperature to ultimately freeze to a global minimum state. There are many ways of controlling the cooling rate or the decrease of the temperature. Geometric cooling schedules are often widely used, which essentially decrease the temperature by a cooling factor $0 < \alpha < 1$ so that T is replaced by αT, or

$$T(t) = T_0 \alpha^t, \quad t = 1, 2, \ldots, t_f, \tag{2.12}$$

where t_f is the maximum number of iterations. The advantage of this method is that $T \to 0$ when $t \to \infty$, and thus there is no need to specify the maximum number of iterations if a tolerance or accuracy is prescribed. Simulated annealling has been applied in a wide range of optimization problems.

2.2.5. *Genetic Algorithms*

Simulated annealing is a trajectory-based algorithm, as it only uses a single agent. Other algorithms such as genetic algorithms use multiple agents or a population to carry out the search, which may have some advantages due to its potential parallelism.

Genetic algorithms are a class of algorithms based on the abstraction of Darwin's evolution of biological systems, pioneered by J. Holland and his collaborators in the 1960s and 1970s (Holland, 1975). Holland was the first to use genetic operators such as crossover and recombination, mutation, and selection, in the study of adaptive and artificial systems. Genetic algorithms have two main advantages over traditional algorithms: the ability to deal with complex problems, and parallelism. Whether the objective function is stationary or transient, linear or nonlinear, continuous or discontinuous, it can be dealt with by genetic algorithms. Multiple genes can be suitable for parallel implementation.

Three main components or genetic operators in genetic algorithms are: crossover, mutation, and selection of the fittest. Each solution is encoded in a string (often binary or decimal), called a chromosome. The crossover of two parent strings produce offsprings (new solutions) by swapping part, or genes, of the chromosomes. Crossover has a higher probability, typically 0.8 to 0.95. On the other hand, mutation is carried out by flipping some digits of a string, which generates new solutions. This mutation probability is typically low, from 0.001 to 0.05. New solutions generated in each generation will be evaluated by their fitness, which is linked to the objective function of the optimization problem. The new solutions are selected according to their fitness — selection of the fittest. Sometimes, in order to make sure that the best solutions remain in the population, the best solutions are passed onto the next generation without much change; this is called elitism.

Genetic algorithms have been applied to almost all areas of optimization, design and applications. There are hundreds of books and thousands of research articles. There are also many variants and hybridization with other algorithms, and interested readers can refer to more advanced literature such as Goldberg (1989) and Michalewicz (1998).

2.2.6. *Differential Evolution*

Differential evolution (DE) was developed by R. Storn and K. Price by their nominal papers in 1996 and 1997 (Storn, 1996; Storn and Price, 1997). It is a vector-based evolutionary algorithm, and can be considered as a further development to genetic algorithms. It is a stochastic search algorithm with a self-organizing tendency and does not use the information of derivatives. Thus, it is a population-based, derivative-free method.

As in genetic algorithms, design parameters in a d-dimensional search space are represented as vectors, and various genetic operators are operated over their bits of strings. However, unlikely genetic algorithms, differential evolution carries out operations over each component (or each dimension of the solution). Almost everything is done in terms of vectors. For example, in genetic algorithms, mutation is carried out at one site or multiple sites of a chromosome, while in differential evolution, a difference vector of two randomly chosen population vectors is used to perturb an existing vector. Such vectorized mutation can be viewed as a self-organizing search, directed towards an optimality.

For a d-dimensional optimization problem with d parameters, a population of n solution vectors are initially generated; we have $\boldsymbol{x}_i$ where $i = 1, 2, \ldots, n$. For each solution $\boldsymbol{x}_i$ at any generation t, we use the conventional notation as

$$\boldsymbol{x}_i^t = (x_{1,i}^t, x_{2,i}^t, \ldots, x_{d,i}^t), \tag{2.13}$$

which consists of d-components in the d-dimensional space. This vector can be considered as the chromosomes or genomes.

Differential evolution consists of three main steps: mutation, crossover, and selection.

Mutation is carried out by the mutation scheme. For each vector $\boldsymbol{x}_i$ at any time or generation t, we first randomly choose three distinct vectors $\boldsymbol{x}_p$, $\boldsymbol{x}_q$, and $\boldsymbol{x}_r$ at t, and then generate a so-called donor vector by the mutation scheme

$$\boldsymbol{v}_i^{t+1} = \boldsymbol{x}_p^t + F\big(\boldsymbol{x}_q^t - \boldsymbol{x}_r^t\big), \tag{2.14}$$

where $F \in [0, 2]$ is a parameter, often referred to as the differential weight. This requires that the minimum number of population size is $n \geq 4$. In principle, $F \in [0, 2]$, but in practice, a scheme with $F \in [0, 1]$ is more efficient and stable.

The crossover is controlled by a crossover probability $C_r \in [0, 1]$ and actual crossover can be carried out in two ways: binomial and exponential. Selection is essentially the same as that used in genetic algorithms. It is to select the most fittest, and for the minimization problem, the minimum objective value. Therefore, we have

$$\boldsymbol{x}_i^{t+1} = \begin{cases} \boldsymbol{u}_i^{t+1} & \text{if } f(\boldsymbol{u}_i^{t+1}) \leq f(\boldsymbol{x}_i^t), \\ \boldsymbol{x}_i^t & \text{otherwise.} \end{cases} \tag{2.15}$$

Most studies have focused on the choice of F, C_r, and n as well as the modification of (2.14). In fact, when generating mutation vectors, we can use many different ways of formulating (2.14), and this leads to various schemes with the naming convention: DE/x/y/z where x is the mutation scheme (rand or best), y is the number of difference vectors, and z is the crossover scheme (binomial or exponential). The basic DE/Rand/1/Bin scheme is given in (2.14). Following a similar strategy, we can design various schemes.

In fact, at least ten different schemes have been formulated, and for details, readers can refer to Price *et al.* (2005).

2.2.7. *Particle Swarm Optimization*

Particle swarm optimization (PSO) was developed by Kennedy and Eberhart (1995), based on swarm behavior such as fish and bird schooling in nature. Since then, PSO has generated much wider interests, and forms an exciting, ever-expanding research subject, called swarm intelligence. PSO has been applied to almost every area in optimization, computational intelligence, and design/scheduling applications. There are at least two dozen PSO variants, and hybrid algorithms, created by combining PSO with other existing algorithms, are also increasingly popular.

This algorithm searches the space of an objective function by adjusting the trajectories of individual agents, called particles, as the piecewise paths formed by positional vectors in a quasi-stochastic manner. The movement of a swarming particle consists of two major components: a stochastic component and a deterministic component. Each particle is attracted toward the position of the current global best g^* and its own best location x_i^* in history, while at the same time it has a tendency to move randomly.

Let x_i and v_i be the position vector and velocity for particle i, respectively. The new velocity vector is determined by the following formula:

$$v_i^{t+1} = v_i^t + \alpha\epsilon_1 \odot \left[g^* - x_i^t\right] + \beta\epsilon_2 \odot \left[x_i^* - x_i^t\right], \qquad (2.16)$$

where ϵ_1 and ϵ_2 are two random vectors, and each entry taking the values between 0 and 1. The Hadamard product of two matrices $u \odot v$ is defined as the entrywise product, that is $[u \odot v]_{ij} = u_{ij}v_{ij}$. The parameters α and β are the learning parameters or acceleration constants, which can typically be taken as, say, $\alpha \approx \beta \approx 2$.

The initial locations of all particles should distribute relatively uniformly so that they can sample over most regions, which is especially important for multimodal problems. The initial velocity of a particle can be taken as zero, that is, $v_i^{t=0} = 0$. The new position can then be updated by

$$x_i^{t+1} = x_i^t + v_i^{t+1}. \qquad (2.17)$$

However, v_i can take any values in the range $[0, v_{\max}]$.

There are many variants which extend the standard PSO algorithm (Kennedy *et al.*, 2001; Yang, 2008; Yang, 2010b), and the most noticeable improvement is probably to use inertia function $\theta(t)$ so that $\boldsymbol{v}_i^t$ is replaced by $\theta(t)\boldsymbol{v}_i^t$:

$$\boldsymbol{v}_i^{t+1} = \theta\boldsymbol{v}_i^t + \alpha\boldsymbol{\epsilon}_1 \odot \left[\boldsymbol{g}^* - \boldsymbol{x}_i^t\right] + \beta\boldsymbol{\epsilon}_2 \odot \left[\boldsymbol{x}_i^* - \boldsymbol{x}_i^t\right], \qquad (2.18)$$

where θ takes the values between 0 and 1. In the simplest case, the inertia function can be taken as a constant, typically $\theta \approx 0.5 \sim 0.9$. This is equivalent to introducing a virtual mass to stabilize the motion of the particles, and thus the algorithm is expected to converge more quickly.

2.2.8. *Harmony Search*

Harmony search (HS) is a relatively new heuristic optimization algorithm and it was first developed by Z. W. Geem *et al.* (2001). Harmony search can be explained in more detail with the aid of a discussion of the improvisation process by a musician. When a musician is improvising, he or she has three possible choices: (i) play any famous piece of music (a series of pitches in harmony) exactly from his or her memory, (ii) play something similar to a known piece (thus adjusting the pitch slightly), or (iii) compose new or random notes. If we formalize these three options for optimization, we have three corresponding components: usage of harmony memory, pitch adjusting, and randomization.

The usage of harmony memory is important as it is similar to choosing the best fit individuals in the genetic algorithms. This will ensure the best harmonies will be carried over to the new harmony memory. In order to use this memory more effectively, we can assign a parameter $r_{\text{accept}} \in [0, 1]$, called harmony memory accepting or considering rate. If this rate is too low, only a few best harmonies are selected and it may converge too slowly. If this rate is extremely high (near 1), almost all the harmonies are used in the harmony memory, and consequently other harmonies are not explored well, leading to potentially wrong solutions. Therefore, typically, $r_{\text{accept}} = 0.7 \sim 0.95$.

To adjust the pitch slightly in the second component, we have to use a method such that it can adjust the frequency efficiently. In theory, the pitch can be adjusted linearly or nonlinearly, but in practice, linear adjustment is used. If x_{old} is the current solution (or pitch), then the new solution (pitch) $\boldsymbol{x}_{\text{new}}$ is generated by

$$\boldsymbol{x}_{\text{new}} = \boldsymbol{x}_{\text{old}} + b_p \left(2\,\epsilon - 1\right), \qquad (2.19)$$

where ϵ is a random number drawn from a uniform distribution $[0, 1]$. Here b_p is the bandwidth, which controls the local range of pitch adjustment. In fact, we can see that the pitch adjustment (2.19) is a random walk.

Pitch adjustment is similar to the mutation operator in genetic algorithms. We can assign a pitch-adjusting rate (r_{pa}) to control the degree of the adjustment. If r_{pa} is too low, then there is rarely any change. If it is too high, then the algorithm may not converge at all. Thus, we usually use $r_{pa} = 0.1 \sim 0.5$ in most simulations.

The third component is the randomization, which increases the diversity of the solutions. Although adjusting pitch has a similar role, it is limited to certain local pitch adjustment and thus corresponds to a local search. The use of randomization can drive the system further to explore various regions with high solution diversity so as to find the global optimality. HS has been applied to solve many optimization problems including function optimization, water distribution network, groundwater modeling, energy-saving dispatch, structural design, vehicle routing, and others.

2.2.9. *Firefly Algorithm*

The firefly algorithm (FA) was first developed by Xin-She Yang in 2007 (Yang, 2008; Yang, 2009) and is based on the flashing patterns and behavior of fireflies. In essence, FA uses the following three idealized rules:

(1) Fireflies are unisex so that one firefly will be attracted to other fireflies regardless of their sex.
(2) The attractiveness is proportional to the brightness and they both decrease as their distance increases. Thus for any two flashing fireflies, the less bright one will move towards the brighter one. If there is no brighter firefly than a particular firefly, it will move randomly.
(3) The brightness of a firefly is determined by the landscape of the objective function.

As a firefly's attractiveness is proportional to the light intensity seen by adjacent fireflies, we can now define the variation of attractiveness β with the distance r by

$$\beta = \beta_0 e^{-\gamma r^2}, \tag{2.20}$$

where β_0 is the attractiveness at $r = 0$.

The movement of a firefly i attracted to another more attractive (brighter) firefly j is determined by

$$x_i^{t+1} = x_i^t + \beta_0 e^{-\gamma r_{ij}^2}\left(x_j^t - x_i^t\right) + \alpha\,\epsilon_i^t, \qquad (2.21)$$

where the second term is due to the attraction. The third term is randomization with α being the randomization parameter, and ϵ_i^t is a vector of random numbers drawn from a Gaussian distribution or uniform distribution at time t. If $\beta_0 = 0$, it becomes a simple random walk. Furthermore, the randomization ϵ_i^t can easily be extended to other distributions such as Lévy flights.

The Lévy flight essentially provides a random walk whose random step length is drawn from a Lévy distribution

$$\text{Lévy} \sim u = t^{-\lambda}, \quad (1 < \lambda \le 3), \qquad (2.22)$$

which has an infinite variance with an infinite mean. Here the steps essentially form a random walk process with a power-law step-length distribution with a heavy tail. Some of the new solutions should be generated by Lévy walk around the best solution obtained so far; this will speed up the local search.

A demo version of firefly algorithm implementation, without Lévy flights, can be found at the Mathworks file exchange website (Yang, 2011b). The firefly algorithm has attracted much attention (Sayadi *et al.*, 2010; Apostolopoulos and Vlachos, 2011; Gandomi *et al.*, 2012). A discrete version of FA can efficiently solve non-deterministic polynomial-time hard, or NP-hard, scheduling problems (Sayadi *et al.*, 2010), while a detailed analysis has demonstrated the efficiency of FA over a wide range of test problems, including multobjective load dispatch problems (Apostolopoulos and Vlachos, 2011). Highly nonlinear and non-convex global optimization problems can be solved using FA efficiently (Gandomi *et al.*, 2012; Yang *et al.*, 2012).

2.2.10. *Cuckoo Search*

Cuckoo search (CS) is one of the latest nature-inspired metaheuristic algorithms, developed in 2009 by Xin-She Yang and Suash Deb (Yang and Deb, 2009). Cuckoo search is based on the brood parasitism of some cuckoo species. In addition, this algorithm is enhanced by the so-called Lévy flights

(Pavlyukevich, 2007), rather than by simple isotropic random walks. Recent studies show that CS is potentially far more efficient than PSO and genetic algorithms (Yang and Deb, 2010).

Cuckoos are fascinating birds, not only because of the beautiful sounds they can make, but also because of their aggressive reproduction strategy. Some species such as the *Ani* and *Guira* cuckoos lay their eggs in communal nests, though they may remove others' eggs to increase the hatching probability of their own eggs. Quite a number of species engage in obligate brood parasitism by laying their eggs in the nests of other host birds (often other species).

There are three basic types of brood parasitism: intraspecific brood parasitism, cooperative breeding, and nest takeover. Some host birds can engage in direct conflict with the intruding cuckoos. If a host bird discovers the eggs are not their owns, it will either get rid of the alien eggs or simply abandon its nest and build a new nest elsewhere. Some cuckoo species such as the New World brood-parasitic *Tapera* have evolved in such a way that female parasitic cuckoos are often very specialized in the mimicry in colour and pattern of the eggs of a few chosen host species. This reduces the probability of their eggs being abandoned and thus increases their reproductivity.

In addition, the timing of egg-laying of some species is also amazing. Parasitic cuckoos often choose a nest where the host bird has just laid its own eggs. In general, the cuckoo eggs hatch slightly earlier than their host eggs. Once the first cuckoo chick is hatched, the first instinctive action it will take is to evict the host eggs by blindly propeling the eggs out of the nest, which increases the cuckoo chick's share of food provided by its host bird. Studies also show that a cuckoo chick can also mimic the call of host chicks to gain access to more feeding opportunity.

For simplicity in describing the cuckoo search, we now use the following three idealized rules:

(1) Each cuckoo lays one egg at a time, and dumps it in a randomly chosen nest.
(2) The best nests with high-quality eggs will be carried over to the next generations.
(3) The number of available host nests is fixed, and the egg laid by a cuckoo is discovered by the host bird with a probability $p_a \in [0, 1]$. In this case, the host bird can either get rid of the egg, or simply abandon the nest and build a completely new nest.

As a further approximation, this last assumption can be approximated by a fraction p_a of the n host nests are replaced by new nests (with new random solutions).

For a maximization problem, the quality or fitness of a solution can simply be proportional to the value of the objective function. Other forms of fitness can be defined in a similar way to the fitness function in genetic algorithms.

For the implementation point of view, we can use the following simple representation, that each egg in a nest represents a solution, and each cuckoo can lay only one egg (thus representing one solution), the aim is to use the new and potentially better solutions (cuckoos) to replace a not-so-good solution in the nests. Obviously, this algorithm can be extended to the more complicated case where each nest has multiple eggs representing a set of solutions. For the present work, we will use the simplest approach where each nest has only a single egg. In this case, there is no distinction between egg, nest, or cuckoo, as each nest corresponds to one egg which also represents one cuckoo.

This algorithm uses a balanced combination of a local random walk and the global explorative random walk, controlled by a switching parameter p_a. The local random walk can be written as

$$x_i^{t+1} = x_i^t + \alpha s \otimes H(p_a - \epsilon) \otimes (x_j^t - x_k^t), \qquad (2.23)$$

where x_j^t and x_k^t are two different solutions selected randomly by random permutation, $H(u)$ is a Heaviside function, ϵ is a random number drawn from a uniform distribution, and s is the step size. On the other hand, the global random walk is carried out by using Lévy flights

$$x_i^{t+1} = x_i^t + \alpha L(s, \lambda), \qquad (2.24)$$

where

$$L(s, \lambda) = \frac{\lambda \Gamma(\lambda) \sin(\pi\lambda/2)}{\pi} \frac{1}{s^{1+\lambda}}, \quad (s \gg s_0 > 0). \qquad (2.25)$$

Here $\alpha > 0$ is the step size scaling factor, which should be related to the scales of the problem of interest. In most cases, we can use $\alpha = O(L/10)$, where L is the characteristic scale of the problem of interest, while in some cases $\alpha = O(L/100)$ can be more effective and avoid flying to far. The above equation is essentially the stochastic equation for a random walk. In general, a random walk is a Markov chain whose next status/location only depends on the current location (the first term in the above equation) and the transition probability (the second term). However, a substantial fraction of the new solutions should be generated by far field randomization and their locations should be far enough from the current best solution; this will make sure that the system will not be trapped in a local optimum (Yang and Deb, 2010).

Although the pseudo-code given in many papers is sequential, vectors should be used from the implementation point of view, as vectors are more efficient than loops. A Matlab implementation is given by the author, and can be downloaded (Yang, 2011c).

The literature on cuckoo search is expanding rapidly. Interestingly, cuckoo search was originally published in 2009 and our Matlab program was in the public domain in 2010, while some authors later in 2011 used a different name, cuckoo optimization algorithm, to essentially talk about the same inspiration from cuckoo behavior.

There has been a lot of attention and recent studies using cuckoo search with a diverse range of applications (Gandomi *et al.*, 2011; Walton *et al.*, 2011; Durgun and Yildiz, 2012; Yang and Deb, 2012). Walton *et al.* improved the algorithm by formulating a modified cuckoo search algorithm (Walton *et al.*, 2011), while Yang and Deb extended it to multiobjective optimization problems (Yang and Deb, 2012).

2.2.11. *Other Algorithms*

There are many other metaheuristic algorithms which are equally popular and powerful, and these include Tabu search (Glover and Laguna, 1997), artificial immune system (Farmer *et al.*, 1986), and others (Yang, 2010a; Yang, 2010b; Koziel and Yang, 2011).

The efficiency of metaheuristic algorithms can be attributed to the fact that they imitate the best features in nature, especially the selection of the fittest in biological systems which has evolved by natural selection over millions of years.

Two important characteristics of metaheuristics are intensification and diversification (Blum and Roli, 2003). Intensification intends to search locally and more intensively, while diversification makes sure the algorithm explores the search space globally (hopefully also efficiently).

Furthermore, intensification is also called exploitation, as it typically searches around the current best solutions and selects the best candidates or solutions. Similarly, diversification is also called exploration, as it tends to explore the search space more efficiently, often by large-scale randomization.

A fine balance between these two components is very important to the overall efficiency and performance of an algorithm. Too little exploration and too much exploitation could cause the system to be trapped in local optima, which makes it very difficult or even impossible to find the global optimum. On the other hand, if there is too much exploration but too little exploitation, it may be difficult for the system to converge and thus slows down the overall search performance. A proper balance itself is an optimization problem, and one of the main tasks of designing new algorithms is to find a certain balance concerning this optimality and/or tradeoff.

Furthermore, just exploitation and exploration are not enough. During the search, we have to use a proper mechanism or criterion to select the best solutions. The most common criterion is to use the *survival of the fittest*, that is to keep updating the the current best found so far. In addition, certain elitism is often used, and this is to ensure the best or fittest solutions are not lost, and should be passed onto the next generations.

References

Afshar, A., Haddad, O.B., Marino, M.A. and Adams, B.J. (2007). Honey-bee mating optimization (HBMO) algorithm for optimal reservoir operation, *J. Frank. Inst.*, **344**, 452–462.

Apostolopoulos, T. and Vlachos, A. (2011). Application of the firefly algorithm for solving the economic emissions load dispatch problem, *Int. J. Combinatorics*, **2011**, Article ID 523806. [Online] Available at: http://www.hindawi. com/journals/ijct/2011/523806.html. Date accessed: 14 Jan 2012.

Bandler, J.W., Cheng, Q.S., Dakroury, S.A., Mohamed, A.S., Bakr, M.H., Madsen, K. and Sondergaard, J. (2004). Space mapping: state of the art, *IEEE T. Microw. Theory Tech.*, **52**(1), 337–361.

Blum, C. and Roli, A. (2003). Metaheuristics in combinatorial optimization: overview and conceptual comparision, *ACM Comput. Surv.*, **35**, 268–308.

Conn, A.R., Schneinberg, K. and Vicente, L.N. (2009). *Introduction to Derivative-Free Optimization*, MPS-SIAM Series on Optimization, Society for Industrial and Applied Mathematics, Philadelphia: PA.

Durgun, I. and Yildiz, A.R. (2012). Structural design optimization of vehicle components using cuckoo search algorithm, *Mater. Test.*, **3**, 185–188.

Dorigo, M. and Stütle, T., (2004). *Ant Colony Optimization*, MIT Press, Cambridge: MA.

Farmer, J.D., Packard, N. and Perelson, A. (1986). The immune system, adapation and machine learning, *Physica D*, **2**, 187–204.

Gandomi, A.H., Yang, X.S. and Alavi, A.H. (2011). Cuckoo search algorithm: a meteheuristic approach to solve structural optimization problems, *Eng. Comput.*, **27**, 1–19.

Gandomi, A.H., Yang, X.S., Talatahari, S. and Deb, S. (2012). Coupled eagle strategy and differential evolution for unconstrained and constrained global optimization, *Comput. Math. Appl.*, **63**(1), 191–200.

Geem, Z.W., Kim, J.H. and Loganathan, G.V. (2001). A new heuristic optimization: harmony search, *Simulation*, **76**, 60–68.

Glover, F. and Laguna, M. (1997). *Tabu Search*, Kluwer Academic Publishers, Boston: MA.

Goldberg, D.E. (1989). *Genetic Algorithms in Search, Optimization and Machine Learning*, Addison Wesley, Reading: MA.

Holland, J. (1975). *Adaptation in Natural and Artificial Systems*, University of Michigan Press, Ann Anbor: MI.

Karaboga, D. (2005). An idea based on honey bee swarm for numerical optimization, Tech. Rep. TR06, Erciyes University, Turkey.

Karmarkar, N. (1984). A new polynomial-time algorithm for linear programming, *Combinatorica*, **4** (4), 373–395.

Kennedy, J. and Eberhart, R.C. (1995). Particle swarm optimization, in *Proc. of IEEE International Conference on Neural Networks*, IEEE Service Center, Piscataway: NJ, pp. 1942–1948.

Kennedy, J., Eberhart, R.C. and Shi, Y. (2001). *Swarm Intelligence*, Morgan Kaufmann Publishers, San Francisco: CA.

Kirkpatrick, S., Gelatt, C.D. and Vecchi, M.P. (1983). Optimization by simulated annealing, *Science*, **220** (4598), 671–680.

Koziel, S. and Yang, X.S. (2011). *Computational Optimization, Methods and Algorithms*, Springer-Verlag, Berlin.

Michalewicz, Z. (1998). *Genetic Algorithms + Data Structures = Evolution Programs*, Springer-Verlag, Berlin.

Nakrani, S. and Tovey, C. (2004). On honey bees and dynamic server allocation in internet hosting centers, *Adaptive Behaviour*, **12**(3-4), 223–240.

Nelder J.A. and Mead, R. (1965). A simplex method for function optimization, *Comput. J.*, **7**, 308–313.

Parpinelli, R.S. and Lopes, H.S. (2011). New inspirations in swarm intelligence: a survey, *Int. J. Bio-Inspired Computation*, **3**, 1–16.

Pavlyukevich I. (2007). Lévy flights, non-local search and simulated annealing, *J. Comput. Physics*, **226**, 1830–1844.

Pham, D.T., Ghanbarzadeh, A., Koc, E., Otri, S., Rahim, S. and Zaidi, M. (2006). The bees algorithm: A novel tool for complex optimisation problems, in Pham, D.T., Eldukhri, E.E. and Sovoka, A.J. (eds.), *Proceedings of IPROMS 2006 Conference*, Elsevier, Oxford, pp. 454–461.

Price, K., Storn, R. and Lampinen, J. (2005). *Differential Evolution: A Practical Approach to Global Optimization*, Springer-Verlag, Berlin.

Sayadi, M.K., Ramezanian, R. and Ghaffari-Nasab, N. (2010). A discrete firefly meta-heuristic with local search for makespan minimization in permutation flow shop scheduling problems, *Int. J. Ind. Eng. Comput.*, **1**, 1–10.

Storn, R. (1996). On the usage of differential evolution for function optimization, Biennial Conference of the North American Fuzzy Information Processing Society (NAFIPS), Berkeley: CA, pp. 519–523.

Storn, R. and Price, K. (1997). Differential evolution — a simple and efficient heuristic for global optimization over continuous spaces, *J. Global Optim.*, **11**, 341–359.

Talbi, E.G. (2009). *Metaheuristics: From Design to Implementation*, John Wiley & Sons, Hoboken: NJ.

Walton, S., Hassan, O., Morgan, K. and Brown, M.R. (2011). Modified cuckoo search: a new gradient free optimization algorithm. *Chaos Soliton. Fract.*, **44**(9), 710–718.

Yang, X.S. (2005). Engineering optimization via nature-inspired virtual bee algorithms, in *Artificial Intelligence and Knowledge Engineering Applications: A Bioinspired Approach*, Lecture Notes in Computer Science, **3562**, Springer-Verlag, Berlin, pp. 317–323.

Yang, X.S. (2008). *Nature-Inspired Metaheuristic Algorithms* (1st ed.), Lunver Press, Frome.

Yang, X.S. (2009). Firefly algorithms for multimodal optimization, in Watanabe, O. and Zeugmann, T. (eds.), *Proceedings of the 5th Symposium on Stochastic Algorithms, Foundation and Applications (SAGA 2009)*, Sapparo, Japan, LNCS, **5792**, pp. 169–178.

Yang, X.S. (2010a). *Nature-Inspired Metaheuristic Algoirthms* (2nd ed.), Luniver Press, Frome.

Yang, X.S. (2010b). *Engineering Optimization: An Introduction with Metaheuristic Applications*, Wiley, New York: NY.

Yang, X.S. (2010c). A new metaheuristic bat-inspired algorithm, in Cruz, C., Gonzales, J.R., Krasnogor, N., Pelta, D.A. and Terrazas, G. (eds.), *Nature-Inspired Cooperative Strategies for Optimization* (NICSO 2010), Springer-Verlag, Berlin, SCI **284**, pp. 65–74.

Yang, X.S. (2011a), Bat algorithm for multi-objective optimisation, *Int. J. Bio-Inspired Computation*, **3** (5), 267–274.

Yang, X.S. (2011b). Firefly algorithm. [Online] Available at: http://www.mathworks.com/matlabcentral/fileexchange/29693-firefly-algorithm. Date accessed: 14 Jan 2011.

Yang, X.S. (2011c). Cuckoo search. [Online] Available at: http://www.mathworks.com/matlabcentral/fileexchange/29809-cuckoo-search-cs-algorithm. Date accessed: 14 Jan 2011.

Yang, X.S. and Deb, S. (2009). Cuckoo search via Lévy flights, in Abraham, A., Carvalho, A., Herrara, F. and Pai, V. (eds.), *Proc. of World Congress on Nature & Biologically Inspired Computing* (NaBic 2009). Coimbatore, India, IEEE Publications, Piscataway: NJ, pp. 210–214.

Yang, X.S. and Deb, S. (2010). Engineering optimization by cuckoo search, *Int. J. Math. Modelling Num. Optimisation*, **1**(4), 330–343.

Yang, X.S. and Koziel, S. (2011). *Computational Optimization and Applications in Engineering and Industry*, Springer-Verlag, Berlin.

Yang, X.S., Hossein S. S., Gandomi, A.H. (2012), Firefly algorithm for solving
    non-convex economic dispatch problems with valve loading effect, *Applied
    Soft Computing*, **12**(3), 1180–1186.
Yang, X.S. and Deb, S. (2012). Multiobjective cuckoo search for design optimiza-
    tion, *Comput. Oper. Res.*, accepted October 2011. [Online] Available at:
    doi:10.1016/j.cor.2011.09.026. Date accessed: 14 Nov 2011.

Chapter 3

Surrogate-Based Optimization

Slawomir Koziel, Leifur Leifsson, and Xin-She Yang

3.1. Introduction

Numerical simulations have been used in the engineering community for decades. Initially, they were mostly utilized for design verification, partially because of limitations in computing power and due to lack of reliable software, programming environment, as well as appropriate algorithms. Today, computer simulations have become one of the main design tools. Ever-increasing demand for accuracy, as well as growing complexity of the engineering systems, has resulted in a situation where design-ready theoretical models are neither available nor sufficiently accurate. Consequently, the repetitive use of high-fidelity simulations is, in many cases, the only way to carry out the design process.

Simulation-driven design can be formulated as a nonlinear minimization problem of the following form

$$x^* = \arg\min_{x} f(x), \qquad (3.1)$$

where $f(x)$ is the objective function, and $x \in R^n$ the design variable vector. In many engineering problems, f is of the form $f(x) = U(R_f(x))$, where $R_f \in R^m$ denotes the response vector of the system of interest (in particular, one may have $m > n$ or even $m \gg n$ (Bandler *et al.*, 2004)), whereas U is a given scalar merit function. In particular, U can be defined through a norm that measures the distance between $R_f(x)$ and a (user-selected) target vector y. An optimal design vector is denoted by x^*. In many cases, R_f is obtained through computationally expensive computer simulations. We will refer to it as a high-fidelity or fine model. To simplify the notation, f itself will also be referred to as the high-fidelity (fine) model.

41

Conventional optimization algorithms (e.g., gradient-based schemes with numerical derivatives) require tens, hundreds, or even thousands of objective function calls per run (depending on the number of design variables). Numerical simulations can be very time-consuming, in some instances, the simulation time can be as long as several hours, days, or even weeks per single design. Therefore, attempts to solve (3.1) by embedding the simulator directly in the optimization loop may often be impractical. The presence of massive computing resources is not always translated in computational speedup, due to a growing demand for simulation accuracy, both by including multi-physics and second-order effects, and by using finer discretization of the structure under consideration.

Objective functions coming from computer simulations are often analytically intractable (i.e., discontinuous, nondifferentiable, and inherently noisy) and can complicate simulation-driven design. In addition to this, sensitivity information is frequently unavailable, or too expensive to compute. In some cases it is, however, possible to obtain derivative information inexpensively through adjoint sensitivities (El Sabbagh *et al.*, 2006), but only if detailed knowledge of and access to the simulator source code is available. Recently, adjoint sensitivity has become commercially available, e.g., CST Microwave Studio (CST, 2011).

Surrogate-based optimization (SBO) (Bandler *et al.*, 2004; Queipo *et al.*, 2005; Forrester and Keane, 2009) has been suggested as an effective approach when time-consuming computer simulations are involved. The basic concept of SBO is that the direct optimization of the computationally expensive model is replaced by an iterative process that involves the creation, optimization, and updating of a fast and analytically tractable surrogate model. The surrogate should be a reasonably accurate representation of the high-fidelity model, at least locally. The design obtained through optimizing the surrogate model is verified by evaluating the high-fidelity model. The high-fidelity model data obtained in this verification process is then used to update the surrogate. SBO proceeds in this predictor-corrector fashion iteratively until some termination criterion is met. Because most of the operations are performed on the surrogate model, SBO can often significantly reduce the computational cost of the optimization process when compared to optimizing the high-fidelity model directly, without resorting to any surrogate.

In this chapter, we review the basics of surrogate-based optimization. We briefly recall the SBO concept and discuss a few popular surrogate-based optimization techniques, including space mapping (Bandler *et al.*, 2004; Koziel *et al.*, 2006; Koziel *et al.*, 2008a), approximation model management (Alexandrov and Lewis, 2001a), manifold mapping (Echeverria and Hemker, 2005), and surrogate-management framework (Booker *et al.*, 1999). We review various ways of generating surrogate models, and we emphasize the distinction between models based on function approximations of sampled high-fidelity model data and models constructed from physically-based low-fidelity models. We also discuss other important issues concerning global optimization such as exploration versus exploitation in the relevant part of the main text.

3.2. Surrogate-Based Optimization Concept

In this section, we briefly recall the concept of surrogate-based optimization and discuss its potential advantages over the direct optimization approach.

3.2.1. *Direct Optimization*

It is pertinent to describe the direct optimization process before the surrogate-based one. Direct optimization employs the high-fidelity simulation model directly in the optimization loop as shown in Fig. 3.1(a), and the flow can be described as follows. First, an initial design $x^{(0)}$ is generated and the high-fidelity simulation model is evaluated at that design, yielding values of the objective function and the constraints. Then, the optimization algorithm finds a new design, x, and the high-fidelity simulation model is evaluated at that design and the objective and constraints are recalculated. Based on the improvement or deterioration in the objective function and the values of the constraints (fulfilled, critical, or violated), either the optimizer finds another design to evaluate, or it uses the current design for the ith design iteration to yield design $x^{(i)}$. The high-fidelity model can be evaluated several times during one design iteration. The optimization loop is repeated until a given termination condition is met and an optimized design has been reached. The termination condition could be, for example, based on the change in airfoil shape between two adjacent design iterations $x^{(i)}$ and $x^{(i+1)}$. There are a number of conventional optimization techniques available, both gradient-based (Nocedal and Wright, 2006) and derivative-free (Conn *et al.*, 2009). In general, most of these methods require a large number of objective function evaluations, which may be prohibitive if computationally expensive simulation is involved in the process.

3.2.2. *Surrogate-Based Optimization*

Direct optimization of the high-fidelity model may not work or can be impractical, as mentioned in the introduction. There are several possible reasons, including a high computational cost of each model evaluation, numerical noise, high nonlinearity and/or discontinuities in the cost function. Surrogate-based optimization (SBO) (Queipo *et al.*, 2005; Koziel *et al.*, 2006; Forrester and Keane, 2009) aims at alleviating such problems by using an auxiliary model, the surrogate, that is preferably fast, amenable to optimization, and yet reasonably accurate. Optimization of the surrogate model yields an approximation of the minimizer associated to the high-fidelity model. This approximation is verified by evaluating the high-fidelity model at the predicted high-fidelity model minimizer. Depending on the result of this verification, the optimization process may be terminated. Otherwise, the surrogate model is updated using the newly available high-fidelity model data, and then re-optimized to obtain a new, and hopefully better, approximation of the high-fidelity model minimizer.

The generic SBO process can be summarized as follows (Fig. 3.1(b)):

1. Generate the initial surrogate model.
2. Obtain an approximate solution to (3.1) by optimizing the surrogate.
3. Evaluate the high-fidelity model at the approximate solution computed in Step 2.
4. Update the surrogate model using the new high-fidelity model data.
5. Stop if the termination condition is satisfied; otherwise go to Step 2.

Formally, the SBO process can be written as an iterative procedure (Queipo *et al.*, 2005; Koziel *et al.*, 2006; Forrester and Keane, 2009):

$$x^{(i+1)} = \arg\min_{x} s^{(i)}(x). \tag{3.2}$$

This scheme generates a sequence of designs $x^{(i)}$ that (hopefully) converge to a solution (or a good approximation) of the original design problem (3.1). Each $x^{(i+1)}$ is the optimal design of the surrogate model $s^{(i)}$, which is assumed to be a computationally cheap and sufficiently reliable representation of the fine model f, particularly in the neighborhood of the current design $x^{(i)}$. Under these assumptions, the algorithm (3.2) aims at a sequence of designs to quickly approach x^*. Typically, and for verification purposes, the high-fidelity model is evaluated only once per iteration (at every new design $x^{(i+1)}$). The data obtained from the validation is used to update the surrogate model. Because the surrogate

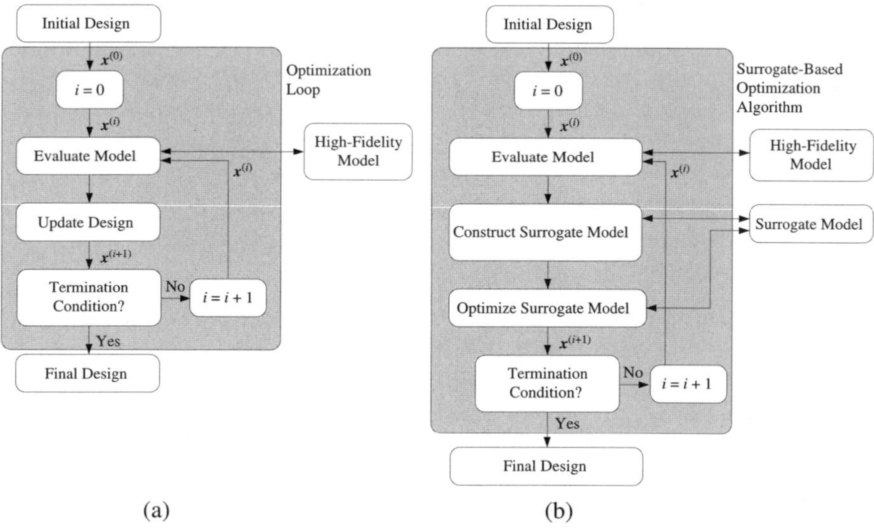

(a) (b)

Fig. 3.1. Direct versus surrogate-based optimization: (a) A flowchart of the direct optimization process with the high-fidelity model evaluation embedded within the loop, (b) a flowchart of the surrogate-based optimization (SBO) process. An approximate high-fidelity model minimizer is obtained iteratively by optimizing the surrogate model. The high-fidelity model is evaluated at each new design for verification purposes. If the termination condition is not satisfied, the surrogate model is updated and the search continues. In most cases the high-fidelity model is evaluated only once per iteration. The number of iterations needed in SBO is often substantially smaller than for conventional (direct) optimization techniques.

model is computationally cheap, the optimization cost associated with (3.2) can — in many cases — be viewed as negligible, so that the total optimization cost is determined by the evaluation of the high-fidelity model. Normally, the number of iterations often needed by the SBO algorithm is substantially smaller than needed by any method that optimizes the high-fidelity model directly (e.g., gradient-based schemes with numerical derivatives) (Koziel *et al.*, 2006).

The SBO algorithm (3.2) is provably convergent to a local optimum of f (Alexandrov *et al.*, 1998) if the surrogate model satisfies zero- and first-order consistency conditions with the high-fidelity model (i.e., $s^{(i)}(x^{(i)}) = f(x^{(i)})$ and $\nabla s^{(i)}(x^{(i)}) = \nabla f(x^{(i)}$ (Alexandrov and Lewis, 2001)), and the surrogate-based algorithm is enhanced by a trust-region mechanism (Conn *et al.*, 2000) (see Section 3.2.2). It should be noted that the satisfaction of the first-order consistency condition requires high-fidelity model sensitivity data. Formally, some standard assumptions concerning the smoothness of the functions involved are also necessary for convergence (Echeverria and Hemker, 2008). Convergence can also be guaranteed if the SBO algorithm is embedded within the framework given in Koziel *et al.* (2006) and Koziel *et al.* (2008) (space mapping), Echeverria and Hemker (2008) (manifold mapping), or Booker *et al.* (1999) (surrogate management framework).

It should be noted that the SBO scheme (3.2), particularly if the surrogate model is first-order consistent, is typically executed as a local search. If the surrogate model is constructed globally over the entire design space, it is of course possible to turn the process (3.2) into a global optimization algorithm, where the surrogate model is optimized using evolutionary algorithms such as genetic algorithms, and is updated using certain statistical infill criteria based on expected improvement of the objective function or minimization of the (global) modeling error. More extensive discussion of these aspects of surrogate-based optimization can be found in Section 3.3.6.

Numerous SBO techniques have been developed over the last decade or so. Some of the popular techniques are described in Section 3.3. Although all of them are based on the same principle, as described above, they can differ in operation and the method of constructing the surrogate model. The latter is a key component of any SBO algorithm. It has to be computationally cheap, preferably smooth, and, at the same time, reasonably accurate, so that it can be used to predict the approximate location of high-fidelity model minimizers. We can clearly distinguish between approximation- and physics-based surrogate models. Approximation-based (or functional) surrogate models are constructed through approximations of the high-fidelity model data obtained by sampling the design space using appropriate design of experiments (DOE) methodologies (Queipo *et al.*, 2005). Some of the strategies for allocating samples (Simpson *et al.*, 2001), generating approximations (Simpson *et al.*, 2001; Queipo *et al.*, 2005; Forrester and Keane, 2009), as well as validating the surrogates, are discussed in Section 3.4. Another modeling approach exploits knowledge about the system under consideration embedded in physics-based low-fidelity models. Here, the surrogates are constructed by applying appropriate correction techniques. This type of surrogate model is discussed in Section 3.5.

3.3. Surrogate-Based Optimization Techniques

We now briefly review a few selected SBO techniques that have gained popularity in the engineering community. More specifically, we describe the approximation model management framework (AMMO) (Alexandrov and Lewis, 2001), space mapping (Bandler *et al.*, 2004), manifold mapping (Echeverria and Hemker, 2008), and the surrogate model framework (SMF) (Booker *et al.*, 1999). We also mention other relevant techniques briefly and discuss the important issue of the trade-off between exploration and exploitation in the surrogate-based optimization process.

3.3.1. *Approximation Model Management Optimization*

The approximation model management optimization (AMMO) algorithm (Alexandrov and Lewis, 2001) is a general approach for controlling the use of variable-fidelity models when solving a nonlinear minimization problem, such as Eq. (3.1). A flowchart of the AMMO algorithm is shown in Fig. 3.2. The optimizer receives the function and constraint values, as well as their sensitivities, from the low-fidelity model. The response of the low-fidelity model is corrected to satisfy at least zero- and first-order

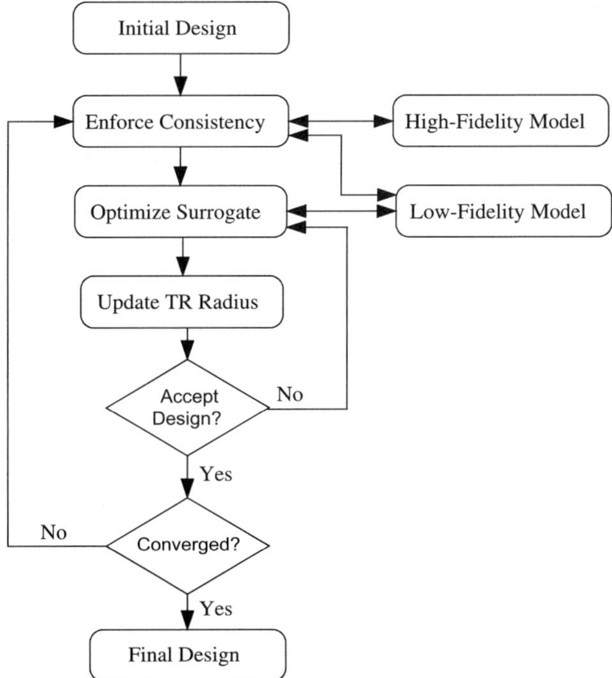

Fig. 3.2. A flowchart of the approximation model management optimization (AMMO) algorithm.

consistency conditions with the high-fidelity model, i.e., agreement between the function values and the first-order derivatives at a given iteration point. The expensive high-fidelity computations are performed outside the optimization loop and serve to re-calibrate the low-fidelity model occasionally, based on a set of systematic criteria. AMMO exploits the trust-region methodology (Conn *et al.*, 2000), which is an adaptive move limit strategy for improving the global behavior of optimization algorithms based on local models. By combining the trust-region approach with the use of the low-fidelity model, satisfying at least first-order consistency conditions, convergence of AMMO to the optimum of the high-fidelity model can be guaranteed.

First-order consistency in variable-fidelity SBO can be insufficient to achieve acceptable convergence rates, which can be similar to those achieved by first-order optimization methods, such as steepest-descent or sequential linear programming (Rao, 1996). More successful optimization methods, such as sequential quadratic programming, use at least approximate second-order information to achieve super-linear or quadratic convergence rates in the neighborhood of the minimum. Eldred *et al.* (2004) present second-order correction methods for variable-fidelity SBO algorithms. The second-order corrections enforce consistency with the high-fidelity model Hessian.

However, since full second-order information is not commonly available in practical engineering problems, consistency can also be enforced to an approximation using finite difference, quasi-Newton, or Gauss–Newton to the high-fidelity Hessian. The results show that all of these approaches outperform the first-order corrections. Then again, the second-order corrections come at a price, since additional function evaluations are required. Additionally, they can become impractical for large design problems, unless adjoint-based gradients are employed. Finally, the issue of how numerical noise affects the second-order corrected SBO process has not been addressed.

3.3.2. *Space Mapping*

The space mapping (SM) paradigm (Bandler *et al.*, 2004; Koziel *et al.*, 2006) was originally developed in microwave engineering optimal design applications, and gave rise to an entire family of surrogate-based optimization approaches. Nowadays, its popularity is spreading across several engineering disciplines (Leary *et al.*, 2001; Bandler *et al.*, 2004; Redhe and Nilsson, 2004). The initial space mapping optimization methodologies were based on input SM (Bandler *et al.*, 2004), i.e., a linear correction of the coarse model design space. This kind of correction is well suited for many engineering problems, particularly in electrical engineering, where the model discrepancy is mostly due to second-order effects (e.g., presence of parasitic components). In these applications the model response ranges are often similar in shape, but slightly distorted and/or shifted with respect to a sweeping parameter (e.g., signal frequency).

Space mapping can be incorporated in the SBO framework (3.2) by identifying the sequence of surrogates with

$$s^{(0)}(\boldsymbol{x}) = U(\boldsymbol{R}_c(\boldsymbol{x})), \qquad (3.3)$$

and

$$s^{(i)}(\boldsymbol{x}) = U(R_s(\boldsymbol{x}; \boldsymbol{p}_{SM}^{(i)})), \qquad (3.4)$$

for $i > 0$. Here, $\boldsymbol{R}_c$ is the physics-based low-fidelity (or coarse) model, whereas $\boldsymbol{R}_s$ is the space mapping surrogate defined as a composition of $\boldsymbol{R}_c$ and some simple (usually linear) transformations parameterized by $\boldsymbol{p}_{SM}$. The parameters $\boldsymbol{p}_{SM}^{(i)}$ are obtained in a parameter extraction process $\boldsymbol{p}_{SM}^{(i)} = \mathrm{argmin}\{\boldsymbol{p} : \Sigma_k \|\boldsymbol{R}_f(\boldsymbol{x}^{(k)}) - \boldsymbol{R}_s(\boldsymbol{x}^{(k)}; \boldsymbol{p})\|\}$ that aim at reducing the misalignment between the high-fidelity model and the surrogate model at the previously considered designs. An example SM transformation is $\boldsymbol{R}_s(\boldsymbol{x}) = \boldsymbol{R}_c(\boldsymbol{x} + \boldsymbol{p})$, which is the so-called input SM (Bandler *et al.*, 2004). Other examples of SM transformations as well as a more detailed exposition of space mapping, can be found in Chapter 4.

The accuracy of the corrected surrogate will clearly depend on the quality of the coarse model (Koziel and Bandler, 2007b). In microwave design applications it has been observed that the number of points p needed for obtaining a satisfactory SM-based corrected surrogate is on the order of the number of optimization variables n (Bandler *et al.*, 2004). Though output SM can be used to obtain both zero- and first-order consistency conditions with $f(\boldsymbol{x})$, many other SM-based optimization algorithms that have been applied in practice do not satisfy those conditions, and in some occasions convergence problems have been identified (Koziel *et al.*, 2008). Additionally, the choice of an adequate SM correction approach is not always obvious (Koziel *et al.*, 2008). However, on multiple occasions and in several different disciplines (Leary *et al.*, 2001; Bandler *et al.*, 2004; Redhe and Nilsson, 2004), space mapping has been reported as a very efficient means for obtaining satisfactory optimal designs.

Convergence properties of space mapping optimization algorithms can be improved when these are safeguarded by a trust region (Koziel *et al.*, 2010a). Similarly to AMMO, the SM surrogate model optimization is restricted to a neighborhood of $x^{(i)}$ as follows:

$$x^{(i+1)} = \arg\min_x s^{(i)}(x) \quad \text{subject to} \quad \|x - x^{(i)}\| \leq \delta^{(i)}, \qquad (3.5)$$

where $\delta^{(i)}$ denotes the trust-region radius at iteration i. The trust region is updated at every iteration as in Conn *et al.* (2000). A number of enhancements for space mapping have been suggested recently in the literature (e.g., zero-order and approximate/exact first-order consistency conditions with $f(x)$ (Koziel *et al.*, 2010a), or adaptively constrained parameter extraction (Koziel *et al.*, 2010b)).

The quality of a surrogate within space mapping can be assessed by means of the techniques described in Koziel *et al.* (2008) and Koziel and Bandler (2007a). These methods are based on evaluating the high-fidelity model at several points (and thus, they require some extra computational effort). With that information, some conditions required for convergence are approximated numerically, and as a result, low-fidelity models can be compared based on these approximate conditions. The quality assessment algorithms presented in Koziel *et al.* (2008) and Koziel and Bandler (2007b) can also be embedded into SM optimization algorithms in order to throw some light on the delicate issue of selecting the most adequate SM surrogate correction.

It should be emphasized that space mapping is not a general-purpose optimization approach. The existence of a computationally cheap and sufficiently accurate low-fidelity model is an important prerequisite for this technique. If such a coarse model does exist, satisfactory designs are often obtained by space mapping after a relatively small number of evaluations of the high-fidelity model. This number is usually on the order of the number of optimization variables n (Koziel *et al.*, 2008), and very frequently represents a dramatic reduction in the computational cost required for solving the same optimization problem with other methods that do not rely on surrogates. In the absence of the above-mentioned low-fidelity model, space mapping optimization algorithms may not perform efficiently.

3.3.3. *Manifold Mapping*

Manifold mapping (MM) (Echeverria and Hemker, 2005; Echeverria, 2007a) is a particular case of output space mapping, that is supported by a convergence theory (Echeverria, 2007a; Echeverria and Hemker, 2008), and does not require the parameter extraction step (cf. Section 3.3.2). Manifold mapping can be integrated in the SBO framework by just considering $s^{(i)}(\boldsymbol{x}) = U(\boldsymbol{R}_s^{(i)}(\boldsymbol{x}))$ with the response correction for $i \geq 0$ defined as

$$\boldsymbol{R}_s^{(i)}(\boldsymbol{x}) = \boldsymbol{R}_f(\boldsymbol{x}^{(i)}) + \boldsymbol{S}^{(i)}(\boldsymbol{R}_c(\boldsymbol{x}) - \boldsymbol{R}_c(\boldsymbol{x}^{(i)})), \tag{3.6}$$

where $\boldsymbol{S}^{(i)}$, for $i \geq 1$, is the following $m \times m$ matrix:

$$\boldsymbol{S}^{(i)} = \Delta \boldsymbol{F} \Delta \boldsymbol{C}^\dagger, \tag{3.7}$$

with

$$\Delta \boldsymbol{F} = [\boldsymbol{R}_f(\boldsymbol{x}^{(i)}) - \boldsymbol{R}_f(\boldsymbol{x}^{(i-1)}) \dots \boldsymbol{R}_f(\boldsymbol{x}^{(i)}) - \boldsymbol{R}_f(\boldsymbol{x}^{(\max\{i-n,0\})})] \tag{3.8}$$

$$\Delta \boldsymbol{C} = [\boldsymbol{R}_c(\boldsymbol{x}^{(i)}) - \boldsymbol{R}_c(\boldsymbol{x}^{(i-1)}) \dots \boldsymbol{R}_c(\boldsymbol{x}^{(i)}) - \boldsymbol{R}_c(\boldsymbol{x}^{(\max\{i-n,0\})})]. \tag{3.9}$$

The matrix $\boldsymbol{S}^{(0)}$ is typically taken as the identity matrix $\boldsymbol{I}_m$. Here, $\dagger$ denotes the pseudoinverse operator defined for $\Delta \boldsymbol{C}$ as

$$\Delta \boldsymbol{C}^\dagger = \boldsymbol{V}_{\Delta C} \Sigma_{\Delta C}^\dagger \boldsymbol{U}_{\Delta C}^T, \tag{3.10}$$

where $\boldsymbol{U}_{\Delta \mathbf{C}}$, $\Sigma_{\Delta \mathbf{C}}$, and $\boldsymbol{V}_{\Delta \mathbf{C}}$ are the factors in the singular value decomposition of $\Delta \boldsymbol{C}$. The matrix $\Sigma_{\Delta \mathbf{C}}^\dagger$ is the result of inverting the nonzero entries in $\Sigma_{\Delta \mathbf{C}}$, leaving the zeroes invariant (Echeverria and Hemker, 2005). Some mild general assumptions on the model responses are made in theory (Echeverria, 2007a) so that every pseudoinverse introduced is well defined.

The response correction $\boldsymbol{R}_s^{(i)}(\boldsymbol{x})$ is an approximation of

$$\boldsymbol{R}_s^*(\boldsymbol{x}) = \boldsymbol{R}_f(\boldsymbol{x}^*) + \boldsymbol{S}^*(\boldsymbol{R}_c(\boldsymbol{x}) - \boldsymbol{R}_c(\boldsymbol{x}^*)), \tag{3.11}$$

with $\boldsymbol{S}^*$ being the $m \times m$ matrix defined as

$$\boldsymbol{S}^* = J_f(\boldsymbol{x}^*) J_c^\dagger(\boldsymbol{x}^*), \tag{3.12}$$

where $J_f(\boldsymbol{x}^*)$ and $J_c(\boldsymbol{x}^*)$ stand for the fine and coarse model response Jacobian, respectively, evaluated at $\boldsymbol{x}^*$. Obviously, neither $\boldsymbol{x}^*$ nor $\boldsymbol{S}^*$ is known beforehand. Therefore, one needs to use an iterative approximation, such as the one in (3.6)–(3.10), in the actual manifold-mapping algorithm.

The manifold-mapping model alignment is illustrated in Fig. 3.3 for the least-squares optimization problem $U(\boldsymbol{R}_f(\boldsymbol{x})) = \|\boldsymbol{R}_f(\boldsymbol{x}) - \boldsymbol{y}\|_2^2$, with $\boldsymbol{y} \in R^m$ being the design specifications given. In that figure the point $\boldsymbol{x}_c^*$ denotes the minimizer corresponding to the coarse model cost function $U(\boldsymbol{R}_c(\boldsymbol{x}))$. We note that, in absence of constraints, the optimality

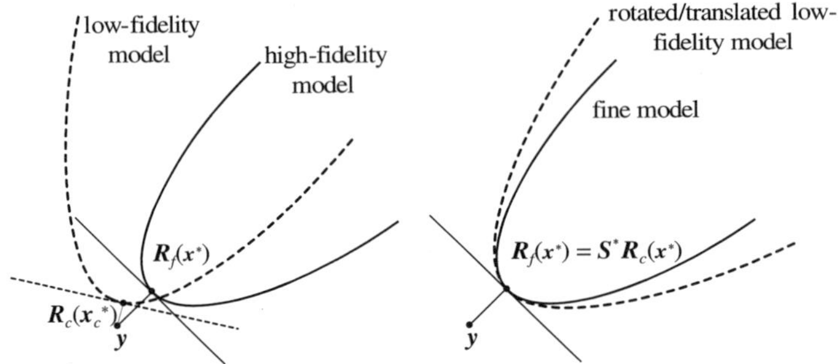

Fig. 3.3. Illustration of the manifold-mapping model alignment for a least-squares optimization problem. The point x_c^* denotes the minimizer corresponding to the coarse model response, and the point y is the vector of design specifications. Thin-solid and dashed-straight lines denote the tangent planes for the fine and coarse model response at their optimal designs, respectively. By the linear correction S^*, the point $R_c(x^*)$ is mapped to $R_f(x^*)$, and the tangent plane for $R_c(x)$ at $R_c(x^*)$ to the tangent plane for $R_f(x)$ at $R_f(x^*)$ (Echeverria and Hemker, 2008).

associated to the least-squares objective function is translated into the orthogonality between the tangent plane for $R_f(x)$ at x^* and the vector $R_f(x^*) - y$.

For least-squares optimization problems, manifold mapping is supported by a mathematically sound convergence theory (Echeverria and Hemker, 2008). In general, manifold-mapping algorithms can be expected to converge for a merit function U sufficiently smoothly. Since the correction in (3.6) does not involve U, if the model responses are smooth enough, and even when U is not differentiable, manifold mapping may still yield satisfactory solutions. The experimental evidence given in Koziel and Echeverria (2010c) for designs based on minimax objective functions indicates that the MM approach can be used successfully in more general situations than those for which theoretical results have been obtained.

The basic manifold-mapping algorithm can be modified in a number of ways. Convergence appears to improve if derivative information is introduced in the algorithm (Echeverria and Hemker, 2008). The incorporation of a Levenberg–Marquardt strategy in manifold mapping (Hemker and Echeverria, 2007) can be seen as a convergence safeguard analogous to a trust-region method (Conn *et al.*, 2000). Manifold mapping can also be extended to designs where the constraints are determined by time-consuming functions, and for which surrogates are available as well (Echeverria, 2007b).

3.3.4. *Surrogate Management Framework*

The surrogate management framework (SMF) algorithm (Booker *et al.*, 1999) is mainly based on pattern search (Kolda *et al.*, 2003), which is a general set of derivative-free optimizers that can be proven to be globally convergent to first-order stationary points. A pattern search optimization algorithm is based on exploring the search space by means of a structured set of points (pattern or stencil) that is modified along iterations.

The pattern search scheme considered in Booker *et al.* (1999) has two main steps in each iteration: SEARCH and POLL. Each iteration starts with a pattern of size Δ centered at $x^{(i)}$. The SEARCH step is optional and is always performed before the POLL step. In the SEARCH step, a (small) number of points are selected from the search space (typically by means of a surrogate), and the cost function $f(x)$ is evaluated at these points. If the cost function for some of them improves on $f(x^{(i)})$, the SEARCH step is declared successful, the current pattern is centered at this new point, and a new SEARCH step is started. Otherwise a POLL (or a local search) step is taken. Polling requires computing $f(x)$ for neighboring points in the pattern. If one of these points is found to improve on $f(x^{(i)})$, the POLL step is declared successful, the pattern is translated to this new point, and a new SEARCH step is performed. Otherwise the whole pattern search iteration is considered unsuccessful and the termination condition is checked. This stopping criterion is typically based on the pattern size Δ (Booker *et al.*, 1999; Marsden *et al.*, 2004). If, after the unsuccessful pattern search iteration another iteration is needed, the pattern size Δ is decreased, and a new SEARCH step is taken with the pattern centered again at $x^{(i)}$. Surrogates are incorporated in the SMF through the SEARCH step. For example, kriging (with Latin hypercube sampling) is considered in the SMF application studied in Marsden *et al.* (2004). The flow of the algorithm is shown in Fig. 3.4.

In order to guarantee convergence to a stationary point, the set of vectors formed by each pattern point and the pattern center should be a generating (or positive spanning) set (Kolda *et al.*, 2003; Marsden *et al.*, 2004). A generating set for R^n consists of a set of vectors whose non-negative linear combinations span R^n. Generating sets are crucial in proving convergence (for smooth objective functions) due to the following property: if a generating set is centered at $x^{(i)}$ and $\nabla f(x^{(i)}) \neq 0$, then at least one of the vectors in the generating set defines a descent direction (Kolda *et al.*, 2003). Therefore, if $f(x)$ is smooth and $\nabla f(x^{(i)}) \neq 0$, we can expect that

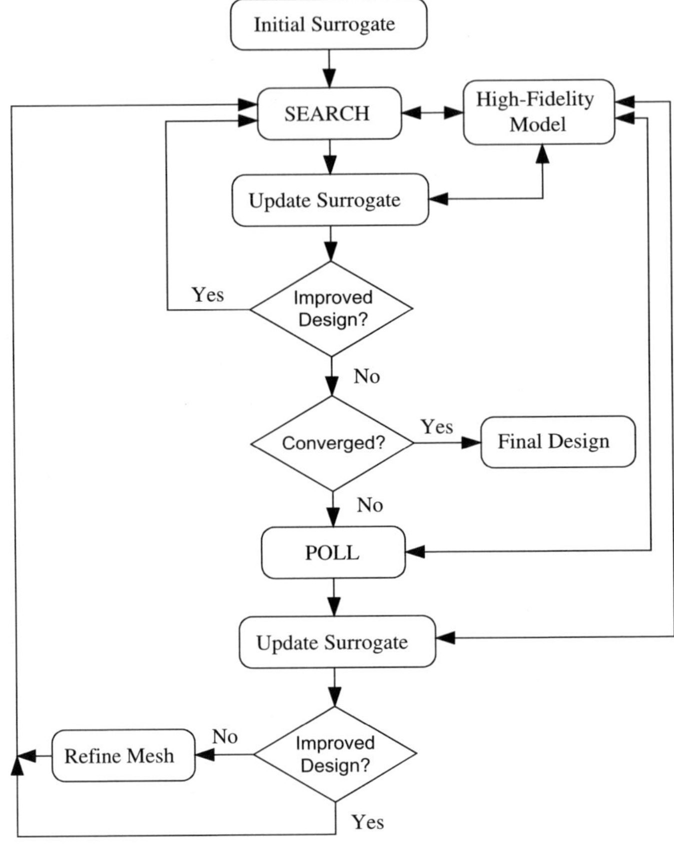

Fig. 3.4.   A flowchart of the surrogate management framework (SMF).

for a pattern size Δ small enough, some of the points in the associated stencil will improve on $f(\boldsymbol{x}^{(i)})$.

Though pattern search optimization algorithms typically require many more function evaluations than gradient-based techniques, the computations in both the SEARCH and POLL steps can be performed in a distributed fashion. On top of that, the use of surrogates, as is the case for the SMF, generally accelerates noticeably the entire optimization process.

3.3.5. *Other Techniques*

Several other SBO techniques are available and they differ mainly in the way the surrogate model is constructed. One particular way is to correct the response of the coarse model to fit the fine model response. This approach can be roughly categorized into parametric techniques, where the response correction function can be represented using explicit formulas, and non-parametric ones, where the response correction is defined implicitly through certain relations between the coarse and fine models which may not have any compact analytical form.

Examples of parametric correction techniques are output space mapping (Koziel *et al.*, 2010a), and multi-point response correction (Koziel, 2010a). Examples of non-parametric correction techniques are adaptive response correction (Koziel *et al.*, 2009), shape-preserving response prediction (Koziel, 2010b), and adaptively adjusted design specifications (Koziel, 2010c). All of these techniques are described in Chapter 5.

3.3.6. *Exploration vs. Exploitation*

The surrogate-based optimization framework starts from an initial surrogate model which is updated using the high-fidelity model data that is accumulated in the optimization process. In particular, the high-fidelity model has to be evaluated for verification at any new design $x^{(i)}$ provided by the surrogate model. The new points at which we evaluate the high-fidelity model are sometimes referred to as infill points (Forrester and Keane, 2009). We reiterate that this data can be used to enhance the surrogate. The selection of the infill points is also known as adaptive sampling (Forrester and Keane, 2009).

Infill points in approximation model management optimization, space mapping, and manifold mapping are in practice selected through local optimization of the surrogate (global optimization for problems with a medium/large number of variables and even relatively inexpensive surrogates can be a time-consuming procedure). The new infill points in the surrogate management framework are taken based only on high-fidelity cost function improvement. As we have seen in this section, the four surrogate-based optimization approaches discussed are supported by local optimality theoretical results. In other words, these methodologies intrinsically aim at the exploitation of a certain region of the design space (the neighborhood of a first-order stationary point). If the surrogate is valid globally, the first iterations of these four optimization approaches can be used to avoid being trapped in unsatisfactory local solutions (i.e., global exploration steps).

The exploration of the design space implies in most cases a global search. If the underlying objective function is non-convex, exploration usually boils down to performing a global sampling of the search space, for example, by selecting those points that maximize some estimation of the error associated to the surrogate considered (Forrester and Keane, 2009). It should be stressed that global exploration is often impractical, especially for computationally expensive cost functions with a medium/large number of optimization variables (more than a few tens). Additionally, pure exploration may not be a good approach for updating the surrogate in an optimization context, since a great amount of computing resources may be spent in modeling parts of the search space that are either infeasible or far from optimality, from the design optimization point of view.

Therefore, it appears that in optimization there should be a balance between exploitation and exploration. As suggested by Forrester and Keane (2009), this tradeoff could be formulated in the context of surrogate-based optimization, for example, by means of a bi-objective optimization problem with a global measure of the error associated to the surrogate as a second objective function.

One formulation is to balance exploitation/exploration of the prediction $\hat{y}(\boldsymbol{x})$ by using the variance $s^2(\boldsymbol{x})$. This can be done by minimizing a *statistical lower bound* (Forrester and Keane, 2009)

$$LB(\boldsymbol{x}) = \hat{y}(\boldsymbol{x}) - As(\boldsymbol{x}), \tag{3.13}$$

where A is a constant that controls the exploitation/exploration balance. From (3.13) we see that pure exploitation, $LB(\boldsymbol{x}) \rightarrow \hat{y}(\boldsymbol{x})$, is attained if $A \rightarrow 0$. On the other hand, pure exploration is attained when $A \rightarrow \infty$. It is not quite clear how to choose the constant A as it is function dependent. A possible solution is to try a number of values and position infill points where there are clusters of minima of (3.13) (Forrester and Keane, 2009).

Another formulation is to maximize the probability of improvement upon the best observed objective function value. Here, the uncertainty in the prediction $\hat{y}(\boldsymbol{x})$ is represented by a Gaussian distribution with variance $s^2(\boldsymbol{x})$ centered on it. Yet another formulation is to maximize the expected improvement given the mean $\hat{y}(\boldsymbol{x})$ and variance $s^2(\boldsymbol{x})$. Another possible approach is to use model selection and information criteria such as Akaike information criterion (AIC) (Yang and Forbes, 2011).

As mentioned above, these hybrid approaches will find difficulties in performing an effective global search in designs with a medium/large number of optimization variables.

Surrogate-based methods that utilize infill criteria as a part of their formulation, i.e., balancing the need for exploration versus exploitation, are often referred to as efficient global optimization (EGO) techniques. A detailed discussion of EGO is given by Jones *et al.* (1998) and by Forrester and Keane (2009).

3.4. Approximation-Based Surrogate Models

In this section, we describe the fundamental steps for constructing approximation-based surrogates. Various design of experiment techniques are presented in Section 3.4.1. The surrogate formulations and the model validations steps are tackled in Section 3.4.2 and Section 3.4.3, respectively.

3.4.1. *Surrogate Construction Overview*

Approximation-based surrogate models (Søndergaard, 2003; Forrester and Keane, 2009):

- can be constructed without previous knowledge of the physical system of interest;
- are generic, and therefore applicable to a wide class of problems;
- are based on (usually simple) algebraic models, and;
- are often very cheap to evaluate but require a considerable amount of data to ensure reasonable general accuracy.

An initial approximation-based surrogate can be generated using high-fidelity model data obtained through sampling of the design space. Figure 3.5 shows the model construction flowchart for an approximation-based surrogate. Design of experiments involves the use of strategies for allocating samples within the design space. The particular choice depends on the number of samples one can afford (in some occasions only a few points may be allowed), but also on the specific modeling technique that will be used to create the surrogate. Though in some cases the surrogate can be found using explicit formulas (e.g., polynomial approximation (Queipo *et al.*, 2005)), in most situations it is computed by means of a separate minimization problem (e.g., when using kriging (Kleijnen, 2009) or neural networks (Rayas-Sanchez, 2004)). The accuracy of the model should be tested in order to estimate its prediction/generalization capability. The main difficulty in obtaining a good functional surrogate lies in keeping a balance between accuracy at the known and at the unknown data (training and testing set, respectively). The surrogate could be subsequently updated

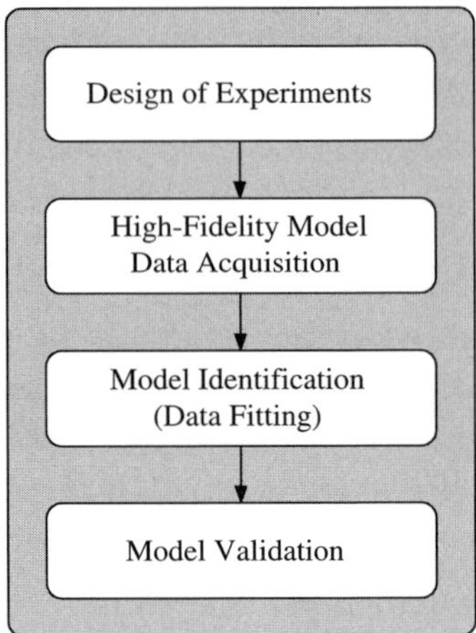

Fig. 3.5.   Approximation-based surrogate model construction flowchart. If the quality of the model is not satisfactory, the procedure can be iterated (more data points will be required).

using new high-fidelity model data that is accumulated during the run of the surrogate-based optimization algorithm.

3.4.2. *Design of Experiments*

Design of experiments (DOE) (Koehler and Owen, 1996; Giunta *et al.*, 2003; Santner *et al.*, 2003) is a strategy for allocating samples (points) in the design space that aims at maximizing the amount of information acquired. The high-fidelity model is evaluated at these points to create the training data set that is subsequently used to construct the functional surrogate model. When sampling, there is a clear tradeoff between the number of points used and the amount of information that can be extracted from these points. The samples are typically spread apart as much as possible in order to capture global trends in the design space.

Factorial designs (Giunta *et al.*, 2003) are classical DOE techniques that, when applied to discrete design variables, explore a large region of the search space. The sampling of all possible combinations is called full factorial design. Fractional factorial designs can be used when model evaluation is expensive and the number of design variables is large (in full factorial design the number of samples increases exponentially with the number of design variables). Continuous variables, once discretized, can be easily analyzed through factorial design. Full factorial two-level and

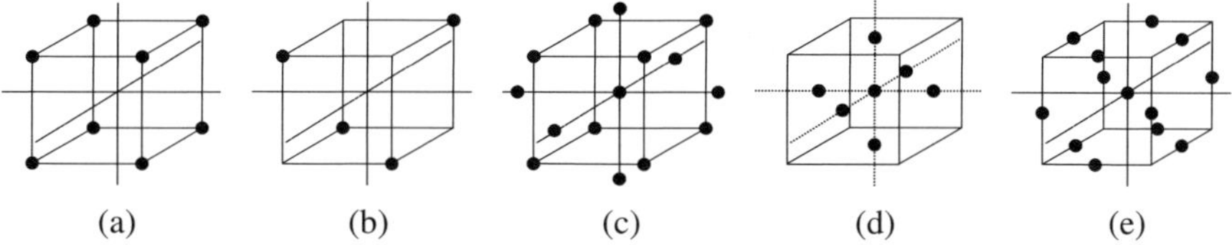

Fig. 3.6. Factorial designs for three design variables ($n = 3$): (a) full factorial design, (b) fractional factorial design, (c) central composite design, (d) star design, and (e) Box–Behnken design.

three-level design (also known as 2^k and 3^k design) allows us to estimate main effects and interactions between design variables, and quadratic effects and interactions, respectively. Figures 3.6(a) and 3.6(b) show examples of full two-level and fractional two-level design, respectively, for three design variables (i.e., $n = 3$). Alternative factorial designs can be found in practice: central composite design (see Fig. 3.6(c)), star design (frequently used in combination with space mapping (McKay *et al.*, 1979); see Fig. 3.6(d)), or Box–Behnken design (see Fig. 3.6(e)).

If no prior knowledge about the objective function is available (typical while constructing the initial surrogate), some recent DOE approaches tend to allocate the samples uniformly within the design space (Queipo *et al.*, 2005). A variety of space-filling designs are available. The simplest ones do not ensure sufficient uniformity (e.g., pseudo-random sampling (Guinta *et al.*, 2003)) or are not practical (e.g., stratified random sampling, where the number of samples needed is on the order of 2^n). One of the most popular DOE for (relatively) uniform sample distributions is Latin hypercube sampling (LHS) (McKay *et al.*, 1979). In order to allocate p samples with LHS, the range for each parameter is divided into p bins, which for n design variables yields a total number of p^n bins in the design space. The samples are randomly selected in the design space so that (1) each sample is randomly placed inside a bin, and (2) for all one-dimensional projections of the p samples and bins, there is exactly one sample in each bin. Figure 3.7 shows a LHS realization of 15 samples for two design variables ($n = 2$). It should be noted that the standard LHS may lead to non-uniform distributions (for example, samples allocated along the design space diagonal satisfy conditions (1) and (2)). Numerous improvements of standard LHS, e.g., Ye (1998), Beachkofski and Palmer and Tsui (2001), Grandhi (2002), and Leary *et al.* (2003), provide more uniform sampling distributions.

                                   Slawomir Koziel et al.

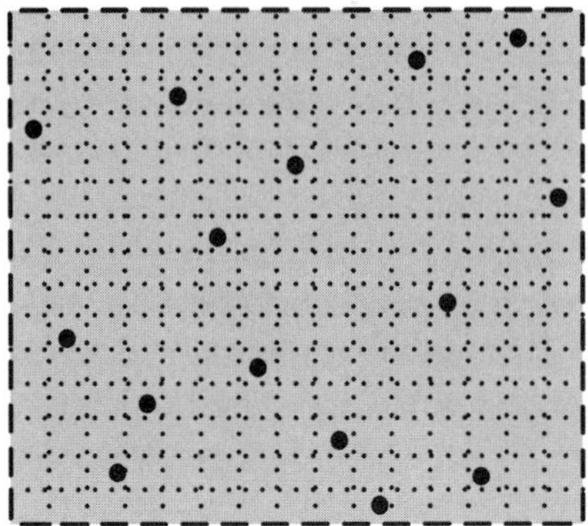

Fig. 3.7.   Latin hypercube sampling realization of 15 samples in a two-dimensional (2D) design space.

Other DOE methodologies commonly used include orthogonal array sampling (Queipo *et al.*, 2005), quasi-Monte Carlo sampling (Guinta *et al.*, 2003), or Hammersley sampling (Guinta *et al.*, 2003). Sample distribution can be improved through the incorporation of optimization techniques that minimize a specific non-uniformity measure, e.g., $\sum_{i=1}^{p}\sum_{j=i+1}^{p} d_{ij}^{-2}$ (Leary *et al.*, 2003), where d_{ij} is the Euclidean distance between samples i and j.

3.4.3. *Polynomial Regression*

Polynomial regression (Queipo *et al.*, 2005) assumes the following relation between the function of interest f and K polynomial basis functions v_j using p samples $f(\boldsymbol{x}^{(i)})$, $i = 1, \ldots, p$:

$$f(\boldsymbol{x}^{(i)}) = \sum_{j=1}^{K} \beta_j v_j(\boldsymbol{x}^{(i)}). \tag{3.14}$$

These equations can be represented in matrix form

$$\boldsymbol{f} = \boldsymbol{X}\boldsymbol{\beta}, \tag{3.15}$$

where $\boldsymbol{f} = [f(\boldsymbol{x}^{(1)})f(\boldsymbol{x}^{(2)})\ldots f(\boldsymbol{x}^{(p)})]^{T}$, $\boldsymbol{X}$ is a $p \times K$ matrix containing the basis functions evaluated at the sample points, and $\boldsymbol{\beta} = [\beta_1\ \beta_2 \ldots \beta_K]^{T}$. The number of sample points p should be consistent with the number of basis functions considered K (typically $p \geq K$). If the sample points and basis function are taken arbitrarily, some columns of $\boldsymbol{X}$ can be linearly dependent. If $p \geq K$ and rank$(\boldsymbol{X}) = K$, a solution of (3.15) in the least-squares sense can be computed through $\boldsymbol{X}^{+}$, the pseudoinverse (or generalized inverse) of $\boldsymbol{X}$ (Golub and Van Loan, 1996):

$$\beta = \boldsymbol{X}^+ = (\boldsymbol{X}^T\boldsymbol{X})^{-1}\boldsymbol{X}^T\boldsymbol{f}. \tag{3.16}$$

The simplest examples of regression models are the first- and second-order order polynomial models

$$s(\boldsymbol{x}) = s([x_1\,x_2\ldots x_n]^T) = \beta_0 + \sum_{j=1}^{n}\beta_j x_j, \tag{3.17}$$

$$s(\boldsymbol{x}) = s([x_1\,x_2\ldots x_n]^T) = \beta_0 + \sum_{j=1}^{n}\beta_j x_j + \sum_{i=1}^{n}\sum_{j\leq i}^{n}\beta_{ij} x_i x_j. \tag{3.18}$$

Polynomial interpolation/regression appears naturally and is crucial in developing robust and efficient derivative-free optimization algorithms. For more details, please refer to Conn *et al.* (2009).

3.4.4. *Radial Basis Functions*

Radial basis function interpolation/approximation (Wild *et al.*, 2008; Forrester and Keane, 2009) exploits linear combinations of K radially symmetric functions ϕ

$$s(\boldsymbol{x}) = \sum_{j=1}^{K}\lambda_j \phi(\|\boldsymbol{x} - \boldsymbol{c}^{(j)}\|), \tag{3.19}$$

where $\lambda = [\lambda_1\lambda_2\ldots\lambda_K]^T$ is the vector of model parameters, and $\boldsymbol{c}^{(j)}$, $j = 1,\ldots,K$, are the (known) basis function centers.

As in polynomial regression the model parameters λ can be computed by

$$\lambda = \boldsymbol{\Phi}^+ = (\boldsymbol{\Phi}^T\boldsymbol{\Phi})^{-1}\boldsymbol{\Phi}^T\boldsymbol{f}, \tag{3.20}$$

where again $\boldsymbol{f} = [f(\boldsymbol{x}^{(1)})\ f(\boldsymbol{x}^{(2)})\ldots f(\boldsymbol{x}^{(p)})]^T$, and the $p \times K$ matrix Φ is defined as

$$\boldsymbol{\Phi} = \begin{bmatrix} \phi(\|\boldsymbol{x}^{(1)} - \boldsymbol{c}^{(1)}\|) & \phi(\|\boldsymbol{x}^{(1)} - \boldsymbol{c}^{(2)}\|) & \cdots & \phi(\|\boldsymbol{x}^{(1)} - \boldsymbol{c}^{(K)}\|) \\ \phi(\|\boldsymbol{x}^{(2)} - \boldsymbol{c}^{(1)}\|) & \phi(\|\boldsymbol{x}^{(2)} - \boldsymbol{c}^{(2)}\|) & \cdots & \phi(\|\boldsymbol{x}^{(2)} - \boldsymbol{c}^{(K)}\|) \\ \vdots & \vdots & \ddots & \vdots \\ \phi(\|\boldsymbol{x}^{(p)} - \boldsymbol{c}^{(1)}\|) & \phi(\|\boldsymbol{x}^{(p)} - \boldsymbol{c}^{(2)}\|) & \cdots & \phi(\|\boldsymbol{x}^{(p)} - \boldsymbol{c}^{(K)}\|) \end{bmatrix}. \tag{3.21}$$

If we select $p = K$ (i.e., the number of basis functions is equal to the number of samples), and if the centers of the basis functions coincide with the data points (and these are all different), Φ is a regular square matrix (and thus, $\boldsymbol{\lambda} = \boldsymbol{\Phi}^{-1}\boldsymbol{f}$).

Typical choices for the basis functions are $\phi(r) = r$, $\phi(r) = r^3$, or $\phi(r) = r^2 \ln r$ (thin plate spline). More flexibility can be obtained by using parametric basis functions such as $\phi(r) = \exp(-r^2/2\sigma^2)$ (Gaussian), $\phi(r) = (r^2 + \sigma^2)^{1/2}$ (multi-quadric), or $\phi(r) = (r^2 + \sigma^2)^{-1/2}$ (inverse multi-quadric).

3.4.5. *Kriging*

Kriging is a popular technique to interpolate deterministic noise-free data (O'Hagan, 1978; Journel and Huijbregts, 1981; Simpson *et al.*, 2001; Kleijnen, 2009). Kriging is a Gaussian process-based modeling method, which is compact and cheap to evaluate (Rasmussen and Williams, 2006). Kriging has been proven to be useful in a wide variety of fields (see, e.g., Forrester and Keane (2009) and Jones *et al.* (1998) for applications in optimization).

In its basic formulation, kriging (Journel and Huijbregts, 1981; Simpson *et al.*, 2001) assumes that the function of interest f is of the form

$$f(\boldsymbol{x}) = \boldsymbol{g}(\boldsymbol{x})^T \boldsymbol{\beta} + Z(\boldsymbol{x}), \tag{3.22}$$

where $\boldsymbol{g}(\boldsymbol{x}) = [g_1(\boldsymbol{x}) g_2(\boldsymbol{x}) \ldots g_K(\boldsymbol{x})]^T$ are known (e.g., constant) functions, $\boldsymbol{\beta} = [\beta_1 \beta_2 \ldots \beta_K]^T$ are the unknown model parameters, and $Z(\boldsymbol{x})$ is a realization of a normally distributed Gaussian random process with zero mean and variance σ^2.

The regression part $\boldsymbol{g}(\boldsymbol{x})^T \boldsymbol{\beta}$ approximates globally the function f, and $Z(\boldsymbol{x})$ takes into account localized variations. The covariance matrix of $Z(\boldsymbol{x})$ is given as

$$Cov[Z(\boldsymbol{x}^{(i)}) Z(\boldsymbol{x}^{(j)})] = \sigma^2 \boldsymbol{R}([R(\boldsymbol{x}^{(i)}, \boldsymbol{x}^{(j)})]), \tag{3.23}$$

where $\boldsymbol{R}$ is a $p \times p$ correlation matrix with $R_{ij} = R(\boldsymbol{x}^{(i)}, \boldsymbol{x}^{(j)})$. Here, $R(\boldsymbol{x}^{(i)}, \boldsymbol{x}^{(j)})$ is the correlation function between sampled data points $\boldsymbol{x}^{(i)}$ and $\boldsymbol{x}^{(j)}$. The most popular choice is the Gaussian correlation function

$$R(\boldsymbol{x}, \boldsymbol{y}) = \exp\left[-\sum_{k=1}^{n} \theta_k |x_k - y_k|^2\right], \tag{3.24}$$

where θ_k are unknown correlation parameters, and x_k and y_k are the kth component of the vectors $\boldsymbol{x}$ and $\boldsymbol{y}$, respectively.

The kriging predictor (Journel and Huijbregts, 1981; Simpson *et al.*, 2001) is defined as

$$s(\boldsymbol{x}) = \boldsymbol{g}(\boldsymbol{x})^T \boldsymbol{\beta} + \boldsymbol{r}^T(\boldsymbol{x}) \boldsymbol{R}^{-1}(\boldsymbol{f} - \boldsymbol{G}\boldsymbol{\beta}), \tag{3.25}$$

where
$\boldsymbol{r}(\boldsymbol{x}) = [R(\boldsymbol{x}, \boldsymbol{x}^{(1)}) \ldots R(\boldsymbol{x}, \boldsymbol{x}^{(p)})]^T$, $\boldsymbol{f} = [f(\boldsymbol{x}^{(1)})f(\boldsymbol{x}^{(2)}) \ldots f(\boldsymbol{x}^{(p)})]^T$, and $\boldsymbol{G}$ is a $p \times K$ matrix with $G_{ij} = g_j(\boldsymbol{x}^{(i)})$.

The vector of model parameters $\boldsymbol{\beta}$ can be computed as

$$\boldsymbol{\beta} = (\boldsymbol{G}^T \boldsymbol{R}^{-1} \boldsymbol{G})^{-1} \boldsymbol{G}^T \boldsymbol{R}^{-1} \boldsymbol{f}. \tag{3.26}$$

An estimate of the variance σ^2 is given by

$$\hat{\sigma}^2 = \frac{1}{p}(\boldsymbol{f} - \boldsymbol{G}\boldsymbol{\beta})^T \boldsymbol{R}^{-1}(\boldsymbol{f} - \boldsymbol{G}\boldsymbol{\beta}). \tag{3.27}$$

Model fitting is accomplished by maximum likelihood for θ_k (Journel and Huijbregts, 1981). In particular, the n-dimensional unconstrained nonlinear maximization problem with cost function

$$-[p\ln(\hat{\sigma}^2) + \ln|\boldsymbol{R}|]/2, \tag{3.28}$$

where the variance σ^2 and $|\boldsymbol{R}|$ are both functions of θ_k, is solved for positive values of θ_k as optimization variables.

It should be noted that, once the kriging-based surrogate has been obtained, the random process $Z(\boldsymbol{x})$ gives valuable information regarding the approximation error that can be used for improving the surrogate (Journel and Huijbregts, 1981; Forrester and Keane, 2009).

3.4.6. *Neural Networks*

The basic structure in a neural network (Minsky and Papert, 1969; Haykin, 1998) is the neuron (or single-unit perceptron). A neuron performs an affine transformation followed by a nonlinear operation (see Fig. 3.8(a)). If the inputs to a neuron are denoted as $x_1, \ldots, x_n$, the neuron output y is computed as

$$y = \frac{1}{1 + \exp(-\eta/T)}, \tag{3.29}$$

where $\eta = w_1 x_1 + \cdots + w_n x_n + \gamma$, with $w_1, \ldots, w_n$ being regression coefficients. Here, γ is the bias value of a neuron, and T is a user-defined (slope) parameter. Neurons can be combined in multiple ways (Haykin,

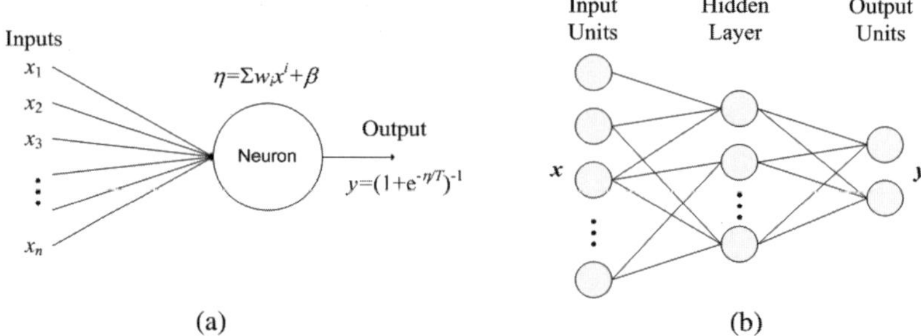

Fig. 3.8. Neural networks: (a) neuron basic structure, (b) two-layer feed-forward neural network architecture.

1998). The most common neural network architecture is the multi-layer feed-forward network (see Fig. 3.8(b)).

The construction of a neural network model requires two main steps: (i) architecture selection, and (ii) network training. The network training can be stated as a nonlinear least-squares regression problem for a number of training points. Since the optimization cost function is nonlinear in all the optimization variables (neuron coefficients), the solution cannot be written using a closed-form expression, as was the case before in (3.5) or in (3.9). A very popular technique for solving this regression problem is the error back-propagation algorithm (Haykin, 1998; Simpson *et al.*, 2001). If the network architecture is sufficiently complex, a neural network can approximate a general set of functions (Simpson *et al.*, 2001). However, in complicated cases (e.g., nonsmooth functions with a large number of variables) the underlying regression problem may be significantly involved.

3.4.7. *Support Vector Regression*

Support vector regression (SVR) (Gunn, 1998) is a relatively recent technique, which is characterized by good generalization capability (Angiulli *et al.*, 2007) and easy training through quadratic programming resulting in a global optimum for the model parameters (Smola and Schölkopf, 2004). SVR is a variant of the support vector machines (SVM) methodology developed by Vapnik (1995), which was originally applied to solve classification problems. SVM exploits the structural risk minimization (SRM) principle, which has been shown to be superior (Gunn, 1998) to the traditional empirical risk minimization (ERM) principle, employed by conventional methods used in empirical data modeling, e.g., neural networks. ERM is based on minimizing an error function for the set of training points. When the model structure is complex (e.g., higher order polynomials), ERM-based surrogates often result in overfitting. SRM incorporates the model complexity in the regression, and therefore yields surrogates that may be more accurate outside of the training set. Support vector regression is currently gaining popularity in the electrical engineering area, e.g., Ceperic and Baric (2004); Rojo-Alvarez *et al.* (2005); Yang *et al.* (2005); Martinez-Ramon and Christodoulou (2006); Meng and Xia (2007); Xia *et al.* (2007).

Here, we formulate the SVR surrogate model for vector-valued functions. Let $\boldsymbol{R}^k = \boldsymbol{R}_f(\boldsymbol{x}^k)$, $k = 1, 2, \ldots, N$, denote the sampled high-fidelity model responses. We want to use support vector regression to approximate $\boldsymbol{R}^k$ at all base points $\boldsymbol{x}^k$, $k = 1, 2, \ldots, N$. We shall also use the notation $\boldsymbol{R}^k = [R_1^k R_2^k \ldots R_m^k]^T$ to denote components of vector $\boldsymbol{R}^k$. In the case of linear regression, we want to approximate a given set of data, in our case, the data pairs $D_j = \{(\boldsymbol{x}^1, R_j^1), \ldots, (\boldsymbol{x}^N, R_j^N)\}$, $j = 1, 2, \ldots, m$, by a linear function $f_j(\boldsymbol{x}) = \boldsymbol{w}_j^T \boldsymbol{x} + b_j$. The optimal regression function is given by the minimum of the functional (Smola and Schölkopf, 2004)

$$\Phi_j(\boldsymbol{w}, \xi) = \frac{1}{2}\|\boldsymbol{w}_j\|^2 + C_j \sum_{i=1}^{N} (\xi_{j.i}^+ + \xi_{j.i}^-) \tag{3.30}$$

where C_j is a user-defined value, and $\xi_{j.i}^+$ and $\xi_{j.i}^-$ are slack variables representing upper and lower constraints on the output of the system. The typical cost function used in support vector regression is the so-called ε-insensitive loss function

$$L_\varepsilon(y) = \begin{cases} 0 & \text{for } |f_j(\boldsymbol{x}) - y| < \varepsilon \\ |f_j(\boldsymbol{x}) - y| & \text{otherwise} \end{cases}. \tag{3.31}$$

The value of C_j determines the trade-off between the flatness of f_j and the amount up to which deviations larger than ε are tolerated (Gunn, 1998).

Here, we describe nonlinear regression employing the kernel approach, in which the linear function $\boldsymbol{w}_j^T \boldsymbol{x} + b_j$ is replaced by the nonlinear function $\Sigma_i \gamma_{j.i} K(\boldsymbol{x}^k, \boldsymbol{x}) + b_j$, where K is a kernel function. Thus, the SVR model is defined as

$$\tilde{\boldsymbol{R}}_s = \begin{bmatrix} \sum_{i=1}^{N} \gamma_{1.i} K(\boldsymbol{x}^i, \boldsymbol{x}) + b_1 \\ \vdots \\ \sum_{i=1}^{N} \gamma_{m.i} K(\boldsymbol{x}^i, \boldsymbol{x}) + b_m \end{bmatrix}, \tag{3.32}$$

with parameters $\gamma_{j.i}$ and b_j, $j = 1, \ldots, m$, $i = 1, \ldots, N$ obtained according to a general support vector regression methodology. In particular, Gaussian kernels of the form

$$K(\boldsymbol{x}, \boldsymbol{y}) = \exp\left(-\frac{\|\boldsymbol{x} - \boldsymbol{y}\|^2}{2c^2}\right), \quad c > 0 \tag{3.33}$$

can be used, where c is the scaling parameter. Both c as well as parameters C_j and ε can be adjusted to minimize the generalization error calculated using a cross-validation method (Queipo *et al.*, 2005) and exponential grid search (Ceperic and Baric, 2004).

3.4.8. *Other Regression Techniques*

Moving least squares (MLS) (Levin, 1998) is a technique particularly popular in aerospace engineering. MLS is formulated as weighted least squares (WLS) (Aitken, 1935). In MLS, the error contribution from each training point $\boldsymbol{x}^{(i)}$ is multiplied by a weight ω_i that depends on the distance between $\boldsymbol{x}$ and $\boldsymbol{x}^{(i)}$. A common choice for the weights is

$$\omega_i(\|\boldsymbol{x} - \boldsymbol{x}^{(i)}\|) = \exp(-\|\boldsymbol{x} - \boldsymbol{x}^{(i)}\|^2). \tag{3.34}$$

MLS is essentially an adapting surrogate, and this additional flexibility can be translated in more appealing designs (especially in computer graphics applications). However, MLS is computationally more expensive than WLS, since computing the approximation for each point $\boldsymbol{x}$ requires solving a new optimization problem.

Gaussian process regression (GPR) (Rasmussen and William, 2006) is a surrogate modeling technique that, as kriging, addresses the approximation problem from a stochastic point of view. From this perspective, and since Gaussian processes are mathematically tractable, it is relatively easy to compute error estimations for GPR-based surrogates in the form of uncertainty distributions. Under appropriate conditions, Gaussian processes can be shown to be equivalent to large neural networks (Rasmussen and William, 2006). Nevertheless, Gaussian process modeling typically requires much fewer regression parameters than neural networks.

Cokriging (Forrester *et al.*, 2007; Toal and Keane, 2011) is a variation of the standard kriging method (described in Section 3.4.5). Cokriging is capable of using multiple levels of information about an objective function, such as data from varying simulation fidelities (Forrester *et al.*, 2007; Toal and Keane, 2011) or even from physical experiments (Forrester, 2010).

Although cokriging was first introduced in the 1980s, it has only recently been applied to microwave engineering design (Couckuyt *et al.*, 2011).

3.4.9. *Model Validation*

Some of the methodologies described above determine a surrogate model together with some estimation of the attendant approximation error (e.g., kriging or Gaussian process regression). Alternatively, there are procedures that can be used in a stand-alone manner to validate the prediction capability of a given model beyond the set of training points. A simple way for validating a model is the split-sample method (Queipo *et al.*, 2005). In this algorithm, the set of available data samples is divided into two subsets. The first subset is called the training subset and contains the points considered for the construction of the surrogate. The second subset is the testing subset and serves purely as a model validation objective. In general, the error estimated by a split-sample method depends strongly on how the set of data samples is partitioned. We also note that in this approach the samples available do not appear to be put to good use, since the surrogate is based on only a subset of them.

Cross-validation (Geisser, 1993; Queipo *et al.*, 2005) is a popular methodology for verifying the prediction capabilities of a model generated from a set of samples. In cross-validation the data set is divided into L subsets, and each of these subsets is sequentially used as a testing set for a surrogate constructed on the other L–1 subsets. If the number of subsets L is equal to the sample size p, the approach is called leave-one-out cross-validation (Queipo *et al.*, 2005). The prediction error can be estimated with all the L error measures obtained in this process (for example, as an average value). Cross-validation provides an error estimation that is less biased than that created with the split-sample method (Queipo *et al.*, 2005). The disadvantage of this method is that the surrogate has to be constructed more than once. However, having multiple approximations may improve the robustness of the whole surrogate generation and validation approach, since all the data available is used with both training and testing purposes.

3.5. Physics-Based Surrogate Models

In this section, we describe the concept and techniques for constructing physics-based surrogate models. Physics-based surrogates exploit underlying low-fidelity models that embed certain knowledge about the system of interest and, therefore, allow us to build surrogates at a low computational cost (in terms of the required amount of high-fidelity model data) and with good generalization capabilities. Detailed discussion of some of the techniques mentioned here can be found in Chapters 4 and 6.

3.5.1. *The Modeling Concept*

Physics-based surrogates are constructed using an underlying low-fidelity (coarse) model and, usually, an appropriate correction. The coarse model is a representation of the system of interest with relaxed accuracy (Bandler *et al.*, 2004). Coarse models are computationally cheaper than the fine models and, in many cases, have better analytical properties. The coarse model can be obtained, for example, from the same simulator as the one used for the fine model but using a coarse discretization (Alexandrov *et al.*, 2000). Alternatively, the coarse model can be based on simplified physics (e.g., by exploiting simplified equations (Bandler *et al.*, 2004), or by neglecting certain second-order effects (Wu *et al.*, 2004)), or on a significantly different physical description (e.g., lumped parameter versus partial differential equation-based models (Bandler *et al.*, 2004)). In some cases, coarse models can be formulated using analytical or semi-empirical formulas (Robinson *et al.*, 2008). The coarse model can be corrected if additional data from the fine model is available (for example, during the course of the optimization).

In microwave engineering, there are several possibilities to obtain the physics-based low-fidelity model. In some cases, models based on explicit analytical or semi-empirical formulas are available. Such models are normally very fast; however, they are usually of limited accuracy. For certain structures, particularly filters, low-fidelity models can be implemented as equivalent circuits, see Fig. 3.9 (Koziel and Bandler, 2007c). Equivalent circuit models are also very fast although not as convenient as analytical ones as an extra simulator has to be used for their evaluation. Their accuracy may also be limited, particularly for more complex structures. The most versatile type of low-fidelity model, available for any microwave structure, can be obtained through coarse-discretization EM simulation (cf. Fig. 3.10). This type of model is usually quite accurate; however, it might be relatively expensive. Therefore, selecting a good tradeoff between the model accuracy and speed might not be obvious. It is worth mentioning that relaxed mesh is not the only way of creating a EM-based low-fidelity model. Other options include the reduced number of cells in the perfectly

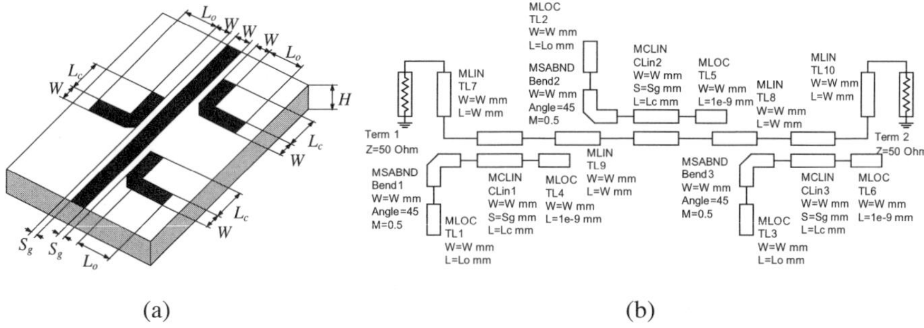

(a)   (b)

Fig. 3.9. Microstrip filter: (a) geometry, and (b) equivalent circuit (Koziel and Bandler, 2007c).

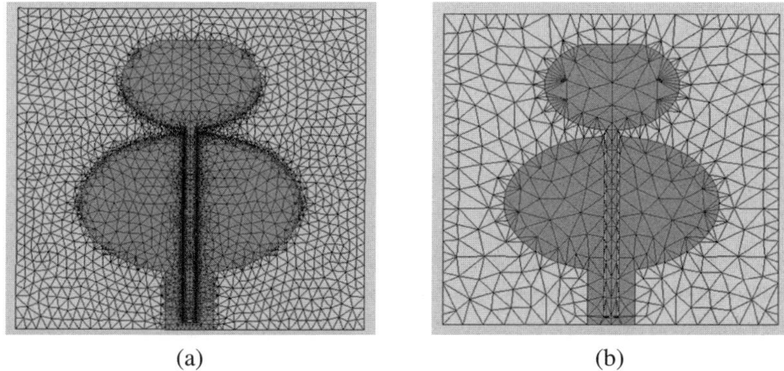

(a)   (b)

Fig. 3.10. Ultrawideband monopole antenna: (a) finely discretized geometry and (b) coarsely discretized geometry.

matched layer absorbing boundary conditions, as well as reduced distance from the simulated structure to the absorbing boundary conditions, as in the case of finite-volume electromagnetic simulators (Taflove and Hagness, 2005); the lower order of the basis functions, as in the case of finite-element and integral equation solvers (Jin, 2002); simplified excitation, e.g., the discrete source of the coarse model versus the waveguide port of the fine model; zero thickness of metallization; the use of perfect electric conductor in place of finite-conductivity metals.

In general, physics-based surrogate models are:

- based on particular knowledge about the physical system of interest,
- dedicated (reuse across different designs is uncommon), and,

- more expensive to evaluate and more accurate (in a global sense) than functional surrogates.

It should be noted that the evaluation of a physics-based surrogate may involve, for example, the numerical solution of partial differential equations, or even actual measurements of the physical system.

The main advantage of physics-based surrogates is that the amount of fine model data necessary for obtaining a given level of accuracy is generally substantially smaller than for approximation-based surrogates (physics-based surrogates inherently embed knowledge about the system of interest) (Bandler *et al.*, 2004). Hence, surrogate-based optimization algorithms that exploit physics-based surrogate models are usually more efficient than those using approximation-based surrogates (in terms of the number of fine model evaluations required to find a satisfactory design (Koziel *et al.*, 2006)).

In the first stages on any surrogate-based optimization procedure, it is desirable to use a surrogate that is valid globally in the search space (Forrester and Keane, 2009) in order to avoid being trapped in local solutions with unacceptable cost function values. Once the search starts becoming local, the global accuracy of the initial surrogate may not be beneficial for making progress in the optimization[1]. For this reason, surrogate correction is crucial within any SBO methodology. We will now describe several correction techniques available in the literature.

3.5.2. *Objective Function Correction*

Most of the objective function corrections used in practice can be identified in one of these three groups: compositional, additive, or multiplicative corrections. We will briefly illustrate each of these categories for correcting the surrogate $s^{(i)}(\boldsymbol{x})$, and discuss if zero- and first-order consistency conditions with $f(\boldsymbol{x})$ (Alexandrov and Lewis, 2001) can be satisfied.

The compositional correction (Søndergaard, 2003)

$$s^{(i+1)}(\boldsymbol{x}) = g(s^{(i)}(\boldsymbol{x})) \tag{3.35}$$

represents a simple scaling of the objective function. Since the mapping g is a real-valued function of a real variable, a compositional correction will not

---

[1] As mentioned in Section 3.2, when solving the original optimization problem in (3.1) using a surrogate-based optimization framework, zero- and first-order local consistency conditions are essential for obtaining convergence to a first-order stationary point.

in general yield first-order consistency conditions. By selecting a mapping g that satisfies

$$g'(s^{(i)}(\boldsymbol{x}^{(i)})) = \frac{\nabla f(\boldsymbol{x}^{(i)})\nabla s^{(i)}(\boldsymbol{x}^{(i)})^T}{\nabla s^{(i)}(\boldsymbol{x}^{(i)})\nabla s^{(i)}(\boldsymbol{x}^{(i)})^T}, \tag{3.36}$$

the discrepancy between $\nabla f(\boldsymbol{x}^{(i)})$ and $\nabla s^{(i+1)}(\boldsymbol{x}^{(i)})$ (expressed in Euclidean norm) is minimized. It should be noted that the correction in (3.36), as many transformations that ensure first-order consistency, requires a high-fidelity gradient, which may be expensive to compute. However, numerical estimates of $\nabla f(\boldsymbol{x}^{(i)})$ may yield in practice acceptable results.

The compositional correction can be also introduced in the parameter space (Bandler *et al.*, 2004)

$$s^{(i+1)}(\boldsymbol{x}) = s^{(i)}(\boldsymbol{p}(\boldsymbol{x})). \tag{3.37}$$

If $f(\boldsymbol{x}^{(i)})$ is not in the range of $s^{(i)}(\boldsymbol{x})$, then the condition $s^{(i)}(\boldsymbol{p}(\boldsymbol{x}^{(i)})) = f(\boldsymbol{x}^{(i)})$ is not achievable. We can overcome that issue by combining both compositional corrections. In that case, the following selection for g and $\boldsymbol{p}$

$$g(t) = t - s^{(i)}(\boldsymbol{x}^{(i)}) + f(\boldsymbol{x}^{(i)}), \tag{3.38}$$

$$\boldsymbol{p}(\boldsymbol{x}) = \boldsymbol{x}^{(i)} + \boldsymbol{J}_p(\boldsymbol{x} - \boldsymbol{x}^{(i)}), \tag{3.39}$$

where $\boldsymbol{J}_\mathbf{p}$ is a $n \times n$ matrix for which $\boldsymbol{J}_\mathbf{p}^T \nabla s^{(i)} = \nabla f(\boldsymbol{x}^{(i)})$, guarantees consistency.

Additive and multiplicative corrections allow obtaining first-order consistency conditions. For the additive case we can generally express the correction as

$$s^{(i+1)}(\boldsymbol{x}) = \lambda(\boldsymbol{x}) + s^{(i)}(\boldsymbol{x}). \tag{3.40}$$

The associated consistency conditions require that $\lambda(\boldsymbol{x})$ satisfies

$$\lambda(\boldsymbol{x}^{(i)}) = f(\boldsymbol{x}^{(i)}) - s^{(i)}(\boldsymbol{x}^{(i)}), \tag{3.41}$$

and

$$\nabla\lambda(\boldsymbol{x}^{(i)}) = \nabla f(\boldsymbol{x}^{(i)}) - \nabla s^{(i)}(\boldsymbol{x}^{(i)}). \tag{3.42}$$

Those requirements can be obtained by the following linear additive correction:

$$\begin{aligned} s^{(i+1)}(\boldsymbol{x}) = {}& f(\boldsymbol{x}^{(i)}) - s^{(i)}(\boldsymbol{x}^{(i)}) \\ & + (\nabla f(\boldsymbol{x}^{(i)}) - \nabla s^{(i)}(\boldsymbol{x}^{(i)}))(\boldsymbol{x} - \boldsymbol{x}^{(i)}) + s^{(i)}(\boldsymbol{x}) \end{aligned} \tag{3.43}$$

Multiplicative corrections (also known as the β-correlation method (Søndergaard, 2003)) can be represented generically by

$$s^{(i+1)}(\boldsymbol{x}) = \alpha(\boldsymbol{x})s^{(i)}(\boldsymbol{x}). \tag{3.44}$$

Assuming that $s^{(i)}(\boldsymbol{x}^{(i)}) \neq 0$, zero- and first-order consistency can be achieved if

$$\alpha(\boldsymbol{x}^{(i)}) = \frac{f(\boldsymbol{x}^{(i)})}{s^{(i)}(\boldsymbol{x}^{(i)})}, \tag{3.45}$$

and

$$\nabla\alpha(\boldsymbol{x}^{(i)}) = [\nabla f(\boldsymbol{x}^{(i)}) - f(\boldsymbol{x}^{(i)})/s^{(i)}(\boldsymbol{x}^{(i)})\nabla s^{(i)}(\boldsymbol{x}^{(i)})]/s^{(i)}(\boldsymbol{x}^{(i)}). \tag{3.46}$$

The requirement $s^{(i)}(\boldsymbol{x}^{(i)}) \neq 0$ is not strong in practice since very often the range of $f(\boldsymbol{x})$ (and thus, of the surrogate $s^{(i)}(\boldsymbol{x})$) is known beforehand, and hence, a bias can be introduced both for $f(\boldsymbol{x})$ and $s^{(i)}(\boldsymbol{x})$ to avoid cost function values equal to zero. In these circumstances the following multiplicative correction,

$$s^{(i+1)}(\boldsymbol{x}) = \left[\frac{f(\boldsymbol{x}^{(i)})}{s^{(i)}(\boldsymbol{x}^{(i)})} + \frac{\nabla f(\boldsymbol{x}^{(i)})s^{(i)}(\boldsymbol{x}^{(i)}) - f(\boldsymbol{x}^{(i)})\nabla s^{(i)}(\boldsymbol{x}^{(i)})}{(s^{(i)}(\boldsymbol{x}^{(i)}))^2}(\boldsymbol{x} - \boldsymbol{x}^{(i)})\right]$$
$$\times s^{(i)}(\boldsymbol{x}), \tag{3.47}$$

is consistent with conditions (3.41) and (3.42).

3.5.3. *Space Mapping*

Space mapping (SM) (Bandler *et al.*, 2004; Koziel and Bandler, 2010e) is a well-known methodology for correcting a given coarse model $\boldsymbol{R}_c$. The idea is to enhance $\boldsymbol{R}_c$ and create a space mapping surrogate model $\boldsymbol{R}_s$ using auxiliary mappings with parameters determined so that $\boldsymbol{R}_s$ matches the high-fidelity (or fine) model $\boldsymbol{R}_f$ as closely as possible at all base (training) points. Because the coarse model is assumed to be physics-based, i.e., describes the same phenomenon as the fine model, we hope that the surrogate model will retain a good match with the fine model over the whole region of interest.

The standard space mapping model is defined as (Koziel *et al.*, 2005)

$$\boldsymbol{R}_s(\boldsymbol{x}) = \bar{\boldsymbol{R}}_s(\boldsymbol{x}, \boldsymbol{p}) \tag{3.48}$$

where the space mapping parameters p are obtained using the parameter extraction process

$$p = \arg\min_{r} \sum_{k=1}^{N} \|R_f(x^k) - \bar{R}_s(x^k, r)\|, \qquad (3.49)$$

while $\bar{R}_s$ is a generic space mapping model, i.e., the coarse model composed with some suitable mappings. Vectors x^k, $k = 1, 2, \ldots, N$ are training points. In many cases it is sufficient to use a quite limited number of training points, e.g., $N = 2n + 1$, with n being the number of design variables, allocated according to so-called star-distribution (Koziel *et al.*, 2005).

The model often used in practice has the form

$$\bar{R}_s(x, p) = \bar{R}_s(x, A, B, c) = A \cdot R_c(B \cdot x + c) + d \qquad (3.50)$$

where $A = diag\{a_1, \ldots, a_m\}$, B is an $n \times n$ matrix, c is an $n \times 1$ vector, and d is an $m \times 1$ vector.

In many cases, both fine and coarse models have parameters that are normally fixed and not used in the optimization process (so-called pre-assigned parameters). These parameters can be used as additional degrees of freedom in the coarse model and adjusted in order to obtain a better match with the fine model, which leads us to implicit space mapping (Bandler *et al.*, 2005). Let us denote the coarse model exploiting the pre-assigned parameters x_p as $R_c(x, x_p)$. The surrogate (3.50) enhanced by implicit space mapping could take the following form:

$$\bar{R}_s(x, p) = \bar{R}_s(x, A, B, c, d, B_p, c_p)$$
$$= A \cdot R_c(B \cdot x + c, B_p \cdot x + c_p) + d, \qquad (3.51)$$

where A, B and c are as in (3.50), while B_p is an $n_p \times n$ matrix, and c_p is an $n_p \times 1$ vector, where n_p is the number of pre-assigned parameters. Here, we use a generalized implicit space mapping (Koziel *et al.*, 2006) in which pre-assigned parameters are dependent on design variables in order to increase the flexibility of the surrogate model.

More detailed information about space mapping in the context of design optimization can be found in Chapter 4.

3.5.4. *Other Techniques*

Certain recently introduced surrogate-based techniques such as shape-preserving response prediction (SPRP) (Koziel, 2010), or adaptive response correction (ARC) (Koziel *et al.*, 2009), see also Section 3.3.5 and Chapter 6, have been specifically introduced in the context of design optimization, where the surrogate model is normally defined locally, in the vicinity of the current iteration point considered during the optimization process. However, both approaches can also be viewed as modeling techniques, which can be obtained by appropriate reformulation and using multiple (not single as in the original versions) training points. This has been done, in particular, for SPRP in Koziel and Szczepanski (2011). The details are beyond the scope of this chapter. Interested readers are referred to more advanced literature such as Koziel and Szczepanski (2011).

3.6. Conclusion

In this chapter, an overview of surrogate-based modeling and optimization has been presented. SBO plays an important and increasing role in contemporary engineering design. The reason is that computer simulations have become a primary design tool in most engineering areas and accurate simulations are time consuming, particularly for time-varying three-dimensional (3D) structures considered in many engineering fields. With evaluation times of several hours, days, or even weeks per design, the direct use of expensive numerical models in conventional optimization procedures (such as gradient-based with approximate derivatives) is often prohibitive. Surrogate-based optimization can be a very useful approach in this context, since, apart from reducing significantly the number of high-fidelity expensive simulations in the whole design process, it also helps in addressing important high-fidelity cost function issues (e.g., presence of discontinuities and/or multiple local optima).

References

Aitken, A.C. (1935). On least squares and linear combinations of observations, *P. Roy. Soc. Edinb. A*, **55**, 42–48.

Alexandrov, N.M., Dennis, J.E., Lewis, R.M. and Torczon, V. (1998). A trust region framework for managing use of approximation models in optimization, *Struct. Multidiscip. O.*, **15**, 16–23.

Alexandrov, N.M., Nielsen, E.J., Lewis, R.M. and Anderson, W.K. (2000). First-order model management with variable-fidelity physics applied to multi-element airfoil optimization, *AIAA/USAF/NASA/ISSMO Symposium on Multidisciplinary Design and Optimization*, Long Beach: CA, paper AIAA 2000-4886.

Alexandrov, N.M. and Lewis, R.M. (2001). An overview of first-order model management for engineering optimization, *Optim. Eng.*, **2**, 413–430.

Angiulli, G., Cacciola, M. and Versaci, M. (2007). Microwave devices and antennas modeling by support vector regression machines, *IEEE T. Magn.*, **43**, 1589–1592.

Bandler, J.W., Cheng, Q.S., Dakroury, S.A., Mohamed, A.S., Bakr, M.H., Madsen, K. and Søndergaard, J. (2004). Space mapping: the state of the art, *IEEE T. Microw. Theory*, **52**, 337–361.

Beachkofski, B. and Grandhi, R. (2002). Improved distributed hypercube sampling, American Institute of Aeronautics and Astronautics, paper AIAA 2002-1274.

Booker, A.J., Dennis, J.E., Frank, P.D., Serafini, D.B., Torczon, V. and Trosset, M.W. (1999). A rigorous framework for optimization of expensive functions by surrogates, *Struct. Optimization*, **17**, 1–13.

Ceperic, V. and Baric, A. (2004). Modeling of analog circuits by using support vector regression machines, *Proc. 11th Int. Conf. Electronics, Circuits, Syst.*, Tel-Aviv, Israel, pp. 391–394.

Couckuyt, I., Koziel, S. and Dhaene, T. (2011). Kriging, Co-Kriging and Space Mapping for Microwave Circuit Modeling, *41st European Microwave Conference (EuMW 2011)*, Manchester, UK, pp. 444–447.

Conn, A.R., Gould, N.I.M. and Toint, P.L. (2000). *Trust Region Methods*, Society for Industrial and Applied Mathematics, Philadelphia: PA.

Conn, A.R., Scheinberg, K. and Vicente, L.N. (2009). *Introduction to Derivative-Free Optimization*, Society for Industrial and Applied Mathematics, Philadelphia: PA.

CST Microwave Studio (2011). CST AG, Bad Nauheimer Str. 19, D-64289 Darmstadt, Germany.

Echeverria, D. and Hemker, P.W. (2005). Space mapping and defect correction, *CMAM*, **5**, 107–136.

Echeverria, D. (2007a). Multi-level optimization: space mapping and manifold mapping. Ph.D. Thesis, Faculty of Science, University of Amsterdam.

Echeverria, D. (2007b). Two new variants of the manifold-mapping technique, *COMPEL*, **26**, 334–344.

Echeverria, D. and Hemker, P.W. (2008). Manifold mapping: a two-level optimization technique, *Computing and Visualization in Science*, **11**, 193–206.

Forrester, A.I.J., Sóbester, A. and Keane, A.J. (2007). Multi-Fidelity Optimization via Surrogate Modeling, *P. Roy. Soc. A-Math. Phy.*, **463**, 3251–3269.

Forrester, A.I.J. and Keane, A.J. (2009). Recent advances in surrogate-based optimization, *Prog. Aerosp. Sci.*, **45**, 50–79.

Forrester, A.I.J. (2010). Black-Box Calibration for Complex Systems Simulation, *Philos. T. R. Soc. A*, **368**, 3567–3579.

Geisser, S. (1993). *Predictive Inference*, Chapman and Hall, New York: NY.

Giunta, A.A., Wojtkiewicz, S.F. and Eldred, M.S. (2003). Overview of modern design of experiments methods for computational simulations, American Institute of Aeronautics and Astronautics, paper AIAA 2003-0649.

Golub, G.H. and Van Loan, C.F. (1996). *Matrix Computations* (3rd ed.), The Johns Hopkins University Press, Baltimore: MD.

Gunn, S.R. (1998). Support vector machines for classification and regression, Tech. Rep., School of Electronics and Computer Science, University of Southampton.

Haykin, S. (1998). *Neural Networks: A Comprehensive Foundation*, (2nd ed.), Upper Saddle River: NJ.

Hemker, P.W. and Echeverría, D. (2007). A trust-region strategy for manifold mapping optimization, *J. Comput. Phys.*, **224**, 464–475.

Jin, J.M. (2002). *The Finite Element Method in Electromagnetics*, (2nd ed.), Wiley–IEEE Press, New York: NY.

Jones, D., Schonlau, M. and Welch, W. (1998). Efficient global optimization of expensive black-box functions, *J. Global Optim.*, **13**, 455–492.

Journel, A.G. and Huijbregts, C.J. (1981). *Mining Geostatistics*, Academic Press, Orlando: FL.

Kleijnen, J.P.C. (2009). Kriging metamodeling in simulation: a review, *Eur. J. Oper. Res.*, **192**, 707–716.

Koehler, J.R. and A.B. Owen (1996). Computer experiments, in Ghosh, S. and Rao, C.R. (eds.), *Handbook of Statistics*, Elsevier Science BV, Amsterdam, pp. 261–308.

Kolda, T.G., Lewis, R.M. and Torczon, V. (2003). Optimization by direct search: new perspectives on some classical and modern methods, *SIAM Rev.*, **45**, 385–482.

Koziel, S., Bandler, J.W., Mohamed, A.S. and Madsen, K. (2005). Enhanced surrogate models for statistical design exploiting space mapping technology, *IEEE MTT-S*, Long Beach: CA, 1609–1612.

Koziel, S., Bandler, J.W. and Madsen, K. (2006). A space mapping framework for engineering optimization: theory and implementation, *IEEE T. Microw. Theory*, **54**. 3721–3730.

Koziel, S. and Bandler, J.W. (2007a). Space-mapping optimization with adaptive surrogate model, *IEEE T. Microw. Theory*, **55**, 541–547.

Koziel, S. and Bandler, J.W. (2007b). Coarse and surrogate model assessment for engineering design optimization with space mapping, *IEEE MTT-S*, Honolulu: HI, 107–110.

Koziel, S. and Bandler, J.W. (2007c). Interpolated coarse models for microwave design optimization with space-mapping, *IEEE T. Microw. Theory.*, **55**, 1739–1746.

Koziel, S., Cheng, Q.S. and Bandler, J.W. (2008a). Space mapping, *IEEE Microw. Mag.*, **9**, 105–122.

Koziel, S., Bandler, J.W. and Madsen, K. (2008b). Quality assessment of coarse models and surrogates for space mapping optimization, *Optim. Eng.*, **9**, 375–391.

Koziel, S., Bandler, J.W. and Madsen, K. (2009). Space mapping with adaptive response correction for microwave design optimization, *IEEE T. Microw. Theory*, **57**, 478–486.

Koziel, S. (2010a). Improved microwave circuit design using multipoint-response-correction space mapping and trust regions, *Int. Symp. Antenna Technology and Applied Electromagnetics, ANTEM 2010*, Ottawa, Canada.

Koziel, S. (2010b). Shape-preserving response prediction for microwave design optimization, *IEEE T. Microw. Theory*, **58**, 2829–2837.

Koziel, S. (2010c). Adaptively adjusted design specifications for efficient optimization of microwave structures, *Prog. Electromagn. Res.*, **21**, 219–234.

Koziel, S., Bandler, J.W. and Cheng, Q.S. (2010a). Robust trust-region space-mapping algorithms for microwave design optimization, *IEEE T. Microw. Theory*, **58**, 2166–2174.

Koziel, S., Bandler, J.W. and Cheng, Q.S. (2010b). Adaptively constrained parameter extraction for robust space mapping optimization of microwave circuits, *IEEE MTT-S*, Anaheim: CA, 2010, 205–208.

Koziel, S. and Echeverría Ciaurri, D. (2010c). Reliable simulation-driven design optimization of microwave structures using manifold mapping, *Prog. Electromagn. Res.*, **26**, 361–382.

Koziel, S. and Bandler, J.W. (2010e). Recent advances in space-mapping-based modeling of microwave devices, *Int. J. Numer. Model. El.*, **23**, 425–446.

Koziel, S. and Szczepanski, S. (2011). Accurate modeling of microwave structures using shape-preserving response prediction, *IET Microw. Antenna P.*, **5**, 1116–1122.

Leary, S.J., Bhaskar, A. and Keane, A.J. (2001). A constraint mapping approach to the structural optimization of an expensive model using surrogates, *Optim. Eng.*, **2**, 385–398.

Leary, S., Bhaskar, A. and Keane, A. (2003). Optimal orthogonal-array-based latin hypercubes, *J. Appl. Stat.*, **30**, 585–598.

Levin, D. (1998). The approximation power of moving least-squares, *Math. Comput.*, **67**, 1517–1531.

Marsden, A.L., Wang, M., Dennis, J.E. and Moin, P. (2004), Optimal aeroacoustic shape design using the surrogate management framework, *Optim. Eng.*, **5**, 235–262.

Martinez-Ramon, M. and Christodoulou, C. (2006). Support vector machines for antenna array processing and electromagnetic, *Synthesis Lectures on Computational Electromagnetics*, **1**, 1–120.

Meng, J. and Xia, L. (2007). Support-vector regression model for millimeter wave transition, *Int. J. Infrared Milli.*, **28**, 413–421.

McKay, M., Conover, W. and Beckman, R. (1979). A comparison of three methods for selecting values of input variables in the analysis of output from a computer code, *Technometrics*, **21**, 239–245.

Minsky, M.I. and Papert, S.A. (1969). *Perceptrons: An Introduction to Computational Geometry*, The MIT Press, Cambridge: MA.

Nocedal, J. and Wright, S. (2006). *Numerical Optimization*, (2nd ed.), Springer-Verlag, Berlin.

O'Hagan, A. (1978). Curve fitting and optimal design for predictions, *J. Roy. Stat. Soc. B Met.*, **40**, 1–42.

Palmer, K. and Tsui, K.-L. (2001). A minimum bias latin hypercube design, *IIE Trans.*, **33**, 793–808.

Queipo, N.V., Haftka, R.T., Shyy, W., Goel, T., Vaidynathan, R. and Tucker, P.K. (2005). Surrogate-based analysis and optimization, *Prog. Aerosp. Sci.*, **41**, 1–28.

Rayas-Sanchez, J.E. (2004). EM-based optimization of microwave circuits using artificial neural networks: the state-of-the-art, *IEEE T. Microw. Theory*, **52**, 420–435.

Rao, S.S. (1996). *Engineering Optimization: Theory and Practice*, (3rd ed.), Wiley, New York: NY.

Rasmussen, C.E. and Williams, C.K.I. (2006). *Gaussian Processes for Machine Learning*, The MIT Press, Cambridge: MA.

Redhe, M. and Nilsson, L. (2004). Optimization of the new Saab 9-3 exposed to impact load using a space mapping technique, *Struct. Multidiscip. O.*, **27**, 411–420.

Robinson, T.D., Eldred, M.S., Willcox, K.E. and Haimes, R. (2008). Surrogate-based optimization using multifidelity models with variable parameterization and corrected space mapping, *AIAA J.*, **46**, 2814–2822.

Rojo-Alvarez, J.L., Camps-Valls, G., Martinez-Ramon, M., Soria-Olivas, E., Navia-Vazquez, A. and Figueiras-Vidal, A.R. (2005). Support vector machines framework for linear signal processing, *Signal Process.*, **85**, 2316–2326.

Santner, T.J., Williams, B. and Notz, W. (2003). *The Design and Analysis of Computer Experiments*, Springer-Verlag, New York: NY.

Simpson, T.W., Peplinski, J., Koch, P.N. and Allen, J.K. (2001). Metamodels for computer-based engineering design: survey and recommendations, *Eng. Comput.*, **17**, 129–150.

Smola, A.J. and Schölkopf, B. (2004). A tutorial on support vector regression, *Stat. Comput.*, **14**, 199–222.

Søndergaard, J. (2003). Optimization using surrogate models — by the space mapping technique. Ph.D. Thesis, Informatics and Mathematical Modelling, Technical University of Denmark, Lyngby.

Taflove, A. and Hagness, S.C. (2005). *Computational Electrodynamics: The Finite-Difference Time-Domain Method*, (3rd ed.), Artech House, Norwood: MA.

Toal, D.J.J. and Keane, A.J. (2011). Efficient Multipoint Aerodynamic Design Optimization via Cokriging, *J. Aircraft*, **48**, 1685–1695.

Vapnik, V.N. (1995). *The Nature of Statistical Learning Theory*, Springer-Verlag, New York: NY.

Wild, S.M., Regis, R.G. and Shoemaker, C.A. (2008). ORBIT: Optimization by radial basis function interpolation in trust-regions, *SIAM J. Sci. Comput.*, **30**, 3197–3219.

Wu, K.-L., Zhao, Y.-J. Wang, J. and Cheng, M.K.K. (2004). An effective dynamic coarse model for optimization design of LTCC RF circuits with aggressive space mapping, *IEEE T. Microw. Theory*, **52**, 393–402.

Xia, L., Xu, R.M. and Yan, B. (2007). Ltcc interconnect modeling by support vector regression, *Prog. Electromagn. Res.*, **69**, 67–75.

Yang, Y., Hu, S.M. and Chen, R.S. (2005). A combination of FDTD and least-squares support vector machines for analysis of microwave integrated circuits, *Microw. Opt. Techn. Let.*, **44**, 296–299.

Yang, X.S. and Forbes, A.B. (2011). Model and feature selection in metrology data approximation, in Georgoulis, E.H., Iske, A. and Levesley, J. (eds.), *Approximation Algorithms for Complex Systems*, Springer Proceedings in Mathematics, 3, Springer, Berlin, pp. 293–307.

Ye, K.Q. (1998). Orthogonal column latin hypercubes and their application in computer experiments, *J. Am. Stat. Assoc.*, **93**, 1430–1439.

Chapter 4

Space Mapping

Slawomir Koziel, Stanislav Ogurtsov, Qingsha S. Cheng,
and John W. Bandler

Electromagnetic (EM) simulation is an essential component of contemporary microwave design. High-fidelity EM simulation allows accurate evaluation of microwave devices; however, it is computationally expensive. As a result, the use of EM solvers for automated design optimization may be prohibitive. This chapter describes space mapping (SM), one of the most popular surrogate-based optimization and modeling techniques in microwave engineering to date. The main concern of SM is to reduce the computational cost of automated EM-simulation-driven design by shifting the optimization burden onto a fast low-fidelity (or coarse) model of the device under consideration, for example, a circuit equivalent. By a proper combination of the knowledge contained in the coarse model, and simple transformations that aim at improving the alignment between the coarse and original high-fidelity EM-simulated (or fine) models, one can create cheap and reliable surrogates that can be used to predict the fine model optimum design and thus to dramatically reduce the cost of the design process. This chapter presents the concept and historical overview of space mapping. It offers various SM formulations, discusses practical issues and open problems, and presents application examples.

4.1. Space Mapping: Concept and Historical Overview

Space mapping (SM) (Bandler *et al.*, 1994; Bandler *et al.*, 2004a; Bandler *et al.*, 2006; Koziel *et al.*, 2008a) is an intelligent methodology for solving engineering optimization problems where the primary issue is the high computational cost of evaluating the system under design (referred to as a high-fidelity or "fine" model). Space mapping aims at reducing the optimization cost through replacing direct optimization of the fine model by iterative updating (and re-optimization, if necessary) of the so-called surrogate model. The surrogate is supposed to be a fast and yet reasonable representation of the fine model. In this respect, space mapping fits within the framework of surrogate-based optimization (SBO) (Queipo *et al.*, 2005; Forrester and Keane, 2009; Koziel *et al.*, 2011a). Practical SBO techniques differ in the way the surrogate model is set up and then used in the optimization process (Koziel *et al.*, 2011a).

There are two main approaches to construct the surrogate model, either by approximating sampled fine model data, or through suitable correction of a physics-based low-fidelity (or "coarse") model. Among a variety of approximation models used by engineers and scientists, the most popular include polynomial regression (Queipo *et al.*, 2005), radial-basis function interpolation (Wild *et al.*, 2008; Forrester and Keane, 2009), kriging (O'Hagan, 1978; Journel and Huijbregts, 1981; Simpson *et al.*, 2001; Koziel *et al.*, 2011a), artificial neural networks (Minsky and Papert, 1969; Haykin, 1998), and support vector regression (Vapnik, 1995; Gunn, 1998; Smola and Schölkopf, 2004). Kriging seems to be the most suitable for implementation of SBO schemes because it provides an estimated error in its predictions (Forrester and Keane, 2009), which can be employed to develop efficient infill criteria and sequential design methods (Gorrisen *et al.*, 2007; Forrester and Keane, 2009; Gorrisen *et al.*, 2009; Gorrisen *et al.*, 2010). In general, the computational cost of creating approximation-based surrogate models may be quite substantial, particularly when the number of design variables is large.

Another way of creating the surrogate, by using the physics-based coarse model, is typically much more efficient (Søndergaard, 2003; Bandler *et al.*, 2004a). The knowledge about the system, embedded in the coarse model, allows the creation of a relatively accurate surrogate model using a limited amount of fine model data (in many cases, through the simulation of a single design). The obvious limitations of physics-based surrogate approaches include the lack of reliable coarse models in general and nontransferability of some specific optimization approaches from one application area to another. On the other hand, physics-based low-fidelity models have been used in engineering design for decades, and in any particular field the experts are normally familiar with their construction.

Space mapping, together with a few other techniques, e.g., Alexandrov and Lewis (2001); Koziel (2010), exploits physics-based surrogates.

A distinguishing feature of space mapping is the parameter extraction (PE) step (Bandler *et al.*, 2004a). Parameter extraction is, in essence, a nonlinear regression procedure that aims at aligning the surrogate with the fine model at one or more points (design variable vectors) by applying suitable transformations to the domain and/or the response of the underlying coarse model (Koziel *et al.*, 2006). Another fundamental component of any SM algorithm, common to other SBO techniques, is surrogate model optimization (SO), where the surrogate is used as a prediction tool to find a good approximation of the fine model optimum. The PE-SO scheme is normally iterated. In the original and aggressive space mapping methods SO is implied (Bandler *et al.*, 1994, Bandler *et al.*, 1995).

Assuming that the underlying coarse model is reasonably accurate, the SM algorithm typically yields a satisfactory solution just after a few iterations. Because both parameter extraction and surrogate model optimization usually require multiple evaluations of the coarse model, the latter should ideally be substantially (at least two orders of magnitude) cheaper than the fine model (Bandler *et al.*, 2004a). As demonstrated in numerous publications, see e.g., Leary *et al.* (2001); Choi *et al.* (2001); Redhe and Nilsson (2002); Ismail *et al.* (2004); Wu *et al.* (2004); Amari *et al.* (2006); Dorica and Giannacopoulos (2006); Rayas-Sánchez and Gutiérrez-Ayala (2006); Crevecoeur *et al.* (2007), space mapping is capable of producing an optimized design at a computational cost corresponding to a few evaluations of the fine model.

The "mystery" of space mapping efficiency lies in one's knowledge about the system under design as contained in the coarse model. In most areas, the art and science of creating low-fidelity models has matured over the years because, before the era of massive parallel computing and highly accurate commercial simulators, simple yet reliable physics-based models were essential for the design process. Today, such models can serve — after applying small corrections — as good and globally accurate surrogates. At the same time, the coarse model is the most critical component of SM algorithms. Wrongly selected or insufficiently accurate, it may lead to the failure of the SM optimization process (Koziel and Bandler, 2007a; Koziel *et al.*, 2008b).

Space mapping was originally developed in microwave engineering in the early 1990s (Bandler *et al.*, 1994) out of competitive necessity. Full-wave electromagnetic simulation had long been accepted for validating microwave designs obtained through circuit models. While the idea of employing electromagnetic solvers for design attracted microwave engineers, electromagnetic solvers are notoriously CPU-intensive. They also suffered from non-differentiable response evaluation and nonparameterized design variables that were often discrete in the parameter space. Such characteristics are unfriendly to classical gradient optimization algorithms.

The original idea of space mapping (Bandler *et al.*, 1994; Bandler *et al.*, 2004a) was to map designs from optimized circuit models to corresponding electromagnetic models. A "parameter extraction" step calibrated the circuit solver against the electromagnetic simulator in order to minimize discrepancies between the two simulations. The circuit model (surrogate or "mapped coarse model") was then implicitly updated (mapped) through extracted parameters and made ready for subsequent iteration. Hence Bandler *et al.* (1994) coined the term "space mapping."

A number of offshoots of space mapping have been developed in the 1990s and early 2000s, including aggressive space mapping (ASM) (Bandler *et al.*, 1995), trust-region aggressive space mapping (TRASM) (Bakr *et al.*, 1998), frequency space mapping (Bandler *et al.*, 1995), and neural space mapping (Devabhaktuni *et al.*, 2003; Rayas-Sánchez *et al.*, 2005; Zhang *et al.*, 2005). Several researchers address theoretical aspects of space mapping, mostly its convergence properties (Vicente, 2003; Madsen and Søndergaard, 2004, Koziel *et al.*, 2006; Koziel *et al.*, 2008b). All initial versions of SM algorithms were based on coarse and fine model alignment through suitable distortion of the coarse model design space (space mapping). This was sufficient mainly because in original microwave engineering applications, the coarse model was typically quite accurate, and, most importantly, the coarse and fine model ranges are similar.

In later years, correction of the coarse model response (output SM) was also introduced (Bandler *et al.*, 2004a). There are many variations of "output" SM that involve both additive and multiplicative corrections. One of the most efficient output SM variations is manifold mapping (MM) (Echeverria and Hemker, 2005; Echeverria, 2007; Echeverria and Hemker, 2008; Koziel and Echeverria, 2010). MM is one of the few SM variations that comes with rigorous convergence theory (Echeverria and Hemker, 2005). Some convergence results are also available for generic space mapping (Koziel *et al.*, 2008b). While these results are of limited usefulness due to the main assumptions being hard to verify in practice, they indicate the importance of the coarse model quality for the performance of SM algorithms.

One of the most characteristic developments of space mapping is implicit SM (Bandler *et al.*, 2003; Bandler *et al.*, 2004b; Koziel *et al.*, 2010), where the coarse and fine model alignment is realized by using physics-based coarse model parameters (so-called implicit or preassigned parameters). These parameters are normally fixed in the fine model, however, they can be freely adjusted in the coarse model in order to align the two (Bandler *et al.*, 2004b).

Recent advances in SM technology include adaptive SM (Koziel and Bandler, 2007a), SM with adaptively constrained parameter extraction (Koziel *et al.*, 2011b), corrected SM (Robinson *et al.*, 2008), as well as tuning space mapping (Koziel *et al.*, 2009a; Cheng *et al.*, 2010a; Cheng *et al.*, 2010b; Cheng *et al.*, 2010c). Tuning space mapping is an invasive version of SM developed for microwave applications where the surrogate model is constructed as a combination of full-wave electromagnetic simulation and circuit model. It has proved to be one of the most efficient SM approaches, where the optimized design is produced after one or two fine model evaluations. See Chapter 5 for details.

Space mapping methodology continues to provide success in many engineering disciplines: electronics, photonics, microwaves, antennas, and magnetic systems; civil, mechanical, and aerospace structure engineering, including automotive crashworthiness design (Leary *et al.*, 2001; Redhe and Nilsson, 2002; Ismail *et al.*, 2004; Wu *et al.*, 2004; Amari *et al.*, 2006; Robinson *et al.*, 2008). Space mapping also inspired development of recent response correction techniques such as adaptive response correction (Koziel *et al.*, 2009b) and shape preserving response prediction (Koziel, 2010; Koziel and Szczepanski, 2011).

4.2. Space Mapping Formulation and Algorithms

As mentioned in the introduction, a variety of algorithms have been proposed since the introduction of space mapping in 1994. In this section, we briefly recall the formulation of space mapping, as well as discuss several SM algorithms, including the aggressive SM (Bandler *et al.*, 1995), input SM (Bandler *et al.*, 2004a), frequency and implicit SM (Bandler *et al.*, 2004a), output SM (Koziel *et al.*, 2006), as well as manifold mapping (MM) (Echeverria and Hemker, 2005).

4.2.1. *Space Mapping Concept*

The original design problem to be solved is formulated as follows:

$$\boldsymbol{x}_f^* = \arg\min_{\boldsymbol{x}_f} U(\boldsymbol{R}_f(\boldsymbol{x}_f)), \qquad (4.1)$$

where $\boldsymbol{R}_f$ is a vector of m responses of the high-fidelity (or fine) model, $\boldsymbol{x}_f$ is the vector of n design parameters, and U is a suitable merit function. In microwave engineering U is typically the minimax objective function with upper and lower specifications; $\boldsymbol{x}_f^*$ is the optimal solution to be determined and is assumed to be unique.

Space mapping assumes the existence of an auxiliary (low-fidelity or coarse) model $\boldsymbol{R}_c$ which is a computationally cheap representation of the fine model. The fundamental SM component is a mapping $\boldsymbol{P}$ relating the fine and coarse model parameters (Bandler *et al.*, 1994):

$$\boldsymbol{x}_c = \boldsymbol{P}(\boldsymbol{x}_f), \tag{4.2}$$

such that

$$\boldsymbol{R}_c(\boldsymbol{P}(\boldsymbol{x}_f)) \approx \boldsymbol{R}_f(\boldsymbol{x}_f) \tag{4.3}$$

at least in some subset of the fine model parameter space.

Having the mapping $\boldsymbol{P}$ known, the direct optimization, i.e., solving the problem (4.1), can be replaced by finding $\bar{\boldsymbol{x}}_f$ given by

$$\bar{\boldsymbol{x}}_f = \boldsymbol{P}^{-1}(\boldsymbol{x}_c^*). \tag{4.4}$$

Here, $\bar{\boldsymbol{x}}_f$ can be considered as a good estimate of $\boldsymbol{x}_f^*$, whereas $\boldsymbol{x}_c^*$ is the optimal solution of the coarse model defined as $\boldsymbol{x}_c^* = \arg\min\{\boldsymbol{x}_c: U(\boldsymbol{R}_c(\boldsymbol{x}_c))\}$.

Using (4.3), the original problem (4.1) can be reformulated as

$$\bar{\boldsymbol{x}}_f = \arg\min_{\boldsymbol{x}_f} U(\boldsymbol{R}_c(\boldsymbol{P}(\boldsymbol{x}_f))), \tag{4.5}$$

where $\boldsymbol{R}_c(\boldsymbol{P}(\boldsymbol{x}_f))$ is an enhanced (mapped) coarse model (also referred to as a surrogate model).

The problem with formulation (4.5) is that space mapping $\boldsymbol{P}$ is not known in an analytical form. It can only be evaluated at any $\boldsymbol{x}_f$ by performing the so-called parameter extraction (PE) procedure

$$\boldsymbol{P}(\boldsymbol{x}_f) = \arg\min_{\boldsymbol{x}_c} \| \boldsymbol{R}_f(\boldsymbol{x}_f) - \boldsymbol{R}_c(\boldsymbol{x}_c) \|. \tag{4.6}$$

In general, the solution to (4.6) may not be unique. This situation has been dealt with in many publications, see e.g., Bandler *et al.* (1995) and Bandler *et al.* (2004a). Here, for the sake of simplicity, we assume the uniqueness of the parameter extraction process. On the other hand, the ranges of the fine and coarse model may not be similar to each other which poses additional difficulties, e.g., Alexandrov and Lewis (2001). For that reason, space mapping based on response correction may be necessary in practice (Koziel *et al.*, 2006).

4.2.2. *Aggressive Space Mapping*

Assuming the uniqueness of the coarse model optimum x_c^*, the solution to (4.5) is equivalent to driving the following residual vector f to zero:

$$f = f(x_f) = P(x_f) - x_c^*. \tag{4.7}$$

The aggressive space mapping (ASM) technique (Bandler *et al.*, 1995) iteratively solves the nonlinear system

$$f(x_f) = 0 \tag{4.8}$$

for x_f. The ASM algorithm starts by finding x_c^*. At jth iteration, the calculation of the error vector $f^{(j)}$ requires an evaluation of $P^{(j)}(x_f^{(j)})$, which is realized by executing the parameter extraction process (4.6), i.e.,

$$P(x_f^{(j)}) = \arg\min_{x_c} \|R_f(x_f^{(j)}) - R_c(x_c)\|. \tag{4.9}$$

The quasi-Newton step in the fine model parameter space is given by

$$B^{(j)} h^{(j)} = -f^{(j)}, \tag{4.10}$$

where $B^{(j)}$ is the approximation of the space mapping Jacobian $J_P = J_P(x_f) = [\partial P^T/\partial x_f]^T = [\partial(x_c^T)/\partial x_f]^T$. Solving (4.10) for $h^{(j)}$ gives the next iterate $x_f^{(j+1)}$

$$x_f^{(j+1)} = x_f^{(j)} + h^{(j)}. \tag{4.11}$$

The algorithm terminates if $\|f^{(j)}\|$ is sufficiently small. The output of the algorithm is an approximation to $\bar{x}_f = P^{-1}(x_c^*)$.

The matrix B can be obtained in several ways. Assuming that the Jacobians of both the coarse and fine models, J_c and J_f, are known, the following expression can be derived (Bakr *et al.*, 1998):

$$B = (J_c^T J_c)^{-1} J_c^T J_f, \tag{4.12}$$

provided that J_c has full rank and $m \geq n$. In practice, the fine model Jacobian is rarely known. The original version of ASM uses a rank one

Broyden update (Broyden, 1965) of the form

$$\boldsymbol{B}^{(j+1)} = \boldsymbol{B}^{(j)} + \frac{\boldsymbol{f}^{(j+1)} - \boldsymbol{f}^{(j)} - \boldsymbol{B}^{(j)}\boldsymbol{h}^{(j)}}{\boldsymbol{h}^{(j)T}\boldsymbol{h}^{(j)}}\boldsymbol{h}^{(j)T}. \tag{4.13}$$

If $\boldsymbol{h}^{(j)}$ is the quasi-Newton step, (4.13) can be simplified using (4.10) to

$$\boldsymbol{B}^{(j+1)} = \boldsymbol{B}^{(j)} + \frac{\boldsymbol{f}^{(j+1)}}{\boldsymbol{h}^{(j)T}\boldsymbol{h}^{(j)}}\boldsymbol{h}^{(j)T}. \tag{4.14}$$

It is also possible to use $\boldsymbol{B}^{(j)} = \boldsymbol{I}$, which corresponds to the mapping between the coarse and fine model spaces being just a shift (Bandler *et al.*, 2004a).

Several variations of ASM have been considered in the literature, including hybrid ASM (Bakr *et al.*, 1999) and trust-region ASM (Bakr *et al.*, 1998), that aimed at addressing issues of the original ASM such as non-uniqueness of the parameter extraction (Bakr *et al.*, 1998).

4.2.3. *Parametric Space Mapping: Input, Implicit, Output, and Others*

The mapping $\boldsymbol{P}$ used by the ASM algorithm is not defined explicitly but only through the parameter extraction process (4.6). In general, $\boldsymbol{P}$ can be assumed to have a certain analytical form. In such a case, the SM algorithm is defined as an iterative process

$$\boldsymbol{x}^{(i+1)} = \arg\min_{\boldsymbol{x}} U(\boldsymbol{R}_s^{(i)}(\boldsymbol{x})), \tag{4.15}$$

where $\boldsymbol{R}_s^{(i)}$ is a surrogate model at iteration i, whereas $\boldsymbol{x}^{(i)}$, $i = 0, 1, \ldots$, is a sequence of approximate solutions to the original problem (4.1).

Chronologically, the first parameterized version of space mapping was the input SM (Bandler *et al.*, 1994), where the SM surrogate model takes the form

$$\boldsymbol{R}_s^{(i)}(\boldsymbol{x}) = \boldsymbol{R}_c(\boldsymbol{B}^{(i)} \cdot \boldsymbol{x} + \boldsymbol{c}^{(i)}), \tag{4.16}$$

where $\boldsymbol{B}^{(i)}$ and $\boldsymbol{c}^{(i)}$ are matrices of suitable dimensions obtained in the parameter extraction process

$$[\boldsymbol{B}^{(i)}, \boldsymbol{c}^{(i)}] = \arg\min_{[\boldsymbol{B}, \boldsymbol{c}]} \sum_{k=0}^{i} w_{i.k} \|\boldsymbol{R}_f(\boldsymbol{x}^{(k)}) - \boldsymbol{R}_c(\boldsymbol{B} \cdot \boldsymbol{x}^{(k)} + \boldsymbol{c})\|. \tag{4.17}$$

Here, $w_{i.k}$ are weighting factors; a common choice of $w_{i.k}$ is $w_{i.k} = 1$ for all i and all k (all previous designs contribute to the parameter extraction process) or $w_{i.1} = 1$ and $w_{i.k} = 0$ for $k < i$ (the surrogate model depends on the most recent design only).

In particular, $\boldsymbol{B}$ can be an identity matrix, in which case the input SM represents the shift of parameters only.

In general, the SM surrogate model is constructed as follows:

$$\boldsymbol{R}_s^{(i)}(\boldsymbol{x}) = \bar{\boldsymbol{R}}_s(\boldsymbol{x}, \boldsymbol{p}^{(i)}), \tag{4.18}$$

where $\bar{\boldsymbol{R}}_s$ is a generic SM surrogate model, i.e., $\boldsymbol{R}_c$ composed with suitable (usually linear) transformations, whereas

$$\boldsymbol{p}^{(i)} = \arg\min_{\boldsymbol{p}} \sum_{k=0}^{i} w_{i.k} \| \boldsymbol{R}_f(\boldsymbol{x}^{(k)}) - \bar{\boldsymbol{R}}_s(\boldsymbol{x}^{(k)}, \boldsymbol{p}) \| \tag{4.19}$$

is a vector of model parameters.

Various SM surrogate models are available (Bandler *et al.*, 2004a; Koziel *et al.*, 2006). They can be categorized into four groups:

- Models based on a (usually linear) distortion of the coarse model parameter space, e.g., the aforementioned input SM of the form $\bar{\boldsymbol{R}}_s(\boldsymbol{x}, \boldsymbol{p}) = \bar{\boldsymbol{R}}_s(\boldsymbol{x}, \boldsymbol{B}, \boldsymbol{c}) = \boldsymbol{R}_c(\boldsymbol{B} \cdot \boldsymbol{x} + \boldsymbol{c})$ (Bandler *et al.*, 1994; Bandler *et al.*, 2004a).
- Models based on a distortion of the coarse model response, e.g., the output SM of the form $\bar{\boldsymbol{R}}_s(\boldsymbol{x}, \boldsymbol{p}) = \bar{\boldsymbol{R}}_s(\boldsymbol{x}, \boldsymbol{d}) = \boldsymbol{R}_c(\boldsymbol{x}) + \boldsymbol{d}$ or $\bar{\boldsymbol{R}}_s(\boldsymbol{x}, \boldsymbol{p}) = \bar{\boldsymbol{R}}_s(\boldsymbol{x}, \boldsymbol{A}) = \boldsymbol{A} \cdot \boldsymbol{R}_c(\boldsymbol{x})$ (Bandler *et al.*, 2003; Koziel *et al.*, 2006).
- Implicit space mapping, where the parameters used to align the surrogate with the fine model are separate from the design variables, i.e., $\bar{\boldsymbol{R}}_s(\boldsymbol{x}, \boldsymbol{p}) = \bar{\boldsymbol{R}}_s(\boldsymbol{x}, \boldsymbol{x}_p) = \boldsymbol{R}_{c.i}(\boldsymbol{x}, \boldsymbol{x}_p)$, with $\boldsymbol{R}_{c.i}$ being the coarse model dependent on both the design variables $\boldsymbol{x}$ and so-called preassigned parameters $\boldsymbol{x}_p$ (e.g., dielectric constant, substrate height) that are normally fixed in the fine model but can be freely altered in the coarse model (Bandler *et al.*, 2003; Bandler *et al.*, 2004b; Koziel *et al.*, 2010a; Koziel *et al.*, 2011b).
- Custom models exploiting characteristic parameters of a given design problem. The most commonly used characteristic parameter is frequency. Frequency SM exploits a surrogate model of the form $\bar{\boldsymbol{R}}_s(\boldsymbol{x}, \boldsymbol{p}) = \bar{\boldsymbol{R}}_s(\boldsymbol{x}, \boldsymbol{F}) = \boldsymbol{R}_{c.f}(\boldsymbol{x}, \boldsymbol{F})$ (Bandler *et al.*, 2003), where $\boldsymbol{R}_{c.f}$ is a frequency-mapped coarse model. Here, the coarse model is evaluated at frequencies different from the original frequency sweep for the fine model, according to the mapping $\omega \rightarrow f_1 + f_2\omega$, with $\boldsymbol{F} = [f_1\ f_2]^T$.

The basic SM types can be combined, e.g., a surrogate model employing both the input, output, and frequency SM types would be as follows: $\bar{R}_s(x, p) = \bar{R}_s(x, c, d, F) = R_{c.f}(x + c, F) + d$. The rationale for this is that a properly chosen mapping may significantly improve the performance of the SM algorithm, however, the optimal selection of the mapping type for a given design problem is not trivial (Koziel and Bandler, 2007a; Koziel and Bandler, 2008b; Koziel *et al.*, 2008b). Figure 4.1 illustrates the basic types of SM surrogate models.

The space mapping optimization algorithm using a parametric surrogate can be described as follows:

1. Set $i = 0$; choose the initial design solution $x^{(0)}$;
2. Evaluate the fine model to find $R_f(x^{(i)})$;
3. Obtain the surrogate model $R_s^{(i)}$ using (4.18) and (4.19);
4. Given $x^{(i)}$ and $R_s^{(i)}$, obtain $x^{(i+1)}$ using (4.15);
5. If the termination condition is not satisfied go to Step 2; else terminate the algorithm;

Typically, $x^{(0)} = \arg\min\{x : U(R_c(x))\}$, i.e., it is the optimal solution of the coarse model: the best initial design is normally available. The algorithm is usually terminated when it converges (i.e., $||x^{(i)} - x^{(i-1)}||$ is smaller than some user-defined value) or when the maximum number of iterations (or, more often, the number of R_f evaluations) is exceeded.

4.2.4. *Space Mapping Illustration*

In order to illustrate the operation of a basic SM algorithm we consider the so-called 1D wedge-cutting problem (Bandler *et al.*, 2004a) reformulated as follows. Given wedges as shown in Fig. 4.2(a), cut a piece of length x so that the corresponding area, $R_f(x)$ is equal to $A_0 = 100$. We assume that the wedge area can only be measured after cutting (fine model simulation) and we have a number of sufficiently large identical raw wedges available. Our goal then is to cut a piece to specification (fine model optimum) with a small number of cuts.

Our fine model is R_f, which, for the sake of this example, is assumed to be given by $R_f(x) = x(5 + x/16)$. We also consider a simplified representation of the wedge, a rectangle of height H shown in Fig. 4.2(b). The coarse model is the area of a piece of this rectangle, determined by the length x, so that we have $R_c(x) = Hx$. Here, we assume that $H = 5$.

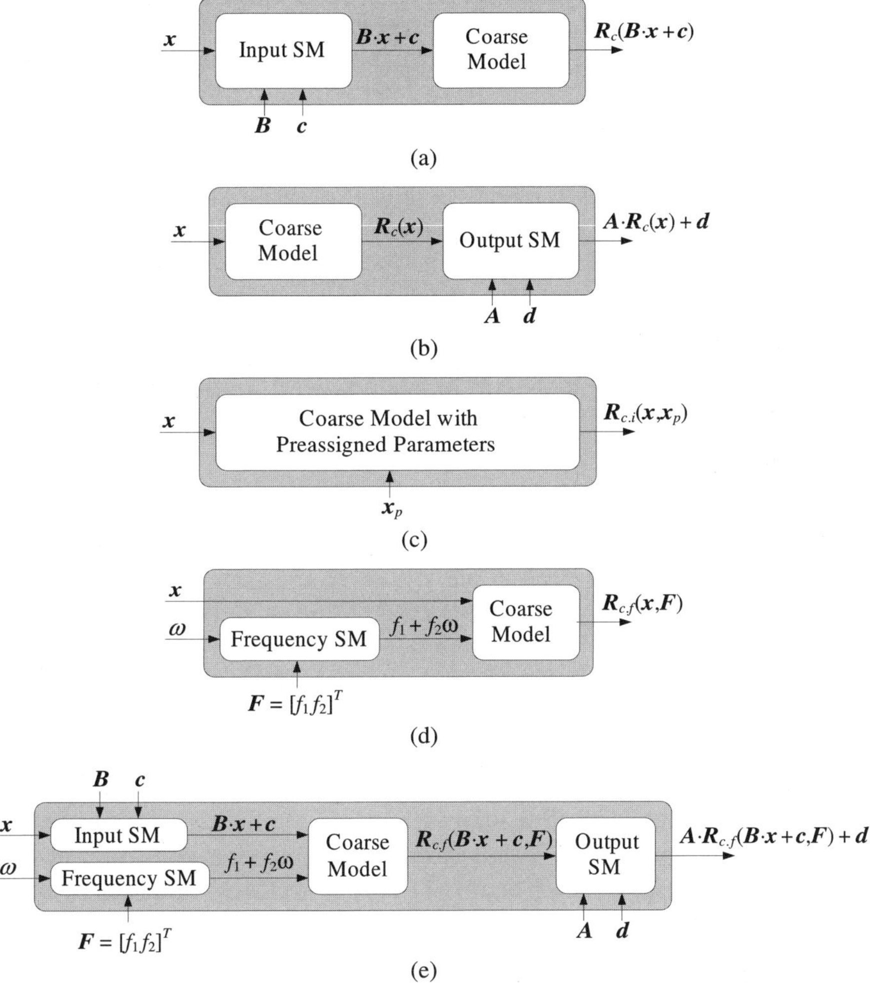

Fig. 4.1. Basic space mapping surrogate model types: (a) input SM, (b) output SM, (c) implicit SM, (d) frequency SM, (e) example of a composed surrogate model using input, output, and frequency SM. In case of input SM, the coarse model design space is subjected to an affine transformation (shift + scaling), for output SM, the coarse model response is scaled and/or shifted. Implicit SM assumes that the coarse model has additional degrees of freedom, so-called preassigned parameters (x_p) that can be used to adjust the model response. The frequency SM is a special type of mapping where the coarse model is evaluated at frequencies different than the original frequency sweep of interest, according to the linear transformation determined by the vector F.

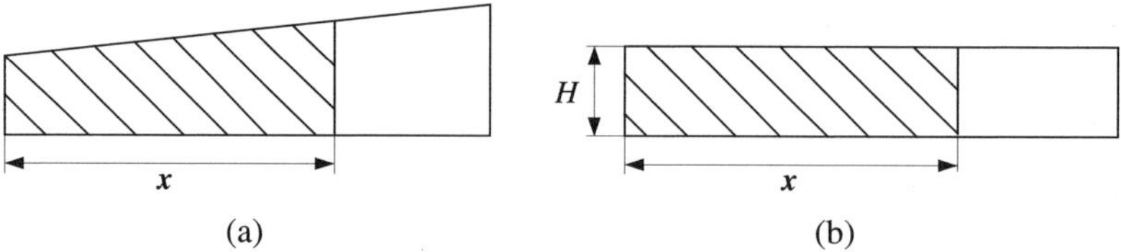

Fig. 4.2. Wedge-cutting problem (Bandler *et al.*, 2004a): (a) the fine model, and (b) the coarse model.

The starting point of SM optimization is a coarse model optimal solution $x^{(0)} = 20$. The fine model at $x^{(0)}$ is $R_f(x^{(0)}) = 125$. For illustration purposes we will solve our problem using the simplest version of the input space mapping and then using implicit space mapping.

4.2.4.1. *Wedge-cutting problem solved using input space mapping*

We use the following setup for the input SM approach. The generic surrogate model is given by $\bar{R}_s(x, p) = \bar{R}_s(x, c) = R_c(x + c)$. The weighting factors in the parameter extraction process (4.19) are given by $w_{i.k} = 1$ for $k = i$ and $w_{i.k} = 0$ otherwise. Thus, the surrogate model can be $R_s^{(i)}(x) = R_c(x+c^{(i)})$, where $c^{(i)} = \arg\min\{c : \|R_f(x^{(i)})-R_c(x^{(i)}+c)\|\}$. In this simple case, the parameter extraction problem has an analytical solution given by $c^{(i)} = R_f^{(i)}/H - x^{(i)}$. Figure 4.3 shows the first four iterations of the SM algorithm solving the wedge-cutting problem.

This particular input SM approach is both simple and direct, yet it converges to an acceptable result (from an engineering point of view) in a remarkably small number of iterations. It is clearly knowledge-based, since the coarse model is a physical approximation to the fine model, and the iteratively updated coarse model attempts to align itself with the fine model. The optimization process mimics a learning process derived from intuition.

4.2.4.2. *Wedge-cutting problem solved using implicit space mapping*

We use the following setup for the implicit SM approach. The generic surrogate model is given by $\bar{R}_s(x, p) = \bar{R}_s(x, H) = R_{c_i}(x, H)$, where $R_{c_i}(x, H) = Hx$. The weighting factors in the parameter extraction process (4.19) are, as before, $w_{i.k} = 1$ for $k = i$ and $w_{i.k} = 0$ otherwise. The surrogate model can be restated as: $R_s^{(i)}(x) = R_{c_i}(x, H^{(i)}) = H^{(i)}x$, where $H^{(i)} = \arg\min\{H: \|R_f(x^{(i)}) - R_{c_i}(x^{(i)}, H)\|\}$. In this simple case,

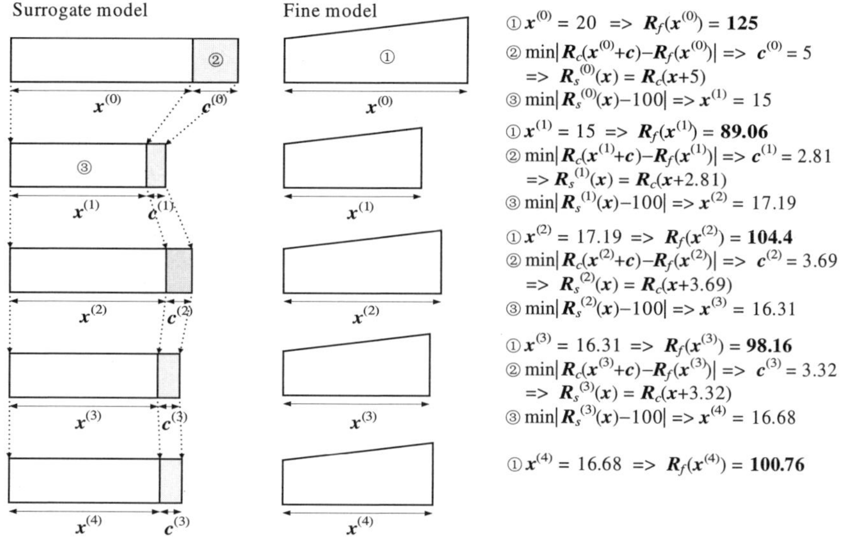

① fine model evaluation ② parameter extraction ③ surrogate optimization

Fig. 4.3. Input space mapping iterations for solving the wedge-cutting problem (Koziel and Bandler, 2010).

The right-hand column of the figure reads:

$$① \; x^{(0)} = 20 \;=>\; R_f(x^{(0)}) = 125$$
$$② \; \min| R_c(x^{(0)}+c) - R_f(x^{(0)})| \;=>\; c^{(0)} = 5$$
$$=> \; R_s^{(0)}(x) = R_c(x+5)$$
$$③ \; \min| R_s^{(0)}(x) - 100| \;=>\; x^{(1)} = 15$$

$$① \; x^{(1)} = 15 \;=>\; R_f(x^{(1)}) = 89.06$$
$$② \; \min| R_c(x^{(1)}+c) - R_f(x^{(1)})| \;=>\; c^{(1)} = 2.81$$
$$=> \; R_s^{(1)}(x) = R_c(x+2.81)$$
$$③ \; \min| R_s^{(1)}(x) - 100| \;=>\; x^{(2)} = 17.19$$

$$① \; x^{(2)} = 17.19 \;=>\; R_f(x^{(2)}) = 104.4$$
$$② \; \min| R_c(x^{(2)}+c) - R_f(x^{(2)})| \;=>\; c^{(2)} = 3.69$$
$$=> \; R_s^{(2)}(x) = R_c(x+3.69)$$
$$③ \; \min| R_s^{(2)}(x) - 100| \;=>\; x^{(3)} = 16.31$$

$$① \; x^{(3)} = 16.31 \;=>\; R_f(x^{(3)}) = 98.16$$
$$② \; \min| R_c(x^{(3)}+c) - R_f(x^{(3)})| \;=>\; c^{(3)} = 3.32$$
$$=> \; R_s^{(3)}(x) = R_c(x+3.32)$$
$$③ \; \min| R_s^{(3)}(x) - 100| \;=>\; x^{(4)} = 16.68$$

$$① \; x^{(4)} = 16.68 \;=>\; R_f(x^{(4)}) = 100.76$$

the parameter extraction problem has an analytical solution given by $H^{(i)} = R_f^{(i)}/x^{(i)}$. Figure 4.4 shows the first four iterations of the SM algorithm for solving the wedge-cutting problem.

This indirect, implicit SM approach is as simple as the input SM approach, and it also converges to an acceptable result (from an engineering point of view) in a few iterations. Since our target is area, the H and the x in the Hx of the rectangle are of equal significance as design parameters in the coarse model. The physical approximation remains valid and this optimization process also mimics a learning process derived from intuition. In effect, in both processes, we are recalibrating the coarse model against measurements of the fine model after each change to the fine model.

4.2.5. *Practical Issues and Open Problems*

It is easy to notice that the SM algorithm (4.15)–(4.19) has the following two features. First of all, the consistency conditions between the fine and surrogate models are not necessarily satisfied. In particular, it is not required that the surrogate model matches the fine model with respect to the value and first-order derivative at any of the iteration points. Second,

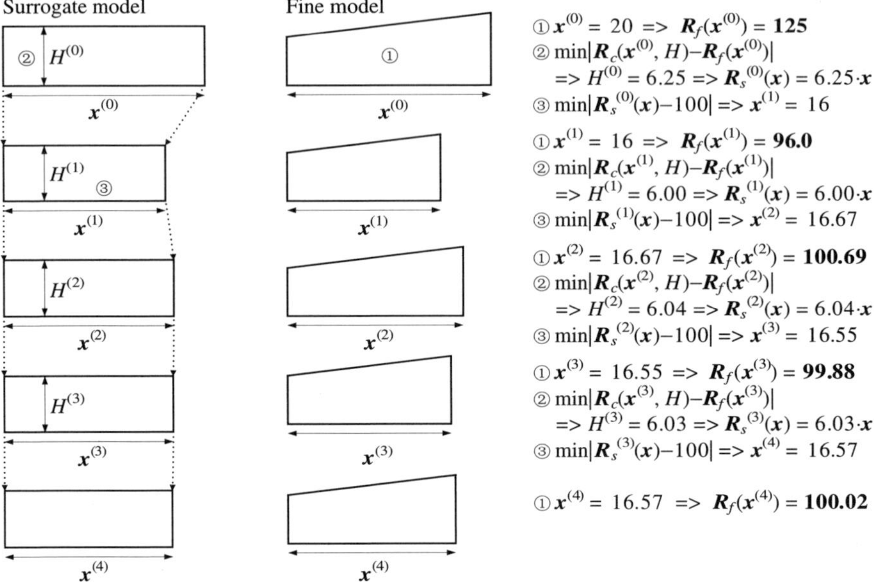

Fig. 4.4.   Implicit SM solving the wedge-cutting problem (Koziel and Bandler, 2010).

subsequent iterations are accepted regardless of the objective function improvement. As a consequence, convergence of the SM algorithm is not guaranteed in general, and the choice of an optimal SM approach for a given problem is not obvious (Koziel *et al.*, 2008b).

Several methods for assessing the quality of the coarse/surrogate model have been proposed that are based on information obtained from the fine model at a set of test points (Koziel and Bandler, 2007a; Koziel and Bandler, 2007b; Koziel *et al.*, 2008b; Koziel *et al.*, 2010a). This information is used to estimate certain conditions in the convergence results and allows us to predict whether a given model might be successfully used in SM optimization. Using these methods one can also compare the quality of different coarse models, or choose the proper type of space mapping which would suit a given optimization problem (Koziel and Bandler, 2007b; Koziel *et al.*, 2008b). Assessment methods can also be embedded into the SM optimization algorithm so that the most suitable surrogate model is selected in each iteration of the algorithm out of a given candidate model set (Koziel and Bandler, 2007a).

Another way of improving reliability of space mapping is to impose suitable (and adaptively adjusted) constraints on the parameter extraction process, surrogate model optimization process, or both of them. As demonstrated in Koziel *et al.* (2011b) and Koziel (2011), this approach can lead to substantial improvement of the SM algorithm performance. Another, more conventional way of improving algorithm convergence is through trust-region approach (Conn *et al.*, 2000) as described in the next section.

4.2.6. *Trust-Region Space Mapping*

One of the common problems of the SM algorithm (4.15) is possible convergence issues. Also, the new design $x^{(i+1)}$ is accepted regardless of the corresponding objective function value so that whether the algorithm finds a satisfactory design or not heavily depends on the quality of the coarse model (Koziel and Bandler, 2007a; Koziel *et al.*, 2008b). Convergence properties of the SM algorithm can be improved using the trust-region approach (Conn *et al.*, 2000; Koziel *et al.*, 2010b). In particular, the surrogate optimization process (4.15) can be constrained to a neighborhood of $x^{(i)}$, defined with $||x - x^{(i)}|| \leq \delta^{(i)}$, as follows

$$x^{(i+1)} = \arg \min_{x,\, ||x - x^{(i)}|| < \delta^{(i)}} U\big(R_s^{(i)}(x)\big), \tag{4.20}$$

where $\delta^{(i)}$ is a trust-region radius at iteration i. The trust-region radius is reduced if the improvement of the fine model objective function is not sufficient, i.e., if $(U(R_f(x^{(i+1)})) - U(R_f(x^{(i)})))/$ $(U(R_s^{(i)}(x^{(i+1)})) - U(R_s^{(i)}(x^{(i)})))$ is smaller than a user-defined threshold, or if $U(R_f(x^{(i+1)})) \geq U(R_f(x^{(i)}))$, in which case the new design is rejected. Typically, the standard trust-region radius updating rules (Conn *et al.*, 2000) are used.

If, for all $i = 0, 1, 2, \ldots$, the surrogate model satisfies the zero- and first-order consistency conditions of the form

$$R_s^{(i)}(x^{(i)}) = R_f(x^{(i)}) \tag{4.21}$$

$$J_{R_s^{(i)}}(x^{(i)}) = J_{R_f}(x^{(i)}), \tag{4.22}$$

where J denotes the Jacobian of the respective model. Then, under mild assumptions about the smoothness of the models, algorithm (4.15) is convergent to the local fine model optimum (Alexandrov *et al.*, 1998). The fundamental reason is that (4.21) and (4.22) ensure that $U(\boldsymbol{R}_f(\boldsymbol{x}^{(i+1)})) < U(\boldsymbol{R}_f(\boldsymbol{x}^{(i)}))$ if the trust-region radius is sufficiently small.

Unfortunately, (4.21) and (4.22) are not necessarily satisfied by the SM surrogate model. In particular, the SM optimization process may become stuck at some point as the reduction of the trust-region radius $\delta^{(i)}$ does not bring any improvement to the fine model objective function, which results in the termination of the algorithm. In other words, the trust-region method applied to the SM algorithm as in (4.20) is a heuristic procedure that may improve robustness of the algorithm but still does not ensure sufficient performance.

On the other hand, an important prerequisite of SM algorithms is that $\boldsymbol{R}_c$ is physics-based so that the surrogate model reflects the general features of $\boldsymbol{R}_f$; in particular, the local behavior of both models is similar. This, in combination with the multi-point parameter extraction (4.19) ensures that (4.21) and (4.22) may be satisfied approximately. Moreover, condition (4.21) can be easily enforced by means of the output SM (Koziel *et al.*, 2006) using the surrogate

$$\boldsymbol{R}_s^{(i)}(\boldsymbol{x}) = \bar{\boldsymbol{R}}_s(\boldsymbol{x}, \boldsymbol{p}^{(i)}) + \boldsymbol{d}^{(i)}, \tag{4.23}$$

with $\boldsymbol{p}^{(i)}$ obtained by (4.19) and

$$\boldsymbol{d}^{(i)} = \boldsymbol{R}_f(\boldsymbol{x}^{(i)}) - \bar{\boldsymbol{R}}_s(\boldsymbol{x}^{(i)}, \boldsymbol{p}^{(i)}). \tag{4.24}$$

Relations (4.21) and (4.22) can be enforced if the SM surrogate explicitly uses fine model sensitivity information, e.g., as in the following model (Koziel *et al.*, 2010b):

$$\boldsymbol{R}_s^{(i)}(\boldsymbol{x}) = \bar{\boldsymbol{R}}_s(\boldsymbol{x}, \boldsymbol{p}^{(i)}) + \boldsymbol{d}^{(i)} + \boldsymbol{E}^{(i)} \cdot (\boldsymbol{x} - \boldsymbol{x}^{(i)}), \tag{4.25}$$

where $\bar{\boldsymbol{R}}_s(\boldsymbol{x}, \boldsymbol{p}^{(i)})$ is any SM surrogate with parameters $\boldsymbol{p}^{(i)}$ obtained with (4.19) and

$$\boldsymbol{d}^{(i)} = \boldsymbol{R}_f(\boldsymbol{x}^{(i)}) - \bar{\boldsymbol{R}}_s(\boldsymbol{x}^{(i)}, \boldsymbol{p}^{(i)}) \tag{4.26}$$

$$\boldsymbol{E}^{(i)} = \boldsymbol{J}_{\boldsymbol{R}_f}(\boldsymbol{x}^{(i)}) - \boldsymbol{J}_{\bar{\boldsymbol{R}}_s(\cdot, \boldsymbol{p}^{(i)})}(\boldsymbol{x}^{(i)}). \tag{4.27}$$

Convergence of algorithm (4.15) with surrogate model (4.25)–(4.27) is guaranteed under standard assumptions concerning the smoothness of the fine and coarse models (Alexandrov *et al.*, 1998). Note, however, that the computational cost of SM optimization may substantially increase in this case because of the necessity of calculating the fine model Jacobian at all iterations. Jacobian estimation using finite differences exploits extra n fine model evaluations per estimation, unless some other techniques are used, such as adjoint sensitivity (Nikolova *et al.*, 2006). Another option is to approximate the Jacobian using, e.g., a rank-one Broyden update formula (Broyden, 1965).

4.3. Application Example

Let us consider an interdigital filter design (Bandler *et al.*, 1997). The fine model, shown in Fig. 4.5, is implemented in the electromagnetic (EM) simulator Sonnet *em* (2010). The substrate height and the dielectric constant are 15 mils and 9.8, respectively. The shielding cover height is 75 mils. The cell size is set at 1 mil $\times$ 1 mil. The design specifications are $|S_{21}| \leq -30\,\mathrm{dB}$ for $4.0\,\mathrm{GHz} \leq \omega \leq 4.5\,\mathrm{GHz}$ and for $5.45\,\mathrm{GHz} \leq \omega \leq 6.0\,\mathrm{GHz}$, and $|S_{11}| \leq -0.1\,\mathrm{dB}$ for $4.9\,\mathrm{GHz} \leq \omega \leq 5.3\,\mathrm{GHz}$, where S_{21} and S_{11} are so-called transmission and reflection coefficients, respectively (Pozar, 2004).

The initial coarse model shown in Fig. 4.6 is a circuit equivalent implemented in the circuit simulator Agilent ADS (Agilent Technologies, 2008). A circuit theory model is a simpler way of approximating the same device than solving the Maxwell equations in an EM solver. The EM fine model is accurate, but its evaluation at any design point takes more than an hour. The less accurate circuit model needs only a few milliseconds to

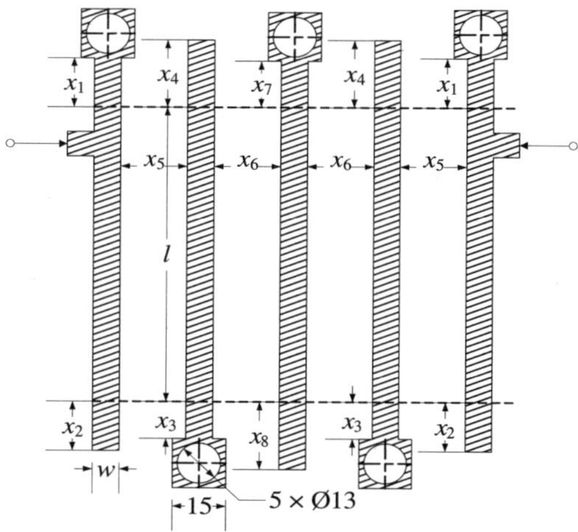

Fig. 4.5.   Microstrip interdigital filter fine model structure and dimensions (Bandler *et al.*, 1997). The dimensions are in mils (Koziel *et al.*, 2008a).

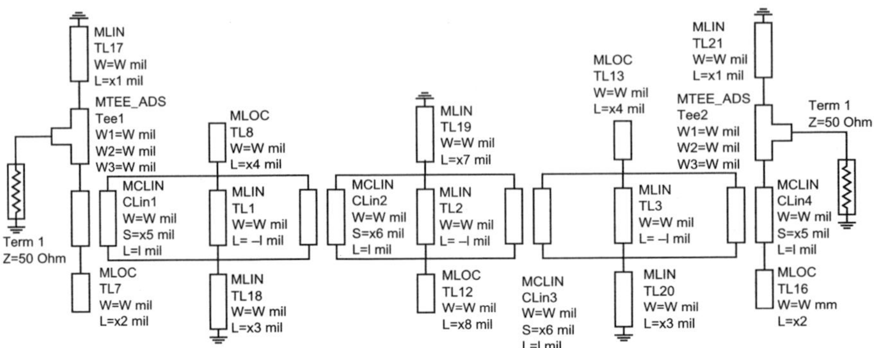

Fig. 4.6. Initial ADS implementation of the microstrip interdigital filter (Koziel *et al.*, 2008a).

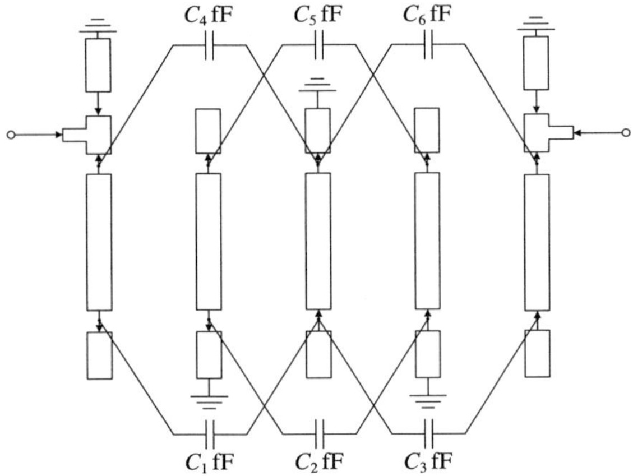

Fig. 4.7. Microstrip interdigital filter coarse model with capacitors inserted between non-adjacent transmission lines (Koziel *et al.*, 2008a).

evaluate. To increase the accuracy of the coarse model, we insert capacitors between non-adjacent microstrip lines as in Fig. 4.7.

The SM approach we adopt is the input and implicit SM. In the interdigital filter case, the lengths of the microstrip lines (x_1, x_2, x_3, x_4, x_7 and x_8) and the gaps (x_5 and x_6) are defined as designable parameters. The implicit SM explores the preassigned parameters in order to align the details of the surrogate with the fine model. For the filter, the preassigned parameter set consists of ε_r (substrate dielectric constant) and capacitors C_1 to C_6 added between non-adjacent conductors (cf. Fig. 4.7). The parameters for the input SM are the shifts of design parameters, i.e., δx_i, $i = 1, \ldots, 8$. The input and implicit SM are combined in one parameter extraction step.

The starting point is the optimal design from Bandler *et al.* (1997). The specification is not satisfied after the first iteration, but the parameter extraction using input and implicit SM yields a good match between the coarse and fine models as shown in Fig. 4.8(a). The coarse model is then

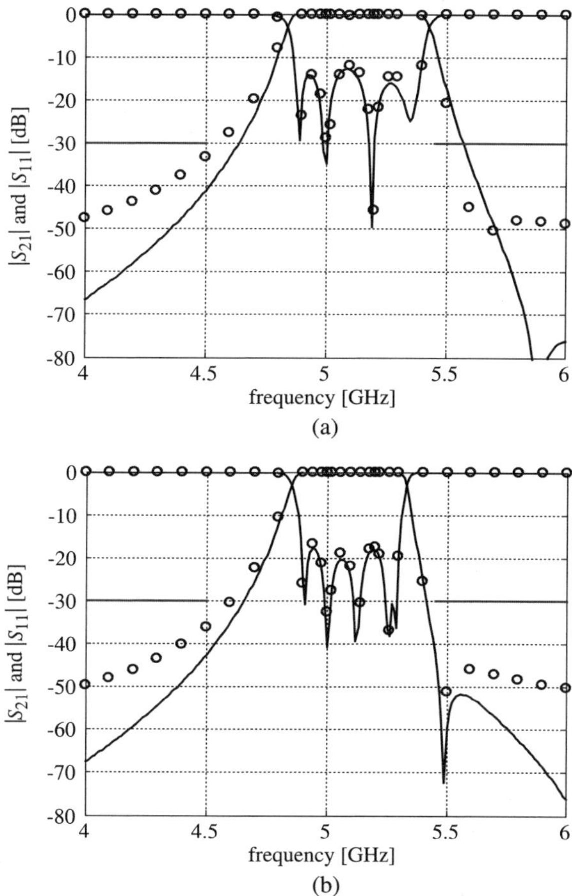

Fig. 4.8. Microstrip interdigital filter: (a) initial design using parameters from (Bandler *et al.*, 1997): the initial fine model responses (o) and the coarse model responses after the first parameter extraction (—), (b) the final fine model responses (o) and the corresponding surrogate model responses (—) (Koziel *et al.*, 2008a). Design specifications marked using thick horizontal lines.

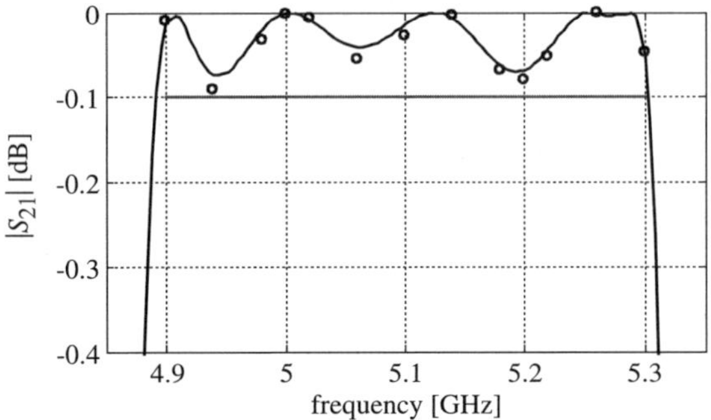

Fig. 4.9. The microstrip interdigital filter passband details of the final fine model simulation (o) and the corresponding surrogate model simulation (—) (Koziel *et al.*, 2008a). Design specifications marked using horizontal line.

(re)optimized and the on-grid solutions are exhaustively searched near the optimum. In just two SM iterations, a good fine model solution is obtained (Figs. 4.8(b) and 4.9). An accurate surrogate is also obtained. The on-grid fine solution is $x_1 = 12$, $x_2 = 30$, $x_3 = 15$, $x_4 = 22$, $x_5 = 31$, $x_6 = 36$, $x_7 = 16$ and $x_8 = 21$ (dimensions in mil).

4.4. Conclusions

In this chapter, we reviewed the original 1994 (Bandler *et al.*) space mapping concept and several of its offshoots, the ensuing 1995 (Bandler *et al.*) aggressive space mapping method, and also the more recent parametric space mapping techniques. We illustrated the algorithms with a simple wedge-cutting problem. We reviewed a trust-region space mapping method that improves convergence. An interdigital filter design demonstrates the input and implicit space mapping approaches.

Acknowledgement

This work was supported in part by the Icelandic Centre for Research (RANNIS) Grant 110034021, and the Natural Sciences and Engineering Research Council of Canada under Grants RGPIN7239-11 and STPGP 381153-09, and by Bandler Corporation. The authors thank Sonnet Software Inc. for *em*, and Agilent Technologies, Santa Rosa, CA, for making ADS available.

References

Agilent ADS (2008). Version 2008, Agilent Technologies, 1400 Fountaingrove Parkway, Santa Rosa: CA 95403-1799.

Alexandrov, N.M., Dennis, J.E., Lewis, R.M. and Torczon, V. (1998). A trust region framework for managing use of approximation models in optimization, *Struct. Multidiscip. O.*, **15**, 16–23.

Alexandrov, N.M. and Lewis, R.M. (2001). An overview of first-order model management for engineering optimization, *Optim. Eng.*, **2**, 413–430.

Amari, S., LeDrew, C. and Menzel, W. (2006). Space-mapping optimization of planar coupled-resonator microwave filters, *IEEE T. Microw. Theory*, **54**, 2153–2159.

Bakr, M.H., Bandler, J.W., Biernacki, R.M., Chen, S.H. and Madsen, K. (1998). A trust region aggressive space mapping algorithm for EM optimization, *IEEE T. Microw. Theory*, **46**, 2412–2425.

Bakr, M.H., Bandler, J.W., Georgieva, N.K. and Madsen, K. (1999). A hybrid aggressive space-mapping algorithm for EM optimization, *IEEE T. Microw. Theory*, **47**, 2440–2449.

Bandler, J.W., Biernacki, R.M., Chen, S.H., Grobelny, P.A. and Hemmers, R.H. (1994). Space mapping technique for electromagnetic optimization, *IEEE T. Microw. Theory*, **4**, 536–544.

Bandler, J.W., Biernacki, R.M., Chen, S.H., Hemmers, R.H. and Madsen, K. (1995). Electromagnetic optimization exploiting aggressive space mapping, *IEEE T. Microw. Theory*, **43**, 2874–2882.

Bandler, J.W., Biernacki, R.M., Chen, S.H. and Huang, Y.F. (1997). Design optimization of interdigital filters using aggressive space mapping and decomposition, *IEEE T. Microw. Theory*, **45**, 761–769.

Bandler, J.W., Cheng, Q.S., Gebre-Mariam, D.H., Madsen, K., Pedersen, F. and Søndergaard, J. (2003). EM-based surrogate modeling and design exploiting implicit, frequency and output space mappings, *IEEE Int. Microwave Symp. Digest*, Philadelphia: PA, pp. 1003–1006.

Bandler, J.W., Cheng, Q.S., Dakroury, S.A., Mohamed, A.S., Bakr, M.H., Madsen, K. and Søndergaard, J. (2004a). Space mapping: the state of the art, *IEEE T. Microw. Theory*, **52**, 337–361.

Bandler, J.W., Cheng, Q.S., Nikolova, N.K. and Ismail, M.A. (2004b). Implicit space mapping optimization exploiting preassigned parameters, *IEEE T. Microw. Theory*, **52**, 378–385.

Bandler, J.W., Koziel, S. and Madsen, K. (2006). Space mapping for engineering optimization, *SIAG/Optimization Views-and-News Special Issue on Surrogate/Derivative-free Optimization*, **17**, 19–26.

Broyden, C.G. (1965). A class of methods for solving nonlinear simultaneous equations, *Math. Comput.*, **19**, 577–593.

Cheng, Q.S., Rautio, J.C., Bandler, J.W. and Koziel, S. (2010a). Progress in simulator-based tuning — the art of tuning space mapping, *IEEE Microw. Mag.*, **11**, 96–110.

Cheng, Q.S., Bandler, J.W. and Koziel, S. (2010b). Space mapping design framework exploiting tuning elements, *IEEE T. Microw. Theory*, **58**, 136–144.

Cheng, Q.S., Bandler, J.W., Koziel, S., Bakr, M.H. and Ogurtsov, S. (2010c). The state of the art of microwave CAD: EM-based optimization and modeling, *Int. J. RF Microw. CE*, **20**, 475–491.

Choi, H.-S., Kim, D.H., Park, I.H. and Hahn, S.Y. (2001). A new design technique of magnetic systems using space mapping algorithm, *IEEE T. Magn.*, **37**, 3627–3630.

Conn, A.R., Gould, N.I.M. and Toint, P.L. (2000). *Trust Region Methods*, MPS-SIAM Series on Optimization, SIAM, Philadelphia: PA.

Crevecoeur, G., Dupre, L. and Van de Walle, R. (2007). Space mapping optimization of the magnetic circuit of electrical machines including local material degradation, *IEEE T. Magn.*, **43**, 2609–2611.

Devabhaktuni, V.K., Chattaraj, B., Yagoub, M.C.E. and Zhang, Q.-J. (2003). Advanced microwave modeling framework exploiting automatic model generation, knowledge neural networks, and space mapping, *IEEE T. Microw. Theory*, **51**, 1822–1833.

Dorica, M. and Giannacopoulos, D.D. (2006). Response surface space mapping for electromagnetic optimization, *IEEE T. Magn.*, **42**, 1123–1126.

Echeverria, D. and Hemker, P.W. (2005). Space mapping and defect correction, *CMAM*, **5**, 107–136.

Echeverría, D. (2007). Multi-level optimization: space mapping and manifold mapping. Ph.D. Thesis, Faculty of Science, University of Amsterdam.

Echeverría, D. and Hemker, P.W. (2008). Manifold mapping: a two-level optimization technique, *Computing and Visualization in Science*, **11**, 193–206.

em^{TM} Version 12.54 (2010). Sonnet Software, Inc., 100 Elwood Davis Road, North Syracuse: NY 13212, USA.

Forrester, A.I.J. and Keane, A.J. (2009). Recent advances in surrogate-based optimization, *Prog. Aerosp. Sci.*, **45**, 50–79.

Gorissen, D., Crombecq, K., Hendrickx, W. and Dhaene, T. (2007). Adaptive distributed metamodeling. *High Performance Computing for Computational Science, VECPAR 2006*, pp. 579–588.

Gorissen, D., Tommasi, L.D., Crombecq, K. and Dhaene, T. (2009). Sequential modeling of a low noise amplifier with neural networks and active learning, *Neural Comput. Appl.*, **18**, 485–494.

Gorissen, D., Crombecq, K., Couckuyt, I., Dhaene, T. and Demeester, P. (2010). A surrogate modeling and adaptive sampling toolbox for computer based design, *J. Mach. Learn. Res.*, **11**, 2051–2055.

Gunn, S.R. (1998). Support vector machines for classification and regression. Tech. Rep., School of Electronics and Computer Science, University of Southampton.

Haykin, S. (1998). *Neural Networks: A Comprehensive Foundation*, (2nd ed.), Prentice Hall, Upper Saddle River: NJ.

Ismail, M.A., Smith, D., Panariello, A., Wang, Y. and Yu, M. (2004). EM-based design of large-scale dielectric-resonator filters and multiplexers by space mapping, *IEEE T. Microw. Theory*, **52**, 386–392.

Journel, A.G. and Huijbregts, C.J. (1981). *Mining Geostatistics*, Academic Press, Suffolk.

Koziel, S., Bandler, J.W. and Madsen, K. (2006). A space mapping framework for engineering optimization: theory and implementation, *IEEE T. Microw. Theory*, **54**, 3721–3730.

Koziel, S. and Bandler, J.W. (2007a). Space-mapping optimization with adaptive surrogate model, *IEEE T. Microw. Theory*, **55**, 541–547.

Koziel, S. and Bandler, J.W. (2007b). Coarse and surrogate model assessment for engineering design optimization with space mapping, *IEEE Int. Microwave Symp. Digest*, Honolulu: HI, pp. 107–110.

Koziel, S., Cheng, Q.S. and Bandler, J.W. (2008a). Space mapping, *IEEE Microw. Mag.*, **9**, 105–122.

Koziel, S., Bandler, J.W. and Madsen, K. (2008b). Quality assessment of coarse models and surrogates for space mapping optimization, *Optim. Eng.*, **9**, 375–391.

Koziel, S., Meng, J., Bandler, J.W., Bakr, M.H. and Cheng, Q.S. (2009a). Accelerated microwave design optimization with tuning space mapping, *IEEE T. Microw. Theory*, **57**, 383–394.

Koziel, S., Bandler, J.W. and Madsen, K. (2009b). Space mapping with adaptive response correction for microwave design optimization, *IEEE T. Microw. Theory*, **57**, 478–486.

Koziel, S. (2010). Shape-preserving response prediction for microwave design optimization, *IEEE T. Microw. Theory*, **58**, 2829–2837.

Koziel, S. and Echeverría Ciaurri, D. (2010). Reliable simulation-driven design optimization of microwave structures using manifold mapping, *Prog. Electromagn. Res.*, **26**, 361–382.

Koziel, S., Cheng, Q.S. and Bandler, J.W. (2010a). Implicit space mapping with adaptive selection of preassigned parameters, *IET Microw. Antenna P*, **4**, 361–373.

Koziel, S., Bandler, J.W. and Cheng, Q.S. (2010b). Robust trust-region space-mapping algorithms for microwave design optimization, *IEEE T. Microw. Theory*, **58**, 2166–2174.

Koziel, S. and Bandler, J.W. (2010). Knowledge-based variable-fidelity optimization of expensive objective functions through space mapping, in Tenne, Y. and Goh, C.-K. (eds.), *Computational Intelligence in Expensive Optimization Problems, (Adaptation, Learning, and Optimization)*, Springer-Verlag, Berlin/Heidelberg, pp. 85–109.

Koziel, S. (2011). Role of constraints in surrogate-based design optimization of microwave structures, *IET Microw. Antenna P.*, **5**, 588–595.

Koziel, S. and Szczepanski, S. (2011). Accurate modeling of microwave structures using shape-preserving response prediction, *IET Microw. Antenna P.*, **5**, 1116–1122.

Koziel, S., Echeverría-Ciaurri, D. and Leifsson, L. (2011a). Surrogate-based methods, in Koziel, S. and Yang, X.S. (eds.), *Computational Optimization, Methods and Algorithms*, Series: Studies in Computational Intelligence, Springer-Verlag, New York: NY, pp. 33–60.

Koziel, S., Bandler, J.W. and Cheng, Q.S. (2011b). Constrained parameter extraction for microwave design optimization using implicit space mapping, *IET Microw. Antenna P.*, **5**, 1156–1163.

Leary, S.J., Bhaskar, A. and Keane, A.J. (2001). A constraint mapping approach to the structural optimization of an expensive model using surrogates, *Optim. Eng.*, **2**, 385–398.

Madsen, K. and Søndergaard, J. (2004). Convergence of hybrid space mapping algorithms, *Optim. Eng.*, **5**, 145–156.

Minsky, M.I. and Papert, S.A. (1969). *Perceptrons: An Introduction to Computational Geometry*, The MIT Press, Cambridge: MA.

Nikolova, N.K., Li, Y., Li, Y. and Bakr, M.H. (2006). Sensitivity analysis of scattering parameters with electromagnetic time-domain simulators, *IEEE T. Microw. Theory*, **54**, 1598–1610.

O'Hagan, A. (1978). Curve fitting and optimal design for predictions, *J. Roy. Stat. Soc. B*, **40**, 1–42.

Pozar, D.M. (2004). *Microwave Engineering*, (3rd ed.), Wiley, Hoboken: NJ.

Queipo, N.V., Haftka, R.T., Shyy, W., Goel, T., Vaidynathan, R. and Tucker, P.K. (2005). Surrogate-based analysis and optimization, *Prog. Aerosp. Sci.*, **41**, 1–28.

Rayas-Sánchez, J.E., Lara-Rojo, F. and Martinez-Guerrero, E. (2005). A linear inverse space-mapping (LISM) algorithm to design linear and nonlinear RF and microwave circuits, *IEEE T. Microw. Theory*, **53**, 960–968.

Rayas-Sánchez, J.E. and Gutiérrez-Ayala, V. (2006). EM-based Monte Carlo analysis and yield prediction of microwave circuits using linear-input neural-output space mapping, *IEEE T. Microw. Theory*, **54**, 4528–4537.

Redhe, M. and Nilsson, L. (2002). Using space mapping and surrogate models to optimize vehicle crashworthiness design, *9th AIAA/ISSMO Multidisciplinary Analysis and Optimization Symposium*, Atlanta: GA, Paper AIAA-2002-5536.

Robinson, T.D., Eldred, M.S., Willcox, K.E. and Haimes, R. (2008). Surrogate-based optimization using multifidelity models with variable parameterization and corrected space mapping, *AIAA J.*, **46**, 2814–2822.

Simpson, T.W., Peplinski, J., Koch, P.N. and Allen, J.K. (2001). Metamodels for computer-based engineering design: survey and recommendations, *Eng. Comput.*, **17**, 129–150.

Smola, A.J. and Schölkopf, B. (2004). A tutorial on support vector regression, *Stat. Comput.*, **14**, 199–222.

Søndergaard, J. (2003). Optimization using surrogate models — by the space mapping technique. Ph.D. Thesis, Informatics and Mathematical Modelling, Technical University of Denmark, Lyngby.

Vapnik, V.N. (1995). *The Nature of Statistical Learning Theory*, Springer-Verlag, New York: NY.

Vicente, L.N. (2003). Space mapping: models, sensitivities, and trust-regions methods, *Optim. Eng.*, **4**, 159–175.

Wild, S.M., Regis, R.G. and Shoemaker, C.A. (2008). ORBIT: Optimization by radial basis function interpolation in trust-regions, *SIAM J. Sci. Comput.*, **30**, 3197–3219.

Wu, K.-L., Zhao, Y.-J., Wang, J. and Cheng, M.K.K. (2004). An effective dynamic coarse model for optimization design of LTCC RF circuits with aggressive space mapping, *IEEE T. Microw. Theory*, **52**, 393–402.

Zhang, L., Xu, J., Yagoub, M.C.E., Ding, R. and Zhang, Q.-J. (2005). Efficient analytical formulation and sensitivity analysis of neuro-space mapping for nonlinear microwave device modeling, *IEEE T. Microw. Theory*, **53**, 2752–2767.

Chapter 5

Tuning Space Mapping

Qingsha S. Cheng, John W. Bandler, and Slawomir Koziel

The widely accepted space mapping technology of Chapter 4 facilitates optimal circuit design with electromagnetic (EM) accuracy and circuit-optimization speed. The so-called tuning space mapping approach combines the intuition of the space mapping concept and EM-simulator-based tuning. In this chapter, to begin with, we review the classical computer-based tuning technology. Then, through "tuning space mapping" procedures, we explain microwave design optimization using various types of tuning models as surrogates. By means of simple examples, we demonstrate the implementation of these surrogates in design optimization. We provide illustrations utilizing commercial simulation software. Our purpose is to explain and clarify the differences among the tuning space mapping techniques and to inspire new implementations and applications.

5.1. Tuning Technology

Tuning (or postproduction tuning) refers to a traditional and often essential process in the manufacturing or post-manufacturing operation of engineering devices in general and electrical circuits in particular. For electrical circuits, tolerances on components, parasitic electromagnetic effects, electrical mismatches, and uncertainties in circuit models cause deviations in the manufactured circuit outcomes, and violations of the design specifications may result. Therefore, postproduction tuning is often included in the final stages of a production process to readjust the manufactured circuit performance in an effort to meet the original design specifications. Of course, the facility to tune is also designed into devices that are intended to be tunable by the end user.

Pinel's computer-oriented tuning strategy (Pinel, 1971) exploits response sensitivities to component variations to select a best component at each stage of tuning, its direction of change and the size of the adjustment. This strategy combined with Monte Carlo techniques was used in filter design evaluations to verify that production units could indeed be tuned to obtain desired production yields as well as determine the effects of aging and temperature.

Bandler *et al.* (1976) formally considered tunable elements as an integral part of a design centering and optimal tolerance assignment process. A suitable objective function determines one or more tunable components to further increase tolerances so as to reduce manufacturing cost or to make otherwise unrealistically toleranced solutions more attractive.

A general, interactive postproduction tuning technique (Bandler *et al.*, 1981) harnessed linear programming iteratively for estimating necessary tuning amounts. The algorithm is applicable to reversible and irreversible tuning processes at the manufacturing stage. By eliminating the then common trial and error approach the process optimally exploits circuit response measurements of the manufactured outcome under test.

The approach to postproduction tuning presented by Bandler and Salama (1983, 1985) addresses the determination of a suitable set of sampling measurements, the specification of the circuit response of interest, the determination of the parameters to be tuned, and the tuning algorithm itself. It utilizes a nonlinear programming notation applicable to circuit design problems. The aims were to relax manufacturing tolerances and compensate for uncertainties in the model parameters. The required changes in the tunable parameters are computed using two functional tuning algorithms. Minimax optimization identifies the tuning frequencies, and least-one optimization is employed to minimize the number of tunable parameters. Worst-case analysis is utilized to reduce the size of the problem.

To reduce the complexity of the filter tuning problem, group delay (Ness, 1998), time-domain (Dunsmore, 1999) and fuzzy logic (Miraftab and Mansour, 2008) techniques have been developed for the design and postproduction tuning of coupled resonator filters. They rely on both the phase and/or magnitude quantities of the reflection coefficient. Miraftab and Mansour (2008) introduce modeling of human experience by linguistic if-then rules in terms of fuzzy logic controllers for tuning RF/Microwave filters. The approach could be used for any circuit or system tuning problem that involves human expert information provided that the expert information could be described in terms of linguistic if-then rules. The tuning is done in two stages of fuzzy controllers. The first stage uses the phase response of the filter, the second uses the magnitude response of the filter for adjustment of the tuning screws. An automated experimental setup involves high resolution motors with flexible leads to make the tuning possible. Three-pole and seven-pole Chebyshev waveguide filters demonstrate the idea which is validated by measurements.

Parameter-extraction-based tuning methods (Hsu *et al.*, 2002; Pepe *et al.*, 2004; Meng and Wu, 2009) employ pole and zero identification of the filter transfer function from phase data and eventual transcription to the coupling matrix model for tuning. Meng and Wu (2009) present an approach to extract the coupling matrix of a narrowband Chebyshev bandpass filter with losses. It reveals the concept of phase loading for computer-aided tuning of a microwave bandpass filter. The proposal is applicable to general coupled resonator filters with losses and can be used in computer-aided tuning of high-order filters with cross couplings.

A transmission-based tuning technique (Zahirovic *et al.*, 2010) has been developed for on-board and on-chip automatic tuning of tunable filters without the use of a vector network analyzer and with minimal additional hardware. A coupling matrix-based model of the tuning algorithm predicts the performance of the algorithm. The tuning model is verified with the automated tuning algorithm operating on a manufactured tunable filter. The authors (Zahirovic *et al.*, 2010) present a low-cost hardware prototype for scalar transmission measurement for standalone implementation of their algorithm.

Yu and Tang (2003) developed a robotic system for the postproduction tuning of microwave filters using phase and magnitude data measured by a vector network analyzer.

The terms "tuning" and "optimization" are often interchangeably used by the engineering community when referring to a design process or a parameter extraction process. In Agilent ADS and other simulation software packages, "tuning" means that a user can change one or more design parameter values and quickly see its effect on the output without re-simulating the entire design.

The terms "tuning" and "space mapping" (see Chapter 4) have been mentioned together (Cheng *et al.*, 2008; Rasmita *et al.*, 2009) in the context where a coarse model or a surrogate is "tuned" within a space mapping-based algorithm. In these particular cases, the coarse models are purely circuit-based surrogates.

5.2. EM-Simulator-Based Tuning

While the use of accurate but CPU-intensive EM simulation is now taken for granted by the microwave industry and computers are increasingly more powerful, microwave engineers are still faced with ever larger design problems, tighter specifications, and shorter closure time requirements. Thus, design optimization processes based on full-wave EM simulations and validations will continue to remain challenging.

The EM-simulator-based tuning method (Swanson and Wenzel, 2001; Rautio, 2006) is an effective tuning or design approach that combines EM accuracy with circuit speed. The approach allows easy tuning and instantaneous visualization of the EM-simulated responses of a structure. The heart of the method is the hybridization of EM simulation and circuit simulation in one structure (in our terminology a "tuning model"). It is achieved by replacing part of the EM-simulated structure (high-fidelity or "fine" model) with physics-based equivalent circuit models or by suitable numerical approximations. Predefined physics-based models are preferable for their minimal sample data requirements.

5.3. Introduction to Tuning Space Mapping

As presented in Chapter 4, space mapping techniques effectively exploit surrogates to model or to iteratively optimize a fine model (Bandler *et al.*, 1994; Bandler *et al.*, 1995; Bandler *et al.*, 2004; Koziel *et al.*, 2008). Surrogate models themselves are constructed from a so-called "coarse" model, such as an equivalent circuit. They are of low fidelity but are much cheaper to evaluate than the fine model. Conventional space mapping techniques such as described in Chapter 4 typically use purely circuit-based models or interpolated coarse-mesh EM models as surrogates.

Inspired by the EM-simulator-based tuning method, we formulate a tuning space mapping technology. Our tuning space mapping algorithms exploit tuning models (EM-simulated model data or S-parameter files augmented by tuning elements) as surrogates. The surrogates approximate the full-wave EM simulation responses in an iterative space mapping updating process, essentially in the same way as do the conventional space mapping approaches of Chapter 4.

Tuning space mapping algorithms involve a fine model (e.g., a full-wave EM simulation), auxiliary fine models (fine models with tuning ports), a tuning model, and a calibration scheme that sometimes exploits an extra calibration (coarse) model. We may induce infinitesimal tuning gaps (negligible compared to the wavelength) into the fine model. We may then define tuning ports at the gap edges or at the boundary of the fine model to form candidate auxiliary fine models. We simulate the auxiliary models in an EM simulator and save the resulting multi-port S-parameter file. Within a circuit simulator, we insert or attach suitable tuning elements to the tuning ports. Preferably, the tuning elements are distributed circuit elements with physical dimensions corresponding to those of the fine model. This new tunable model bearing the EM-based S-parameter file data and tuning elements, namely, this new surrogate, is called a tuning model.

After a simple alignment procedure (also known as parameter extraction or PE), we match the tuning model with the fine model. Some of the fine-model couplings are preserved (or represented through the S-parameter file) in the tuning model. We normally obtain a good surrogate of the fine model. In the next stage, the tuning model is optimized by varying the tuning parameter values of the tuning elements to satisfy given design specifications. Following a translation process, the resulting design parameter values constitute our next fine model iterate.

In this chapter, we review categorized types (Cheng *et al.*, 2012) of tuning space mapping. Swanson and Wenzel (2001) use a tuning method we consider as Type 0– tuning space mapping. Hirtenfelder (2007) and Edquist (2009) revisited the idea. Type 0 tuning is first described in Koziel *et al.* (2009) and more examples are demonstrated in Cheng *et al.* (2010b) and Thoma (2010). Type 1 tuning space mapping is demonstrated in Cheng *et al.* (2010a) and is compared to Type 0 in Cheng *et al.* (2010b). The Type 1 tuning method is simplified in Koziel *et al.* (2010); Koziel *et al.* (2011). Type 2 is presented in various papers including Hirtenfelder (2007); Koziel and Bandler (2011); Koziel (2011).

5.4. General Tuning Space Mapping Algorithm

The design optimization problem we are concerned with is to find the optimal solution $\boldsymbol{x}_f^*$ of a fine model of the device of interest, namely,

$$\boldsymbol{x}_f^* = \arg\min_{\boldsymbol{x}} U(\boldsymbol{R}_f(\boldsymbol{x})), \tag{5.1}$$

where $\boldsymbol{R}_f \in R^m$ denotes the response vector, U is a suitable objective function, and $\boldsymbol{x}$ is the vector of designable parameters.

A typical tuning space mapping algorithm consists of several characteristic steps. Various EM simulations include the fine model and corresponding auxiliary fine models. A tuning model alignment step updates the tuning model parameters to match the tuning model with the fine model. The tuning model is optimized to meet the original design specifications. A translation process determines the corresponding fine model design parameter values. In the ith iteration, firstly, based on data from an auxiliary fine model (for example, a fine model with so-called co-calibrated ports) at the current design $\boldsymbol{x}^{(i)}$, an initial tuning model $\boldsymbol{R}_t^{(i)}$ is built on the auxiliary fine model data (usually S-parameter files) augmented with appropriate tuning elements. The current tuning model is supplied with appropriate (initial) tuning parameters (parameter values $\boldsymbol{t}^{(i)}$) determined from microwave engineering knowledge. The tuning model is augmented by an additional set of parameters $\boldsymbol{p}$ (such as input mapping parameters and/or preassigned parameters) similar to those found in the surrogate of the traditional space mapping techniques (see, for example, Chapter 4 of this book and Koziel *et al.* (2008)). Our initial tuning model response may not agree with the response of the original fine model at $\boldsymbol{x}^{(i)}$. We align these models by

$$\boldsymbol{p}^{(i)} = \arg \min_{\boldsymbol{p}} \|\boldsymbol{R}_f(\boldsymbol{x}^{(i)}) - \boldsymbol{R}_t^{(i)}(\boldsymbol{t}^{(i)}, \boldsymbol{p})\|, \qquad (5.2)$$

where $\boldsymbol{R}_t \in R^m$ denotes the response vector of the tuning model. After $\boldsymbol{p}^{(i)}$ is extracted, we optimize $\boldsymbol{R}_t^{(i)}$ to have it meet the design specifications w.r.t. the selected tuning parameters $\boldsymbol{t}$, giving

$$\boldsymbol{t}_{\text{opt}}^{(i)} = \arg \min_{\boldsymbol{t}} U(\boldsymbol{R}_t^{(i)}(\boldsymbol{t}, \boldsymbol{p}^{(i)})). \qquad (5.3)$$

The new design parameter values $\boldsymbol{x}^{(i+1)}$ can be calculated either directly from $\boldsymbol{x}^{(i)} + \boldsymbol{t}_{\text{opt}}^{(i)} - \boldsymbol{t}^{(i)}$ (when the design parameters and tuning parameters have a corresponding physical meaning and units) or by substituting $\boldsymbol{t}_{\text{opt}}^{(i)}$, $\boldsymbol{t}^{(i)}$, and $\boldsymbol{x}^{(i)}$ into a suitable calibration model (Koziel *et al.*, 2009; Cheng *et al.*, 2010b). Note that the current initial tuning parameters $\boldsymbol{t}^{(i)}$ are related to how the current auxiliary fine model and tuning model are set up; their values may or may not be directly related to the previous optimal tuning parameter values $\boldsymbol{t}_{\text{opt}}^{(i-1)}$.

The tuning space mapping algorithm can be described as follows.

Step 1   Assign initial design parameters $\boldsymbol{x}^{(i)}$, $i = 0$, to the fine model.
Comment: The initial fine model design may be obtained from an optimized circuit-based coarse model.
Step 2   Simulate the fine model at $\boldsymbol{x}^{(i)}$.
Step 3   Stop if the design criterion is satisfied.
Comment: For example, the fine model responses $\boldsymbol{R}_f(\boldsymbol{x}^{(i)})$ should satisfy certain specifications or the iteration count should exceed a certain number.
Step 4   Create an auxiliary fine model.
Comment: The auxiliary fine model is created by cutting into or partitioning the given fine model and adding suitable tuning ports.
Step 5   Simulate this auxiliary fine model.
Comment: The multi-port auxiliary fine model is simulated by an EM-based simulator. The results are saved in S-parameter files.
Step 6   Create a tuning model.
Comment: Augment the resulting multi-port S-parameter file with tuning elements in a circuit simulator to create a tunable model.
Step 7   Update the parameter values of the tuning model.
Comment: Match the tuning model responses to the fine model using (5.2). This is also known as the parameter extraction process.
Step 8   Optimize the tuning model w.r.t. the tuning parameters.
Comment: Find the optimal tuning model responses using (5.3).
Step 9   Translate the optimized tuning parameters to the original design parameters
Step 10 $i = i + 1$. Go to *Step 2*.

5.5. Types of Tuning Model

Since the tuning model is based on auxiliary fine models and appropriate tuning elements, there are many ways to create it. We now discuss useful types of tuning model suitable for tuning space mapping algorithms. Conceptual images of the fine model, the auxiliary fine model and the tuning model for each type are shown in Table 5.1. The algorithm outlined in the previous section applies to several variations of tuning space mapping. However, each variation utilizes its own specific auxiliary fine model and tuning model.

Table 5.1. Variations of tuning space mapping.

	Type 0–	Type 0 and 1	Type 1d (fast)	Type 2
Fine model	fine model	fine model	fine model	fine model
Auxiliary fine model*	auxiliary fine model			
Tuning model†	auxiliary fine model	Type 1 Type 0		

*Auxiliary fine model: fine model with tuning ports or split fine model components.
†Tuning model: tuning components are added to the auxiliary fine model.

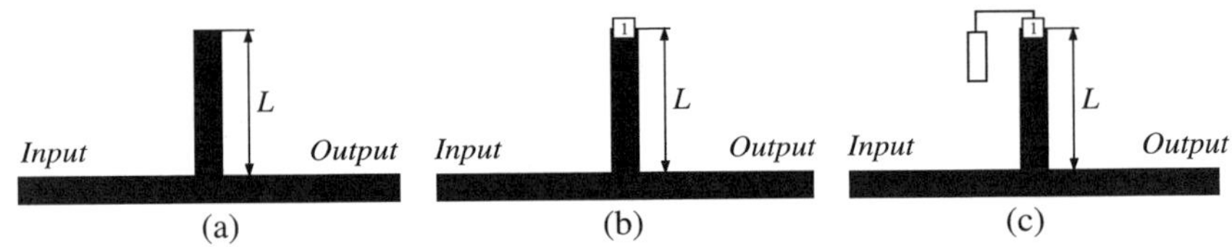

Fig. 5.1. Illustration of TSM with Type 0– tuning of a simple bandstop microstrip filter: (a) fine model, (b) auxiliary fine model with tuning port (Port 1), (c) tuning model (tuning elements are attached to Port 1). Note: the dark area in (c) is a representation (S-parameter file) of the auxiliary fine model.

We illustrate Type 0– tuning by the simple bandstop microstrip filter example in Fig. 5.1. Tuning ports are defined at the boundary of the EM model (fine) to form an auxiliary fine model (see Fig. 5.1(b)). The auxiliary fine model is then simulated. The resulting multi-port S-parameter file assembled with tuning elements (at the tuning ports) in a circuit simulator becomes a tuning model.

We demonstrate Type 0 and Type 1 tuning using the coupled microstrip line example in Fig. 5.2 and Fig. 5.3, respectively. In Type 0 and 1 tuning, we utilize internal tuning ports associated with infinitesimal gaps (see Section 5.3). The auxiliary fine model is formed by defining such ports within the fine model. This multi-port auxiliary model is then simulated in

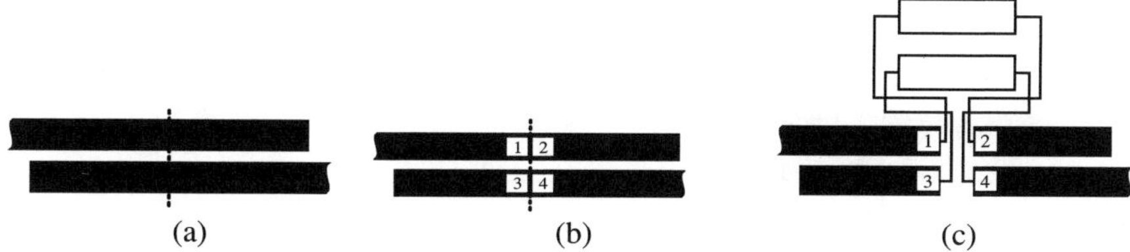

(a)   (b)   (c)

Fig. 5.2. Illustration of TSM with Type 0 tuning of a coupled microstrip line: (a) fine model, (b) auxiliary fine model with tuning ports (Ports 1–4), (c) tuning model (tuning elements are inserted between Ports 1–4). Note: the dark area in (c) is a representation (S-parameter file) of the auxiliary fine model.

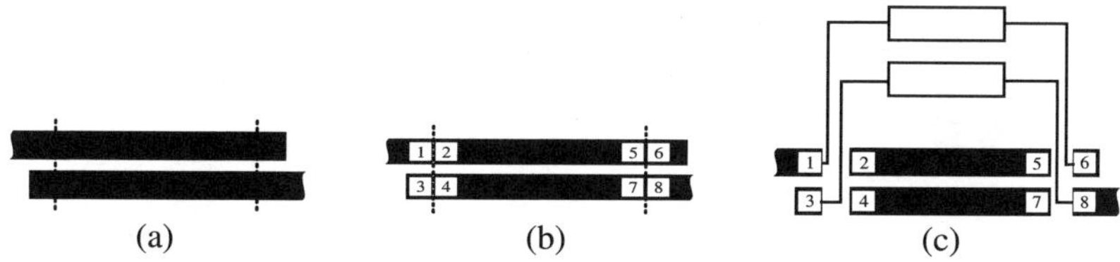

(a)   (b)   (c)

Fig. 5.3. Illustration of TSM with Type 1 tuning of a coupled microstrip line: (a) fine model, (b) auxiliary fine model with tuning ports (Ports 1–8), (c) tuning model (tuning elements are inserted between Ports 1, 3, 6, and 8). Note: the dark area in (c) is a representation (S-parameter file) of the auxiliary fine model.

an EM simulator. The resulting S-parameter file is augmented by tuning elements in a circuit simulator to form a tuning model.

Type 0 tuning and Type 1 tuning differ in the way tuning elements are attached. We attach tuning elements to the adjacent tuning ports in the same tuning gap for Type 0 tuning (see Fig. 5.2(c)). For Type 1 tuning, we add tuning elements across tuning ports belonging to different gaps (see Fig. 5.3(c)). Here, the effects of unused ports and associated couplings are deembedded or compensated for by the subsequent parameter extraction process. The Type 1d (Cheng *et al.*, 2012), also known as fast (Koziel *et al.*, 2011), tuning space mapping approach is based on Type 1 tuning but uses a simplified or reduced auxiliary model (Fig. 5.4).

Type 2 tuning differs from the Type 0–, 0, and 1 in that it breaks the fine model into fragments. Each piece (a submodel) is simulated separately in the EM simulator. The auxiliary fine model in this case is a collection of submodels. The submodel S-parameter files obtained are connected with tuning elements in a circuit simulator to form a tuning model. See Fig. 5.5.

                    Qingsha S. Cheng et al.

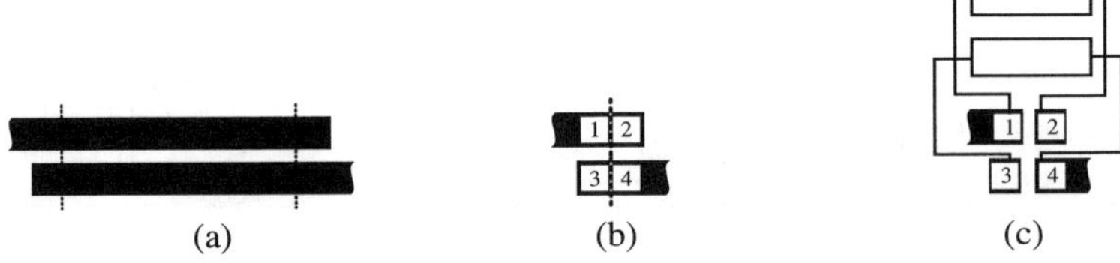

Fig. 5.4. Illustration of TSM with Type 1d (fast) tuning of a coupled microstrip line: (a) fine model, (b) auxiliary fine model with tuning ports (Ports 1–4), (c) tuning model (tuning elements are inserted between Ports 1–4). Note: the dark area in (c) is a representation (S-parameter file) of the auxiliary fine model.

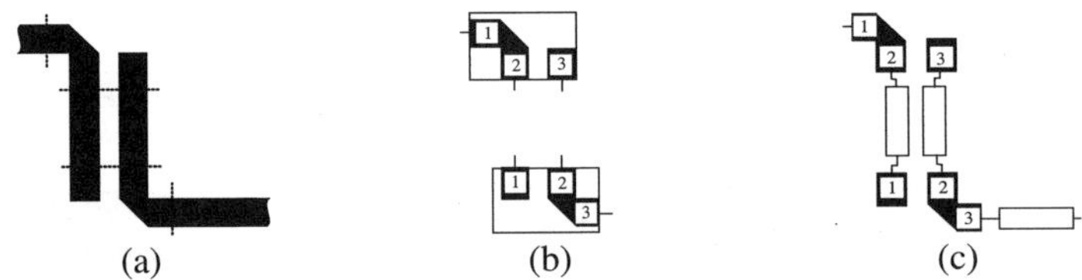

Fig. 5.5. Illustration of TSM with Type 2 tuning of a coupled microstrip line with bends: (a) fine model, (b) auxiliary fine model sections (submodels) with tuning ports (Ports 1–3) — the sections are usually simulated separately in the EM simulator, (c) tuning model (tuning elements connect the submodels). Note: the dark area in (c) is a representation (S-parameter file) of the auxiliary fine model.

5.6. Examples

In this section we select typical examples from published papers to illustrate Type 0−, Type 0, Type 1, and Type 2 tuning. Interested readers should refer to the relevant paper for details.

5.6.1. *Type 0− Tuning*

Using multiport S-parameter data and lumped capacitors at the ports, Swanson and Wenzel (2001) tuned a filter in a circuit simulator. They then optimize their combline filter by mapping the "coarse" circuit model to their "fine" FEM model. With a good starting point, this optimization is shown to converge in a single iteration. The method can be viewed as a Type 0− tuning process (tuning elements added at the boundary). In another Type 0− tuning example (Hirtenfelder, 2007), a seven-post bandpass filter is tuned using a tuning model constructed by adding tuning elements (tuning capacitors and inductors) to the tuning ports at each resonator. The couplings between the resonators are also tuned by tuning elements (transmission lines) added between the resonators. A similar procedure is demonstrated in Edquist (2009).

5.6.2. *Type 0 Tuning*

A microstrip bandpass filter, a second-order tapped-line microstrip filter, and a high-temperature superconducting filter are designed with Type 0 tuning space mapping in Koziel *et al.* (2009). A similar technique is applied to an interdigitated filter using Sonnet (Cheng *et al.*, 2010b) and CST (Thoma, 2010).

Here we show the design of a narrowband 62 GHz interdigitated filter using Type 0 tuning space mapping (Cheng *et al.*, 2010b). The design parameters are $[Offset\ L_1\ L_2\ L_3\ S_1\ S_2]^T$ mm as shown in Fig. 5.6. The specifications are $|S_{11}| \leq -20$ dB for $61\,\text{GHz} \leq \omega \leq 63\,\text{GHz}$, $|S_{21}| \leq -40 + (\omega - 44) \cdot 30/15$ dB for $44\,\text{GHz} \leq \omega \leq 59\,\text{GHz}$, and $|S_{21}| \leq -10 - (\omega - 65) \cdot 30/11$ dB for $65\,\text{GHz} \leq \omega \leq 76\,\text{GHz}$.

The initial design values are $[8.5\ 11.5\ 8.5\ 4\ 5\ 4.75]^T$. After the internal ports are inserted (to form an auxiliary fine model), an EM simulation in Sonnet *em* is conducted. In the circuit schematic (Microwave Office), the tuning model (surrogate) is constructed by importing the EM simulated S-parameters and embedding suitable tuning elements. The tuning elements are capacitors (tuning gaps $[S_1\ S_2]^T$) and microstrip lines (tuning lengths and offset $[Offset\ L_1\ L_2\ L_3]^T$) as shown in Fig. 5.7.

The tuning model is adjusted to satisfy the design specifications. The fine model is based on the standard EM simulation of the entire original filter. The accuracy of such EM simulations has been verified in prior publications (Rautio, 1994; Rautio, 2005). Once tuned to meet the specifications, this filter layout is the one that would actually be fabricated. The small differences between the fine and tuning model responses (i.e., the result with tuning ports and circuit theory tuning

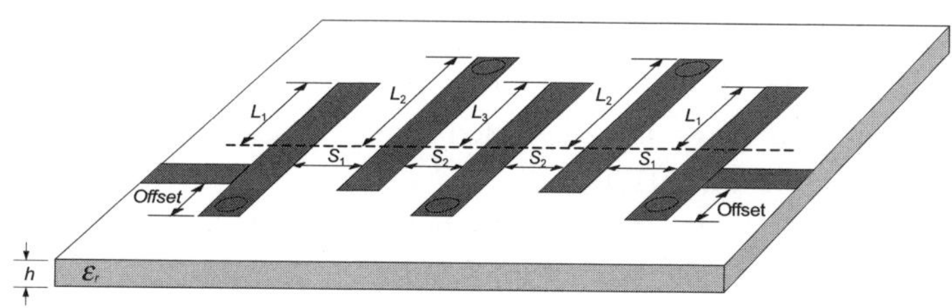

Fig. 5.6.   Narrowband 62 GHz interdigitated filter (Cheng *et al.*, 2010b).

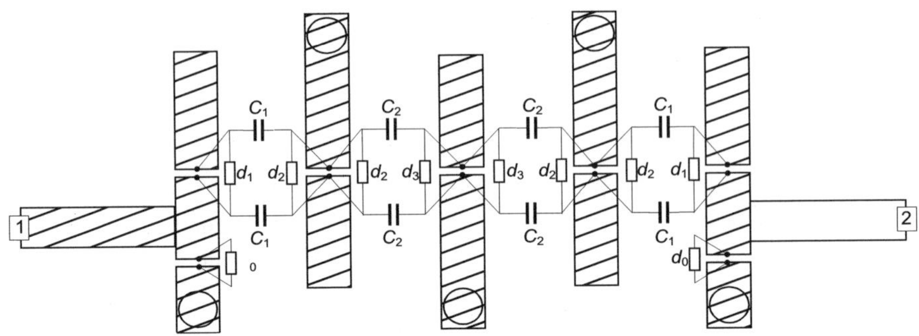

Fig. 5.7.   Narrowband 62 GHz interdigitated filter surrogate (Cheng *et al.*, 2010b). Inserted capacitors are the tuning elements for the separations and microstrip lines are tuning elements for the lengths.

elements set to correspond to the original fine model dimensions) are due to the circuit theory models of the transmission lines being near the limits of their validity. There is enough accuracy in the circuit theory that they can still be used to tune the filter even without a parameter extraction step.

A set of optimal tuning parameter values is obtained to satisfy the specifications: capacitances $[C_1 \; C_2]^T$ and lengths $[d_0 \; d_1 \; d_2 \; d_3]^T$. Their values are now translated to design parameter values as follows. The microstrip lengths $[Offset \; L_1 \; L_2 \; L_3]^T$ are obtained by directly adding lengths $[d_0 \; d_1 \; d_2 \; d_3]^T$ to the initial lengths. The capacitance must be translated into new separations $[S_1, \; S_2]^T$ using a calibration step (Cheng *et al.*, 2010b). After repeating the entire process (including the calibration step) one more time, the specifications are satisfied.

5.6.3. *Type 1 Tuning*

Three designs demonstrate Type 1 tuning space mapping (Cheng *et al.*, 2010a): a three-section transformer, an open-loop ring resonator bandpass filter, and a compact multi-layer low-temperature co-fired ceramics (LTCC) second-order bandpass filter (Yeung and Wu, 2003). The open-loop ring resonator bandpass filter is redesigned using mixed Type 1 and Type 0 tuning in (Cheng *et al.*, 2010b).

Here we review one of the examples, namely, the LTCC filter. Its auxiliary fine model is shown in Fig. 5.8.

The design parameters are $\boldsymbol{x} = [L_c \; L \; L_{cc1} \; L_{cc2} \; L_{c1} \; L_{c2} \; L_1 \; L_2 \; s]^T$ mil (0.001 inch). The specifications are $|S_{21}| \leq -20\,\text{dB}$ for $1\,\text{GHz} \leq \omega \leq 2\,\text{GHz}$,

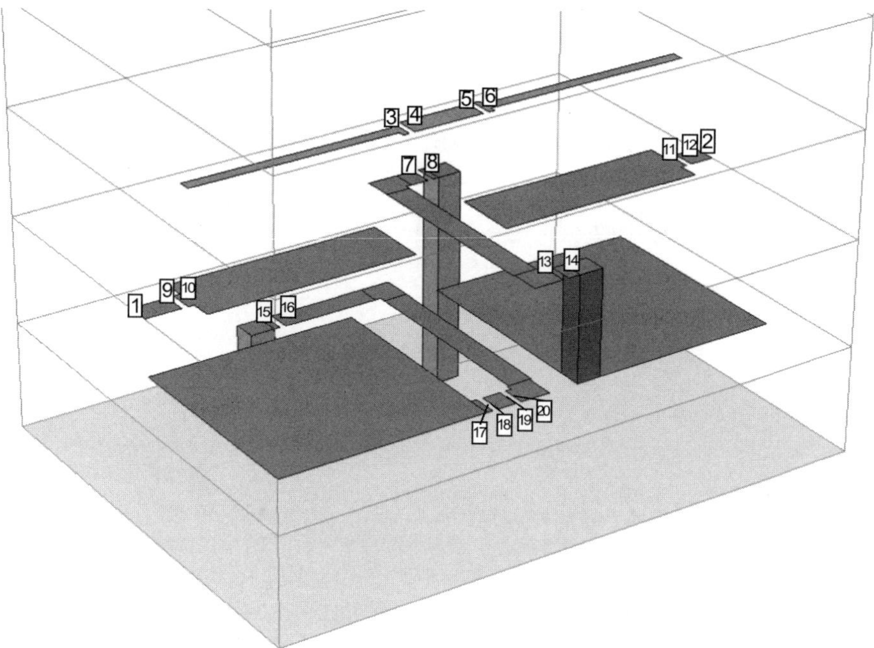

Fig. 5.8. The auxiliary fine model of the LTCC second-order bandpass filter (Cheng *et al.*, 2010a). The input and output ports are labeled 1 and 2. The co-calibrated ports are ports 3–20.

$|S_{21}| \leq -20\,\mathrm{dB}$ for $3\,\mathrm{GHz} \leq \omega \leq 4\,\mathrm{GHz}$, and $|S_{21}| \geq -2\,\mathrm{dB}$ for $2.35\,\mathrm{GHz} \leq \omega \leq 2.65\,\mathrm{GHz}$.

To accommodate tuning elements, we add 18 co-calibrated tuning ports to the structure, as shown in Fig. 5.8.

We now import the 20-port S-parameter file S20P into Agilent ADS and replace the light gray areas (bottom plate excluded) by tuning elements (Fig. 5.9). To accommodate the couplings between the layers, we use multi-layer coupled lines as tuning elements to replace the coupled multi-layer structure.

The tuning model (Fig. 5.10) in Agilent ADS is ready to be optimized. First we extract tuning parameters and preassigned parameters in an alignment process. The selected preassigned parameters are the dielectric constants and substrate heights of the substrate materials between the non-empty layers. They are not uniformly varied in all the components. We divide the tuning elements into three groups which are shaded by different patterns as in Fig. 5.10. Each group has its own multi-layer

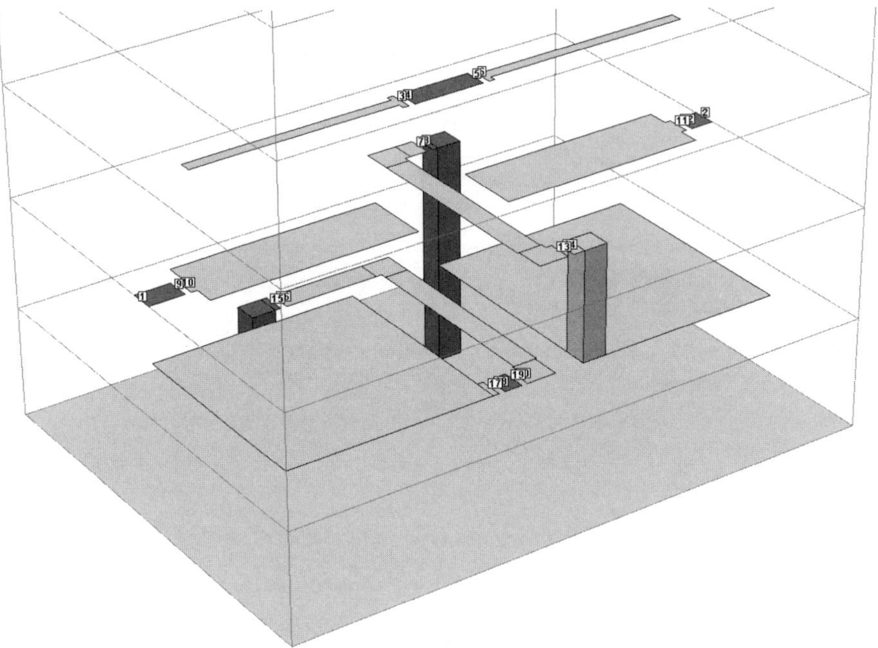

Fig. 5.9.   The auxiliary fine model of the LTCC second-order bandpass filter illustrating Type 1 tuning (Cheng *et al.*, 2010a). The microstrips and patches denoted light gray (bottom plate excluded) are to be replaced by tuning elements.

substrate property that can be varied during the alignment process. The initial guess of the design parameters $x^{(0)} = [12\ 72\ 20\ 20\ 58\ 58\ 58\ 26\ -10]^T$ mil. In four iterations, a good fine model response (Cheng *et al.*, 2010a) is obtained with $x^{(5)} = [5\ 65\ 23.5\ 24\ 84\ 67.6\ 52.5\ 39.5\ -8]^T$.

5.6.4. *Type 1d (Fast) Tuning*

A third-order Chebyshev filter, a second-order ring resonator bandpass filter, and a coupled microstrip bandpass filter are designed with the Type 1d (Cheng *et al.*, 2012; fast (Koziel *et al.*, 2011)) tuning.

The coupled microstrip bandpass filter (Lee and Tsai, 2005) (fine model) is simulated in Sonnet *em* with a grid of $0.1\,\text{mm} \times 0.02\,\text{mm}$ (evaluation time 54 minutes). The structures and the locations of the co-calibrated ports used to construct the auxiliary fine models for the Type 1 and Type 1d (fast) tuning are shown in Figs. 5.11(a) and 5.11(b), respectively.

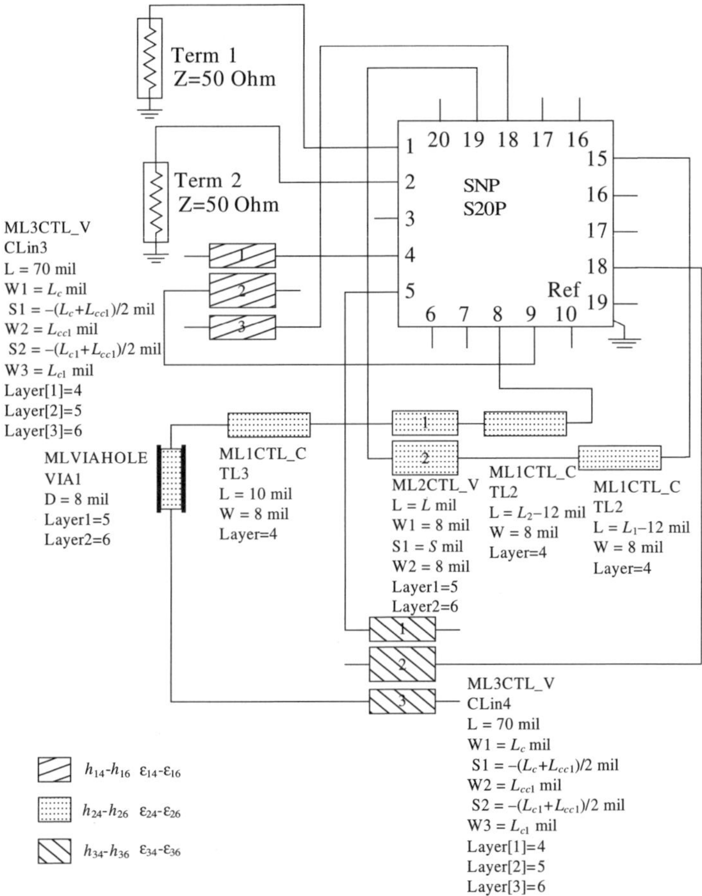

Fig. 5.10. The tuning model of the LTCC second-order bandpass filter is implemented in Agilent ADS and ready to be tuned and designed (Cheng *et al.*, 2010a).

The design parameters are $x = [L_1\, L_2\, L_3\, L_4\, S_1\, S_2]^T$ mm. The design specifications are $|S_{21}| \geq -1$ dB for $2.35\,\text{GHz} \leq \omega \leq 2.45$ GHz, and $|S_{21}| \leq -20$ dB for $1.5\,\text{GHz} \leq \omega \leq 2.2\,\text{GHz}$ and $2.6\,\text{GHz} \leq \omega \leq 3.3\,\text{GHz}$.

The tuning model for the Type 1d (fast) tuning is shown in Fig. 5.11(c). The simulation time of the structure in Fig. 5.11(b) is 1.5 hours, while that the structure in Fig. 5.11(a) is 18 hours.

The initial design is $x^{(0)} = [22.1\, 5.94\, 12.6\, 14.8\, 0.10\, 0.16]^T$ mm. After one iteration of Type 1d (fast) tuning, we obtain $x^{(1)} = [23.0\, 6.12\, 10.4\, 16.9\, 0.10\, 0.24]^T$ mm. As before, the additive perturbations of the design

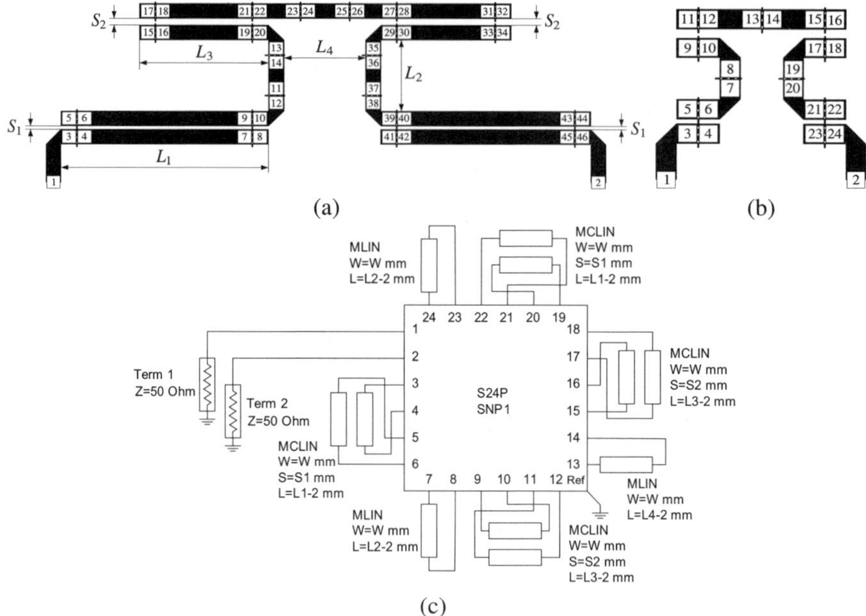

Fig. 5.11. Coupled microstrip bandpass filter (Koziel *et al.*, 2011): (a) auxiliary fine model geometry and the places (dashed lines) for inserting the tuning ports for Type 1 tuning, (b) the reduced structure auxiliary fine model and the places (dashed lines) for inserting the tuning ports for Type 1d (fast) tuning, (c) the tuning model for Type 1d (fast) tuning (Agilent ADS).

variables (here, dL_1 to dL_4, dS_1, and dS_2) are used in the alignment procedure (5.2).

The optimization results are summarized in Koziel *et al.* (2011). While the quality of the final designs are similar, the computational cost corresponding to Type 1d (fast) tuning is significantly less than that for Type 1 (Koziel *et al.*, 2011).

5.6.5. *Type 2 Tuning*

In Koziel and Bandler (2011) Type 2 tuning is illustrated through the design of a coupled-line bandpass microstrip filter and a wideband bandstop filter, two microstrip filters simulated in an EM simulator (FEKO). Tuning models can be constructed by simulating the EM model sections (we call submodels) separately and then connecting them with the tuning components using a "co-simulation" process (Koziel and Bandler, 2011). This facilitates tuning space mapping with any EM simulator. Response misalignment between the original structure and the tuning model is reduced using classical space mapping.

Consider the coupled-line bandpass filter (Lee and Tsai, 2005) again. The fine model is simulated in FEKO. A schematic of the co-simulation-based tuning model is shown in Fig. 5.12(a). Submodels marked black

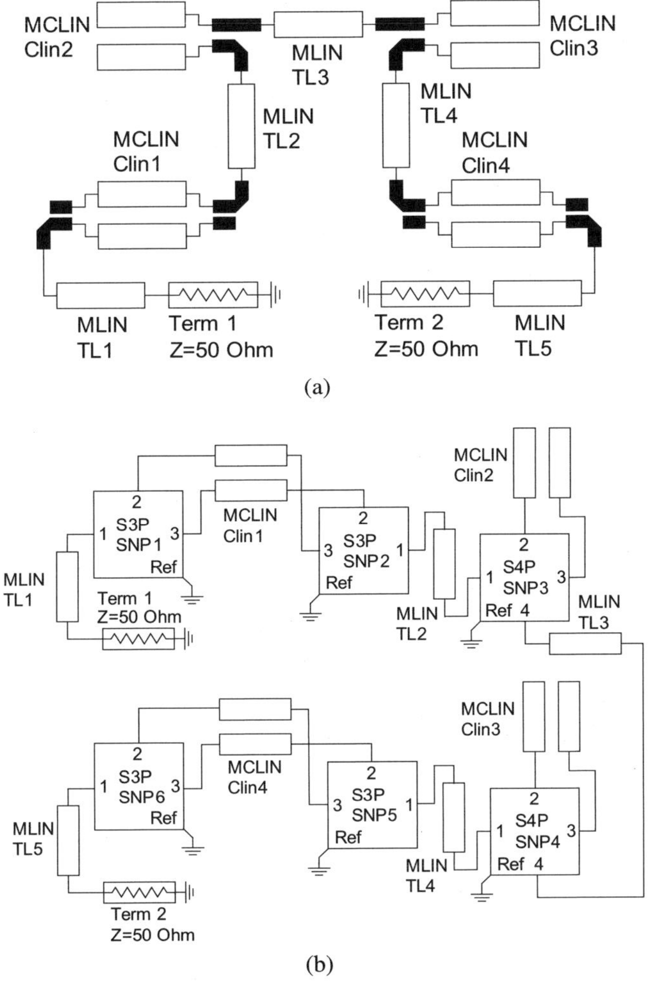

Fig. 5.12. Coupled-line bandpass filter co-simulation-based (Type 2) tuning model (Koziel and Bandler, 2011): (a) Type 2 (co-simulation) tuning model with FEKO-simulated submodels (in black) connecting designable tuning elements, (b) circuit simulator (Agilent ADS) implementation of the tuning model: S-parameters of the EM-simulated submodels are stored in the S3P and S4P data components SNP1 to SNP6.

are simulated in FEKO. Due to symmetry, only two submodels need independent evaluation. The tuning model is handled by the circuit simulator Agilent ADS (Fig. 5.12(b)). All the designable parameters (microstrip lengths, widths, and coupled-line gaps) are associated with the circuit-theory-based tuning elements allowing fast optimization of the tuning model. On the other hand, simulating parts of the filter using the EM solver maintains accuracy and predictability of the tuning model. The alignment procedure uses dielectric constants and substrate heights of the tuning elements corresponding to the design variables.

After two iterations of Type 2 tuning, the specification is satisfied (see Fig. 5.13). In others tests in Koziel and Bandler (2011), both the traditional space mapping (Koziel *et al.*, 2006) and a gradient-based search (Matlab, 2008) fail to satisfy the specifications. The design obtained using a pattern search (Kolda *et al.*, 2003) is slightly better than that obtained by the Type 2 tuning, however, the design cost is substantially higher (Koziel and Bandler, 2011).

In another example of Type 2 tuning (Hirtenfelder, 2007), the author tunes each resonator of a dual band duplexer for 400 MHz using a tuning model comprised of connected submodels in the EM/circuit simulation suite CST Microwave Studio. Tuning elements are attached to the resonators. Similarly, another software vendor shows the technique in filter design

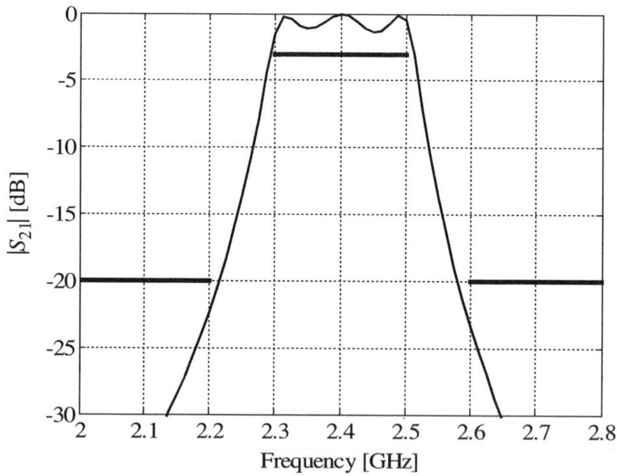

Fig. 5.13. Coupled-line bandpass filter: the fine model response at the final design obtained in two Type 2 iterations (Koziel and Bandler, 2011). Design specifications marked using thick horizontal lines.

(Fernandest, 2008) using its EM (HFSS) and circuit (Ansoft Designer) simulators.

5.6.6. *Discussion*

Tuning space mapping permits rapid design with satisfactory results obtained after a few iterations (typically, one to four). Variants differ in the construction of the tuning model as well as in the calibration procedure. While the calibration process for Type 0 generally requires an auxiliary calibration model, it is likely robust because tuning elements are inserted with minimal or no disturbance to the fine model. Also, Type 0 tuning requires a smaller number of tuning ports than Type 1, thus the simulation time of the auxiliary fine model is longer for Type 1 than for Type 0.

Type 1 tuning involves simple calibration and implementation and allows us to tune cross-sectional parameters. However, the alignment process can be difficult because the insertion of the tuning components requires deembedding of substantial parts of the original structure: more iterations are normally necessary to yield a satisfactory design.

Type 1d (fast) tuning (Koziel *et al.*, 2010) seems simpler and more efficient than Type 1 tuning. However, reducing a complex structure may result in deterioration of the generalization capability of the tuning model, leading to an increased number of iterations.

Type 0− and Type 2 tuning are suited to waveguide structures. Type 2 tuning allows parallel processing, keeping in mind, however, that the couplings between the submodels are not accounted for.

5.7. Conclusions

We have reviewed engineering tuning techniques, and categorized and illustrated tuning space mapping procedures. The several variations of tuning space mapping are generally robust since the tuning elements and subsequent parameter extraction procedures usually compensate for misalignments. The main considerations for choosing among the variations are the engineer's knowledge, the available software, the anticipated difficulties in implementation, and the simulation cost of the auxiliary fine model.

Acknowledgement

This work was supported in part by the Icelandic Centre for Research (RANNIS) Grant 110034021, and the Natural Sciences and Engineering

Research Council of Canada under Grants RGPIN7239-11 and STPGP 381153-09, and by Bandler Corporation. The authors thank Sonnet Software Inc. for *em*, and Agilent Technologies, Santa Rosa, CA, for making ADS available. The authors thank J.C. Rautio, Sonnet, Syracuse, NY, for the interdigitated filter example.

References

Agilent ADS, Agilent Technologies, 1400 Fountaingrove Parkway, Santa Rosa: CA 95403-1799.

Ansoft Designer/Nexxim, Ansoft Corporation, Pittsburgh: PA.

Ansoft HFSS, Ansoft Corporation, Pittsburgh: PA.

Bandler, J.W., Cheng, Q.S., Dakroury, S.A., Mohamed, A.S., Bakr, M.H., Madsen, K. and Søndergaard, J. (2004). Space mapping: the state of the art, *IEEE T. Microw. Theory*, **52**, 337–361.

Bandler, J.W., Liu, P.C. and Tromp, H. (1976). A nonlinear programming approach to optimal design centering, tolerancing and tuning, *IEEE T. Circuits Syst.*, **23**, 155–165.

Bandler, J.W., Rizk, M.R.M. and Salama, A.E. (1981). An interactive optimal postproduction tuning technique utilizing simulated sensitivities and response measurements, *IEEE Int. Microwave Symp. Digest*, Los Angeles: CA, pp. 63–65.

Bandler, J.W. and Salama, A.E. (1983). Integrated approach to microwave post production tuning, *IEEE Int. Microwave Symp. Digest*, Boston: MA, pp. 415–417.

Bandler, J.W. and Salama, A.E. (1985). Functional approach to microwave postproduction tuning, *IEEE T. Microw. Theory*, **33**, 302–310.

Bandler, J.W., Biernacki, R.M., Chen, S.H., Grobelny, P.A., and Hemmers, R.H. (1994) Space mapping technique for electromagnetic optimization, *IEEE T. Microw. Theory*, **42**, 536–544.

Bandler, J.W., Biernacki, R.M., Chen, S.H., Hemmers, R.H. and Madsen, K. (1995) Electromagnetic optimization exploiting aggressive space mapping, *IEEE T. Microw. Theory*, **43**, 2874–2882.

Cheng, Q.S., Bandler, J.W. and Koziel, S. (2012). Tuning space mapping: the state of the art, *Int. J. RF Microw. CE*, **22**, 639–651.

Cheng, Q.S., Bandler, J.W. and Rayas-Sánchez, J.E. (2008). Tuning-aided implicit space mapping, *Int. J. RF Microw. CE*, **18**, 445–453.

Cheng, Q.S., Bandler, J.W. and Koziel, S. (2010a). Space mapping design framework exploiting tuning elements, *IEEE T. Microw. Theory*, **58**, 136–144.

Cheng, Q.S., Rautio, J.C., Bandler, J.W. and Koziel, S. (2010b). Progress in simulator-based tuning – the art of tuning space mapping, *IEEE Microw. Mag.*, **11**, 96–110.

Dunsmore, J. (1999). Tuning band pass filters in the time domain, *IEEE Int. Microwave Symp. Digest*, Anaheim: CA, pp. 1351–1354.

Edquist, A. (2009). Filter design and Ansoft tools, *HFSS Info Day*, Stockholm, Sweden.

FEKO, EM Software & Systems-S.A. (Pty) Ltd, 32 Techno Lane, Technopark, Stellenbosch, 7600, South Africa.

Fernandest, R. (2008). The accuracy of HFSS with the speed of circuit simulation, *10èmes Journées de Caractérisation Microondes et Matériaux*, Limoges, France. [Online] Available at http://jcmm2008.xlim.fr/hfss_presentation. pdf. Date accessed: 16 June 2012.

Hirtenfelder, F. (2007). Filter tuning techniques. *CST: 3rd European User Group Meeting*, September 13–14, Lake Tegernsee, Germany.

Hsu, H.-T., Yao, H.-W., Zaki, K. and Atia, A. (2002). Computer-aided diagnosis and tuning of cascaded coupled resonators filters, *IEEE T. Microw. Theory*, **50**, 1137–1145.

Kolda, T.G., Lewis, R.M. and Torczon, V. (2003). Optimization by direct search: new perspectives on some classical and modern methods, *SIAM Rev.*, **45**, 385–482.

Koziel, S. (2011). Robust optimization of microwave structures using cosimulation-based surrogate models, *Microw. Opt. Techn. Let.*, **53**, 130–135.

Koziel, S. and Bandler, J.W. (2011). Fast design optimization of microwave structures using co-simulation-based tuning space mapping, *Appl. Comput. Electrom.*, **26**, 631–639.

Koziel, S., Bandler, J.W. and Cheng, Q.S. (2010). Design optimization of microwave circuits through fast embedded tuning space mapping, *Proc. 40th European Microwave Conf.*, Paris, France, pp. 1186–1189.

Koziel, S., Bandler, J.W. and Cheng, Q.S. (2011). Tuning space mapping design framework exploiting reduced EM models, *IET Microw. Antenna P*, **5**, 1219–1226.

Koziel, S., Bandler, J.W. and Madsen, K. (2006). A space mapping framework for engineering optimization: theory and implementation, *IEEE T. Microw. Theory*, **54**, 3721–3730.

Koziel, S., Cheng, Q.S. and Bandler, J.W. (2008). Space mapping, *IEEE Microw. Mag.*, **9**, 105–122.

Koziel, S., Meng, J., Bandler, J.W., Bakr, M.H. and Cheng, Q.S. (2009). Accelerated microwave design optimization with tuning space mapping, *IEEE T. Microw. Theory*, **57**, 383–394.

Lee, H.M. and Tsai, C.M. (2005). Improved coupled-microstrip filter design using effective even-mode and odd-model characteristic impedances, *IEEE T. Microw. Theory*, **53**, 2812–2818.

Matlab^{TM} (2008). Version 7.6, The MathWorks, Inc., 3 Apple Hill Drive, Natick: MA 01760-2098.

Meng, M. and Wu, K.-L. (2009). An analytical approach to computer-aided diagnosis and tuning of lossy microwave coupled resonator filters, *IEEE T. Microw. Theory*, **57**, 3188–3195.

Microwave Office, AWR Corporation, 1960 E. Grand Avenue, Suite 430, El Segundo: CA 90245.

Miraftab, V. and Mansour, R.R. (2008). Fully automated RF/microwave filter tuning by extracting human experience using fuzzy controllers, *IEEE T. Circuits I*, **55**, 1357–1367.

Ness, J. (1998). A unified approach to the design, measurement, and tuning of coupled-resonator filters, *IEEE T. Microw. Theory*, **46**, 343–351.

Pepe, G., Gortz, F.-J. and Chaloupka, H. (2004). Computer-aided tuning and diagnosis of microwave filters using sequential parameter extraction, *IEEE Int. Microwave Symp. Digest*, Fort Worth: TX, pp. 1373–1376.

Pinel, J.F. (1971). Computer-aided network tuning, *IEEE T. Circuits Syst.*, 1, 192–194.

Rasmita, A., Guo, Y.X. and Alphones, A. (2009). Constrained explicit knowledge embedded space mapping using circuit tuning based on physical augmentation as parameter extraction, *Asia Pacific Microwave Conference (APMC)*, Singapore, pp. 1431–1434.

Rautio, J.C. (1994). An ultra-high precision benchmark for validation of planar electromagnetic analyses, *IEEE T. Microw. Theory*, **42**, 2046–2050.

Rautio, J.C. (2005). Deembedding the effect of a local ground plane in electromagnetic analysis, *IEEE T. Microw. Theory*, **53**, 770–776.

Rautio, J.C. (2006). RF design closure – Companion modeling and tuning methods, *IEEE Int. Microwave Symp. Workshop on Microwave Component Design Using Space Mapping Technology*, San Francisco: CA.

Swanson, D.G. and Wenzel, R.J. (2001). Fast analysis and optimization of combline filters using FEM, *IEEE Int. Microwave Symp. Digest*, Boston: MA, pp. 1159–1162.

Thoma, P. (2010). Using coupled simulations to efficiently address challenging EM problems, *Int. Workshop on Advances in Modeling and Optimization of High Frequency Structures*, Reykjavik University, Iceland.

Yeung, L.K. and Wu, K.-L. (2003). A compact second-order LTCC bandpass filter with two finite transmission zeros, *IEEE T. Microw. Theory*, **51**, 337–341.

Yu, M. and Tang, W.-C., (2003). A fully automated filter tuning robot for wireless base station diplexers, *IEEE Int. Microwave Symp. Workshop on Computer Aided Filter Tuning*, Philadelphia: PA.

Zahirovic, N., Mansour, R.R. and Yu, M. (2010). Scalar measurement-based algorithm for automated filter tuning of integrated Chebyshev tunable filters, *IEEE T. Microw. Theory*, **58**, 3749–3759.

Chapter 6

Robust Design Using Knowledge-Based Response Correction and Adaptive Design Specifications

Slawomir Koziel, Stanislav Ogurtsov, and Leifur Leifsson

Simulation-based optimization has become an important design tool in microwave engineering. Yet, employing electromagnetic (EM) solvers in the design process is a challenging task, primarily due to a high computational cost of an accurate EM simulation. This chapter is focused on efficient EM-driven design optimization techniques that utilize physics-based low-fidelity models, normally based on coarse-discretization EM simulations. The presented methods attempt to exploit as much of the knowledge about the system or device of interest embedded in the low-fidelity model as possible, so as to reduce the computational cost of the design process. Unlike most of the approaches contained in other chapters of this book, the techniques discussed here are non-parametric, i.e., they are not based on analytical formulas. The chapter presents several specific methods, including those based on correcting the low-fidelity model response (adaptive response correction and shape-preserving response prediction), as well as on suitable modification of design specifications. Detailed formulations, application examples, and the discussion of advantages and disadvantages of these techniques are also included.

6.1. Electromagnetic-Simulation-Driven Design

Electromagnetic (EM) simulation has become an inherent part of contemporary microwave design. It is used for verification purposes, but also, more and more often, to adjust the designable parameters of the structure under consideration so that given performance specifications are met.

129

EM-simulation-driven design is actually a must for the growing number of devices, where systematic design procedures or design-ready theoretical models are not sufficiently accurate, e.g., for including substrate-integrated circuits (Wu, 2009), planar ultrawideband antennas (Schantz, 2005), and dielectric resonator antennas (Petosa, 2007). EM-based design optimization is a challenging task though. Probably the most challenging obstacle is the high computational cost of accurate full-wave simulations, which makes the use of conventional optimization algorithms prohibitive as these techniques require a large number of EM simulations. Another issue is the numerical noise that is inherent to EM simulations, particularly if the solver uses adaptive meshing. These problems can be alleviated to some extent by using adjoint sensitivities (Chung *et al.*, 2001; Nikolova *et al.*, 2006) that have recently become available in some commercial EM solvers (HFSS, 2010; CST, 2011). In general, robust automated and computationally efficient EM-based optimization is still an open problem. In practice, the most common simulation-driven design approach is to perform parameter sweeps guided by engineering experience. This is, however, tedious and does not guarantee optimal results.

A number of techniques for modeling and simulation-driven design of microwave structures have emerged over the recent years including the methods that exploit artificial neural networks (Kabir *et al.*, 2008; Zhang *et al.*, 2008; Gutiérrez-Ayala and Rayas-Sánchez, 2010; Murphy and Yakovlev, 2010; Tighilt *et al.*, 2011), fuzzy systems (Miraftab and Mansour, 2006), kriging (Couckuyt *et al.*, 2010), or multidimensional Cauchy approximation (Shaker *et al.*, 2009), as well as surrogate-based techniques such as space mapping (SM) (Bandler *et al.*, 2004a; Ismail *et al.*, 2004; Amari *et al.*, 2006; Crevecoeur *et al.*, 2007; Koziel *et al.*, 2006; Koziel *et al.*, 2008; Crevecoeur *et al.*, 2008; Cheng *et al.*, 2010a ; Quynang *et al.*, 2010), simulation-based tuning (Swanson and Macchiarella, 2007; Rautio, 2008), and combination of both (Koziel *et al.* 2009a; Cheng *et al.*, 2010b). The last three approaches allow computationally efficient design optimization where, under certain circumstances, a satisfactory design can be obtained after a few high-fidelity (or fine) EM simulations (Bandler *et al.*, 2004a). SM and tuning SM (TSM) (Cheng *et al.*, 2010b), exploit physics-based low-fidelity (or coarse) model that is a basis to build a surrogate model — a computationally cheap and yet reasonably accurate representation of the fine model. The surrogate is iteratively updated and reoptimized instead of optimizing the fine model directly (Bandler *et al.*, 2004a). SM and TSM are both potentially very efficient, however, these methods may not be straightforward to implement and automate (Koziel *et al.*, 2006; Koziel *et al.*, 2010), which, so far, is a limitation for their widespread use by the researchers and designers.

Probably the simplest way of exploiting physics-based coarse models for efficient design optimization is through correcting the coarse model response in order to obtain zero- and (in some cases) first-order consistency (Alexandrov and Lewis, 2001) between the surrogate and the fine model. The corrected coarse model becomes a reliable prediction tool that can be used to find the approximate optimum of the fine model. Several techniques of this approach have been proposed in microwave engineering recently including manifold mapping (MM) (Echeverria and Hemker, 2005) and multi-point response correction (Koziel, 2010). Formally, output SM (Bandler *et al.*, 2004a) and its variations (Koziel *et al.*, 2010a) also belong to this group, as well as the neuro-space mapping techniques (Zhang *et al.*, 2005; Rayas-Sánchez and V. Gutiérrez-Ayala, 2006; Kabir *et al.*, 2010). A detailed exposition of space mapping and similar techniques can be found in Chapters 3 and 4. All of these techniques can be categorized as parametric response correction methods, because the adjustment of the low-fidelity model is realized by means of explicit formulas, parameters of which can be calculated using available low- and high-fidelity model data.

This chapter is focused on non-parametric approaches that allow us to fully exploit the knowledge about the structure of interest embedded in the low-fidelity model. The techniques, such as adaptive response correction (ARC) (Koziel *et al.*, 2009b) and shape-preserving response prediction (SPRP) (Koziel *et al.*, 2010) do not use explicit parameters. They utilize relationships between the low- and high-fidelity model responses as functions of design variables, not just their values at specific designs, which normally results in much better generalization capability when compared to parametric approaches. A similar idea stays behind another method presented here, the adaptively adjusted design specifications (AADS) technique (Koziel, 2010c). In AADS, however, the relationship between the low- and high-fidelity models is reflected in proper modification of the design specifications without directly adjusting the low-fidelity model, which makes this method extremely simple to implement and yet robust.

In this chapter, we recall the response correction concept as well as the formulation of the optimization algorithm exploiting a corrected low-fidelity model. We focus on non-parametric correction techniques; however, for the sake of the chapter being self-contained, parametric approaches are recalled as well. Additionally, we discuss the adaptively adjusted design specification method. All presented techniques are illustrated using microwave design examples.

6.2. Design Optimization Using Response-Corrected Physics-Based Coarse Models

In this section, we briefly formulate the microwave design problem as well as the concept of the response correction. We also recall the fundamental conditions that ensure convergence of the surrogate-based optimization algorithm with the response-corrected coarse model to converge to the fine model optimum. A distinction between the parametric and non-parametric models is also highlighted.

6.2.1. *Formulation of the Design Problem*

The design optimization task can be formulated as the following minimization problem

$$x_f^* \in \arg\min_x U(R_f(x)), \qquad (6.1)$$

where $R_f \in R^m$ denotes the response vector of a fine model of the device of interest evaluated through high-fidelity EM simulation; x is a vector of design parameters. The response $R_f(x)$ might be, e.g., the modulus of the reflection coefficient $|S_{11}|$ evaluated at m different frequencies. U is a given scalar merit function, e.g., a norm, or a minimax function with upper and lower specifications; x_f^* is the optimal design to be determined. It is assumed that R_f is evaluated through computationally expensive EM simulation so that solving (6.1) directly may be impractical.

6.2.2. *Response Correction: The Concept*

Let $R_c(x) \in R^m$ be the response vector of the coarse model of the structure of interest, i.e., a computationally cheap but less accurate representation of the fine model. It is also assumed that the coarse model is physics-based so that it "embeds" some knowledge about the fine model R_f, by using similar (although usually simplified) physical description of the device/system under consideration. Typical examples include equivalent circuits or coarse-discretization EM models.

The coarse model can be used to construct a surrogate R_s of the fine model so that an approximate solution to (6.1) can be obtained by optimizing R_s instead of R_f. In particular, the surrogate can be obtained as follows:

$$R_s(x) = C(R_c(x)), \tag{6.2}$$

where $C : R^m \to R^m$ is a response correction function.

Typically, the optimization process involving surrogates is performed in an iterative manner so that the sequence $x^{(1)}$, $x^{(2)}, \ldots$, of (hopefully) better and better approximations to x_f^* are obtained as

$$x^{(i+1)} = \arg\min_x U(R_s^{(i)}(x)), \tag{6.3}$$

with $R_s^{(i)}$ being the surrogate model at iteration i. Here, $R_s^{(i)}(x) = C^{(i)}(R_c(x))$, where $C^{(i)}$ is the correction function at iteration i. For surrogates constructed using response correction, we typically require that, at least, zero-order consistency between the surrogate and the fine model is satisfied, i.e., $R_s^{(i)}(x^{(i)}) = R_f(x^{(i)})$. It can be shown (Alexandrov *et al.*, 1998) that satisfaction of first-order consistency, i.e., $J[R_s^{(i)}(x^{(i)})] = J[R_f(x^{(i)})]$ (here, $J[\cdot]$ denotes the Jacobian of the respective model), guarantees convergence of $\{x^{(i)}\}$ to a local optimum of R_f assuming that (6.3) is enhanced by the trust-region mechanism (Conn *et al.*, 2000) and the functions involved are sufficiently smooth.

6.2.3. *Parametric and Non-Parametric Response Correction*

Response correction approaches can be roughly categorized into parametric techniques, where the response correction function (6.2) can be represented using explicit formulas, and non-parametric ones, where the response correction is defined implicitly through certain relations between the coarse and fine models which may not have any compact analytical form. Output space mapping (Bandler *et al.*, 2004a; Koziel *et al.*, 2010), manifold mapping (Echeverria and Hemker, 2005), and multi-point response correction (Koziel, 2010a) belong to the first group. We describe them briefly in Section 6.3. The second group includes adaptive response correction (Koziel *et al.*, 2009b) and shape-preserving response prediction (Koziel, 2010b) which are presented in detail in Section 6.4.

6.3. Parametric Response Correction Techniques for Microwave Optimization

In this section, we briefly review several parametric response correction techniques that have been recently developed and applied in simulation-driven microwave optimization. More specifically, we discuss output space mapping (OSM) (Koziel *et al.*, 2010), manifold mapping (MM) (Echeverria and Hemker, 2005), and multi-point response correction (Koziel, 2010a). More information about OSM and MM can be found in Chapters 4 and 3, respectively.

6.3.1. *Output Space Mapping*

The simplest realization of the response correction is output SM (Bandler *et al.*, 2004a), where

$$R_s^{(i)}(x) = R_c(x) + d^{(i)}. \tag{6.4}$$

Here, $d^{(i)} = R_f(x^{(i)}) - R_c(x^{(i)})$, which ensures zero-order consistency. Note that this way of creating the surrogate model is only based on the fine model response at the most current iteration point $x^{(i)}$. Typically, output SM is used as a supplement for other SM techniques (Koziel *et al.*, 2008). Provided that the coarse model is sufficiently accurate, output SM can, also, work as a stand-alone technique (Koziel *et al.*, 2008).

6.3.2. *Output Space Mapping with Sensitivity*

The model (6.4) can be enhanced as follows:

$$R_s^{(i)}(x) = R_c(x) + d^{(i)} + E^{(i)} \cdot (x - x^{(i)}), \tag{6.5}$$

where $d^{(i)}$ is as in (6.4) and $E^{(i)} = J[R_f(x^{(i)}) - R_c(x^{(i)})]$, which ensures first-order consistency. Alternatively, Jacobian of $R_f(x^{(i)}) - R_c(x^{(i)})$ can be replaced by a suitable approximation obtained using, e.g., Broyden update (Broyden, 1965). As before, the model (6.5) is normally used as a supplement but it can, also, work as a stand-alone technique, particularly when the algorithm (6.3) is enhanced by a trust-region approach (Conn *et al.*, 2000; Koziel *et al.*, 2010).

6.3.3. *Manifold Mapping*

The response correction model can be defined by explicitly using most of the available fine model data. According to the manifold mapping (MM) approach (Echeverria and Hemker, 2005), the surrogate model is defined as

$$R_s^{(i)}(x) = R_f(x^{(i)}) + S^{(i)}(R_c(x) - R_c(x^{(i)})), \tag{6.6}$$

with $\boldsymbol{S}^{(i)}$ being the $m \times m$ correction matrix

$$\boldsymbol{S}^{(i)} = \boldsymbol{\Delta F} \cdot \boldsymbol{\Delta C}^{\dagger}, \tag{6.7}$$

where

$$\boldsymbol{\Delta F} = [\boldsymbol{R}_f(\boldsymbol{x}^{(i)}) - \boldsymbol{R}_f(\boldsymbol{x}^{(i-1)}) \cdots \boldsymbol{R}_f(\boldsymbol{x}^{(i)}) - \boldsymbol{R}_f(\boldsymbol{x}^{(\max\{i-n,0\})})] \tag{6.8}$$

$$\boldsymbol{\Delta C} = [\boldsymbol{R}_c(\boldsymbol{x}^{(i)}) - \boldsymbol{R}_c(\boldsymbol{x}^{(i-1)}) \cdots \boldsymbol{R}_c(\boldsymbol{x}^{(i)}) - \boldsymbol{R}_c(\boldsymbol{x}^{(\max\{i-n,0\})})]. \tag{6.9}$$

The pseudoinverse, denoted by †, is defined as $\boldsymbol{\Delta C}^{\dagger} = \boldsymbol{V}_{\boldsymbol{\Delta C}}(\sum_{\boldsymbol{\Delta C}})^{\dagger}$ $(\boldsymbol{U}_{\boldsymbol{\Delta C}})^T$, where $\boldsymbol{U}_{\boldsymbol{\Delta C}}$, $\sum_{\boldsymbol{\Delta C}}$, and $\boldsymbol{V}_{\boldsymbol{\Delta C}}$ are the factors in the singular value decomposition of $\boldsymbol{\Delta C}$. The matrix $\sum_{\boldsymbol{\Delta C}}^{\dagger}$ is the result of inverting the nonzero entries in $\sum_{\boldsymbol{\Delta C}}$, leaving the zeroes invariant (Echeverria and Hemker, 2005).

Although MM does not explicitly use sensitivity information, the surrogate and the fine model Jacobians become more and more similar to each other towards the end of the MM optimization process (i.e., when $||\boldsymbol{x}^{(i)} - \boldsymbol{x}^{(i-1)}|| \to 0$) so that the surrogate (approximately) satisfies both zero- and first-order consistency conditions (Alexandrov and Lewis, 2001). This allows for a more accurate location of the fine model optimum.

6.3.4. *Multi-Point Response Correction*

Multi-point response correction (Koziel, 2010a) is another way of generalizing output SM in order to exploit as much of available fine model data as possible. The surrogate model is formulated as

$$\boldsymbol{R}_s^{(i)}(\boldsymbol{x}) = \boldsymbol{R}_c^{(i)}(\boldsymbol{x}) + \boldsymbol{\Delta R}_g^{(i)}(\boldsymbol{x}) \tag{6.10}$$

with the design-variable-dependent correction term $\boldsymbol{\Delta R}_g^{(i)}$ defined to have the following features:

(1) $\boldsymbol{\Delta R}_g^{(i)}(\boldsymbol{x}^{(i)}) = \boldsymbol{R}_f(\boldsymbol{x}^{(i)}) - \boldsymbol{R}_c^{(i)}(\boldsymbol{x}^{(i)})$, i.e., $\boldsymbol{R}_s^{(i)}(\boldsymbol{x}^{(i)}) = \boldsymbol{R}_f(\boldsymbol{x}^{(i)})$,
(2) $\boldsymbol{R}_s^{(i)}(\boldsymbol{x}^{(k)}) = \boldsymbol{R}_f(\boldsymbol{x}^{(k)})$, i.e., $\boldsymbol{\Delta R}_g^{(i)}(\boldsymbol{x}^{(k)}) = \boldsymbol{R}_f(\boldsymbol{x}^{(k)}) - \boldsymbol{R}_c^{(i)}(\boldsymbol{x}^{(k)})$ for as many $k < i$ as possible,
(3) $\boldsymbol{\Delta R}_g^{(i)}$ is a linear function of $\boldsymbol{x}$ (so that it does not degrade the ability of $\boldsymbol{R}_c$ to model higher order nonlinearities of $\boldsymbol{R}_f$).

Let $S^{(i)}$ be a set of all affinely independent subsets of $X^{(i)} = \{\boldsymbol{x}^{(1)}, \boldsymbol{x}^{(2)}, \ldots, \boldsymbol{x}^{(i)}\}$ containing $\boldsymbol{x}^{(i)}$. Let $s^{(i)} \in S^{(i)}$ be such that: (1) $s^{(i)}$ is maximal, i.e., $|s^{(i)}| = n_i = \max\{|s| : s \in S^{(i)}\}$, and (2) $s^{(i)}$ is the smallest in the sense of its diameter, i.e., $d(s^{(i)}) = \min\{d(s) : s \in S^{(i)} \wedge |s| = n_i\}$, where $d(s) \equiv \max\{\|\boldsymbol{x} - \boldsymbol{y}\| : \boldsymbol{x}, \boldsymbol{y} \in s\}$.

We have $|s^{(i)}| \leq \min(i, n+1)$. Let $\boldsymbol{x}^k$, $k = 1, 2, \ldots, n_i$, denote the elements of $s^{(i)}$ (here, $\boldsymbol{x}^1 = \boldsymbol{x}^{(i)}$). Define $\boldsymbol{R}_k = \boldsymbol{R}_f(\boldsymbol{x}^k) - \boldsymbol{R}_c^{(i)}(\boldsymbol{x}^k)$, $k = 1, 2, \ldots, n_i$. Let us also define the $m \times n_i$ matrix $\boldsymbol{R}^{(i)} = [\boldsymbol{R}_1 \boldsymbol{R}_2 \ldots \boldsymbol{R}_{ni}]$.

Let $M^{(i)} \subseteq R^n$ be a linear manifold generated by $s^{(i)}$, i.e.,

$$M^{(i)} = \left\{ \boldsymbol{x} \in R^n : \ \boldsymbol{x} = \sum_{k=1}^{n_i} \alpha_k \boldsymbol{x}^k, \sum_{k=1}^{n_i} \alpha_k = 1 \right\}. \tag{6.11}$$

$M^{(i)}$ is well defined because $s^{(i)}$ is affinely independent. Let $\boldsymbol{B}^{(i)} = \{\boldsymbol{b}_1, \ldots, \boldsymbol{b}_{ni-1}\}$, where $\boldsymbol{b}_k = \boldsymbol{x}^{k+1} - \boldsymbol{x}^1$, $k = 1, 2, \ldots, n_i - 1$. Let $Y^{(i)} \subseteq R^n$ be a linear subspace generated by $B^{(i)}$, i.e.,

$$Y^{(i)} = \left\{ \boldsymbol{x} \in R^n : \ \boldsymbol{x} = \sum_{k=1}^{n_i-1} \beta_k \boldsymbol{b}_k, \beta_k \in R, k = 1, 2, \ldots, n_i - 1 \right\}. \tag{6.12}$$

Clearly, $Y^{(i)} = M^{(i)} - \boldsymbol{x}^{(i)}$. For any $\boldsymbol{x} \in R^n$ define $P_i(\boldsymbol{x})$ as the orthogonal projection of $\boldsymbol{x} - \boldsymbol{x}^{(i)}$ onto $Y^{(i)}$. For any $\boldsymbol{y} \in M^{(i)}$ define $\lambda_i(\boldsymbol{y}) = [\lambda_1 \lambda_2 \ldots \lambda_{ni}]^T$ so that $\boldsymbol{y} = \lambda_1 \boldsymbol{x}^1 + \lambda_2 \boldsymbol{x}^2 + \cdots + \lambda_{ni} \boldsymbol{x}^{ni}$. λ_i is uniquely determined and $\lambda_1 + \lambda_2 + \cdots + \lambda_{ni} = 1$.

We can now define $\Delta \boldsymbol{R}_g^{(i)}(\boldsymbol{x}), \boldsymbol{x} \in R^n$, as

$$\Delta \boldsymbol{R}_g^{(i)}(\boldsymbol{x}) = \begin{cases} \boldsymbol{R}_f(\boldsymbol{x}^{(i)}) - \boldsymbol{R}_c^{(i)}(\boldsymbol{x}^{(i)}) & \text{for } i = 1 \\ \boldsymbol{R}^{(i)} \cdot \boldsymbol{\lambda}_i(P_i(\boldsymbol{x} - \boldsymbol{x}^{(i)}) + \boldsymbol{x}^{(i)}) & \text{for } i > 1 \end{cases}. \tag{6.13}$$

In (6.13), we distinguished two cases because some of the above objects (e.g., $Y^{(i)}$) are not defined for $i = 1$. Vector $\lambda_i(\boldsymbol{y})$ can be determined as $\lambda_i = (\boldsymbol{T}^{(i)T} \boldsymbol{T}^{(i)})^{-1} \boldsymbol{T}^{(i)T} \boldsymbol{y}^{(i)}$, where $\boldsymbol{T}^{(i)}$ is an $(n+1) \times n_i$ matrix depending on $\boldsymbol{B}^{(i)}$, and $\boldsymbol{y}^{(i)} = [(P_i(\boldsymbol{x} - \boldsymbol{x}^{(i)}) + \boldsymbol{x}^{(i)})^T 1]^T$ (Koziel, 2010a).

It can be shown that $\Delta \boldsymbol{R}_g^{(i)}$ satisfies the properties 1–3 listed above. Also, $\Delta \boldsymbol{R}_g^{(i)}$ contains some information about the Jacobian of $\boldsymbol{R}_f(\boldsymbol{x}^{(i)}) - \boldsymbol{R}_c^{(i)}(\boldsymbol{x}^{(i)})$. In particular, the Jacobian of $\Delta \boldsymbol{R}_g^{(i)}$ is the approximation of the Jacobian reduced to the subspace $Y^{(i)}$: it can be shown that if $d(s^{(i)}) \to 0$, then $\boldsymbol{J}[\Delta \boldsymbol{R}_g^{(i)}(\boldsymbol{x}^{(i)})] \to \boldsymbol{J}[(\boldsymbol{R}_f(\boldsymbol{x}^{(i)}) - \boldsymbol{R}_c^{(i)})(\boldsymbol{x}^{(i)})]$ on $Y^{(i)}$. In particular, if $|s^{(i)}| = n + 1$, then the Jacobian of $\Delta \boldsymbol{R}_g^{(i)}$ becomes a good approximation of the Jacobian of $\boldsymbol{R}_f(\boldsymbol{x}^{(i)}) - \boldsymbol{R}_c^{(i)}(\boldsymbol{x}^{(i)})$ on R^n, provided that $d(s^{(i)})$ is sufficiently small. Some practical issues related to conditioning of the matrix $\boldsymbol{T}^{(i)}$ are discussed in Koziel (2010a). To improve the algorithm performance, a trust-region approach (Conn *et al.*, 2000) can be used as a convergence safeguard.

6.4. Exploiting Maximum Knowledge: Non-Parametric Response Correction Techniques

In this section, we describe non-parametric response correction techniques where the surrogate model is defined implicitly through certain relations between the coarse and fine models. On the other hand, non-parametric correction is able — in many cases — to exploit the knowledge about the system embedded in the coarse model better than the parametric one. We discuss the following approaches: adaptive response correction (Koziel *et al.*, 2009b), and shape-preserving response prediction (Koziel, 2010b).

6.4.1. *Adaptive Response Correction*

Adaptive response correction (ARC) is a generalization of output SM, which makes the correction term design-variable-dependent so that we have

$$R_s^{(i)}(x) = R_c(\ x) + \Delta_r(x, x^{(i)}), \qquad (6.14)$$

where

$$\Delta_r(x^{(i)}, x^{(i)}) = \ d^{(i)} = R_f(x^{(i)}) - R_c(x^{(i)}), \qquad (6.15)$$

i.e., a zero-order consistency condition is satisfied at $x^{(i)}$. An important feature of ARC is that the correction term is determined in such a way that this modification reflects the changes of R_c during the process of surrogate model optimization. In particular, if the response of R_c shifts or changes its shape with respect to a free parameter, e.g., frequency, the response correction term should track these changes.

The concept of ARC is best explained using an example. Consider a wideband bandstop microstrip filter (Hsieh and Wang, 2005) shown in Fig. 6.1. Figures 6.2 and 6.3 show the fine and coarse (in fact, here we use a space mapped coarse model response (Koziel *et al.*, 2009)) at two different designs, as well as the corresponding output SM terms d. It can be observed that the (frequency-wise) changes of the output SM correction term correspond to the changes of the coarse/fine model responses. These changes can be tracked by performing suitable frequency scaling of the coarse model response and then apply it to the original correction term. This is illustrated in Figs. 6.4 and 6.5, where the original output SM correction

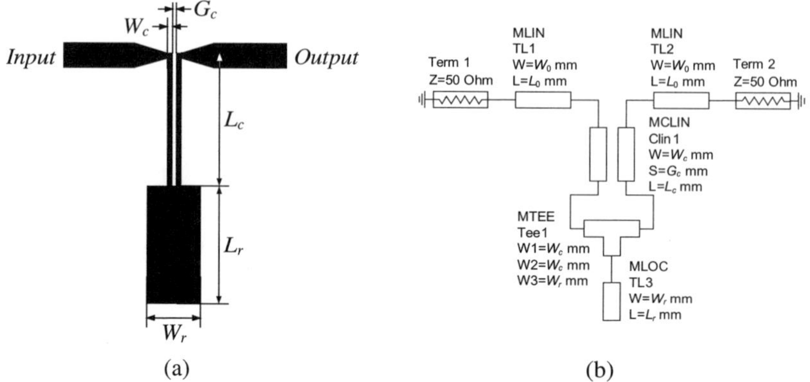

Fig. 6.1.   Wideband bandstop filter (Hsieh and Wang, 2005): (a) geometry, (b) coarse model (Agilent ADS).

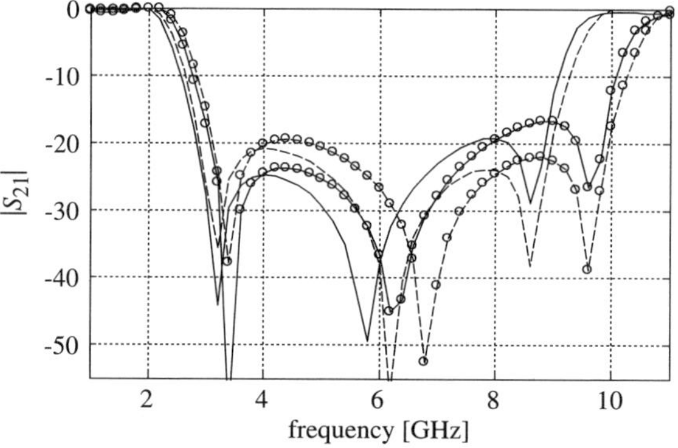

Fig. 6.2.   Wideband bandstop filter: fine (solid line) and coarse model (dashed line) responses at certain design, as well as the fine (solid line with circle markers) and coarse model (dashed line with circle markers) responses at a different design.

term (at certain design $x^{(0)}$), as well as the predicted one (at another design $x^{(1)}$) obtained using the frequency scaling function determined by tracking the coarse model response changes, are shown. The rigorous formulation of ARC can be found in Koziel *et al.* (2009); it is omitted here for the sake of brevity. Figure 6.6 shows the results of optimizing the surrogate model using the output SM correction (Fig. 6.6(a)) and ARC (Fig. 6.6(b)). The optimized surrogate model design in Fig. 6.6(a) is not as good as that in Fig. 6.6(b) because the constant output SM correction terms causes a significant distortion of the model response while moving away from $x^{(0)}$.

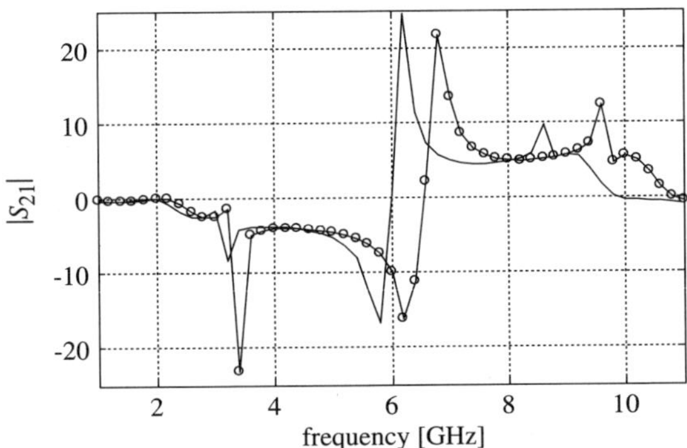

Fig. 6.3. Wideband bandstop filter: standard output SM correction terms corresponding to responses shown in Fig. 6.2.

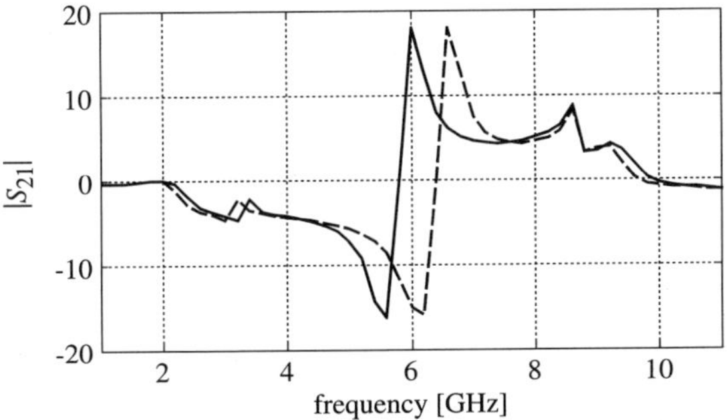

Fig. 6.4. Correction terms at certain design $x^{(0)}$, $\Delta_r(x^{(0)}, x^{(0)})$, (solid line), and at the other design $x^{(1)}$, $\Delta_r(x^{(1)}, x^{(0)})$ (dashed line). Plots obtained for the adaptive response correction method.

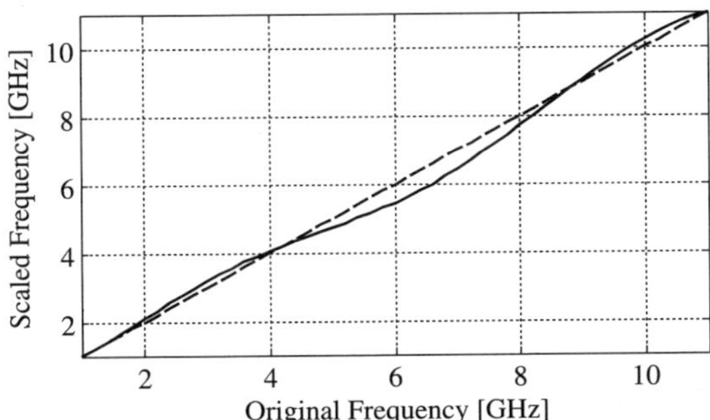

Fig. 6.5. The frequency scaling function used to obtain the dashed line in Fig. 6.4 (solid line). The scaling accounts for the changes (in frequency) of the coarse model response while going from $x^{(0)}$ to $x^{(1)}$. The identity function plot is shown as a dashed line.

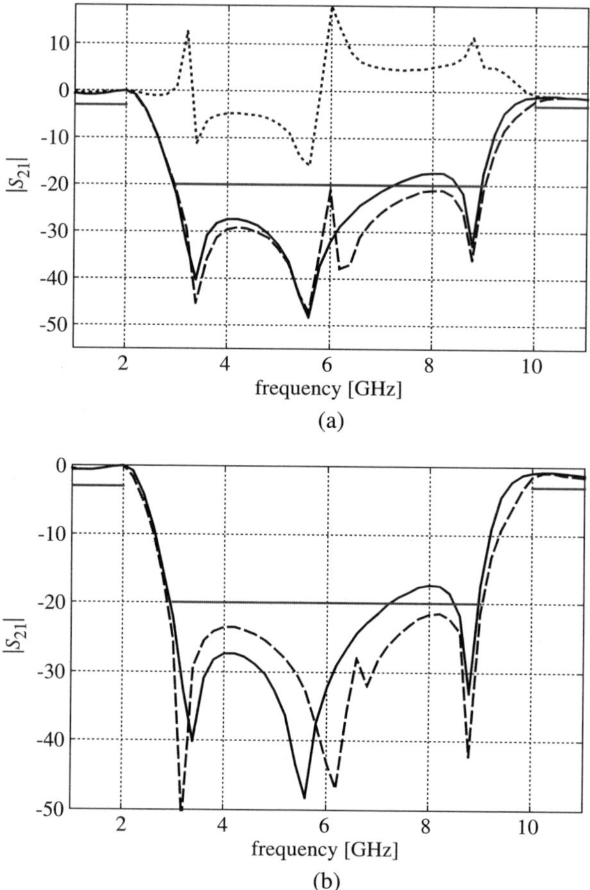

Fig. 6.6. Wideband bandstop filter: (a) initial (solid line) and optimized (dashed line) surrogate model response when output SM is used to create the surrogate, (b) initial (solid line) and optimized (dashed line) surrogate model response when ARC is used to create the surrogate. The plots in (b) correspond to $x^{(0)}$ and $x^{(1)}$ in Fig. 6.5. Design specifications marked using horizontal lines.

Table 6.1 shows the results of optimization of the bandstop filter in Fig. 6.1 using the surrogate model based on output SM and ARC. The latter is able to yield a better result after just three fine model evaluations, whereas the optimization cost with output SM is eight fine model evaluations. Figure 6.7 shows the fine model response at the final design obtained using ARC.

Table 6.1.   Optimization results for the wideband bandstop microstrip filter.

Surrogate Model	Final Specification Error [dB]	Number of Fine Model Evaluations*
Standard Output Space Mapping	−1.1	8
Adaptive Response Correction	−2.1	3

*Excludes the fine model evaluation at the starting point.

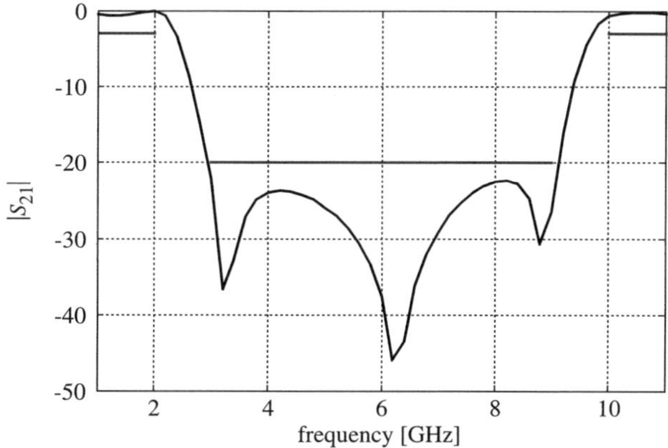

Fig. 6.7.   Wideband bandstop filter: fine model response at the design found by the space mapping optimization algorithm with the adaptive response correction. Design specifications marked using horizontal lines.

6.4.2. *Shape-Preserving Response Prediction*

Shape-preserving response prediction (SPRP) (Koziel, 2010b) is one of the most recent non-parametric response correction techniques. SPRP constructs the surrogate model assuming that the change of the fine model response due to the adjustment of the design variables can be predicted using the actual changes of the coarse model response. Therefore, it is critically important that the coarse model is physics-based, which ensures that the effect of the design parameter variations on the model response is similar for both the fine and coarse models. The change of the coarse model response is described by the translation vectors corresponding to a certain (finite) number of characteristic points of the model's response. These translation vectors are subsequently used to predict the change of the fine model response with the actual response of $\boldsymbol{R}_f$ at the current iteration point, $\boldsymbol{R}_f(\boldsymbol{x}^{(i)})$, used as a reference.

Figure 6.8(a) shows an example of coarse model response, $|S_{21}|$ in the frequency range $8\,\text{GHz}$ to $18\,\text{GHz}$, at the design $\boldsymbol{x}^{(i)}$, as well as a coarse model response at some other design $\boldsymbol{x}$. The responses come from the double-folded stub bandstop filter example considered in Koziel (2010b). Circles denote characteristic points of $\boldsymbol{R}_c(\boldsymbol{x}^{(i)})$, here, selected to represent $|S_{21}| = -3\,\text{dB}$, $|S_{21}| = -20\,\text{dB}$, and the local $|S_{21}|$ maximum (at about $13\,\text{GHz}$). Squares denote corresponding characteristic points for $\boldsymbol{R}_c(\boldsymbol{x})$, while line segments represent the translation vectors ("shift") of the characteristic points of $\boldsymbol{R}_c$ when changing the design variables from $\boldsymbol{x}^{(i)}$ to $\boldsymbol{x}$.

Because the coarse model is physics-based, the fine model response at a given design, $\boldsymbol{x}$, can be predicted using the same translation vectors applied to the corresponding characteristic points of the fine model response at $\boldsymbol{x}^{(i)}$, $\boldsymbol{R}_f(\boldsymbol{x}^{(i)})$. This is illustrated in Fig. 6.8(b). Figure 6.8(c) shows the predicted fine model response at $\boldsymbol{x}$ as well as the actual response, $\boldsymbol{R}_f(\boldsymbol{x})$, with a good agreement between both curves. Note that SPRP surrogate model does not explicitly use any parameters, which makes it easy to implement.

The shape-preserving response prediction can be rigorously formulated as follows. Let $\boldsymbol{R}_f(\boldsymbol{x}) = [R_f(\boldsymbol{x}, \omega_1) \ldots R_f(\boldsymbol{x}, \omega_m)]^T$ and $\boldsymbol{R}_c(\boldsymbol{x}) = [R_c(\boldsymbol{x}, \omega_1) \ldots R_c(\boldsymbol{x}, \omega_m)]^T$, where ω_j, $j = 1, \ldots, m$, is the frequency sweep (it can be assumed without loss of generality that the model responses are parameterized by frequency). Let $\boldsymbol{p}_j^f = [\omega_j^f \; r_j^f]^T$, $\boldsymbol{p}_j^{c0} = [\omega_j^{c0} \; r_j^{c0}]^T$, and $\boldsymbol{p}_j^c = [\omega_j^c \; r_j^c]^T$, $j = 1, \ldots, K$, denote the sets of characteristic points of $\boldsymbol{R}_f(\boldsymbol{x}^{(i)})$, $\boldsymbol{R}_c(\boldsymbol{x}^{(i)})$, and $\boldsymbol{R}_c(\boldsymbol{x})$, respectively. Here, ω and r denote the frequency and magnitude components of the respective point. The translation vectors of the coarse model response are defined as $\boldsymbol{t}_j = [\omega_j^t r_j^t]^T$, $j = 1, \ldots, K$, where $\omega_j^t = \omega_j^c - \omega_j^{c0}$ and $r_j^t = r_j^c - r_j^{c0}$.

The shape-preserving response prediction surrogate model is defined as follows

$$\boldsymbol{R}_s^{(i)}(\boldsymbol{x}) = [R_s^{(i)}(\boldsymbol{x}, \omega_1) \;\; \cdots \;\; R_s^{(i)}(\boldsymbol{x}, \omega_m)]^T, \qquad (6.16)$$

where

$$R_s^{(i)}(\boldsymbol{x}, \omega_j) = \bar{R}_f(\boldsymbol{x}^{(i)}, F(\omega_j, \{-\omega_k^t\}_{k=1}^K)) + R(\omega_j, \{r_k^t\}_{k=1}^K), \qquad (6.17)$$

for $j = 1, \ldots, m$. $\bar{R}_f(\boldsymbol{x}, \omega)$ is an interpolation of $\{R_f(\boldsymbol{x}, \omega_1), \ldots, R_f(\boldsymbol{x}, \omega_m)\}$ onto the frequency interval $[\omega_1, \omega_m]$. The scaling function F interpolates the data pairs $\{\omega_1, \omega_1\}$, $\{\omega_1^f, \omega_1^f - \omega_1^t\}, \ldots, \{\omega_K^f, \omega_K^f - \omega_K^t\}$, $\{\omega_m, \omega_m\}$, onto the frequency interval $[\omega_1, \omega_m]$. The function R does a similar interpolation

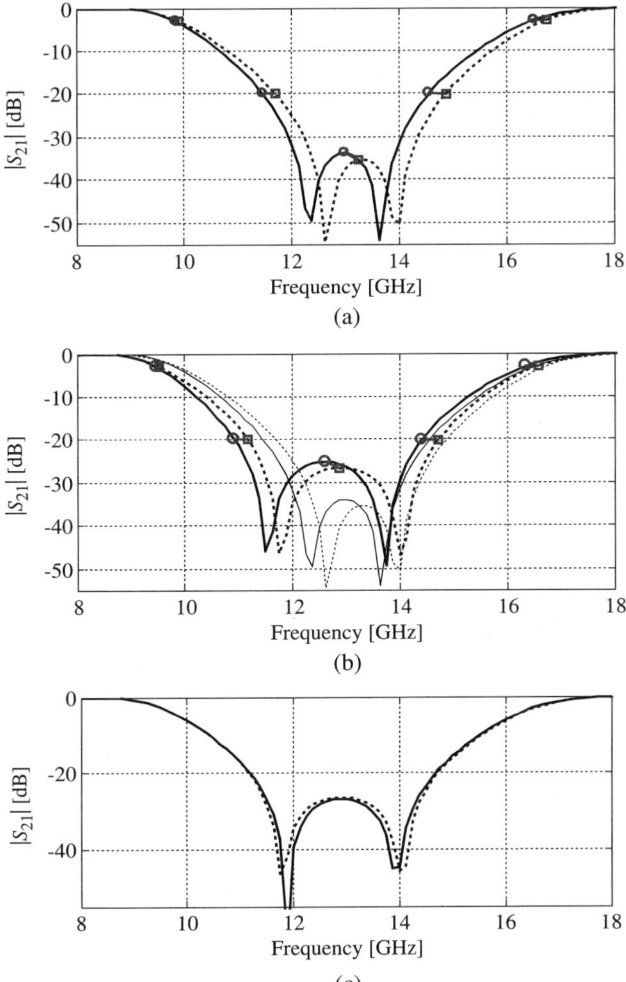

Fig. 6.8. SPRP concept: (a) example coarse model response at the design $x^{(i)}$, $R_c(x^{(i)})$ (solid line), the coarse model response at x, $R_c(x)$ (dotted line), characteristic points of $R_c(x^{(i)})$ (circles) and $R_c(x)$ (squares), and the translation vectors (short lines), (b) fine model response at $x^{(i)}$, $R_f(x^{(i)})$ (solid line) and the predicted fine model response at x (dotted line) obtained using SPRP based on characteristic points of (a); characteristic points of $R_f(x^{(i)})$ (circles) and the translation vectors (short lines) were used to find the characteristic points (squares) of the predicted fine model response; coarse model responses $R_c(x^{(i)})$ and $R_c(x)$ are plotted using thin-solid and dotted line, respectively (Koziel, 2010b), (c) fine model response at x, $R_f(x)$ (solid line), and the fine model response at x obtained using the shape-preserving prediction (dotted line). Good agreement between both curves is observed, particularly in the areas corresponding to the characteristic points of the response.

for data pairs $\{\omega_1, r_1\}$, $\{\omega_1^f, r_1^f - r_1^t\}, \ldots, \{\omega_K^f, r_K^f - r_K^t\}$, $\{\omega_m, r_m\}$; here $r_1 = R_c(\boldsymbol{x}, \omega_1) - R_c(\boldsymbol{x}^r, \omega_1)$ and $r_m = R_c(\boldsymbol{x}, \omega_m) - R_c(\boldsymbol{x}^r, \omega_m)$.

In other words, the function F translates the frequency components of the characteristic points of $\boldsymbol{R}_f(\boldsymbol{x}^{(i)})$ to the frequencies at which they should be located according to the translation vectors $\boldsymbol{t}_j$, while the function R adds the necessary magnitude component. The interpolation onto $[\omega_1, \omega_m]$ is necessary because the original frequency sweep is a discrete set.

Formally, both the translation vectors $\boldsymbol{t}_j$ and their components should have an additional index (i) indicating that they are determined at iteration i of the optimization algorithm (6.3), however, this was omitted for the sake of simplicity.

Figure 6.9 shows the plots of the functions F and R corresponding to the fine/coarse model response of Fig. 6.8(a) and (b). The interpolation of $\{R_f(\boldsymbol{x}, \omega_1), \ldots, R_f(\boldsymbol{x}, \omega_m)\}$, F and R is implemented using cubic splines.

SPRP has some similarities with the adaptive response correction (Koziel, 2010a) (cf. Section 6.4.1). The important similarity is that the actual change of the coarse model response, which is described in terms of the characteristic points of the model response, is used as a prediction tool. On the other hand, the fundamental difference is that ARC adaptively adjusts the output space mapping correction term (Bandler *et al.*, 2004a), whereas SPRP directly affects the fine model response. This has important implications: (i) ARC does not have a shape preserving capability,

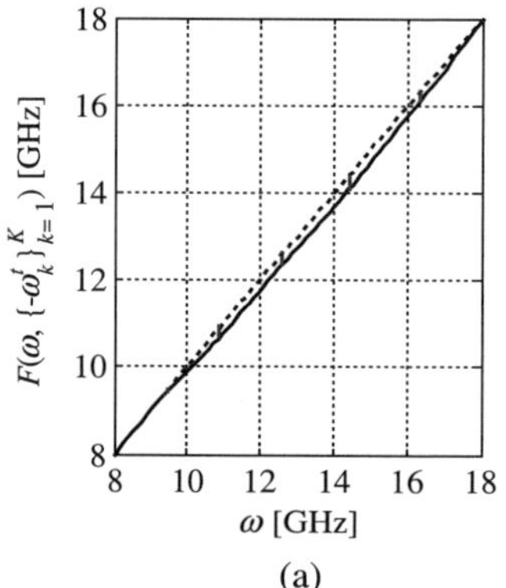
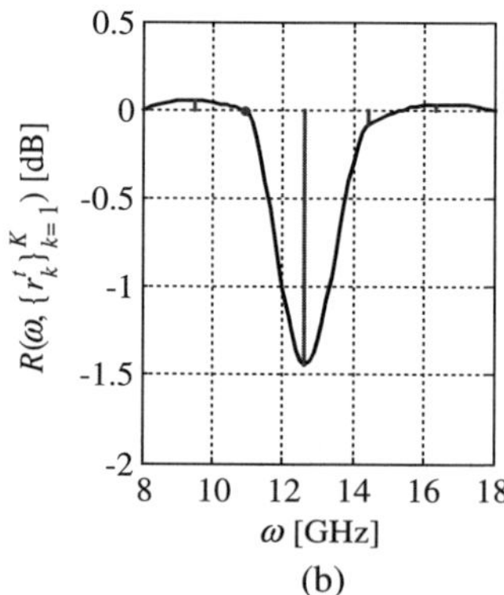

(a)   (b)

Fig. 6.9. Interpolating functions corresponding to the fine/coarse model plots in Figs. 6.8(a) and 6.8(b): (a) function F (solid line); the identity function is denoted using the dotted line, the frequency components of the translation vectors are denoted as short solid lines, (b) function R (solid line); the magnitude components of the translation vectors are denoted using short solid lines (Koziel, 2010b).

(ii) formulation and implementation of ARC is more complicated than that of SPRP. Consequently, the optimization process carried out using SPRP is expected to be more stable and perform better in comparison to the ARC approach. However, ARC is not restricted by the fundamental assumption of SPRP: the one-to-one correspondence between the characteristic points of the low- and high-fidelity model responses. More details regarding this limitation of SPRP as well as on possible remedies can be found in Koziel (2010b).

Consider the dual-band bandpass filter (Guan *et al.*, 2008) (Fig. 6.10(a)). The design parameters are $x = [L_1\, L_2\, S_1\, S_2\, S_3\, d\, g\, W]^T$ mm.

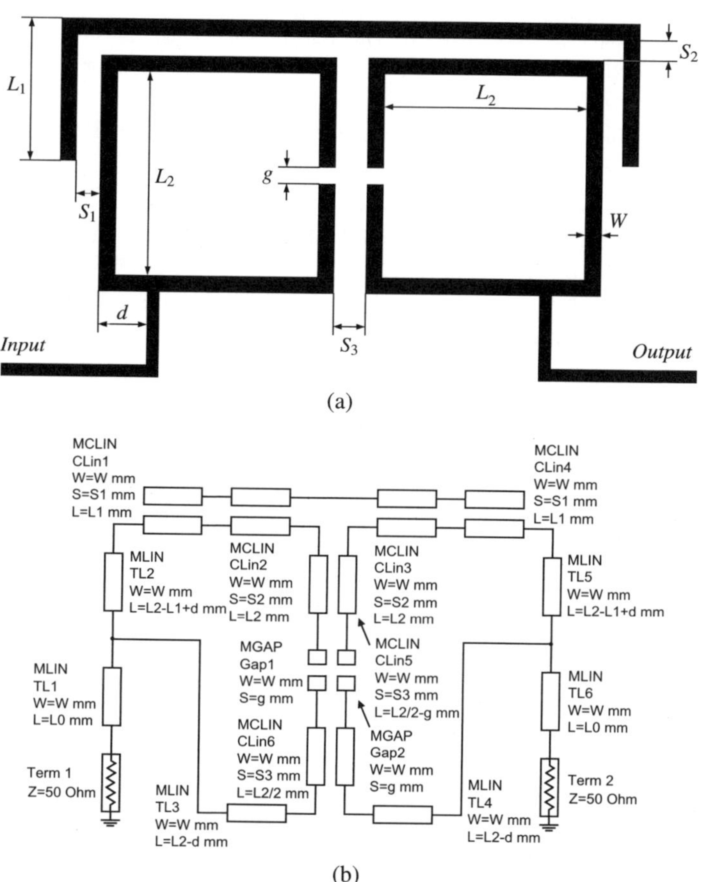

Fig. 6.10. Dual-band bandpass filter: (a) geometry (Guan *et al.*, 2008), (b) coarse model (Agilent ADS).

Table 6.2. Optimization results for dual-band bandpass filter.

Algorithm	Final Specification Error [dB]	Number of Fine Model Evaluations[*]
Output SM[†]	−1.4	8
Output SM^2 + ISM[‡]	−1.9	8
ARC (Koziel *et al.*, 2009)	−1.8	3
ARC (Koziel *et al.*, 2009) + ISM[‡]	−1.5	4
SPRP	−2.0[§]	3
SPRP + ISM[‡]	−1.9[¶]	2

[*]Excludes the fine model evaluation at the starting point.
[†]The surrogate model is of the form $R_s^{(i)}(x) = R_c(x) + [R_f(x^{(i)}) - R_c(x^{(i)})]$.
[‡]The surrogate model is of the form $R_s^{(i)}(x) = R_c(x + c^{(i)})$; $c^{(i)}$ found using parameter extraction (Bandler *et al.*, 2004a).
[§]Design specifications satisfied after the first iteration (spec. error −1.2 dB).
[¶]Design specifications satisfied after the first iteration (spec. error −1.0 dB).

The fine model is simulated in Sonnet *em* (Sonnet, 2010). The design specifications are $|S_{21}| \geq -3$ dB for 0.85 GHz $\leq \omega \leq 0.95$ GHz and 1.75 GHz $\leq \omega \leq 1.85$ GHz, and $|S_{21}| \leq -20$ dB for 0.5 GHz $\leq \omega \leq 0.7$ GHz, 1.1 GHz $\leq \omega \leq 1.6$ GHz and 2.0 GHz $\leq \omega \leq 2.2$ GHz. The coarse model is implemented in Agilent ADS (Agilent, 2008) (Fig. 10(b)). The initial design is $x^{(0)} = [16.14\,17.28\,1.16\,0.38\,1.18\,0.98\,0.98\,0.20]^T$ mm (the optimal solution of R_c). The following characteristic points are selected to set up the SPRP model: four points for which $|S_{21}| = -20$ dB, four points with $|S_{21}| = -5$ dB, as well as six additional points located between −5 dB points.

For the purpose of optimization, the coarse model was enhanced by tuning the dielectric constants and the substrate thickness of the microstrip models corresponding to the design variables L_1, L_2, d and g (original values of ε_r and H were 10.2 and 0.635 mm, respectively). The filter was optimized using the SPRP, SM, and ARC techniques.

Table 6.2 shows the optimization results. SPRP was compared to output SM as well as ARC. All of the considered methods were able to yield solutions satisfying the design specifications. However, the quality of the designs produced by SPRP is better than that of the other methods. Also, computational cost of SPRP is lower than the cost of the other approaches, particularly of output SM. Figure 6.11 shows the initial fine model response as well as the fine model response at the design obtained using SPRP.

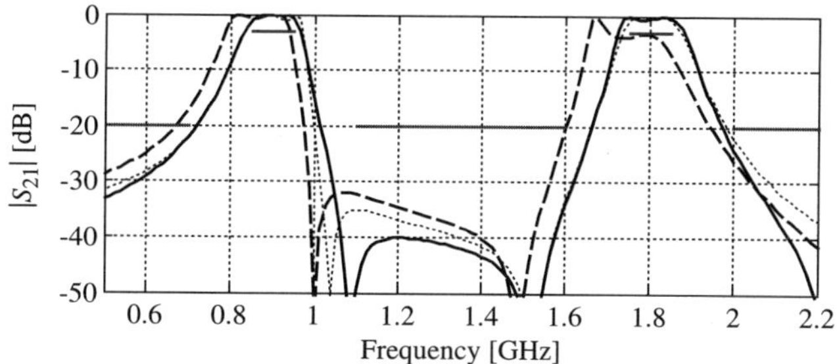

Fig. 6.11. Dual-band bandpass filter: fine model (dashed line) and coarse model (thin-dashed line) response at $x^{(0)}$, and the optimized fine model response (solid line) at the design obtained using shape-preserving response prediction. Design specifications marked using horizontal lines.

6.5. Exploiting Maximum Knowledge: Adaptively Adjusted Design Specifications

It is not necessary to remove the discrepancies between the low- and the high-fidelity models by correcting the low-fidelity model. Another way is to "absorb" the model misalignment by proper adjustment of the design specifications. In microwave engineering, most of the design tasks can be formulated as minimax problems with upper and lower specifications and it is easy to implement modifications by, for example, shifting the specification levels. This approach is exploited by the adaptively adjusted design specifications (AADS) technique (Koziel, 2010c), and it is described in this section.

6.5.1. *Adaptively Adjusted Design Specifications: Optimization Procedure*

Adaptively adjusted design specifications (AADS) technique (Koziel, 2010c) consists of the following two simple steps that can be iterated if necessary:

1. Modify the original design specifications in order to take into account the difference between the responses of R_f and R_c at their characteristic points.
2. Obtain a new design by optimizing the coarse model with respect to the modified specifications.

Characteristic points of the responses should correspond to the design specification levels. They should also include local maxima/minima of the respective responses. Figure 6.12(b) shows characteristic points of $\boldsymbol{R}_f$ and $\boldsymbol{R}_c$ for the bandstop filter example. The points correspond to –3 dB and –30 dB levels as well to the local maxima of the responses. As one can observe in Fig. 6.12(b) the selection of points is rather straightforward.

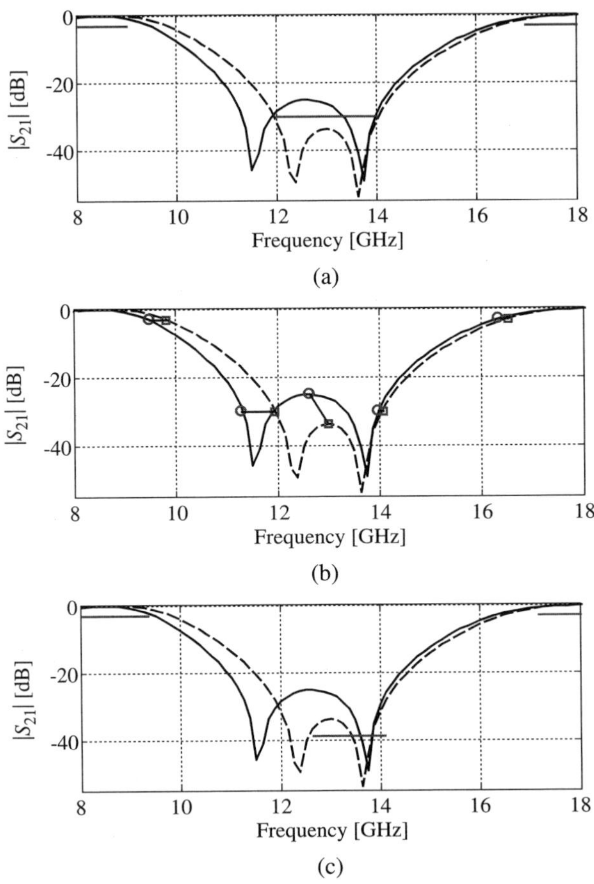

Fig. 6.12. Bandstop filter example (responses of $\boldsymbol{R}_f$ and $\boldsymbol{R}_c$ are marked with solid and dashed line, respectively): (a) fine and coarse model responses at the initial design (optimum of $\boldsymbol{R}_c$) as well as the original design specifications, (b) characteristic points of the responses corresponding to the specification levels (here, –3 dB and –30 dB) and to the local response maxima, (c) fine and coarse model responses at the initial design and the modified design specifications. Design specifications marked using horizontal lines.

In the first step of the AADS optimization procedure, the design specifications are modified (or mapped) so that the level of satisfying/violating the modified specifications by the coarse model response corresponds to the satisfaction/violation levels of the original specifications by the fine model response.

More specifically, for each edge of the specification line, the edge frequency is shifted by the difference of the frequencies of the corresponding characteristic points, e.g., the left edge of the specification line of $-30\,$dB is moved to the right by about $0.7\,$GHz, which is equal to the length of the line connecting the corresponding characteristic points in Fig. 6.12(b). Similarly, the specification levels are shifted by the difference between the local maxima/minima values for the respective points, e.g., the $-30\,$dB level is shifted down by about $8.5\,$dB because of the difference of the local maxima of the corresponding characteristic points of $\boldsymbol{R}_f$ and $\boldsymbol{R}_c$. Modified design specifications are shown in Fig. 6.12(c).

The coarse model is subsequently optimized with respect to the modified specifications and the new design obtained this way is treated as an approximated solution to the original design problem (i.e., optimization of the fine model with respect to the original specifications). Steps 1 and 2 can be repeated if necessary. As demonstrated later, substantial design improvement is typically observed after the first iteration, however, additional iterations may bring further enhancement.

An important prerequisite of AADS is that the coarse model is physics-based; in particular, the adjustment of the design variables has a similar effect on both $\boldsymbol{R}_f$ and $\boldsymbol{R}_c$. In such a case the coarse model design that is obtained in the second stage of the AADS procedure (i.e., optimal with respect to the modified specifications) will be (almost) optimal for $\boldsymbol{R}_f$ with respect to the original specifications.

Figure 6.13 illustrates an iteration of AADS used for design of a CBCPW-to-SIW transition (Deslandes and Wu, 2005). One can observe that the absolute matching between the low- and high-fidelity models is not as important as the shape similarity.

If the similarity between the fine and coarse model response is not sufficient the AADS technique may not work well. In many cases, however, using different reference design for the fine and coarse models may help. In particular, $\boldsymbol{R}_c$ can be optimized with respect to the modified specifications starting not from $\boldsymbol{x}^{(0)}$ (the optimal solution of $\boldsymbol{R}_c$ with respect to the original specifications), but from another design, say $\boldsymbol{x}_c^{(0)}$, at which the response of $\boldsymbol{R}_c$ is as similar to the response of $\boldsymbol{R}_f$ at $\boldsymbol{x}^{(0)}$ as possible. Such

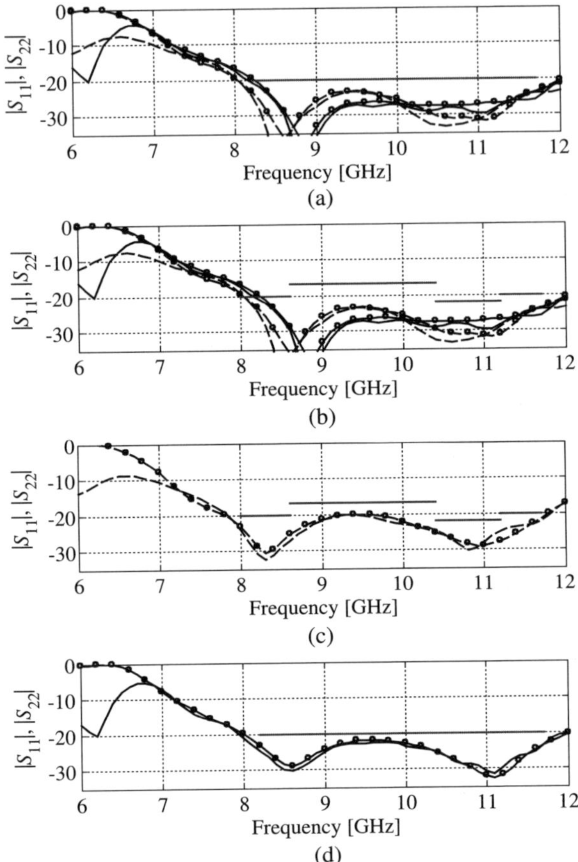

Fig. 6.13. Adaptively adjusted design specification technique applied to optimize CBCPW-to-SIW transitions. High- and low-fidelity model response denoted as solid and dashed lines, respectively. $|S_{22}|$ distinguished from $|S_{11}|$ using circles. Design specifications denoted by thick horizontal lines. (a) High- and low-fidelity model responses at the beginning of the iteration as well as original design specifications; (b) High- and low-fidelity model responses and modified design specifications that reflect the differences between the responses; (c) Low-fidelity model optimized to meet the modified specifications; (d) High-fidelity model at the low-fidelity model optimum shown versus original specifications. Thick horizontal lines indicate the design specifications.

a design can be obtained as follows (Koziel, 2010c):

$$x_c^{(0)} = \arg\min_{z} ||\boldsymbol{R}_f(\boldsymbol{x}^{(0)}) - \boldsymbol{R}_c(\boldsymbol{z})||. \tag{6.18}$$

At iteration i of AADS optimization procedure, the optimal design of the coarse model $\boldsymbol{R}_c$ with respect to the modified specifications, $x_c^{(i)}$, has to be translated to the corresponding fine model design, $x^{(i)}$, as follows: $x^{(i)} = x_c^{(i)} + (x^{(0)} - x_c^{(0)})$. Note that the preconditioning procedure (6.18) is performed only once for the entire optimization process.

The idea of coarse model preconditioning is borrowed from space mapping (more specifically, from the original space mapping concept (Bandler *et al.*, 2004a)). In practice, the coarse model can be "corrected" to reduce its misalignment with the fine model using any available degrees of freedom, for example, preassigned parameters as in implicit space mapping (Bandler *et al.*, 2004b).

6.5.2. *Adaptively Adjusted Design Specifications: Design Example*

To demonstrate operation and efficiency of AADS consider the second example of a bandpass microstrip filter with an open stub inverter (Lee *et al.*, 2000) (Fig. 6.14). The design parameters are $x = [L_1\, L_2\, L_3\, S_1\, S_2\, W_1]^T$. The fine model is simulated in FEKO (FEKO, 2010) The design specifications are $|S_{21}| \leq -20\,\mathrm{dB}$ for $1.5\,\mathrm{GHz} \leq \omega \leq 1.8\,\mathrm{GHz}$, $|S_{21}| \geq -3\,\mathrm{dB}$ for $1.95\,\mathrm{GHz} \leq \omega \leq 2.05\,\mathrm{GHz}$, and $|S_{21}| \leq -20\,\mathrm{dB}$ for $2.2\,\mathrm{GHz} \leq \omega \leq 2.5\,\mathrm{GHz}$. The coarse model is implemented in Agilent ADS (Agilent, 2008) (Fig. 6.15). The initial AADS design is the coarse model optimal solution $x^{(0)} = [25.00\, 5.00\, 1.221\, 0.652\, 0.187\, 0.100]^T\,\mathrm{mm}$ (specification error $+15.7\,\mathrm{dB}$).

The first iteration of AADS already yields a design satisfying the specifications, $x^{(1)} = [23.79\, 5.00\, 1.00\, 0.694\, 0.192\, 0.10]^T\,\mathrm{mm}$ (specification

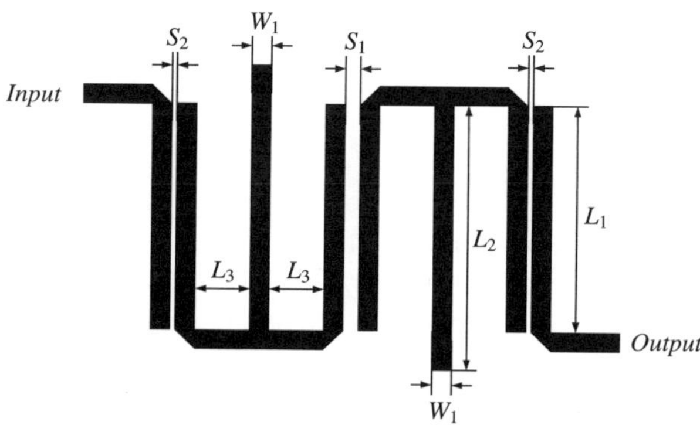

Fig. 6.14.   Bandpass filter with open stub inverter: geometry (Lee *et al.*, 2000).

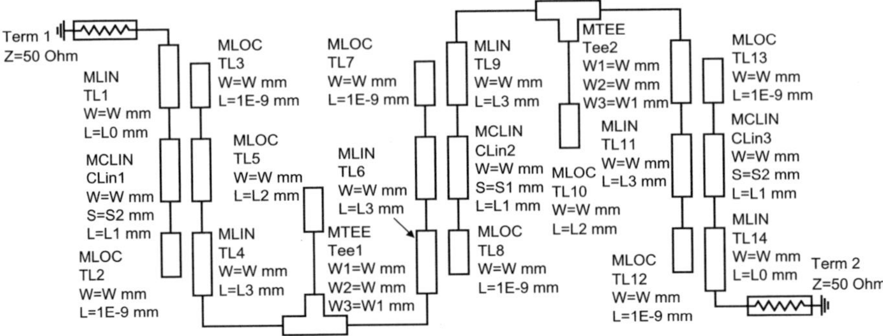

Fig. 6.15.   Bandpass filter with open stub inverter: coarse model (Agilent, 2008).

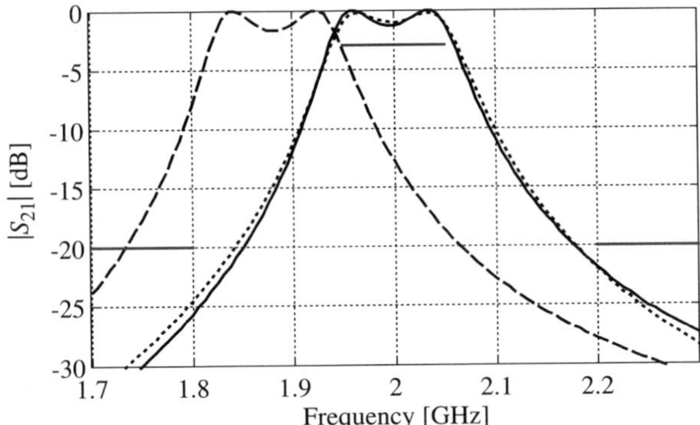

Fig. 6.16.   Bandpass filter with open stub inverter: R_f response (solid line) at the final design obtained after two iterations of our optimization procedure; fine (dashed line) and coarse (dotted line) model responses at the initial design. Design specifications marked using horizontal lines.

error -0.6 dB). After the second iteration, the design was further improved to $x^{(2)} = [23.68\,5.00\,1.00\,0.717\,0.193\,0.10]^T$ mm (specification error -1.7 dB). Figure 6.16 shows the fine and coarse model responses at $x^{(0)}$ and the fine model response at the final design.

For the sake of comparison, the filter was also optimized using the frequency SM algorithm. The design obtained in three iterations, $[23.66\,5.00\,1.00\,0.654\,0.188\,0.100]^T$ mm, satisfies the design specifications, however, it is not as good at the one obtained using the procedure proposed in this work (specification error -0.8 dB).

6.6. Discussion and Recommendations

Here, we attempt to give a qualitative comparison of knowledge-based design techniques presented in this chapter, as well as formulate some guidelines and recommendations for the readers interested in applying these methods in their research and/or design work. The main factors are complexity of implementation, computational efficiency, robustness, as well as range of applications.

In terms of implementation, the parametric response correction techniques are very straightforward. Output space mapping, manifold mapping, and multi-point response correction construct the surrogate model by using analytical formulas, therefore, they are easy to implement. Non-parametric methods may be more involved because the surrogate models are constructed by considering some auxiliary quantities such as scaling functions (adaptive response correction) or characteristic points and translation vectors (SPRP), which generally depend on the response shape. Also, in SPRP, the user is responsible for defining the characteristic points on a case-by-case basis. On the other hand, the AADS approach described in Section 6.4 is probably the easiest to implement out of all the methods considered in this chapter.

In terms of computational complexity, non-parametric methods prove to be more efficient. The reason is that both ARC and SPRP, as well as AADS, exploit the knowledge embedded in the low-fidelity model to a larger extent than the parametric methods which only do a local model alignment. As indicated by the examples, ARC and SPRP are capable of yielding a satisfactory design after two or three iterations, whereas output SM or MM typically require more iterations. On the other hand, parametric techniques tend to be more robust, particularly when embedded in the trust-region framework which improves their convergence properties. Also, the use of (even approximated) fine model sensitivity improves the technique's ability to accurately locate the fine model optimum. Parametric techniques are also more generic than ARC, SPRP, and AADS. For the latter, some considerations are necessary in order to select a proper realization of the scaling function (ARC), the definition of the characteristic points (SPRP), or the analysis of the model responses to properly modify design specifications (AADS). Also, the SPRP technique assumes one-to-one correspondence between characteristic points of the coarse and fine models at all considered designs, which may be difficult to satisfy for certain problems.

Based on the above remarks, parametric methods, particularly output SM, are recommended for less experienced users, and when the underlying coarse model is relatively fast (e.g., equivalent circuit). More experienced users are encouraged to try either ARC or SPRP, particularly if the available coarse model is relatively expensive (e.g., obtained from coarse-discretization EM simulation). AADS, because of its simplicity, can be readily used by both experienced and novice users; in general, it tends to be more reliable when the coarse model is more accurate (e.g., obtained through coarse-discretization EM simulation). In either case, the quality of the coarse model is one of the key factors, therefore its preconditioning using any space mapping transformation is recommended.

References

Agilent ADS (2008). Agilent Technologies, 1400 Fountaingrove Parkway, Santa Rosa: CA 95403-1799.

Alexandrov, N.M., Dennis, J.E., Lewis, R.M. and Torczon, V. (1998). A trust region framework for managing use of approximation models in optimization, *Struct. Multidiscip. O.*, **15**, 16–23.

Alexandrov, N.M. and Lewis, R.M. (2001). An overview of first-order model management for engineering optimization, *Optim. Eng.*, **2**, 413–430.

Amari, S., LeDrew, C. and Menzel, W. (2006). Space-mapping optimization of planar coupled-resonator microwave filters, *IEEE T. Microw. Theory*, **54**, 2153–2159.

Bandler, J.W., Cheng, Q.S., Dakroury, S.A., Mohamed, A.S., Bakr, M.H., Madsen, K. and Sondergaard, J. (2004a). Space mapping: the state of the art, *IEEE T. Microw. Theory*, **52**, 337–361.

Bandler, J.W., Cheng, Q.S., Nikolova, N.K. and Ismail, M.A. (2004). Implicit space mapping optimization exploiting preassigned parameters, *IEEE T. Microw. Theory*, **52**, 378–385.

Broyden, C.G. (1965). A class of methods for solving nonlinear simultaneous equations, *Math. Comp.*, **19**, 577–593.

Cheng, Q.S., Bandler, J.W. and Rayas-Sánchez, J.E. (2010a). Tuning-aided implicit space mapping, *Int. J. RF Microw. CE*, **18**, 445–453.

Cheng, Q.S., Rautio, J.C., Bandler, J.W. and Koziel, S. (2010b). Progress in simulator-based tuning — the art of tuning space mapping, *IEEE Microw. Mag.*, **11**, 96–110.

Chung, Y.S., Cheon, C., Park, I.H. and Hahn, S.Y. (2001). Optimal design method for microwave device using time domain method and design sensitivity analysis-part II: FDTD case, *IEEE T. Magn.*, **37**, 3255–3259.

Conn, A.R., Gould, N.I.M. and Toint, P.L. (2000). *Trust Region Methods*, Society for Industrial and Applied Mathematics, Philadelphia: PA.

Couckuyt, I., Declercq, F., Dhaene, T., Rogier, H. and Knockaert, L. (2010). Surrogate-based infill optimization applied to electromagnetic problems, *Int. J. RF Microw. CE*, **20**, 492–501.

Crevecoeur, G., Dupre, L. and Van de Walle, R. (2007). Space mapping optimization of the magnetic circuit of electrical machines including local material degradation, *IEEE T. Magn.*, **43**, 2609–2611.

Crevecoeur, G., Sergeant, P., Dupre, L. and Van de Walle, R. (2008). Two-level response and parameter mapping optimization for magnetic shielding, *IEEE T. Magn.*, **44**, 301–308.

CST Microwave Studio (2011). CST AG, Bad Nauheimer Str. 19, D-64289 Darmstadt, Germany.

Deslandes, D. and Wu, K. (2005). Analysis and design of current probe transition from grounded coplanar to substrate integrated rectangular waveguides, *IEEE T. Microw. Theory*, **53**, 2487–2494.

Echeverria, D. and Hemker, P.W. (2005). Space mapping and defect correction, *CMAM*, **5**, 107–136.

FEKO (2010). Suite 6.0, EM Software & Systems-S.A. (Pty) Ltd, 32 Techno Lane, Technopark, Stellenbosch, 7600, South Africa.

Guan, X., Ma, Z., Cai, P., Anada, T. and Hagiwara, G. (2008). A microstrip dual-band bandpass filter with reduced size and improved stopband characteristics, *Microw. Opt. Tech. Let.*, **50**, 618–620.

Gutiérrez-Ayala, V. and Rayas-Sánchez, J.E. (2010). Neural input space mapping optimization based on nonlinear two-layer perceptrons with optimized nonlinearity, *Int. J. RF Microw. CE*, **20**, 512–526.

HFSS (2010), release 13.0, ANSYS, http://www.ansoft.com/products/hf/hfss/.

Hsieh, M.Y. and Wang, S.M. (2005). Compact and wideband microstrip bandstop filter, *IEEE Microw. Wirel. Co.*, **15**, 472–474.

Ismail, M.A., Smith, D., Panariello, A., Wang, Y. and Yu, M. (2004). EM-based design of large-scale dielectric-resonator filters and multiplexers by space mapping, *IEEE T. Microw. Theory*, **52**, 386–392.

Kabir, H., Wang, Y., Yu, M. and Zhang, Q.J. (2008). Neural network inverse modeling and applications to microwave filter design, *IEEE T. Microw. Theory*, **56**, 867–879.

Kabir, H., Yu, M. and Zhang, Q.J. (2010). Recent advances of neural network-based EM-CAD, *Int. J. RF Microw. CE*, **20**, 502–511.

Koziel S., Bandler J.W. and Madsen, K. (2006). A space mapping framework for engineering optimization: theory and implementation, *IEEE T. Microw. Theory*, **54**, 3721–3730.

Koziel, S., Cheng, Q.S. and Bandler, J.W. (2008). Space mapping, *IEEE Microw. Mag.*, 9, 105–122.

Koziel, S., Meng, J., Bandler, J.W., Bakr, M.H. and Cheng, Q.S. (2009a). Accelerated microwave design optimization with tuning space mapping, *IEEE T. Microw. Theory*, **57**, 383–394.

Koziel, S., Bandler, J.W. and Madsen, K. (2009b). Space mapping with adaptive response correction for microwave design optimization, *IEEE T. Microw. Theory*, **57**, 478–486.

Koziel, S. (2010a). Improved microwave circuit design using multipoint-response-correction space mapping and trust regions, *Int. Symp. Antenna Technology and Applied Electromagnetics, ANTEM 2010*, Ottawa, Canada.

Koziel, S. (2010b). Shape-preserving response prediction for microwave design optimization, *IEEE T. Microw. Theory*, **58**, 2829–2837.

Koziel, S. (2010c). Adaptively adjusted design specifications for efficient optimization of microwave structures, *Prog. Electromagn. Res.*, **21**, 219–234.

Koziel, S., Bandler, J.W. and Cheng, Q.S. (2010). Robust trust-region space-mapping algorithms for microwave design optimization, *IEEE T. Microw. Theory*, **58**, 2166–2174.

Lee, J.R., Cho, J.H. and Yun, S.W. (2000). New compact bandpass filter using microstrip $\lambda/4$ resonators with open stub inverter, *IEEE Microw. Guided W.*, **10**, 526–527.

Miraftab, V. and Mansour, R.R. (2006). EM-based microwave circuit design using fuzzy logic techniques, *IEE Proc-H.*, **153**, 495–501.

Murphy, E.K. and Yakovlev, V.V. (2010). Neural network optimization of complex microwave structures with a reduced number of full-wave analyses, *Int. J. RF Microw. CE*, **21**, 279–287.

Nikolova, N.K., Li, Y. and Bakr, M.H. (2006). Sensitivity analysis of scattering parameters with electromagnetic time-domain simulators, *IEEE T. Microw. Theory*, **54**, 1598–1610.

Petosa, A. (2007). *Dielectric Resonator Antenna Handbook*, Artech House, Norwood: MA.

Quyang, J., Yang, F., Zhou, H., Nie, Z. and Zhao, Z. (2010). Conformal antenna optimization with space mapping, *J. Electromagn. Wave.*, **24**, 251–260.

Rautio, J.C. (2008). Perfectly calibrated internal ports in EM analysis of planar circuits, *IEEE MTT-S.*, Atlanta: GA, 1373–1376.

Rayas-Sánchez, J.E. and Gutiérrez-Ayala, V. (2006). EM-based Monte Carlo analysis and yield prediction of microwave circuits using linear-input neural-output space mapping, *IEEE T. Microw. Theory*, **54**, 4528–4537.

Schantz, H. (2005). *The Art and Science of Ultrawideband Antennas*, Artech House, Norwood: MA.

Shaker, G.S.A., Bakr, M.H., Sangary, N. and Safavi-Naeini, S. (2009). Accelerated antenna design methodology exploiting parameterized Cauchy models, *Prog. Electromagn. Res.*, **18**, 279–309.

Sonnet Software, Inc. (2010), em^{TM} Version 12.54, 100 Elwood Davis Road, North Syracuse: NY 13212, USA.

Swanson, D. and Macchiarella, G. (2007). Microwave filter design by synthesis and optimization, *IEEE Microw. Mag.*, **8**, 55–69.

Tighilt, Y., Bouttout, F. and Khellaf, A. (2011). Modeling and design of printed antennas using neural networks, *Int. J. RF Microw. CE*, **21**, 228–233.

Wu, K. (2009). Substrate Integrated Circuits (SiCs) — A new paradigm for future Ghz and Thz electronic and photonic systems, *IEEE T. Circuits Syst. Newsl.*, **3**, 1.

Zhang, L., Xu, J.J., Yagoub, M.C.E., Ding, R.T. and Zhang, Q.J. (2005). Efficient analytical formulation and sensitivity analysis of neuro-space mapping for nonlinear microwave device modelling, *IEEE T. Microw. Theory*, **53**, 2752–2767.

Zhang, L., Zhang, Q.J. and Wood, J. (2008). Statistical neuro-space mapping technique for large-signal modeling of nonlinear devices, *IEEE T. Microw. Theory*, **56**, 2453–2467.

Chapter 7

Simulation-Driven Design of Broadband Antennas Using Surrogate-Based Optimization

Slawomir Koziel and Stanislav Ogurtsov

Contemporary antenna design strongly relies on electromagnetic (EM) simulations. Accurate reflection and radiation responses of many antenna geometries, e.g., composite broadband antennas, can only be obtained with discrete full-wave EM simulation. On the other hand, the direct use of high-fidelity EM simulations in the design process, particularly for automated parameter optimization, may result in high computational costs, often prohibitively so. We demonstrate that numerically efficient designs of broadband antennas can be realized using automated surrogate-based optimization (SBO) methodology. Applied SBO techniques include response surface approximation implemented using kriging and combined with space mapping, adaptive design specification technique, variable-fidelity simulation-driven optimization, and space mapping combined with adjoint sensitivities. The essence of these techniques resides in shifting the optimization burden to a fast surrogate of the antenna structure, and the use of coarse-discretization EM models to configure the surrogate. A properly created and handled surrogate serves as a reliable prediction tool so that satisfactory designs can be found at the cost of a few simulations of the high-fidelity antenna model. We also demonstrate the effect of the coarse-discretization model fidelity on the final design quality and the computational cost of the design process. Finally, we give an example of automatic management of the coarse model quality. Recommendations concerning application of specific SBO techniques to antenna design are also presented.

7.1. Introduction

Antenna design is a challenging task that includes, among others, the adjustment of geometry and material parameters so as to obtain the required antenna response with respect to certain figures of interest such as input impedance, radiation pattern, etc. (Schantz, 2005; Petosa, 2007; Volakis, 2007). Computationally inexpensive analytical models of antennas can only be used — in many cases — to yield initial designs. For accurate responses the models should account for the installation fixture, feeding circuit, etc. Such detailed models are normally evaluated with discrete electromagnetic (EM) solvers applied at meshes of fine discretization. Contemporary computational techniques — implemented in commercial simulation packages — are capable of obtaining quite accurate reflection and radiation antenna responses. However, full-wave simulations of realistic models are computationally expensive, and simulation even for a single combination of design parameters may take up to several hours. Such computational cost poses a significant problem for antenna design.

The task of automated adjustment of antenna parameters can be formulated as an optimization problem with the objective function supplied by an EM solver (Special issue, 2007). However, most of the conventional optimization techniques, both gradient-based (Wright and Nocedal, 1999), e.g., conjugate-gradient, quasi-Newton, sequential quadratic programming, etc., and derivative-free (Kolda *et al.*, 2003), e.g., Nelder–Mead, pattern-search techniques, etc., require a large number of design simulations, each of which is already computationally expensive. As a consequence, the direct use of the EM solver evaluating the high-fidelity antenna model in the optimization loop often turns out to be impractical due to the unacceptably high computational cost. Other obstacles for successful application of conventional optimization techniques to antenna design originate from poor analytical properties of simulation-based objective functions (discontinuity, numerical noise, etc.). As a result, the practice of simulation-driven antenna design relies on repetitive parameter sweeps guided by engineering experience. While this approach can be more reliable than brute-force antenna optimization, it is very laborious, time consuming, and it does not guarantee optimal results. Also, only antenna designs with a limited number of parameters can be handled this way.

Some of the problems mentioned above can be alleviated by the use of adjoint sensitivity (Director and Rohrer, 1969), which is a computationally cheap way to obtain derivatives of the system response with respect to its design parameters. Adjoint sensitivities can substantially speed up microwave design optimization while using gradient-based algorithms (Bandler and Seviora 1972; Chung *et al.*, 2001); however, adjoint sensitivities are not yet widespread in commercial EM solvers. Only CST (CST Microwave Studio, 2011) and HFSS (HFSS, 2010) have recently implemented this feature. Also, the use of adjoint sensitivities is limited by numerical noise of the response.

Computationally efficient simulation-driven antenna design can be realized using surrogate-based optimization (SBO) (Koziel and Ogurtsov, 2011e). In SBO, the computational burden is shifted to a surrogate model, a computationally cheap representation of the optimized structure (Bandler *et al.*, 2004; Queipo *et al.*, 2005; Koziel *et al.*, 2006; Koziel and Ogurtsov, 2011a). One can distinguish two types of SBO approaches that use approximation models and physics-based surrogates. Approximation models are constructed by sampling the design space and approximating or interpolating the corresponding high-fidelity simulation data (Simpson *et al.*, 2001). If the design space is sampled sufficiently dense the resulting model becomes reliable so that the optimal antenna design can be found by optimizing the surrogate. There are many techniques available to build an approximation model, including artificial neural networks (Haykin, 1998), radial basis functions (Wild *et al.*, 2008), kriging (Forrester *et al.*, 2009), support vector machines (Smola and Schölkopf, 2004), etc. Because of high costs of sampling the design space, this approach is suitable for building multiple-use library models, whereas it may not be suitable for ad-hoc antenna design optimization.

Other types of surrogates, the physics-based ones, can be constructed from underlying low-fidelity (or coarse) models or the respective structures. Because the low-fidelity models inherit some knowledge of the system under consideration, usually a small number of high-fidelity simulations are sufficient to configure a reliable surrogate. The most popular SBO approaches using physics-based surrogates that proved to be successful in microwave engineering are space mapping (SM) (Bandler *et al.*, 2004), tuning, and tuning SM (Koziel *et al.*, 2009; Cheng *et al.*, 2010). Unfortunately, these techniques are of limited use for antennas. SM normally relies on a fast coarse model, typically a circuit equivalent. Such circuit equivalents are either unavailable for many types of antennas, e.g., for dielectric resonator antennas, or they cannot provide the antenna radiation response with acceptable accuracy, e.g., as for microstrip antennas.

A generic way to build a physics-based surrogate model is to use data obtained from EM simulations at the coarse discretization meshes with a subsequent correction of these data relative to the high-fidelity model (Zhu *et al.*, 2007). A properly created and handled antenna surrogate serves as a predictor for the optimization process, i.e., it allows us to find the high-fidelity model optimum. This approach, once developed, can be applied to different antennas as a quite universal design tool.

It is worth clarifying, in the context of antenna design through simulation-driven optimization, the place of SBO techniques relative to metaheuristic optimization approaches (Yang, 2010), which include, among others, genetic algorithms (GA) (Back *et al.*, 2000), particle swarm optimizers (PSO) (Kennedy, 1997), differential evolution (DE) (Storn and Price, 1997), ant colony optimization (Dorigo and Gambardella, 1997). The essence of the SBO approaches is in numerically efficient generation of the reliable model response through the use of surrogate models whereas the strength of the metaheuristic optimization approaches is in biologically inspired intelligent handling of the objective functions to alleviate some optimization problems, e.g., the problem of multiple local optima (Yang, 2010). Probably the most successful application of the metaheuristic algorithms in antenna design resides so far in the antenna array optimization problems (Ares-Pena *et al.*, 1999; Bevelacqua and Balanis, 2007; Grimaccia *et al.*, 2007; Haupt, 2007; Jin and Rahmat-Samii, 2007; Pantoja *et al.*, 2007; Petko and Werner, 2007; Rajo-Iglesias and Quevedo-Teruel, 2007; Jin and Rahmat-Samii, 2008; Li *et al.*, 2008; Selleri *et al.*, 2008; Roy *et al.*, 2011) where the cost of the single element response is not the primary concern or where the response of a single element is already available, e.g., with a preassigned array element. In contrast, numerical efficiency of the SBO approaches has been demonstrated for design of antennas requiring high-fidelity EM simulations of the whole antenna (Koziel and Ogurtsov, 2011b). In such cases, the use of metaheuristic optimization is not practical because the corresponding computational would be tremendous: a typical GA, PSO, or DE algorithm requires hundreds or thousands of objective function evaluations to yield a solution (Ares-Pena *et al.*, 1999; Bevelacqua and Balanis, 2007; Grimaccia *et al.*, 2007; Haupt, 2007; Jin and Rahmat-Samii, 2007; Pantoja *et al.*, 2007; Petko and Werner, 2007; Rajo-Iglesias and Quevedo-Teruel, 2007; Jin and Rahmat-Samii, 2008; Li *et al.*, 2008; Selleri *et al.*, 2008; Roy *et al.*, 2011).

In this chapter we will review the state of the art of broadband antenna design using physics-based SBO techniques, where the surrogates of the antenna under design originate from the coarse-mesh antenna models.

Challenges common to all presented design examples are the following:

- Simple theoretical models of the antennas are inaccurate or even unavailable — so they can be used only to set up an initial design.
- Accurate high-fidelity simulations are quite expensive, whereas a significant number of accurate simulations might be needed to arrive at an improved design.
- A number of modes/resonances can be found within the impedance bandwidth of interest; the modes are often of unknown (prior simulation) field distributions and effect on antenna response.
- The effect of the particular design variable is design dependent so that an improvement through a one-by-one parameter sweep is hardly feasible to achieve; therefore automated optimization/tuning is the only reliable option to improve the design.
- The simulated antenna response is the net effect of the radiator, fixture, and feeding circuitry; therefore, a realistic antenna model should include all these antenna parts and be simulated as a whole structure at once.

For details concerning background, formulation, and implementation of the SBO algorithms the interested reader is encouraged to see the listed references or refer to Chapters 3, 4, and 6 of this book.

7.2. Broadband Antenna Designs Using SBO Methodology

In this section, we consider a number of broadband antenna design examples. For every example we describe the antenna structure under design, formulate the design problem, and outline the SBO technique that seems to be the most suitable to handle the particular antenna structure of interest. Results, as well as design computational costs, are provided.

In all examples, the SBO techniques have been implemented within the SMF system (Koziel and Bandler, 2007), a user-oriented software package for surrogate modeling and optimization. In particular, SMF has a capability to perform space mapping-based constrained optimization, modeling, and statistical analysis. It implements existing SM approaches, including input, output, implicit, and frequency SM. It contains drivers for various commercial EM simulators that allow linking electromagnetic models of different fidelity to the optimization algorithm so that the design process is fully automatic.

7.2.1. *Dielectric Resonator Antenna*

Consider a rectangular dielectric resonator antenna (DRA) shown in Fig. 7.1. The DRA comprises: a suspended rectangular dielectric resonator (DR) estimated to operate at the perturbed $TE_{\delta 11}$ mode (Petosa, 2007), supporting RO4003C slabs, and polycarbonate housing. The housing is fixed to the circuit board with four M1 through-bolts. The DRA is energized with a 50 ohm microstrip through a slot made in the metal ground. Substrate is 0.5 mm thick RO4003C.

Design specifications are $|S_{11}| \leq -15$ dB for 5.1-to-5.9 GHz; also the DRA is required to have antenna gain better than 5 dBi for the zero zenith angle over the bandwidth of interest.

There are nine design variables: $\boldsymbol{x} = [a_x a_y a_z a_{y0} u_s w_s y_s g_1 y_1]^T$, where a_x, a_y, and a_z are dimensions of the DR brick; a_{y0} stands for the offset of the DR center relative to the slot center marked by a black dot in Fig. 7.1(a); u_s

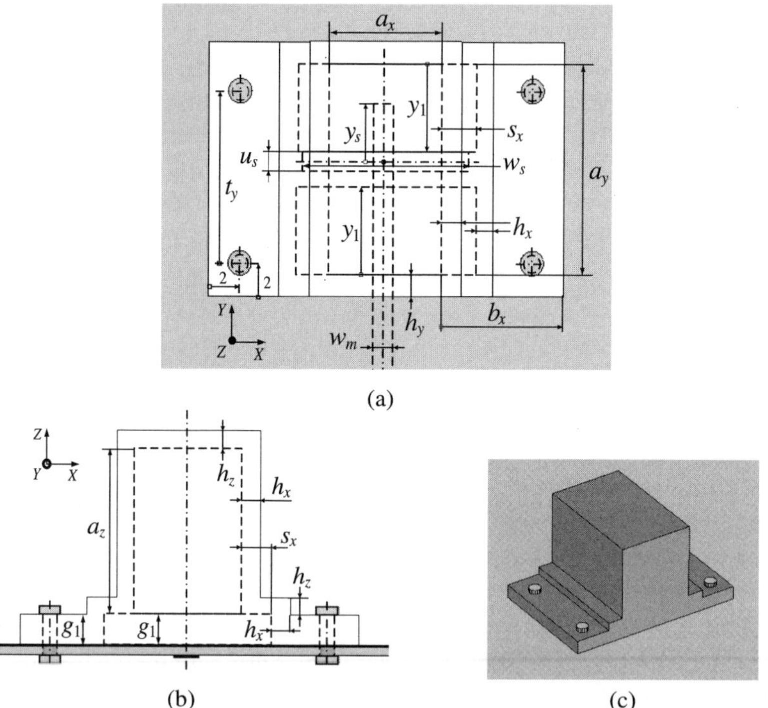

Fig. 7.1. DRA: (a) top view, (b) front view, and (c) 3D view (Koziel and Ogurtsov, 2011c).

and w_s are the slot dimensions; y_s is the length of the open-ended microstrip stub; and g_1 and y_1 are dimensions of the supporting slabs. Relative permittivity and loss tangent of the DR are 10 and 1e-4 respectively at 6.5 GHz. The width of the microstrip signal trace is 1.15 mm. Metallization of the trace and ground is with 50 μm copper. Relative permittivity and loss tangent of the polycarbonate housing are 2.8 and 0.01 at 6.5 GHz respectively. DRA models are defined with the CST MWS (CST Microwave Studio, 2011), and the single-pole Debye model is used for all dielectrics to describe their dispersion properties. Other dimensions are fixed as follows: $h_x = h_y = h_z = 1$, $b_x = 7.5$, $s_x = 2$, and $t_y = a_y - a_{y0} - 1$, all in mm.

The initial design is $x^{init} = [8.00\ 14.00\ 9.00\ 0\ 1.75\ 10.00\ 3.00\ 1.500\ 6.00]^T$ mm. Here we use two EM models: the fine model R_f (1,099,490 mesh cells at x^{in}) and the coarse-discretization model R_{cd} (26,796 mesh cells at x^{in}). Both EM models are evaluated with CST MWS transient solver: R_f in 2,175 seconds, and R_{cd} in 42 seconds.

In this example, there is a significant discrepancy between the fine model, R_f, and the coarse-discretization model, R_{cd}. Because of that significant discrepancy, shown in Fig. 7.2, at the optimum of R_{cd}, which is also observed for other designs, we adopt space mapping as the surrogate modeling technique. SM is more flexible in aligning models of different fidelity than many other SBO methods; however, the coarse-discretization model R_{cd} is too expensive to be used in the SM-based optimization process. To alleviate this difficulty, SM is combined with the response

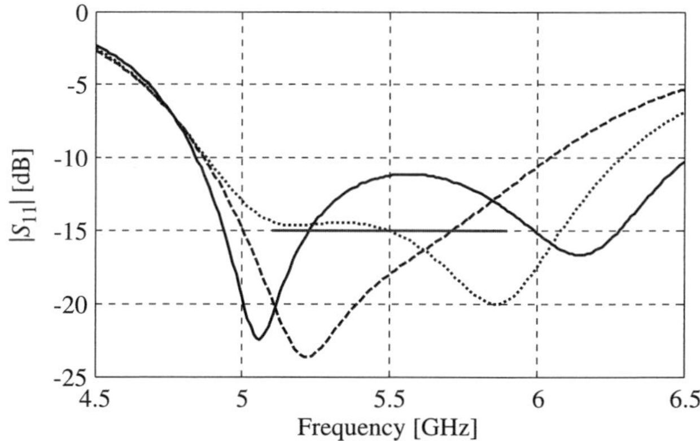

Fig. 7.2. DRA, $|S_{11}|$ versus frequency: Fine model R_f at the initial design (- - -), optimized coarse-discretization model R_{cd} ($\cdots$), and R_f at the optimum of R_{cd} (—) (Koziel and Ogurtsov, 2011c). Design specifications marked using horizontal line.

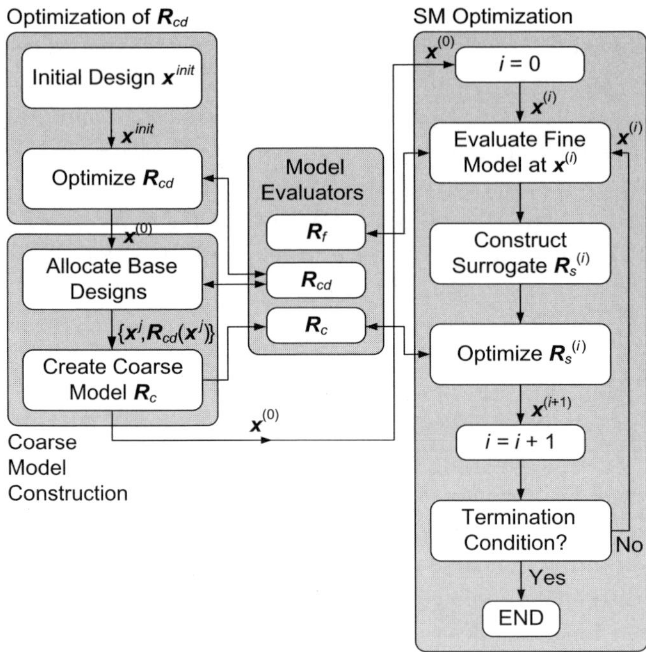

Fig. 7.3. Flowchart of the design optimization procedure exploiting a response-surface-approximation-based coarse model and space mapping (Koziel and Ogurtsov, 2011c).

surface approximation technique, here kriging, which allows us to configure a reliable DRA surrogate R_s without using excessive number of R_{cd} evaluations.

The implemented design optimization procedure, also outlined with a flowchart shown in Fig. 7.3, can be summarized as follows:

1. Take an initial design x^{init}.
2. Find a starting point $x^{(0)}$ for the SM algorithm by optimizing the coarse-discretization model R_{cd}.
3. Allocate N base designs, $X_B = \{x^1, \ldots, x^N\}$.
4. Evaluate R_{cd} at each design x^j, $j = 1, 2, \ldots, N$.
5. Build the coarse model R_c as a kriging interpolation of data pairs $\{(x^j, R_{cd}(x^j))\}_{j=1,\ldots,N}$.
6. Set $i = 0$.
7. Evaluate the fine model R_f at $x^{(i)}$.
8. Construct the surrogate model $R_s^{(i)}$ from R_c with SM correction.

9. Find a new design $x^{(i+1)}$ as the surrogate model optimum: $x^{(i+1)} = \mathrm{argmin}\{x : U(R_s^{(i)}(x))\}$.
10. Set $i = i + 1$.
11. If the termination condition is not satisfied go to 7.
12. END.

The first phase of the design process is to find an optimized design of the coarse-discretization model R_{cd}. The optimum of R_{cd} is usually the best design we can get at a reasonably low computational cost. This cost can be further reduced by relaxing tolerance requirements while searching for $x^{(0)}$: due to a limited accuracy of R_{cd} it is sufficient to find only a rough approximation of its optimum. Steps 3–5 describe the construction of the kriging-based coarse model R_c. Some details concerning the allocation of the base points are given in Beachkofski and Grandhi (2002). Steps 6–12 describe the flow of the SM algorithm with particular operations detailed in Bandler *et al.* (2004); Koziel *et al.* (2006).

It is worth mentioning that the mesh density of model R_{cd} should be adjusted so that its evaluation time is substantially smaller than that of model R_f while providing decent accuracy. Typically, if R_{cd} is set up so that it is 20 to 60 times faster than R_f its accuracy is acceptable for the purpose of the design procedure.

Following the procedure where $|S_{11}|$ is treated as the objective function, whereas the design requirements for the gain at the zero zenith angle are handled through a constraint function, a R_{cd} optimum is found at $x^{(0)} = [7.444\ 13.556\ 9.167\ 0.250\ 1.750\ 10.500\ 2.500\ 1.500\ 6.000]^T$ mm. Figure 7.2 shows the fine model reflection response at the initial design x^{init} as well as that of the fine and coarse-discretization model R_{cd} at $x^{(0)}$. The kriging coarse model R_c is set up using 200 samples of R_{cd} allocated in the vicinity of $x^{(0)}$ of the size $[0.5\ 0.5\ 0.5\ 0.25\ 0.5\ 0.25\ 0.25\ 0.25\ 0.5]^T$ mm.

The final design, $x^{(4)} = [7.556\ 13.278\ 9.630\ 0.472\ 1.287\ 10.593\ 2.667\ 1.722\ 6.482]^T$ mm, is obtained after four SM iterations. The reflection response is in Fig. 7.4. The far-field response of the final design is shown in Fig. 7.5. For the bandwidth of interest, the peak gain is above $5\,\mathrm{dBi}$, and the back radiation level is below $-14\,\mathrm{dB}$ (relative to the maximum). All responses shown include the effect of the $25\,\mathrm{mm}$ input microstrip. The surrogate model used by the optimization algorithm exploited input and output space mapping of the form $R_s(x) = R_c(x + c) + d$. Optimization costs are summarized in Table 7.1. The total design time corresponds to about 11 evaluations of the fine model.

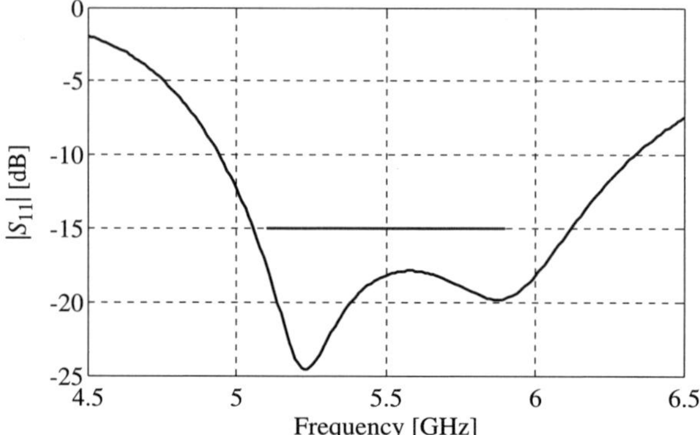

Fig. 7.4. DRA, $|S_{11}|$ versus frequency: $\boldsymbol{R}_f$ at the final design. The design specification is shown with the solid horizontal line (Koziel and Ogurtsov, 2011c).

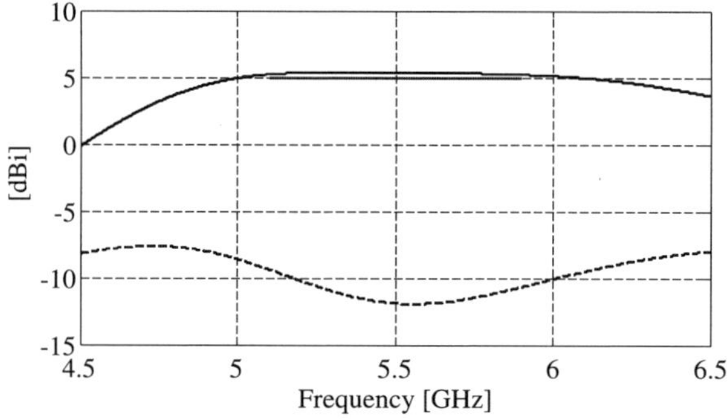

Fig. 7.5. DRA, realized gain versus frequency: (—) is for the zero zenith angle ($\theta = 0^0$); (- - -) is back radiation for $\theta = 180^0$ (Koziel and Ogurtsov, 2011c). Here, only θ-polarization ($\phi = 90^0$) contributes to the gain for the listed directions. The design constrain is shown with the solid horizontal line.

7.2.2. *Broadband Composite Microstrip Antenna*

Consider a composite microstrip antenna (Chen, 2008) shown in Fig. 7.6. Design variables are $\boldsymbol{x} = [l_1 l_2 l_3 l_4 w_2 w_3 d_1 s]^T$. Multilayer substrate is $l_s \times l_s$ ($l_s = 30\,\text{mm}$). The antenna stack comprises: metal ground, RO4003 dielectric, signal trace, RO3006 dielectric with a through via connecting the trace to the driven patch, the driven patch, RO4003 dielectric, and four extra patches. The signal trace is terminated with an open-end stub. Feeding is with a 50 ohm SMA connector. The stack is fixed with four through bolts at the corners.

Table 7.1. DRA: optimization cost (Koziel and Ogurtsov, 2011c).

Algorithm Component	Number of Model Evaluations	CPU Time Absolute	CPU Time Relative to R_f
Optimization of R_{cd}	$150 \times R_{cd}$	105 min	2.9
Setting up R_c	$200 \times R_{cd}$	140 min	3.9
Evaluation of R_f	$4 \times R_f$	145 min	4.0
Total cost	N/A	390 min	**10.8**

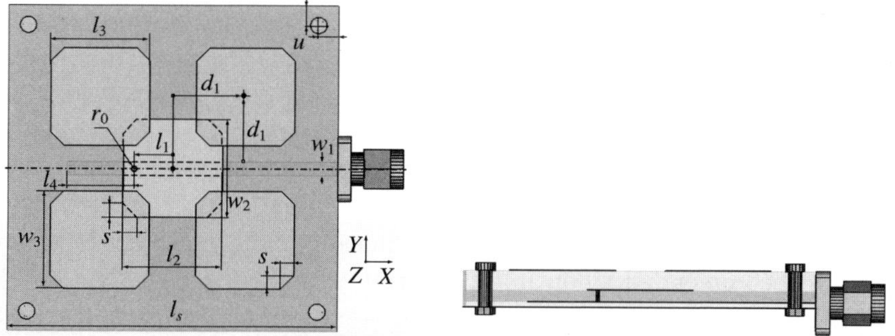

Fig. 7.6. Microstrip broadband antenna (Chen, 2008): top and side views, substrates shown transparent.

The final design is required to have $|S_{11}| \leq -10\,\mathrm{dB}$ for 3.1-to-4.8 GHz. IEEE gain is required to be not less than 5 dB for the zero zenith angle over the whole frequency band of interest.

In this example, the antenna under design is of relatively complex composition so that the choice of the mesh density for the coarse-discretization model as well as other settings of the EM solver, here the CST MWS transient solver, strongly affect the total design optimization time. On the other hand the computational cost of the model and its accuracy can be easily controlled by changing the discretization density. This feature has been exploited in the multi-fidelity optimization algorithm introduced in Koziel and Ogurtsov (2010).

The multi-fidelity optimization is based on a family of coarse-discretization models $\{R_{c.j}\}$, $j = 1, \ldots, K$, all evaluated by the same EM solver. Discretization of the model $R_{c.j+1}$ is finer than that of the model $R_{c.j}$, which results in better accuracy but also longer evaluation time. In practice, the number of coarse-discretization models, K, is two or three.

Having the optimized design $x^{(K)}$ of the finest coarse-discretization model $R_{c.K}$, the model is evaluated at all perturbed designs around $x^{(K)}$, i.e., at $x_k^{(K)} = [x_1^{(K)} \ldots x_k^{(K)} + \mathrm{sign}(k) \cdot d_k \ldots x_n^{(K)}]^T$, $k = -n, -n + 1, \ldots, n - 1, n$. These data can be used to refine the final design without directly optimizing R_f. Instead, an approximation model is set up and optimized in the neighborhood of $x^{(K)}$ defined as $[x^{(K)} - d, \; x^{(K)} + d]$, where $d = [d_1 d_2 \ldots d_n]^T$, and $R^{(k)}$ stands for $R_{c.K}(x_k^{(K)})$. The size of the neighborhood can be selected based on sensitivity analysis of $R_{c.1}$ (the cheapest of the coarse-discretization models); usually d equals 2 to 5 per cent of $x^{(K)}$. A reduced quadratic model $q(x) = [q_1 q_2 \ldots q_m]^T$ is used for approximation where

$$q_j(x) = q_j([x_1 \ldots x_n]^T)$$
$$= \lambda_{j.0} + \lambda_{j.1} x_1 + \cdots + \lambda_{j.n} x_n + \lambda_{j.n+1} x_1^2 + \cdots + \lambda_{j.2n} x_n^2. \qquad (7.1)$$

Coefficients $\lambda_{j.r}$, $j = 1, \ldots, m$, $r = 0, 1, \ldots, 2n$, can be uniquely obtained by solving the linear regression problem.

In order to account for unavoidable misalignment between $R_{c.K}$ and R_f, it is recommended to optimize a corrected model $q(x) + [R_f(x^{(K)}) - R_{c.K}(x^{(K)})]$ that ensures a zero-order consistency (Alexandrov *et al.*, 1998) between $R_{c.K}$ and R_f. The refined design can be then found as

$$x^* = \arg \min_{x^{(K)} - d \leq x \leq x^{(K)} + d} U(q(x) + [R_f(x^{(K)}) - R_{c.K}(x^{(K)})]). \qquad (7.2)$$

This kind of correction is also known as output space mapping (Bandler *et al.*, 2004). If necessary, the step (7.2) can be performed a few times starting from a refined design where each iteration requires only one evaluation of R_f.

The multi-fidelity optimization procedure can be summarized as follows (input arguments are: the initial design $x^{(0)}$ and the number of coarse-discretization models K):

1. Set $j = 1$.
2. Optimize the coarse-discretization model $R_{c.j}$ to obtain a new design $x^{(j)}$ using $x^{(j-1)}$ as a starting point.
3. Set $j = j + 1$; if $j < K$ go to 2.
4. Obtain a refined design x^* as in (7.2).
5. END.

Note that the original model R_f is only evaluated at the final stage (Step 4). Operation of the algorithm in illustrated in Fig. 7.7. Coarse-discretization models can be optimized using any available algorithm.

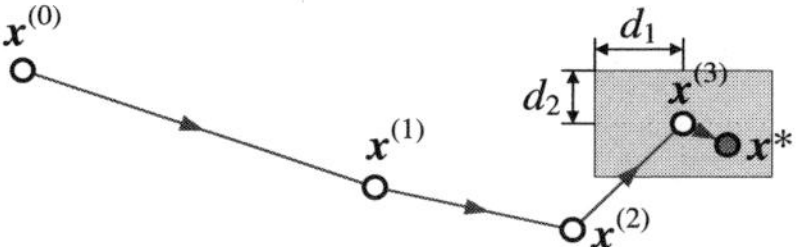

Fig. 7.7. Operation of the multi-fidelity design optimization procedure for three coarse-discretization models ($K = 3$) (Koziel and Ogurtsov, 2011a). The design $x^{(j)}$ is the optimal solution of the model $R_{c.j}$, $j = 1, 2, 3$. A reduced second-order model q is set up in the neighborhood of $x^{(K)}$ (gray area). The final design x^* is obtained by optimizing the model q as in (7.2).

Application of the multi-fidelity optimization methodology to this example can be outlined as follows. The initial design is set to $x^{(0)} = [15$ 15 15 15 20 –4 2 2$]^T$ mm. Two coarse-discretization models are used: $R_{c.1}$ (122,713 mesh cells at $x^{(0)}$) and $R_{c.2}$ (777,888 mesh cells). The evaluation times for $R_{c.1}$, $R_{c.2}$ and R_f (2,334,312 mesh cells) are 3 min, 18 min, and 160 min at $x^{(0)}$, respectively. $|S_{11}|$ is the objective function with the goal of $|S_{11}| \leq -10\,\mathrm{dB}$ for 3.1–4.8 GHz. IEEE gain not less than 5 dB for the zero elevation angle over the band is implemented as an optimization constrain.

Figure 7.8(a) shows the responses of $R_{c.1}$ at $x^{(0)}$ and at its optimal design $x^{(1)}$. Figure 7.8(b) shows the responses of $R_{c.2}$ at $x^{(1)}$ and at its optimized design $x^{(2)}$. Figure 7.8(c) shows the responses of R_f at $x^{(0)}$, at $x^{(2)}$ and at the refined design $x^* = [14.87\ 13.95\ 15.4\ 13.13\ 20.87\ –5.90$ 2.88 0.68$]^T$ mm ($|S_{11}| \leq -11.5\,\mathrm{dB}$ for 3.1 GHz to 4.8 GHz) obtained in two iterations of the refinement step (7.2).

The design cost, shown in Table 7.2, corresponds to about 12 runs of the high-fidelity model R_f. Antenna gain at the final design is shown in Fig. 7.9.

7.2.3. *Broadband Composite DRA*

Consider a rotationally symmetric DRA (Shum and Luk, 1995) shown in Fig. 7.10. It comprises: two annular ring dielectric resonators (DR) with relative permittivity, ε_{r1}, of 36; two supporting Teflon rings; a probe; and cylindrical Teflon filling. The inner radius of the filling is the radius of the probe, 1.27 mm. The probe is an extension (h_0 above the ground) of the inner conductor of the input 50 ohms coax. The radius of each supporting ring equals that of the DR above them. All metal parts are modeled as perfect electric conductors (PEC). The coax is also filled by Teflon. The ground is of infinite extends.

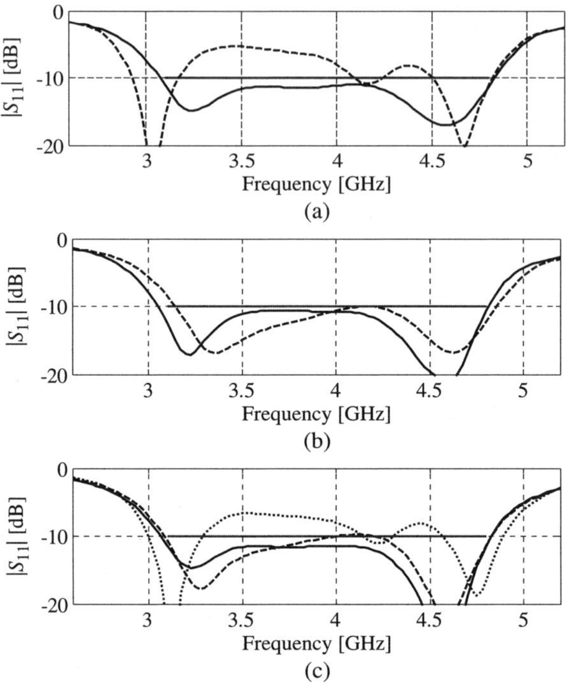

Fig. 7.8. Microstrip antenna: (a) model $R_{c.1}$ at the initial design $x^{(0)}$ (- - -) and at the optimized design $x^{(1)}$ (—), (b) model $R_{c.2}$ at $x^{(1)}$ (- - -) and at its optimized design $x^{(2)}$ (—), (c) model R_f at $x^{(0)}$ (· · · ·), at $x^{(2)}$ (- - -) and at the refined final design x^* (—) (Koziel and Ogurtsov, 2011d). Design specifications marked using horizontal lines.

Table 7.2. Microstrip antenna: optimization cost (Koziel and Ogurtsov, 2011d).

Design Step	Model Evaluations*	Computational Cost	
		Absolute [hours]	Relative to R_f
Optimization of $R_{c.1}$	$125 \times R_{c.1}$	6.3	2.6
Optimization of $R_{c.2}$	$48 \times R_{c.2}$	14.4	5.4
Setup of model q	$17 \times R_{c.2}$	5.1	1.9
Evaluation of R_f	$2 \times R_f$	5.3	2.0
Total design time	N/A	31.1	**11.9**

*Excludes R_f evaluation at the initial design.

Design variables are the inner and outer radii of the DRs, the heights of the DRs and the supporting rings, and the probe length, namely, $x = [a_1 a_2 b_1 b_2 h_1 h_2 g_1 g_2 h_0]^T$. The design objective is $|S_{11}| \leq -20\,\text{dB}$ for $4\,\text{GHz}$ to $6\,\text{GHz}$. Broadside gain not less than $5\,\text{dBi}$ is an optimization constraint.

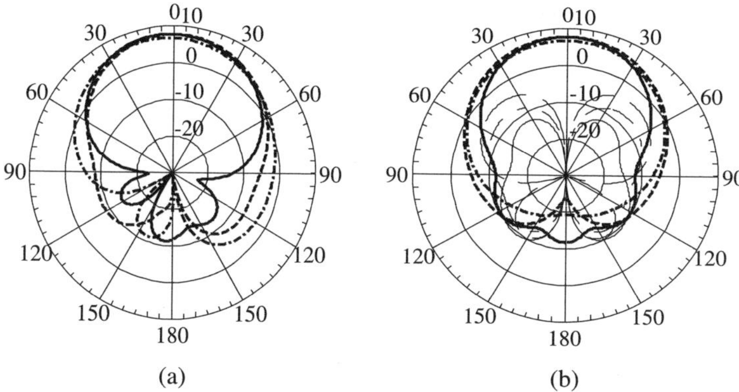

Fig. 7.9.   Microstrip antenna gain [dBi] of the final design at 3.5 GHz ($\bullet\!-\!\bullet$), 4.0 GHz ($--$), and 4.5 GHz ($-$): (a) co-pol. in the E-plane (XOZ), and connector is at 90° on the right, (b) x-pol., primary (thick lines), and co-pol. (thin lines) in the H-plane (Koziel and Ogurtsov, 2011e).

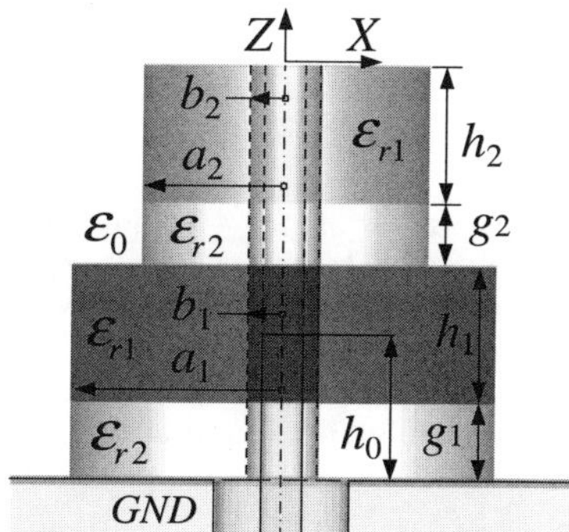

Fig. 7.10.   Annular ring dielectric resonator antenna (Shum and Luk, 1995): side view.

Here, the overall shape of the low- and high-fidelity model responses is quite similar; therefore we use the adaptively adjusted design specifications (AADS) technique (Koziel, 2010) that allows us to account for the misalignment between the models without actually adjusting the low-fidelity one. AADS consists of the following two steps that can be iterated if necessary:

1. Modify the original design specifications to account for the difference between the responses of the high-fidelity model $\boldsymbol{R}_f$ and the coarse-discretization model $\boldsymbol{R}_{cd}$ at their characteristic points.

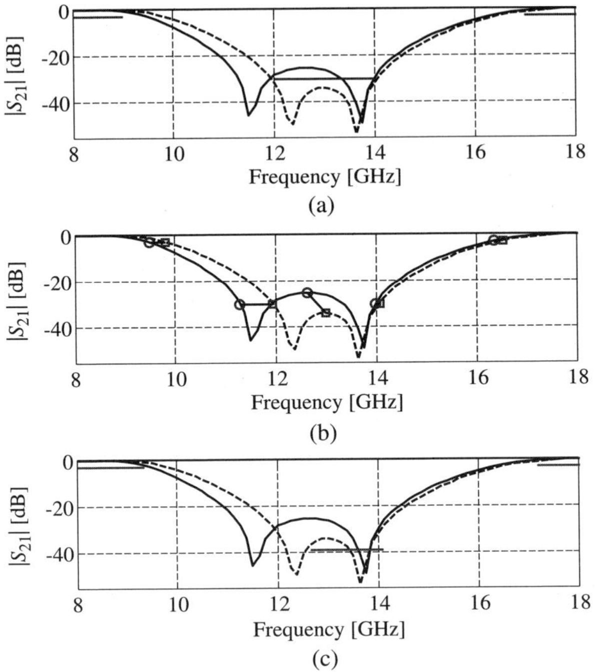

Fig. 7.11. AADS concept (responses of $\boldsymbol{R}_f$ (—) and $\boldsymbol{R}_{cd}$ (- - -)) (Koziel, 2010): (a) responses at the initial design and the original design specifications, (b) characteristic points of the responses corresponding to the specification levels (here, $-3\,\mathrm{dB}$ and $-30\,\mathrm{dB}$) and to the local maxima, (c) responses at the initial design as well as the modified design specifications. The modification accounts for the discrepancy between the models so that optimizing $\boldsymbol{R}_{cd}$ with respect to the modified specifications approximately corresponds to optimizing $\boldsymbol{R}_f$ with respect to the original specifications. Design specifications marked using horizontal lines.

2. Obtain a new design by optimizing the low-fidelity model $\boldsymbol{R}_{cd}$ with respect to the modified specifications.

As $\boldsymbol{R}_{cd}$ is much faster than $\boldsymbol{R}_f$, the design process can be performed at a low cost compared to direct optimization of $\boldsymbol{R}_f$. Figure 7.11 explains the idea of AADS using a bandstop filter example (Koziel, 2010). The design specifications are adjusted using so-called characteristic points which should correspond to the design specification levels. They should also include local maxima/minima of the responses at which the specifications may not be satisfied.

It should be emphasized again that for the AADS technique there is no surrogate model configured from $\boldsymbol{R}_{cd}$ — discrepancy between $\boldsymbol{R}_{cd}$ and $\boldsymbol{R}_f$ is "absorbed" by the modified design specifications.

Figure 7.11(b) shows characteristic points of R_f and R_{cd}. The design specifications are modified (mapped) so that the level of satisfying/violating the modified specifications by the R_{cd} response corresponds to the satisfaction/violation levels of the original specifications by the R_f response (Fig. 7.11(b) and 7.11(c)). R_{cd} is subsequently optimized with respect to the modified specifications and the new design obtained this way is treated as an approximated solution to the original design problem. Typically, a substantial design improvement is observed after first iteration. Additional iterations can bring further improvement.

Here the initial design is $x^{init} = [6.9\ 6.9\ 1.05\ 1.05\ 6.2\ 6.2\ 2.0\ 2.0\ 6.80]^T$. The high- and low-fidelity models are evaluated using CST Microwave Studio (R_f: 829,000 meshes at x^{init}, evaluation time 58 min, R_{cd}: 53,000 meshes at x^{init}, evaluation time 2 min).

The optimized design is found to be $x^* = [5.9\ 5.9\ 1.05\ 1.55\ 7.075\ 7.2\ 4.5\ 1.0\ 8.05]^T$. It is obtained with three iterations of the AADS procedure. Significant improvement of the DRA's bandwidth is observed, namely: the 48% fractional bandwidth at –20 dB is shown in Fig. 7.12. The far-field response of the optimized DRA, shown in Fig. 7.13 at selected frequencies,

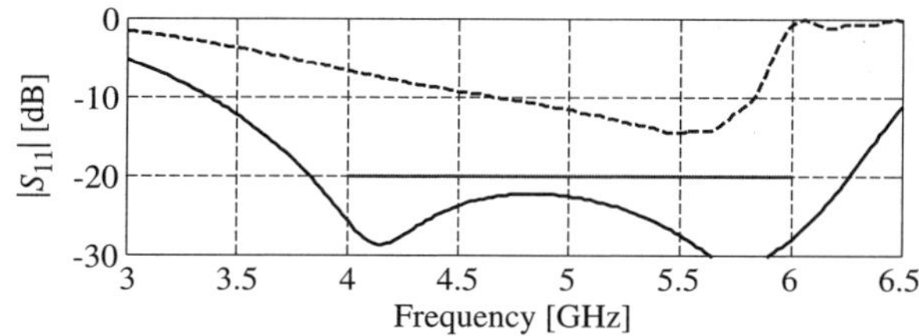

Fig. 7.12.   DRA: fine model response at the initial (- - -) and the optimized design (—).

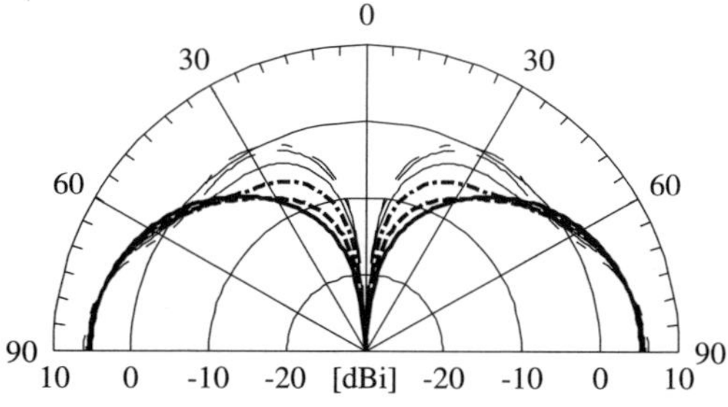

Fig. 7.13.   DRA at the optimal design: gain in the elevation plane at 3.5 GHz (thick −), 4 GHz (thick − −), 4.5 GHz (thick ·−·), 5 GHz (thin −), 5.5 GHz (thin − −), and 6 GHz (thin ·−·).

Table 7.3. DRA: design optimization costs.

Design Step	Model Evaluations*	Computational Cost	
		Absolute [Hours]	Relative to R_f
Optimization of R_{cd}	$235 \times R_{cd}$	7.8	8.1
Evaluation of R_f	$3 \times R_f$	2.9	3.0
Total design time	N/A	10.7	**11.1**

*Excludes R_f evaluation at the initial design.

is of $TM_{01\delta}$ DRA mode behavior over the 60% bandwidth (on $-10\,$dB level). The total design cost is equivalent to about 11 evaluations of the high-fidelity DRA model. The design cost budget is listed in Table 7.3.

7.2.4. *UWB Monopole Antenna*

Consider a monopole antenna in a housing shown in Fig. 7.14. The design variables are $x = [l_1 l_2 l_3 w_1]^T$. Other antenna dimensions are fixed. The monopole is fed through an edge mount SMA connector. The high-fidelity and low-fidelity models of the monopole, R_f and R_c, also include the feeding coaxial cable. Simulation time of R_c (156,000 mesh cells) is 1 min, and that of R_f (1,992,060 mesh cells) is 40 min (both at the initial design). The models are evaluated using the transient solver of CST Microwave Studio. The design specifications for reflection are $|S_{11}| \leq -12\,$dB for 3.1 GHz to 10.6 GHz. The initial design is $x^{init} = [20\ 2\ 25\ 0]^T$ mm.

For this design project we combine space mapping (SM) with adjoint sensitivity. This is done in a comprehensive way so that SM is utilized for: (i) speeding up the surrogate model optimization, (ii) speeding up the

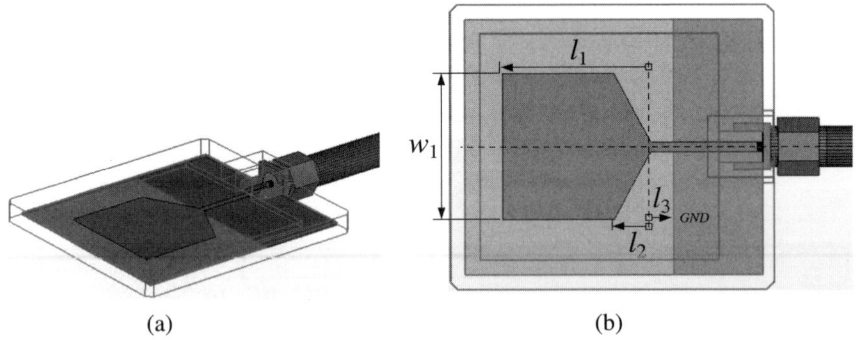

Fig. 7.14. UWB monopole: (a) 3D view, (b) top view. The housing is shown transparent.

parameter extraction process, and (iii) improving the matching between the surrogate and the high-fidelity model. Due to (i) and (ii), both parameter extraction and surrogate model optimization can be performed using a limited number of low-fidelity model evaluations, which allows us to utilize coarse-discretization EM models. This increases the range of SM applications and improves the SM performance because low-fidelity SM models are normally quite accurate. Adjoint sensitivity is also used to match both the response and first-order derivatives of the surrogate and the high-fidelity model, and it improves the performance and convergence of the SM algorithm. The algorithm requires a remarkably small number of both coarse and fine model evaluations to complete the optimization task. Details of the applied procedures are explained below.

Design Problem and the Space Mapping Algorithm. We aim at solving the following problem:

$$x_f^* = \arg\min_x U(\boldsymbol{R}_f(\boldsymbol{x})), \tag{7.3}$$

where $\boldsymbol{R}_f(\boldsymbol{x}) \in R^m$ is the response vector of the high-fidelity model; U is an objective function, e.g., minimax (Bandler *et al.*, 2004); $\boldsymbol{x} \in R^n$ is a vector of design variables. Space mapping replaces solving of (7.3) by the iterative process

$$\boldsymbol{x}^{(i+1)} = \arg\min_x U(\boldsymbol{R}_s^{(i)}(\boldsymbol{x})), \tag{7.4}$$

where $\boldsymbol{x}^{(i)}$, $i = 0, 1, \ldots$, is the sequence of approximate solutions to (7.3), whereas $\boldsymbol{R}_s^{(i)}$ is the surrogate model at iteration i.

Low-Fidelity and Surrogate Models. The SM surrogate is constructed using the underlying low-fidelity (or coarse) model $\boldsymbol{R}_c$. The most versatile type of the coarse model is the one obtained through simulation at coarse-meshes. Coarse-mesh EM models are relatively expensive; therefore, the SM surrogate exploited here is based on the input and output SM of the form (Koziel *et al.*, 2006):

$$\boldsymbol{R}_s^{(i)}(\boldsymbol{x}) = \boldsymbol{R}_c(\boldsymbol{x} + \boldsymbol{c}^{(i)}) + \boldsymbol{d}^{(i)} + \boldsymbol{E}^{(i)}(\boldsymbol{x} - \boldsymbol{x}^{(i)}). \tag{7.5}$$

Here, only the input SM vector $\boldsymbol{c}^{(i)}$ is obtained through a nonlinear parameter extraction process:

$$\boldsymbol{c}^{(i)} = \arg\min_c \|\boldsymbol{R}_f(\boldsymbol{x}^{(i)}) - \boldsymbol{R}_c(\boldsymbol{x}^{(i)} + \boldsymbol{c})\|. \tag{7.6}$$

Output SM parameters are calculated as

$$d^{(i)} = R_f(x^{(i)}) - R_c(x^{(i)} + c^{(i)}) \tag{7.7}$$

and

$$E^{(i)} = J_{R_f}(x^{(i)}) - J_{R_c}(x^{(i)} + c^{(i)}), \tag{7.8}$$

where J denotes the Jacobian of the respective model obtained using adjoint sensitivity. Formulation (7.4)–(7.8) ensures zero- and first-order consistency between the surrogate and the fine model, i.e., $R_s^{(i)}(x^{(i)}) = R_f(x^{(i)})$ and $J_{Rs^{(i)}}(x^{(i)}) = J_{R_f}(x^{(i)})$, and substantially improves the ability of the SM algorithm to quickly locate the high-fidelity model optimum (Koziel *et al.*, 2010). Here, the algorithm (7.4) is embedded in the trust-region (TR) framework (Conn *et al.*, 2000), i.e., we have $x^{(i+1)} = \mathrm{argmin}\{\|x - x^{(i)}\| \leq \delta^{(i)} : U(R_s^{(i)}(x))\}$, where the TR radius $\delta^{(i)}$ is updated using classical rules (Conn *et al.*, 2000; Koziel *et al.*, 2010). This, assuming the first-order consistency and the smoothness of $R_s^{(i)}(x)$, ensures convergence to the local R_f optimum.

Fast Parameter Extraction Using Adjoint Sensitivity. To speed up the parameter extraction process (7.6) we exploit adjoint sensitivity. We use a simple trust-region (TR)-based algorithm, where the approximate solution $c^{(i.k+1)}$ of $c^{(i)}$ is found as

$$c^{(i.k+1)} = \arg \min_{\|c-c^{(i.k)}\| \leq \delta_{PE}^{(k)}} \|R_f(x^{(i)}) - L_{c.c}^{(i.k)}(c)\|, \tag{7.9}$$

where $L_{c.c}^{(i.k)}(c) = R_c(x^{(i)} + c^{(i.k)}) + J_{Rc}(x^{(i)} + c^{(i.k)}) \cdot (c - c^{(i.k)})$ is a linear approximation of $R_c(x^{(i)} + c)$ at $c^{(i.k)}$. The TR radius $\delta_{PE}^{(k)}$ is updated according to the standard rules (Conn *et al.*, 2000). Parameter extraction is terminated upon convergence or exceeding the maximum number of coarse model evaluations. The limit is set to five because parameter extraction accuracy is not critical.

Low-Cost Surrogate Optimization Using Adjoint Sensitivity. Adjoint sensitivity is also utilized to lower the cost of surrogate model optimization. We use a TR-based algorithm that produces a sequence of approximations $x^{(i+1.k)}$ of the solution $x^{(i+1)}$ as follows:

$$x^{(i+1.k+1)} = \arg \min_{\|x-x^{(i+1.k)}\| \leq \delta_{SO}^{(k)}} U(L_{c.x}^{(i.k)}(x)), \tag{7.10}$$

where $L_{c.x}^{(i.k)}(x) = R_s^{(i)}(x^{(i+1.k)} + c^{(i)}) + J_{\mathbf{R}s^{(i)}}(x^{(i+1.k)} + c^{(i)}) \cdot (x - x^{(i+1.k)})$ is a linear approximation of $R_s^{(i)}(x + c^{(i)})$ at $x^{(i+1.k)}$. The TR radius $\delta_{SO}^{(k)}$ is updated according to the standard rules (Conn *et al.*, 2000). Due to the use of adjoint sensitivity, surrogate model optimization requires only a few evaluations of the coarse model R_c. Note that sensitivity of the surrogate model can be calculated using the sensitivity of both coarse and fine model as follows:

$$J_{\mathbf{R}s^{(i)}}(x + c^{(i)}) = J_{\mathbf{R}c}(x + c^{(i)}) + [J_{\mathbf{R}f}(x^{(i)}) - J_{\mathbf{R}c}(x^{(i)} + c^{(i)})].$$

The antenna was optimized using the above described algorithm. Figure 7.15(a) shows the responses of R_f and R_c at x^{init}. Figure 7.15(b) shows the response of the fine model at the final design $x^{(3)} = [20.29\ 19.63\ 2.27\ 0.058]^T$ obtained only after three SM iterations. Table 7.4 summarizes the computational cost. Note that using adjoint sensitivity allows us to greatly reduce the number of both fine and coarse model evaluations in

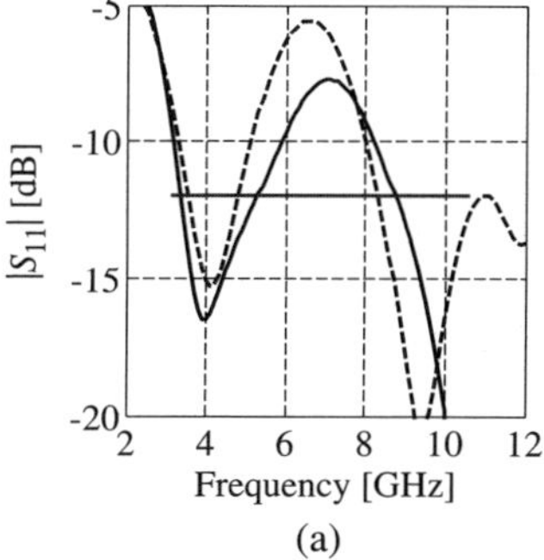
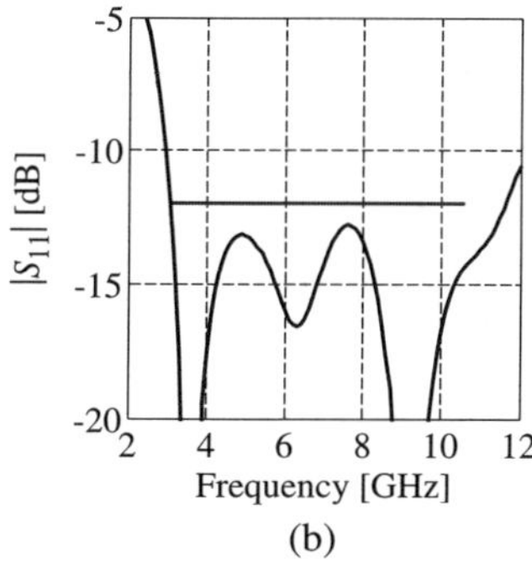

Fig. 7.15. UWB monopole: (a) responses of R_f (—) and R_c (- - -) at the initial design x^{init}, (b) response of R_f (—) at the final design. Design specifications marked using horizontal lines.

Table 7.4. UWB monopole: optimization costs.

Algorithm Component	Number of Model Evaluations*	CPU Time Absolute	CPU Time Relative to R_f
Evaluation of R_c	45	45 min	1.1
Evaluation of R_f	4	160 min	4.0
Total cost*	N/A	205 min	**5.1**

*Excludes R_f evaluation at the initial design.

the design process. The total cost of parameter extraction and surrogate optimization processes is only about five evaluations of $\boldsymbol{R}_c$.

7.2.5. *Design of a Microstrip Antenna Using Coarse-Discretization Models of Different Fidelity*

Consider a coax-fed microstrip antenna shown in Fig. 7.16. Design variables are $\boldsymbol{x} = [a\ b\ c\ d\ e\ l_0\ a_0\ b_0]^T$. The antenna is on 3.81 mm thick Rogers TMM4 ($\varepsilon_1 = 4.5$ at 10 GHz). The TMM4 lateral dimensions are $l_x = l_y = 6.75$ mm. The ground plane is of infinite extends. The feed probe diameter is 0.8 mm. The connector's inner conductor is 1.27 mm in diameter. Design specifications are $|S_{11}| \leq -10$ dB for 5 GHz to 6 GHz. The high-fidelity model $\boldsymbol{R}_f$ is evaluated with CST MWS transient solver (704,165 mesh cells, evaluation time 60 min).

Here we consider three coarse models: $\boldsymbol{R}_{c1}$ (41,496, 1 min), $\boldsymbol{R}_{c2}$ (96,096, 3 min), and $\boldsymbol{R}_{c3}$ (180,480, 6 min). The initial design is $\boldsymbol{x}^{(0)} = [6\ 12\ 15\ 1\ 1\ 1\ 1\ -4]^T$ mm. We investigate the performance of the SBO algorithm working with these models in terms of the computational cost and the quality of the final design.

Figure 7.17(a) shows responses of all the models at the approximate optimum of $\boldsymbol{R}_{c1}$. The major misalignment between the responses is due to the frequency shift so that the surrogate is created here using frequency scaling as well as output SM (Bandler *et al.*, 2004). The results, Table 7.5 and Fig. 7.17(b), indicate that the model $\boldsymbol{R}_{c1}$ is too inaccurate and the SBO design process fails to find a satisfactory design. The designs found

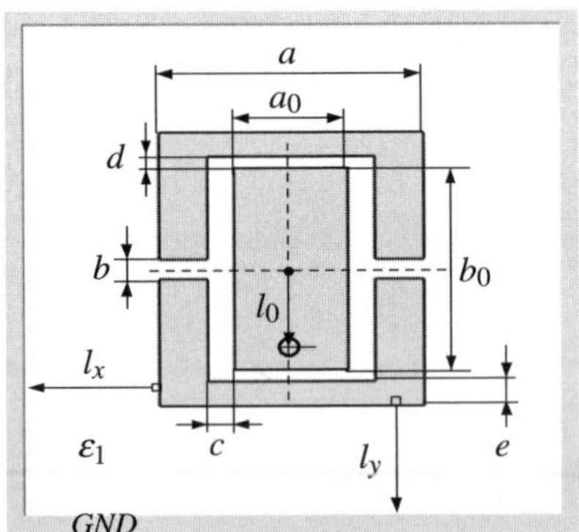

Fig. 7.16.   Microstrip antenna: top view.

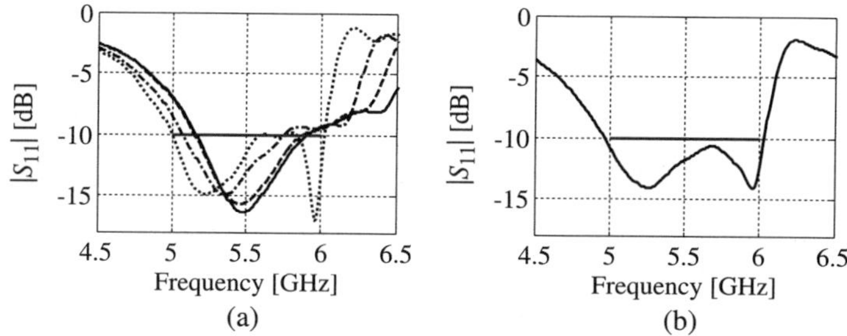

Fig. 7.17. Microstrip antenna: (a) model responses at the approximate optimum of R_{c1}: R_{c1} ($\cdots$), R_{c2} ($-\cdot-$), R_{c3} (- - -), and R_f (—); (b) the high-fidelity model R_f response at the final design found using the low-fidelity model R_{c3}. Design specifications marked using horizontal lines.

Table 7.5. Microstrip antenna: design results and costs.

| Low-Fidelity Model | Design Cost: The Number of Model Evaluations* | | Relative Design Cost† | $\max \|S_{11}\|$ for 2-to-8 GHz at Final Design |
	R_c	R_f		
R_{c1}	385	6	12.4	$-8.0\,$dB
R_{c2}	185	3	12.3	$-10.0\,$dB
R_{c3}	121	2	14.1	$-10.7\,$dB

*Number of R_f evaluations is equal to the number of the SBO algorithm iterations.
†Equivalent number of R_f evaluations includes evaluation at the initial design.

with models R_{c2} and R_{c3} satisfy the specifications and the cost of the SBO process with R_{c2} is slightly lower than that with R_{c3}.

7.2.6. *Design of a Hybrid Antenna with the Coarse-Discretization Model Management*

Consider a hybrid DRA shown in Fig. 7.18. The DRA is fed by a 50 ohm microstrip terminated with an open-ended section. Microstrip substrate is 0.787 mm thick Rogers RT5880. The design variables are $x = [h_0\ r_1\ h_1\ ul_1\ r_2]^T$. Other dimensions are fixed: $r_0 = 0.635$, $h_2 = 2$, $d = 1$, $r_3 = 6$, all in mm. Permittivity of the DRA core is 36, and the loss tangent is 10^{-4}, both at 10 GHz. The DRA support material is Teflon ($\varepsilon_2 = 2.1$). The radome is of polycarbonate ($\varepsilon_3 = 2.7$ and $\tan\delta = 0.01$). The radius of the ground plane opening, shown in Fig. 7.18(b), is 2 mm. The high-fidelity antenna model $R_f(x)$ is evaluated using the time-domain solver of CST Microwave Studio ($\sim$1,400,000 meshes, evaluation time 60 min).

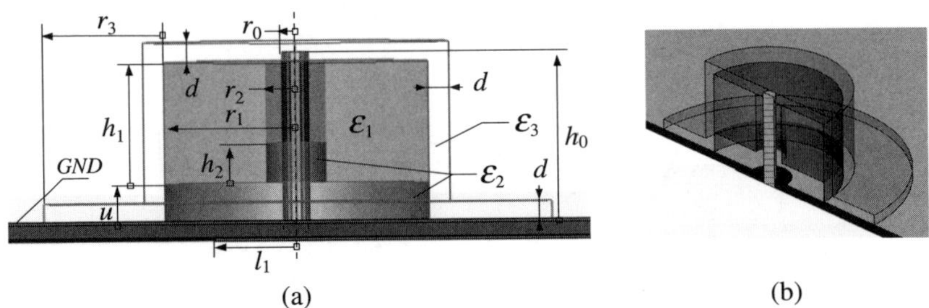

Fig. 7.18.   Hybrid DRA: (a) side view, (b) 3D-cut view.

The design goal is to adjust geometry parameters so that the following specifications are met: $|S_{11}| \leq -12\,$dB for 5.15 GHz to 5.8 GHz. This design problem can be formulated in the form of $\boldsymbol{x}^* = \mathrm{argmin}\{\boldsymbol{x} : U(\boldsymbol{R}_f(\boldsymbol{x}))\}$, where U is an objective function that implements the above specifications. The initial design is $\boldsymbol{x}^{(0)} = [7.0\ 7.0\ 5.0\ 2.0\ 2.0\ 2.0]^T$ mm.

Here, we consider two models of different fidelity, $\boldsymbol{R}_{c1}$ ($\sim$45,000 meshes, evaluation time 1 min), and $\boldsymbol{R}_{c2}$ ($\sim$300,000 meshes, evaluation time 3 min). We investigate the algorithm using either one of these models or both ($\boldsymbol{R}_{c1}$ at the initial state and $\boldsymbol{R}_{c2}$ in the later stages).

Neither of the low-fidelity models is sufficiently accurate to simply replace $\boldsymbol{R}_f$ in the design process. Therefore, we create the surrogate model $\boldsymbol{R}_s^{(i)}$ by aligning the low- and high-fidelity models using frequency scaling and output space mapping (OSM) (Bandler *et al.*, 2004). This particular choice comes from the fact that the major misalignment between the models is due to a frequency shift as illustrated in Fig. 7.19.

This misalignment can be reduced by applying a suitable frequency scaling of the low-fidelity model response. Let $\boldsymbol{R}_c(\boldsymbol{x}) = \boldsymbol{R}_c(\boldsymbol{x}; \Omega)$, where Ω

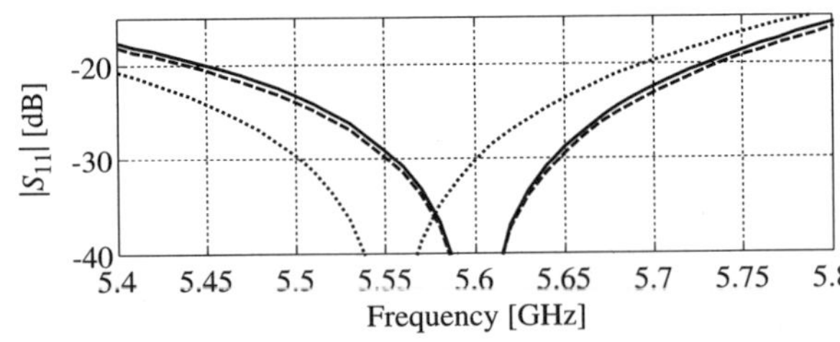

Fig. 7.19.   Hybrid DRA: high- (——) and low-fidelity model response at certain design before ($\cdots$) and after (- - -) applying the frequency scaling.

is a frequency band of interest (here, R_c represents either R_{c1} or R_{c2}). The frequency-scaled model is defined as $R_c(x; F^{(i)}(\Omega))$ where $F^{(i)}$ is a scaling function determined to minimize misalignment between $R_f(x^{(i)})$ and $R_c(x^{(i)}; F^{(i)}(\Omega))$. Here, we use a polynomial scaling $F^{(i)}(\omega) = f_0^{(i)} + f_1^{(i)}\omega$, and $[f_0^{(i)}, f_{1(i)}] = \mathrm{argmin}\{[f_0, f_1] : \|R_f(x^{(i)}) - R_c(x; f_0 + f_1 \cdot \Omega)\|\}$. Figure 7.19 shows the scaled low-fidelity model. The surrogate model is then constructed by reducing the remaining misalignment at $x^{(i)}$ through output SM as follows:

$$R_s^{(i)}(x) = R_c(x; F^{(i)}(\Omega)) + [R_f(x^{(i)}) - R_c(x^{(i)}; F^{(i)}(\Omega))]. \qquad (7.11)$$

The DRA has been optimized using the algorithm and the surrogate model (7.11). We have considered three cases:

1. The surrogate constructed using R_{c1} — cheaper but less accurate.
2. The surrogate constructed using R_{c2} — more expensive but more accurate.
3. The surrogate constructed with R_{c1} at the first iteration and with R_{c2} for subsequent iterations.

The last option allows us to faster locate the approximate optimum of the high-fidelity model and then refine it using the more accurate model.

The number of surrogate model evaluations was limited to 100 in the first iteration that involves the largest design change, and to 50 in the subsequent iterations, where design modifications are smaller.

Table 7.6 shows results for all three cases. Figure 7.20 shows the high-fidelity model response at the final design obtained using the SBO algorithm working with the low-fidelity model R_{c2}. The quality of the final designs is the same in all cases. However, the SBO algorithm using the low-fidelity

Table 7.6.   Design results.

Case	Number of Iterations	Number of Model Evaluations*			Total Design Cost†	Max$|S_{11}|$ for 5.15 GHz to 5.8 GHz at Final Design
		R_{c1}	R_{c2}	R_f		
1	4	250	0	4	8.2	−12.6 dB
2	2	100	50	2	6.2	−12.6 dB
3	2	0	150	2	9.5	−12.6 dB

*Number of R_f evaluations is equal to the number of the SBO iterations.
†Equivalent number of R_f evaluations.

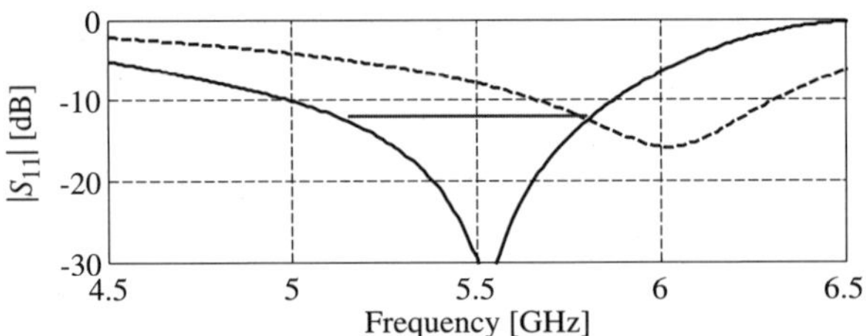

Fig. 7.20. Hybrid DRA: high-fidelity model response at the initial design (- - -) and at the final design obtained using the SBO algorithm using with the low-fidelity model R_{c2} (—). Design specifications marked using a horizontal line.

model R_{c1}, Case 1, requires more iterations than the algorithm using the model R_{c2}, Case 3. This is because the model R_{c1} is more accurate than R_{c2}. In this particular case, the overall computational cost of the design process is still lower for R_{c1} than for R_{c2}. On the other hand, the cheapest approach is Case 2 when the model R_{c1} is utilized in the first iteration that requires the largest number of EM analyses, whereas the algorithm switches to R_{c2} in the second iteration, which allows us to both reduce the number of iterations and number of evaluations of R_{c2} at the same time. The total design cost is the lowest overall. This demonstrates that by proper management of the models involved in the design process one can lower the overall optimization cost without compromising the final design quality.

7.3. SBO for Antenna Design: Discussion and Recommendations

The SBO techniques presented in Section 7.2 have proven to be computationally efficient for design of composite broadband antennas. Typical computational cost of the design process expressed in terms of the number of equivalent high-fidelity model evaluations is comparable to the number of design variables, as demonstrated through examples. In this section, we attempt to qualitatively compare these methods and give some recommendations for the readers interested in using them in their research and design work.

Space mapping with kriging, applied in Section 7.2.1, is a quite generic method. In particular, it is able to work even if the low-fidelity model is rather inaccurate. On the other hand, SM requires some experience in selecting the proper type of the surrogate model, as well as in selecting the region for setting up the kriging coarse model. More implementation details and application examples of this technique to antenna design can be found in Koziel and Ogurtsov (2011c).

Multi-fidelity optimization applied in Section 7.2.2 is one of the most robust techniques, which is yet simple to implement. The only drawback is that it requires at least two low-fidelity models of different discretization density and some initial study of the model accuracy versus computational complexity. While the multi-fidelity technique will work with practically any setup, careful selection of mesh density can reduce the computational cost of the optimization process considerably. See Koziel and Ogurtsov (2011d) and Koziel *et al.* (2012) for more implementation details and application examples of this technique to antenna design.

The simplest method for implementation is definitely AADS, applied in Section 7.2.3, as it does not require any correction of the low-fidelity model. Therefore, AADS can even be executed within any EM solver by modifying the design requirements and using built-in optimization capabilities. On the other hand, AADS only works with minimax-like design specifications. Also, AADS requires that the low-fidelity model is relatively accurate so that the possible discrepancies between the low- and high-fidelity models can be accounted for by design specifications adjustment. More implementation details and application examples of this technique to antenna design can be found in Ogurtsov and Koziel (2011a) and Ogurtsov and Koziel (2011b).

Adjoint sensitivity, as demonstrated in the literature, e.g., Chung *et al.* (2001), can significantly reduce the computational cost of the EM-driven design process while using conventional gradient-based algorithms. At the same time, as shown in Section 7.2.4, it can also be exploited to speed up and improve the robustness of the surrogate-based optimization algorithm, in this case, space mapping. Therefore, it is recommended to use this technology when it is available to enhance the SBO design schemes.

As mentioned before, the low-fidelity model accuracy may be critical for the performance of the SBO algorithms. Using finer, i.e., more expensive but, at the same time, more accurate models, generally reduces the number of SBO iterations necessary to find a satisfactory design; however, each iteration is typically more time consuming. For coarser models, the cost of each iteration is lower but the number of iterations may be larger and, for models that are too coarse, the surrogate-based optimization process may simply fail. The proper selection of the low-fidelity model "coarseness" may not be obvious beforehand as indicated in Section 7.2.5. In most cases, it is recommended to use finer models rather than coarser ones for the sake of ensuring good algorithm performance, even at the cost of some extra computational overhead.

The problem discussed in the previous paragraph can be considered in the wider context of model management. As demonstrated in Section 7.2.6, it may be beneficial to change the low-fidelity model coarseness during the SBO algorithm run. Typically, one starts from the coarser model in order to find an approximate location of the optimum design and switches to the finer model to increase the accuracy of the local search process without compromising the computational efficiency. Proper management of the model fidelity may result in further reduction of the design cost.

7.4. Conclusion

Surrogate-based techniques for simulation-driven design of broadband antennas have been discussed. It was demonstrated that optimized designs can be found at a low computational cost corresponding to a few high-fidelity EM simulations of the antenna structure. Further progress of the considered SBO techniques can be expected with their full automation, combination, and hybridization with adjoint sensitivities as well as with metaheuristic algorithms.

References

Alexandrov, N.M., Dennis, J.E., Lewis, R.M. and Torczon, V. (1998). A trust region framework for managing use of approximation models in optimization, *Struct. Multidiscip. O.*, **15**, 16–23.

Ares-Pena, F.J., Rodriguez-Gonzalez, A., Villanueva-Lopez, E. and Rengarajan, S.R. (1999). Genetic algorithms in the design and optimization of antenna array patterns, *IEEE T. Antenn. Propag.*, **47**, 506–510.

Back, T., Fogel, D.B. and Michalewicz, Z. (eds.) (2000). *Evolutionary Computation 1: Basic Algorithms and Operators*, Taylor & Francis Group, London.

Bandler, J.W. and Seviora, R.E. (1972). Wave sensitivities of networks, *IEEE T. Microw. Theory*, **20**, 138–147.

Bandler, J.W., Cheng, Q.S., Dakroury, S.A., Mohamed, A.S., Bakr, M.H., Madsen, K. and Søndergaard, J. (2004). Space mapping: the state of the art, *IEEE T. Microw. Theory*, **52**, 337–361.

Beachkofski B. and Grandhi, R. (2002). Improved distributed hypercube sampling. American Institute of Aeronautics and Astronautics, paper AIAA 2002-1274.

Bevelacqua, P.J. and Balanis, C.A. (2007). Optimizing antenna array geometry for interference suppression, *IEEE T. Antenn. Propag.*, **55**, 637–641.

Chen, Z.N. (2008). Wideband microstrip antennas with sandwich substrate, *IET Microw. Ant. P.*, **2**, 538–546.

Cheng, Q.S., Rautio, J.C., Bandler, J.W. and Koziel, S. (2010). Progress in simulator-based tuning — the art of tuning space mapping, *IEEE Microw. Mag.*, **11**, 96–110.

Chung, Y.S., Cheon, C., Park, I.H. and Hahn, S.Y. (2001). Optimal design method for microwave device using time domain method and design sensitivity analysis-part II: FDTD case, *IEEE T. Magn.*, **37**, 3255–3259.

Conn, A.R., Gould, N.I.M. and Toint, P.L. (2000). *Trust Region Methods*, Cambridge University Press, Cambridge.

CST Microwave Studio (2011), CST AG, Bad Nauheimer Str. 19, D-64289 Darmstadt, Germany.

Director, S.W. and Rohrer, R.A. (1969). The generalized adjoint network and network sensitivities, *IEEE T. Circuits Syst.*, **16**, 318–323.

Dorigo, M. and Gambardella, L.M. (1997). Ant colony system: a cooperative learning approach to the traveling salesman problem, *IEEE T. Evolut. Comput.*, **1**, 53–66.

Forrester, A.I.J. and Keane, A.J. (2009). Recent advances in surrogate-based optimization, *Prog. Aerosp. Sci.*, **45**, 50–79.

Grimaccia, F., Mussetta, M. and Zich, R.E. (2007). Genetical swarm optimization: self-adaptive hybrid evolutionary algorithm for electromagnetic, *IEEE T. Antenn. Propag.*, **55**, 781–785.

Haupt, R.L. (2007). Antenna design with a mixed integer genetic algorithm, *IEEE T. Antenn. Propag.*, **55**, 577–582.

Haykin, S. (1998). *Neural Networks: A Comprehensive Foundation*, (2nd ed.), Prentice Hall, Upper Saddle River, NJ.

HFSS (2010), release 13.0, ANSYS, http://www.ansoft.com/products/hf/hfss/.

Jin, N. and Rahmat-Samii, Y. (2007). Advances in particle swarm optimization for antenna designs: real-number, binary, single-objective and multiobjective implementations, *IEEE T. Antenn. Propag.*, **55**, 556–567.

Jin, N. and Rahmat-Samii, Y. (2008). Analysis and particle swarm optimization of correlator antenna arrays for radio astronomy applications, *IEEE T. Antenn. Propag.*, **56**, 1269–1279.

Kennedy, J. (1997). The particle swarm: social adaptation of knowledge, *Proc. 1997 Int. Conf. Evolutionary Computation*, Indianapolis: IN, pp. 303–308.

Kolda, T.G., Lewis, R.M. and Torczon, V. (2003). Optimization by direct search: new perspectives on some classical and modern methods, *SIAM Rev.*, **45**, 385–482.

Koziel, S. and Bandler, J.W. (2007). SMF: a user-friendly software engine for space-mapping-based engineering design optimization, *Proc. Int. Symp. Signals, Systems and Electronics, URSI ISSSE* 2007, Montreal, Canada, pp. 157–160.

Koziel, S., Bandler, J.W. and Madsen, K. (2006). Space mapping framework for engineering optimization: theory and implementation, *IEEE Trans. Microw. Theory*, **54**, 3721–3730.

Koziel, S., Bandler, J.W. and Cheng, Q.S. (2010). Robust trust-region space-mapping algorithms for microwave design optimization, *IEEE T. Microw. Theory*, **58**, 2166–2174.

Koziel, S. and Ogurtsov, S. (2010). Robust multi-fidelity simulation-driven design optimization of microwave structures, *IEEE MTT-S.*, Anaheim: CA, 201–204.

Koziel, S., Meng, J., Bandler, J.W., Bakr, M.H. and Cheng, Q.S. (2009). Accelerated microwave design optimization with tuning space mapping, *IEEE T. Microw. Theory*, **57**, 383–394.

Koziel, S. (2010). Efficient optimization of microwave structures through design specifications adaptation, *Proc. IEEE Antennas Propag. Soc. International Symposium (APSURSI)*, Toronto, Canada, 153–178.

Koziel, S. and Ogurtsov, S. (2011a). Simulation-driven design in microwave engineering: methods, in Koziel, S., and Yang, X.S. (eds.), *Computational Optimization, Methods and Algorithms*, Springer-Verlag, Berlin, pp. 1–4.

Koziel, S. and Ogurtsov, S. (2011b). Simulation-driven design in microwave engineering: application case studies, in Yang, X.S. and Koziel, S. (eds.), *Computational Optimization and Applications in Engineering and Industry*, Springer-Verlag, Berlin, pp. 57–98.

Koziel, S. and Ogurtsov, S. (2011c). Rapid design optimization of antennas using space mapping and response surface approximation models, *Int. J. RF Microw. CE*, **21**, 611–621.

Koziel, S. and Ogurtsov, S. (2011d). Antenna design through variable-fidelity simulation-driven optimization, *Loughborough Antennas & Propagation Conference, LAPC 2011, IEEEXplore*.

Koziel, S. and Ogurtsov, S. (2011e). Computationally efficient simulation-driven antenna design using coarse-discretization electromagnetic models, *IEEE Antennas Prop.*, 2928–2931.

Koziel, S., Ogurtsov, S. and Leifsson, L. (2012). Variable-fidelity simulation-driven design optimisation of microwave structures, *IJMNO*, **3**, 64–81.

Li, W.T., Shi, X.W., Hei, Y.Q., Liu, S.F. and Zhu, J. (2008). A hybrid optimization algorithm and its application for conformal array pattern synthesis, *IEEE T. Antenn. Propag.*, **58**, 3401–3406.

Ogurtsov, S. and Koziel, S. (2011a). Optimization of UWB planar antennas using adaptive design specifications, *Proc. the 5th European Conference on Antennas and Propagation, EuCAP*, pp. 2216–2219.

Ogurtsov, S. and Koziel, S. (2011b). Design optimization of a dielectric ring resonator antenna for matched operation in two installation scenarios, *Proc. International Review of Progress in Applied Computational Electromagnetics, ACES*, pp. 424–428.

Ogurtsov, S. and Koziel, S. (2011c). Simulation-driven design of dielectric resonator antenna with reduced board noise emission, *IEEE MTT-S*, Anaheim: CA, 1–4.

Pantoja, M.F., Meincke, P. and Bretones, A.R. (2007). A hybrid genetic algorithm space-mapping tool for the optimization of antennas, *IEEE T. Antenn. Propag.*, **55**, 777–781.

Petko J.S. and Werner, D.H. (2007). An autopolyploidy-based genetic algorithm for enhanced evolution of linear polyfractal arrays, *IEEE T. Antenn. Propag.*, **55**, 583–593.

Petosa, A. (2007). *Dielectric Resonator Antenna Handbook*, Artech House, Norwood: MA.

Queipo, N.V., Haftka, R.T., Shyy, W., Goel, T., Vaidynathan, R. and Tucker, P.K. (2005). Surrogate-based analysis and optimization, *Prog. Aerosp. Sci.*, **41**, 1–28.

Rajo-Iglesias, E. and Quevedo-Teruel, O. (2007). Linear array synthesis using an ant-colony-optimization-based algorithm, *IEEE Antenn. Propag. M.*, **42**, 70–79.

Roy, G.G., Das, S., Chakraborty, P. and Suganthan, P.N. (2011). Design of non-uniform circular antenna arrays using a modified invasive weed optimization algorithm, *IEEE T. Antenn. Propag.*, **59**, 110–118.

Schantz, H. (2005). *The Art and Science of Ultrawideband Antennas*, Artech House, Norwood: MA.

Selleri, S., Mussetta, M., Pirinoli, P., Zich, R.E. and Matekovits, L. (2008). Differentiated meta-PSO methods for array optimization. *IEEE T. Antenn. Propag.*, **56**, 67–75.

Shum, S. and Luk, K. (1995). Stacked anunular ring dielectric resonator antenna excited by axi-symmetric coaxial probe, *IEEE T. Microw. Theory*, **43**, 889–892.

Simpson, T.W., Peplinski, J., Koch, P.N. and Allen, J.K. (2001). Metamodels for computer-based engineering design: survey and recommendations, *Eng. Comput.*, **17**, 129–150.

Smola, A.J. and Schölkopf, B. (2004). A tutorial on support vector regression, *Stat. Comput.*, **14**, 199–222.

Special issue on synthesis and optimization techniques in electromagnetic and antenna system design, (2007). *IEEE T. Antenn. Propag.*, **55**, 518–785.

Storn R. and Price, K. (1997). Differential evolution — a simple and efficient heuristic for global optimization over continuous spaces, *J. Global Optim.*, **11**, 341–359.

Volakis, J.L. (ed.) (2007). *Antenna Engineering Handbook*, (4th ed.), McGraw-Hill, New York: NY.

Wild, S.M., Regis, R.G. and Shoemaker, C.A. (2008). ORBIT: Optimization by radial basis function interpolation in trust-regions, *SIAM J. Sci. Comput.*, **30**, 3197–3219.

Wright, S.J. and Nocedal, J. (1999). *Numerical Optimization*, Springer Verlag, Berlin.

Yang, X.S. (2010). *Engineering Optimization: An Introduction with Metaheuristic Applications*, Wiley, New York: NY.

Zhu, J., Bandler, J.W., Nikolova, N.K. and Koziel, S. (2007). Antenna optimization through space mapping. *IEEE T. Antenn. Propag.*, **55**, 651–658.

Chapter 8

Neural Networks for Radio
Frequency/Microwave Modeling

Chuan Zhang, Lei Zhang, and Qi-Jun Zhang

This chapter presents an introduction to neural networks for radio frequency (RF) and microwave modeling and design. In this chapter, we will present the fundamentals of neural networks, including formulation of neural network and its structures, computational representation of neural networks, and training and validation of neural network models. The step-by-step procedure of neural network model development will be described in detail, and two effective neural network-based modeling techniques, namely automatic model generation (AMG) and parallel AMG (PAMG), will be discussed with reference to the full advantages of automated model generation and minimal human interaction in modeling. Towards the end of this chapter, examples of using neural networks to model RF/microwave components and circuits will be provided to demonstrate the capability of neural networks in addressing different RF/microwave modeling problems.

8.1. Introduction to Neural Networks

Neural networks, also called artificial neural networks (ANNs), are information processing systems whose design is inspired by the studies of the ability of the human brain to learn from observations and to generalize by abstractions (Zhang and Gupta, 2000). The fact that neural networks can be trained to learn any nonlinear input–output relationship from corresponding data has led to their use in a number of areas such as pattern recognition, speech processing, control, and biomedical engineering. In the last decade, neural networks have become a popular alternative in computer-aided design (CAD) of RF/microwave components and circuits,

191

as such high-quality CAD models are urgently needed to enable efficient design and optimization of large and complicated circuits and systems. The neural network model learns component data, with the resulting model being as fast as the equivalent circuit model and as accurate as the detailed EM/physics-based model. With such speed and accuracy advantage, use of neural network models can significantly speed up circuit design, while retaining EM/physics-level accuracy in high-level circuit simulation. Furthermore, ANN formulations are flexible enough to be realized in both frequency-domain and time-domain for different modeling requirements. Conventional multilayer neural networks are formulated to be compatible with existing CAD tools, and can be used in complex simulation and design optimization. There is also lots of research work being conducted on knowledge-based neural networks, which smartly combine existing RF/ microwave knowledge with neural networks to achieve enhanced model accuracy and reduced modeling cost (Zhang *et al.*, 2005).

8.2. Formulation of Neural Networks

8.2.1. *Basic Components*

A typical neural network structure has two types of basic components, namely, the processing elements and the interconnections between them. The processing elements are called neurons, and the connections between the neurons are known as links or synapses. Every link has a corresponding weight parameter associated with it. Each neuron receives stimuli from other neurons connected to it, processes the information, and produces an output. Neurons that receive stimuli from outside the network are called input neurons; neurons whose outputs are externally used are called output neurons; neurons that receive stimuli from other neurons and whose outputs are stimuli for other neurons in the network are called hidden neurons. Different neural network structures can be constructed by using different types of neurons and by connecting them differently (Zhang *et al.*, 2005). These definitions will be further explained in the following sections.

8.2.2. *External View of Neural Networks*

Let n and m represent the number of input and output neurons of a neural network. Let x be an n-vector containing the external inputs to the neural network, y be an m-vector containing the outputs from the output neurons, and w be a vector containing the weight parameters representing

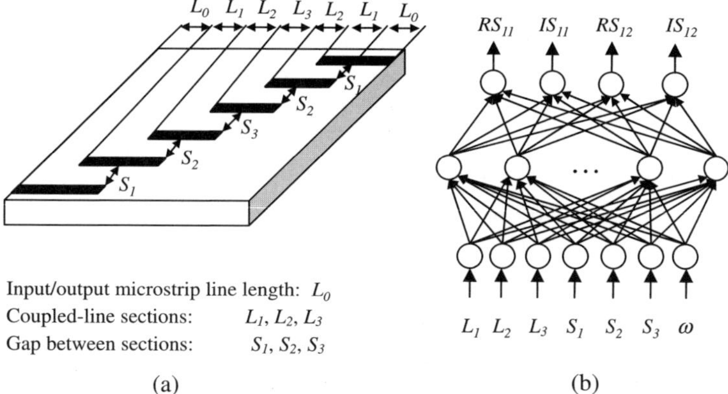

Fig. 8.1. (a) EM-based bandpass microstrip filter to be modeled using a neural network, (b) neural network model.

all interconnections in the neural network. The definition of $\boldsymbol{w}$, and the manner in which $\boldsymbol{y}$ is computed from $\boldsymbol{x}$ and $\boldsymbol{w}$, determine the structure of the neural network.

Consider a bandpass microstrip filter as shown in Fig. 8.1(a). The physical/geometrical parameters of the filter are input variables and any variation in these parameters may affect the electrical responses (or output) of the filter (e.g., S-parameters). Assume that this input–output relationship can be modeled by a neural network, as shown in Fig. 8.1(b), in which:

$$\boldsymbol{x} = [L_1 \ L_2 \ L_3 \ S_1 \ S_2 \ S_3 \ \omega]^T \tag{8.1}$$

$$\boldsymbol{y} = [RS_{11} \ IS_{11} \ RS_{12} \ IS_{12}]^T, \tag{8.2}$$

where ω is the angular frequency. In (8.2), RS_{ij} and IS_{ij} represent real and imaginary parts of the S-parameter S_{ij} $(i, j = 1, 2)$. The superscript T indicates transpose of a vector or matrix. The input parameters in (8.1) are defined in Fig. 8.1(a). The original filter modeling problem can be expressed as

$$\boldsymbol{y} = \boldsymbol{f}(\boldsymbol{x}), \tag{8.3}$$

where $\boldsymbol{f}$ is a function representing the actual input–output relationship of the filter. The neural network model for the filter is defined by

$$\boldsymbol{y} = \boldsymbol{f}_{ANN}(\boldsymbol{x}, \boldsymbol{w}). \tag{8.4}$$

The neural network in (8.4) can represent the filter behavior in (8.3) only after learning the original $\boldsymbol{x}$-$\boldsymbol{y}$ relationship $\boldsymbol{f}$ through an optimization procedure called training. A set of $(\boldsymbol{x}, \boldsymbol{y})$ data pairs, called training data, are required to train neural network model. Such data set can be obtained either from the EM-based simulation or from direct measurements of the filter structure depicted in Fig. 8.1(a). The objective of training is to adjust neural network weights $\boldsymbol{w}$ such that the outputs of the neural model best match those from the training data. A trained neural model can then be used in circuit design to provide instant answers to the task it has learnt. In the filter example, the neural model can be used to provide fast estimation of S-parameters given the filter's physical dimensions.

8.2.3. *Multilayer Perceptron (MLP) Neural Network*

8.2.3.1. *Structure*

Multilayer perceptron (MLP) is a widely used neural network structure. In this structure, the neurons are grouped into various layers. The layer which contains neurons connecting the external inputs is called the input layer. The layer which contains neurons connecting the external outputs is called the output layer. The layers between the input and the output layers are called the hidden layers. The structure of a multilayer perceptron network is illustrated in Fig. 8.2.

Suppose the total number of layers is L. The first layer is the input layer, the Lth layer is the output layer, and layers from 2 to $L - 1$ are hidden layers, where N_l is the number of neurons of each layer.

Information processing, or input–output computation, of each hidden neuron is depicted in Fig. 8.3. Let $\boldsymbol{x}_i$ represent the ith external input to the MLP, z_i^l be the output of the ith neuron of the l the layer, w_{ij}^l be the weight of the link between the jth neuron of the $(l - 1)$th layer and the ith neuron of the lth layer and w_{i0}^l be the additional weight parameter for each neuron representing the bias for the ith neuron of the lth layer.

8.2.3.2. *Feed-forward computation*

Feed-forward computation is an important computational process to calculate the neural network outputs from given inputs. It exists in both neural network training and model evaluation. The overall process is as follows: the external inputs $\boldsymbol{x}$ are applied to the input neurons, and they become the stimuli for the hidden neurons of the second layer. For each hidden neuron, the sum of the weighted stimuli (γ_i^l) is used to trigger the activation function $\sigma(\cdot)$, producing the response z_i^l for the hidden neuron. Continuing this way, the response from second layer neurons become stimuli

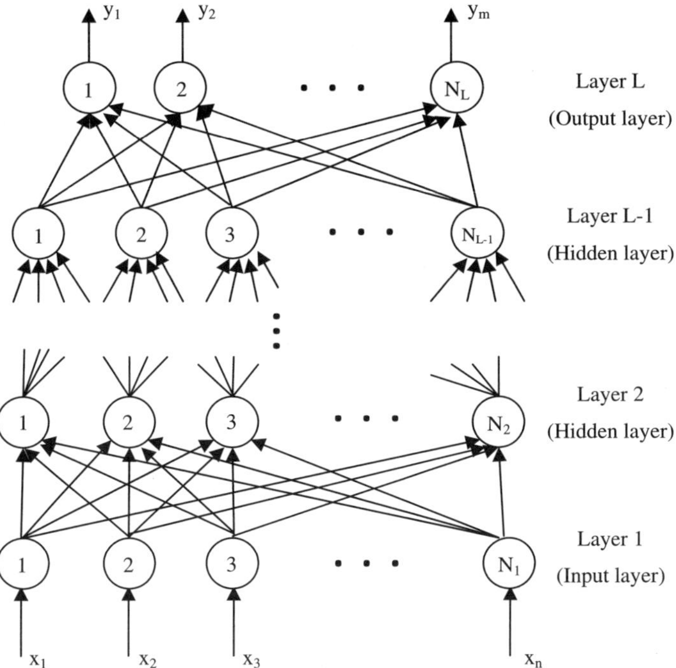

Fig. 8.2. A multilayer perceptron neural network with an input layer, an output layer, and hidden layers.

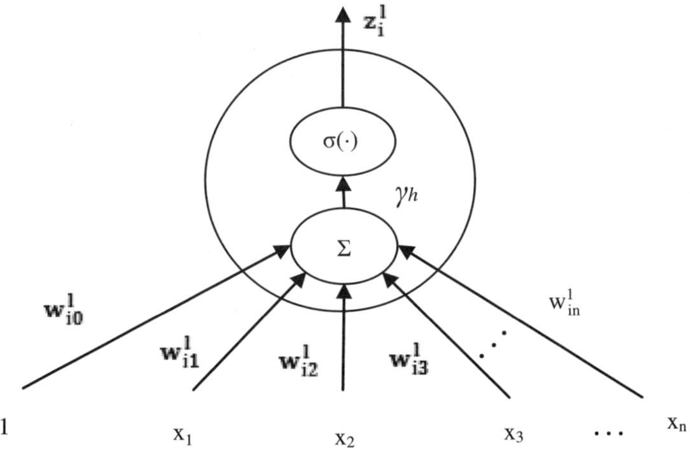

Fig. 8.3. Information processing of the ith hidden neuron at layer l.

for the third layer neurons until the output layer is reached. For each output neuron, the sum of the weighted stimuli becomes the response of the output neuron. Neural network weights $\boldsymbol{w}$, which contain all the weight and bias parameters, remain fixed during feed-forward computation. The layer-to-layer computation from the input to the output is given by:

$$\text{Input layer:} \quad z_i^1 = x_i, \quad i = 1, 2, \ldots, N_1 \quad n = N_1 \tag{8.5}$$

$$\text{Hidden layer:} \quad z_i^l = \sigma(\gamma_i^l), \quad i = 1, 2, \ldots, N_l \quad l = 2, 3, \ldots, L \tag{8.6}$$

$$\text{Output layer:} \quad y_i = z_i^L, \quad i = 1, 2, \ldots, N_L \quad m = N_L, \tag{8.7}$$

where $\sigma(\cdot)$ is called activation function, which is usually the sigmoid function

$$\sigma(\gamma) = \frac{1}{1 + e^{-\gamma}}, \tag{8.8}$$

and

$$\gamma_i^l = \sum_{j=0}^{N_{l-1}} w_{ij}^l z_j^{l-1}. \tag{8.9}$$

There are other activation functions for hidden neurons such as the arctangent function and hyperbolic-tangent function. All of these functions are bounded, continuous, monotonic, and continuously differentiable.

The universal approximation theorem (Hornik *et al.*, 1989) states that MLP always exists with at least one hidden layer which can approximate a general nonlinear, continuous, and multidimensional function to any desired accuracy. To this extent, ANNs can be used to represent a wide variety of modeling problems in RF/microwave area, from modeling individual passive and active components to large nonlinear devices, circuits, and systems.

8.3. Neural Network Modeling Procedure

The neural network cannot predict the behavior of a RF/microwave component or circuit until it is trained with the corresponding data, so the first step in neural modeling is to obtain input–output data pairs (i.e., $\boldsymbol{x} - \boldsymbol{y}$ relationship) of the problem to be modeled. This can be done from either simulation or measurement of the component or circuit. The resulting data set becomes the training data to be used in ANN training. The key steps on neural network modeling are summarized below.

8.3.1. *Inputs and Outputs Definition*

The preliminary step toward developing a neural model is the identification of the inputs and outputs of the problem to be modeled. The model inputs can be circuit input parameters or circuit design variables. Generally, the model outputs are determined based on the purpose of the model. For example, for passive component modeling, the model inputs are physical/geometrical parameters, and the outputs can be the scattering parameters. Some factors that influence the choice of outputs are ease of data generation, ease of incorporation of the neural model into circuit parameters, and so forth.

8.3.2. *Data Generation*

Data generation is the first step for ANN training. In general, data generation is the process to obtain outputs from given inputs. There are two approaches for data generation in practical applications: direct measurement and computer-based simulation. Data generation by measurement is to directly measure the device or circuit to obtain the input–output data. It is accurate to account for the actual device behavior in the real working environment. However, it needs human interaction, thus operation errors may be present in the measurement data. Data generation by simulation is to obtain data by commercial device or circuit simulator. It is a systematic and fast way to approximate the device or circuit behavior, but actual working conditions may not be accurately represented. Both of these methods have their own advantages and drawbacks. For example, measurement data can be obtained even if theory for the device is not available. However, measuring multiple devices with varying physical/geometrical parameters is expensive in terms of human resources. On the other hand, simulation of multiple devices/circuits with different physical dimensions can be set up much easier, but the simulators are normally under certain theoretical assumptions which may limit the accuracy of the simulated data (Zhang and Gupta, 2000).

8.3.3. *Data Organization*

Neural modeling usually requires three sets of data, namely the training data, the validation data, and the test data (Zhang and Gupta, 2000). Training data is necessary for neural model development. Validation data is used to monitor the quality of the neural network model during training and to validate whether stop criteria is met for the training process. Test data is used to independently examine the final quality of the trained neural model in terms of accuracy and generalization capability. To generate a proper amount of data that sufficiently covers the problem to be modeled, range and distribution of the data samples should also be defined. Suppose the lower and upper boundary of the input parameters over a practical design is $[\boldsymbol{x}_{\min}, \boldsymbol{x}_{\max}]$. Validation data and test data should be generated in the desired model usage range $[\boldsymbol{x}_{\min}, \boldsymbol{x}_{\max}]$, while the range of training data should be slightly larger to ensure good performance of the neural model at the boundary of the specified input parameter space.

8.3.4. *Neural Network Training*

The most important step of neural modeling is neural network training, or learning. As mentioned above, a neural network at the initial stage will not represent any device or circuit behavior unless it is trained with corresponding device or circuit data. The fundamentals of neural network training are discussed below.

8.3.4.1. *Training process*

As mentioned in the previous section, training data is in the form of an input–output pair represented by $(\boldsymbol{x}_k, \boldsymbol{d}_k)$, where $\boldsymbol{x}_k$ and $\boldsymbol{d}_k$ are n- and m-dimensional vectors representing the inputs and desired outputs of the problem to be modeled. In other words, the values of $\boldsymbol{y}$ from data generator are represented by the vector $\boldsymbol{d}$. The objective of neural network training is to minimize the difference between the neural network model and the training data, and the training error is defined as:

$$E_{t_r}(\boldsymbol{w}) = \frac{1}{2} \sum_{k \in t_r} \sum_{j=1}^{m} (\boldsymbol{y}_j(\boldsymbol{x}_k, \boldsymbol{w}) - \boldsymbol{d}_{jk})^2, \qquad (8.10)$$

where $\boldsymbol{d}_{jk}$ is the jth element of $\boldsymbol{d}_k$ at input $\boldsymbol{x}_k$ and $\boldsymbol{y}_j(\boldsymbol{x}_k, \boldsymbol{w})$ is the jth neural network output for input $\boldsymbol{x}_k$.

The purpose of neural network training is to adjust the values of weight parameters $\boldsymbol{w}$ in order to minimize the training error of (8.10) to a given error criterion. Iterative algorithms are often used to explore the $\boldsymbol{w}$-space efficiently. For example, in iterative descent methods, the training starts with an initial guess of $\boldsymbol{w}$ and then interactively updates $\boldsymbol{w}$ with directive change to reduce training error. The new values of $\boldsymbol{w}$, denoted as $\boldsymbol{w}_{next}$, is determined by a step down from the current point $\boldsymbol{w}_{now}$ along a direction vector $\boldsymbol{h}$,

$$\boldsymbol{w}_{next} = \boldsymbol{w}_{now} - \eta\boldsymbol{h}, \tag{8.11}$$

where η is a positive step size regulating the extent to which we can proceed in the direction $\boldsymbol{h}$.

An epoch is an iteration of training after all training data are used in the computation of training error. The neural network training is performed epoch by epoch as described by the following algorithm.

Step 1: $\boldsymbol{w} =$ initial guess
$epoch = 0$

Step 2: If $E_v(epoch) < \varepsilon$ (given accuracy criteria) or $epoch > max\_epoch$ (max number of epochs), stop

Step 3: Calculate $E_{tr}(\boldsymbol{w})$ and $\partial E_{tr}(\boldsymbol{w})/\partial\boldsymbol{w}$ using partial or all training data

Step 4: Use optimization algorithm to find $\Delta\boldsymbol{w}$

$$\boldsymbol{w} \leftarrow \boldsymbol{w} + \Delta\boldsymbol{w} \quad \Delta\boldsymbol{w} = \eta\boldsymbol{h}$$

Step 5: If all training data are used, then
$epoch = epoch + 1$, go to Step 2, else go to Step 3

where $\boldsymbol{h}$ is the direction of the update of $\boldsymbol{w}$, η is the step size of the update of $\boldsymbol{w}$. $\boldsymbol{h}$ is typically determined from the training error and its derivative to neural network weights, and η is determined by line search (Zhang and Gupta, 2000).

8.3.4.2. *Training algorithms*

8.3.4.2.1. Back propagation algorithm

One of the most popular algorithms for neural network training is the back-propagation (BP) algorithm, proposed by Rumelhart, Hinton, and Williams in 1986 (Rumelhart *et al.*, 1986). Back-propagation is a stochastic algorithm based upon the steepest descent principle (Luenberger, 1989), in which the weights of the neural network are updated along the negative gradient direction in the weight space.

The basic back-propagation, however, suffers from slower convergence and possible weight oscillation. The addition of momentum, proposed by Rumelhart *et al.* (1986), provides significant improvements to the basic back-propagation scheme, reducing the weight oscillation.

Other approaches to reduce weight oscillation have also been proposed, such as invoking a correction term that uses the difference of gradients (Ochiai *et al.*, 1994), and the constrained optimization approach where constraints on weights are imposed to achieve better alignment between weight updates in different epochs (Perantonis and Karras, 1995).

A very interesting work is the delta-bar-delta rule proposed by Jacobs (1988). He developed an algorithm based on a set of heuristics where the learning rate for different weights was different and adapted separately during the learning process.

8.3.4.2.2. Training algorithms using other gradient-based optimization techniques

Back-propagation based on the steepest descent principle is relatively easy to implement. However, the error surface of neural network training usually contains planes with a gentle slope due to the squashing functions commonly used in neural networks. Since supervised learning of neural networks can be viewed as a function optimization problem, higher order optimization methods using gradient information can be adopted in neural network training to improve the speed of convergence.

Conjugate gradient method

The conjugate gradient method is originally derived from quadratic minimization, where the minimum of the objective function E_{tr} can be efficiently found within a few iterations. In this method, the descent direction runs along the conjugate direction, which can be accumulated using vector. Thus conjugate gradient method is very efficient and scales well with the neural network size.

Conjugate gradient method is generally faster than BP. As far as the memory requirement is concerned, the algorithm requires the storage of only a few vectors whose size is the number of neural network weights. Therefore, conjugate gradient methods are effective for training large neural networks.

Quasi-Newton method

Similar to the conjugate gradient method, the quasi-Newton training method was also derived from quadratic objective function. The inverse of the Hessian matrix $\boldsymbol{B} = \boldsymbol{H}^{-1}$ is used to bias the gradient direction, following Newton's method.

Standard quasi-Newton methods require N_w^2 storage space to maintain an approximation of the inverse Hessian matrix and a line search is indispensable to calculate a reasonably accurate step length, where N_w is the total number of weights in the neural network structure. Limited memory (LM) or one-step Broyden–Fletcher–Goldfarb–Shanno (BFGS) (Broyden, 1967) is a simplified implementation in which the inverse Hessian approximation is reset to the identity matrix after every iteration, thus avoiding the need to store matrix (Nakano, 1997). Parallel implementation of second-order, gradient-based MLP training algorithms featuring full- and limited-memory BFGS algorithms have also been investigated (McLoone and Irwine, 1997). Through the estimation of an inverse Hessian matrix, the quasi-Newton has a faster convergence rate than the conjugate gradient method.

8.3.4.2.3. Nongradient-based training techniques

Nongradient-based training techniques such as simplex methods also exist in the literature, which are in general much slower than the gradient-based methods. The simplex algorithm can be particularly useful for training neural networks with user-defined neuron activation functions (e.g., empirical functions), where the gradient information may not be available in training (Zhang and Gupta, 2000).

8.3.4.2.4. Global optimization algorithms

Global optimization methods include genetic algorithms and simulated annealing (SA) algorithms, which use the information from training error $E_{tr}(\boldsymbol{w})$ only (without gradient information). They are capable of escaping from the traps of the local minimum and finding the global minimum. The genetic algorithm (Goldberg, 1989) is a stochastic optimization algorithm, derived from the concepts of the biological theory of evolution. The simulated annealing algorithm (Kirkpatrick *et al.*, 1983) is inspired by the idea of the physical annealing process in crystal growth.

8.3.4.2.5. Training algorithms utilizing decomposed optimization

Decomposition is an important way of solving large-scale optimization problems. Several training algorithms that decompose the training process by training the neural network layer by layer have been proposed (Ergezinger and Thomsen, 1995; Wang and Chen, 1996; Lengelle and Denceux, 1996).

Linear programming can be used to solve large-scale linearized optimization problems. Neural network training is linearized and formulated as a constrained linear programming (Lengelle and Denceux, 1996).

A combination of linear and nonlinear programming techniques can reduce the degree of nonlinearity of the error function with respect to the

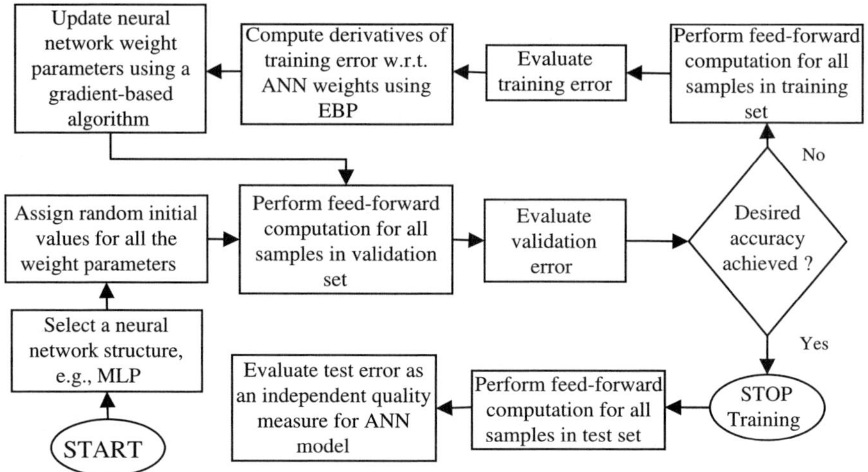

Fig. 8.4.　Major steps in neural network training.

hidden layer weights, and also reduce the probability of being trapped in a local minimum.

8.3.4.3. *More about training*

Figure 8.4 shows a flowchart of the major steps in neural network training. As shown in the figure, validation error is periodically evaluated and the training is terminated once a reasonable validation error E_v is reached during ANN training. While at the end of the training, the quality of the neural network model can be independently assessed by evaluating the test error E_{te}.

As discussed above, training techniques include local optimization methods such as BP, conjugate gradient, quasi-Newton, and global optimization methods including simulated annealing and genetic algorithms.

8.3.4.4. *Generalization, overlearning, and underlearning*

The ability of a trained neural network to estimate output y accurately at a given input x never seen during training is called the generalization ability.

Overlearning is a phenomenon in which the neural network memorizes the training data but cannot generalize well. In other words, the training error E_{tr} is small, but the validation error $E_v \gg E_{tr}$. Possible reasons for overlearning include the presence of too many hidden neurons, or insufficient training data.

Underlearning, on the other hand, is a situation in which the neural network has difficulties even learning the training data itself i.e. $E_{tr} \gg 0$. Possible reasons for underlearning include the presence of insufficient hidden neurons, insufficient training, or training being stuck in a local minimum.

Good learning of a neural network is observed when both training and validation errors have small values and are close to each other.

8.4. Automatic Model Generation

8.4.1. *Introduction*

As mentioned in the preceding text, the trained ANN can offer fast and accurate solutions to RF/microwave problems. However, practical microwave modeling problems are mostly highly nonlinear and multidimensional. The selection of the number of samples and the number of hidden neurons needed for developing a neural model with desired accuracy is not obvious. To enable fast design cycle with minimum modeling burden, a method of solving these problems automatically is urgently needed.

Recently an automatic model generation (AMG) technique was introduced by Devabhaktuni *et al.* (2001), which performs all the subtasks involved in the neural modeling process through a computational algorithm. Starting with zero or minimum amount of data, AMG uses an adaptive data sampling algorithm to automatically drive the EM/physics/circuit simulators to generate new data samples required by ANN training. It also uses a more exhaustive search of different ANN structures to generate the most compact model with the highest possible accuracy. As such, AMG effectively converts the intensive human effort demanded by the conventional step-by-step neural modeling approach into intensive computation by computer algorithms, and realizes the automation of microwave ANN modeling. The integrated process is computerized and is carried out automatically in a stage-wise fashion. Within a stage, the algorithm facilitates periodic communication between various subtasks, thus enabling adjustment or enhancement in the execution of a subtask based on the feedback from other subtasks. As a result, each stage could involve dynamic incremental data generation, neural network size adjustment, neural network training with training data, and neural model testing with validation data.

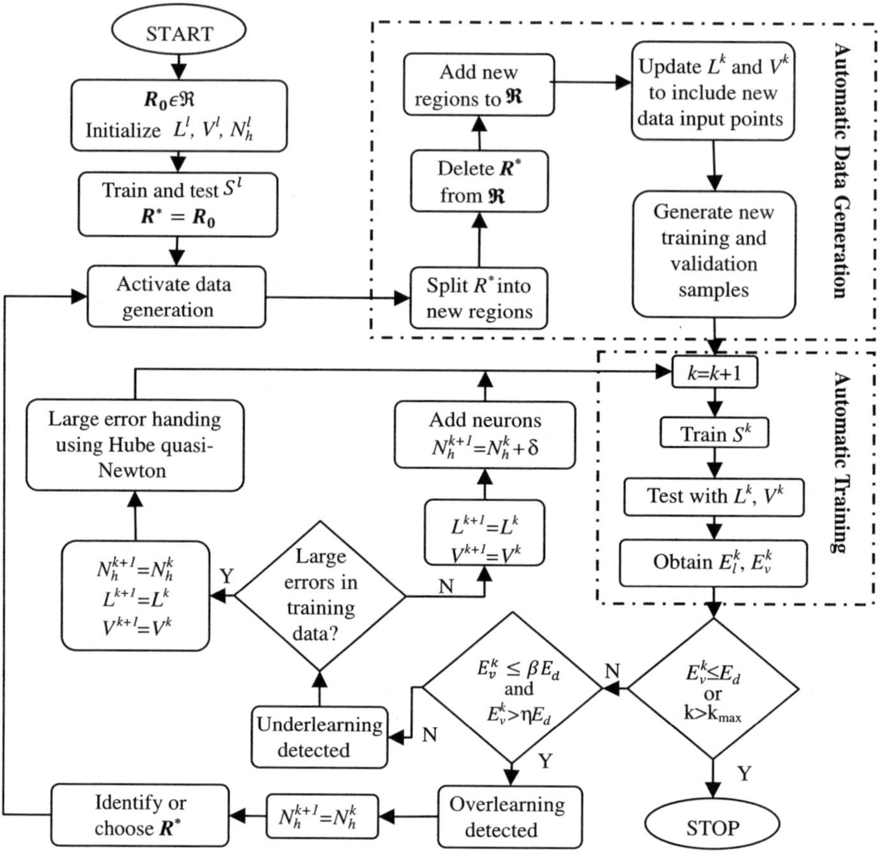

Fig. 8.5.   Automatic neural model development algorithm.

8.4.2. *Key Aspects of the AMG Algorithm*

A framework of the AMG algorithm is shown in Fig. 8.5. We use two disjoint sets of data in AMG, namely, the training data and the validation data. Training data are used to update neural network weights during training, and validation data are used to monitor the quality of the neural model during training.

(1) Automatic sampling and data generation: overlearning of the neural network may be detected, by examining training error and validation error. When overlearning is seen, the AMG algorithm dynamically adds more data samples to the training and validation sets. We developed a unique neural network-oriented technique featuring the growth of training and validation

sets in an intertwined way. Utilizing this technique, the algorithm handles the issues of number of additional samples and their distribution in model input space (Devabhaktuni *et al.*, 2001).

(2) Automatic structure adaptation: underlearning of the neural network may be detected after the kth stage, using training error E_l^k and its gradient $\Delta E_l^k = E_l^{k-1} - E_l^k$. When the algorithm detects underlearning, it dynamically adds more hidden layer neurons. A larger neural network would then provide increased freedom to better learn the nonlinearities in the training data. If the neural network structure S^k has N_h^k hidden neurons, S^{k+1} with $N_h^{k+1} = N_h^k + \delta$, hidden neurons is used by the algorithm in the $(k+1)$th stage. Suggested range for the newly added hidden neurons δ is 10%–20% of N_h^k.

(3) Automatic handling of large errors in generated samples: in general, most of the microwave samples have small measurement/simulation errors and a few samples could even have large errors. A few accidental large errors could occur in training data during dynamic data generation of the automatic model development approach. It is essential to automatically detect these large errors and neglect them during neural network training, because there is no place for human intervention in an automated approach. The large errors are detected as a special case of underlearning. Once large errors are detected, automatic training switches from conventional neural network training algorithms (e.g., quasi-Newton) to Huber norm-based quasi-Newton (HQN) technique (Xi *et al.*, 1999).

Consequently, the Huber norm-based training objective can be robust against both small and large errors in data. When accidental large errors are detected in the kth stage training data, our algorithm switches neural network training process to HQN. The HQN algorithm ensures that the network S^{k+1} learns only the original problem behaviors, neglecting large errors.

(4) Overall automation: at the end of each stage, the algorithm checks for various possible neural network training situations and takes relevant actions, e.g., update data, adjust neural network size, etc. In the subsequent stage, neural network S^{k+1} is trained with samples in L^{k+1} and the neural model is tested with samples in V^{k+1}. We also incorporated a few conservative options to make the algorithm more general: (i) Periodically, after a fixed number of stages, incremental training/validation samples are generated in a randomly chosen sub-region instead of a worst sub-region and, (ii) the algorithm terminates only if $E_v^k \le E_d$ consecutively for a given number of stages.

8.5. Parallel AMG

The central processing unit (CPU)-expensive EM/physics/circuit simulations and the iterative training process are the most computationally intensive parts in AMG. To improve AMG efficiency, the parallel automated model generation (PAMG) approach exploits parallelism to split the processes of the data generation and the training into smaller sections which are simultaneously executed on parallel processors in a multi-processor environment to speed up the computation. Figure 8.6 shows the PAMG algorithm. In this flow chart, $E_{tr}(E_v)$ is the training (validation) error, and $\partial E_{tr}/\partial w$ is the gradient information required by quasi-Newton algorithm training. $T_{si}(i = 1, 2, \ldots, 8)$ represents the CPU time taken by the ith sequential process. T_{sim} and T_{tr} represent the data generation time and training time respectively, for one sample on a single processor.

8.5.1. *Parallel Adaptive-Sampling/Data-Generation*

Parallel adaptive-sampling/data-generation is used to drive several data generators simultaneously to generate multiple training data on different parallel processors. However, due to the sharing violation of the input/output files and simulation environment among multiple simulations, simulators may not allow direct parallel simulations. In order to solve this problem, we set an independent simulation environment in each processor to separate the input/output files in each environment.

Let P be the number of parallel processors, and N_d be the number of training data generated in one adaptive sampling stage. Let $x[i]$ be the inputs of the ith adaptively determined sample according to the AMG criteria, and $y_d[i]$ be the corresponding outputs from the data generator. The PAMG data generation is formulated as:

start: $i = 0$;
while $(i < N_d)$
    parallel executed loop for data generation:
    for (simulation environment $j, j = 1, 2, \ldots, P$)
        inject $x\,[i + j]$ into simulator input file j;
        activate data generator j;
        extract $y_d[i + j]$ from output file j;
end parallel loop;
$i = i + P$,

where P data are generated concurrently on P processors until N_d data generations are finished (Zhang *et al.*, 2007).

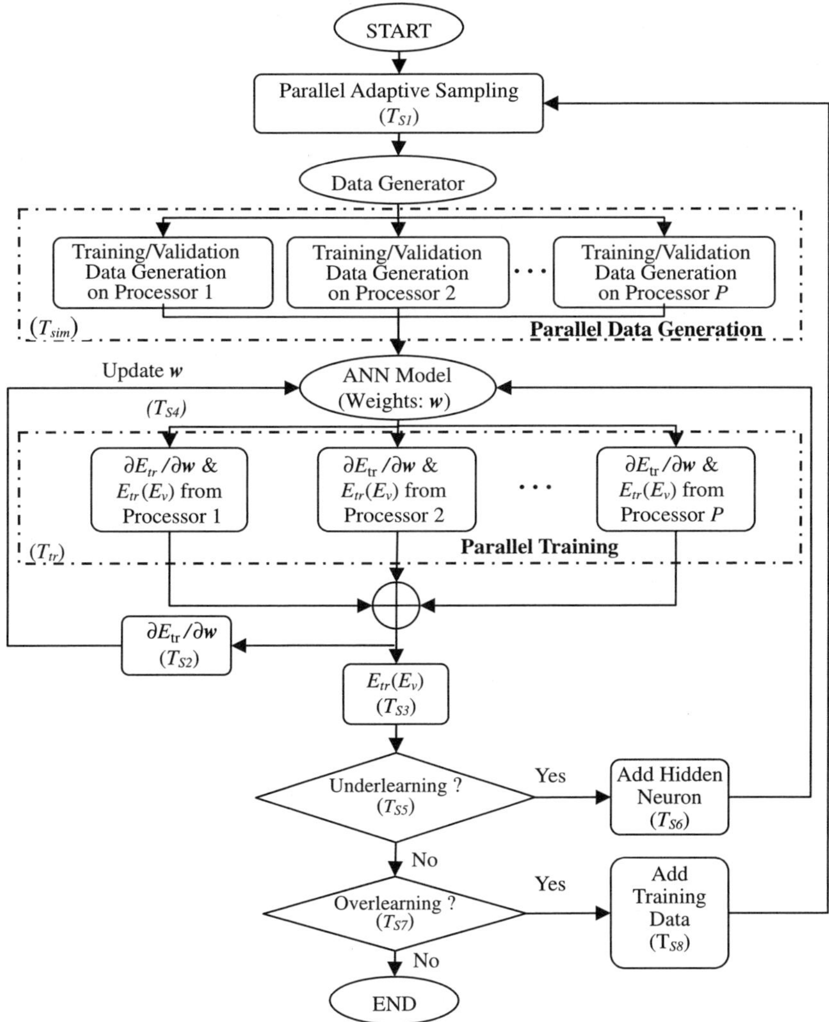

Fig. 8.6.   Framework of the PAMG algorithm.

8.5.2. *Parallel Automatic ANN Training*

The parallel processing technique in ANN training enables the training error computation and error back-propagation to be performed in parallel on different processors.

We implement the parallelism for batch training quasi-Newton (QN) algorithm (Zhang and Gupta, 2000). Let $\boldsymbol{y}_{ANN}(\boldsymbol{x}[i], \boldsymbol{w})$ be an N_y-vector representing the outputs of the ANN model for the ith sample. w is a vector containing neural network synaptic weights. We define $E_{tr}[i]$ as the training error between data $y_d[i]$ and ANN model output $y_{ANN}(x[i], w)$ for the ith sample. Let N_{tr} represent the number of training samples used in AMG training.

The parallel training is formulated below. Let j represent the index of processors in a multiprocessor computing environment. The $(i + j)$th sample is fed into ANN inputs. Feed-forward computations are carried out through hidden neurons to ANN outputs to obtain $y_{ANN}(x[i + j], w)$. $E_{tr}[i + j]$ is then calculated and back-propagated through hidden layers back to the ANN inputs to find the error derivative $\partial E_{tr}[i + j]/\partial w$ used for training. Such feed-forward and back-propagation computations are completely independent between different samples, thus $E_{tr}[i + j]$ and $\partial E_{tr}[i + j]/\partial w$ can be computed concurrently and asynchronously for different j's by the P parallel processors to increase the training efficiency.

start: $i = 0$;
while $(i < N_{tr})$
    parallel executed loop for training:
    for (error computation on processor $j, j = 1, 2, \ldots, P$)
       ANN feed-forward computation of $y^k_{ANN}(x[i + j], w)$;

$$E_{tr}[i + j] = \sum_{k=1}^{N_y}(y^k_d[i + j] - y^k_{ANN}(x[i + j], w))^2/(2 \times N_{tr})$$

       Back-propagate error $E_{tr}[i + j]$ to ANN input;

$$\frac{\partial E_{tr}[i + j]}{\partial w} = \sum_{k=1}^{N_y}(y^k_{ANN}(x[i + j], w) - y^k_d[i + j]) \cdot \frac{\partial y^k_{ANN}x[i + j], w)}{\partial w}/N_{tr}$$

end parallel loop;
$i = i + P$.

8.6. ANN Examples

In previous sections, we introduced the formulation, structures, and training of neural networks. We will now move to application examples to investigate usage of ANN models in microwave/RF modeling.

8.6.1. *Parametric Modeling of Bandstop Microstrip Filter*

In this example, a bandstop microstrip filter is illustrated to show the basic neural network modeling. The training/test data are generated by CST Microwave Studio (2010).

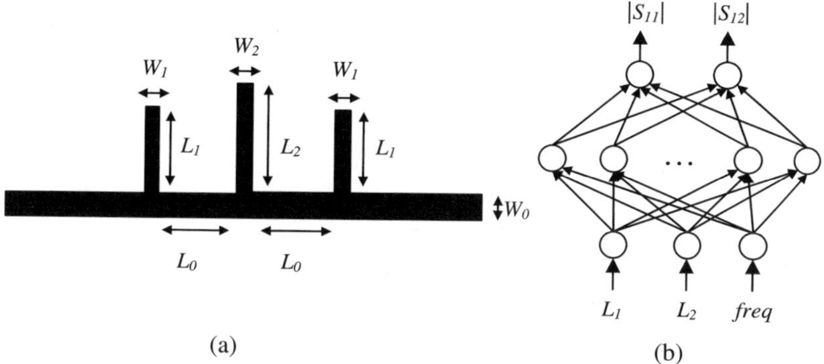

Fig. 8.7.   (a) Structure of a bandstop microstrip filter, (b) structure of neural network for bandstop microstrip filter.

As illustrated in Fig. 8.7, L_1 and L_2 are the open stub lengths, W_1 and W_2 are the width, and L_0 is the length of the middle connected line. This structure is sitting on an alumina substrate with thickness $h = 0.635\,\text{mm}$, width $W_0 = 0.635\,\text{mm}$, and dielectric constant $\varepsilon_r = 9.4$.

Two input geometrical parameters (L_1, L_2) and frequency as the inputs and the magnitude of S-parameter $(|S_{11}|, |S_{12}|)$ as the outputs of the neural network are used to train the neural network; other parameters are fixed for the sake of simplicity. A three-layer MLP structure is used in this example.

Table 8.1 shows the training/test data samples information which is used in neural network processing. As you can see in Figs. 8.8 and 8.9, the neural network modeling result for the magnitude of S-parameter (S_{11} S_{12}) is quite accurate.

The number of hidden neurons is 30. The number of training and test samples are 1600 and 900, respectively. Training error $= 0.89\%$, test error $= 0.80\%$. Both of the two errors are below 1%, which is usually a measure of a good model. CPU time for data generation of samples is 51 min 44 sec. CPU time for training is 12 min 10 sec.

Table 8.1.   Input space for the training data and test data samples.

	Training Data			Test Data		
	Min	Max	Step	Min	Max	Step
L_1 (mm)	1.3	1.42	0.04	1.32	1.4	0.04
L_2 (mm)	1.3	1.42	0.04	1.32	1.4	0.04
freq (GHz)	1.6	2.0	0.00404	1.61	1.99	0.00384

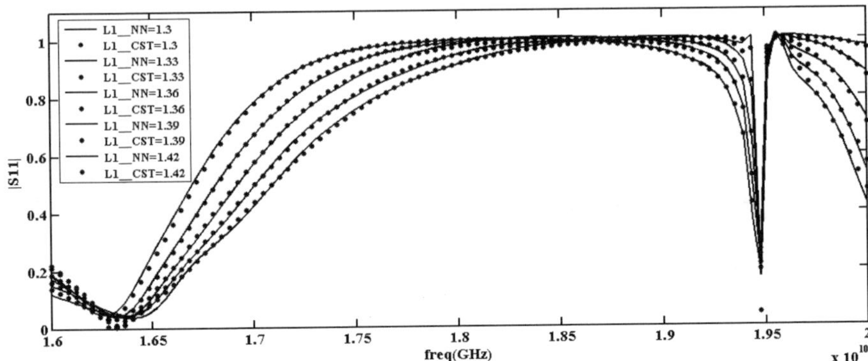

Fig. 8.8. Neural network modeling (line) vs. CST modeling (dot) for magnitude of S-parameter (S_{11}) with five different values of L_1.

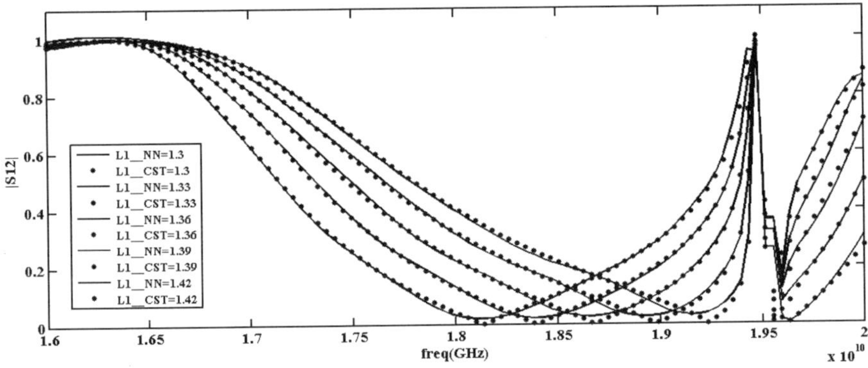

Fig. 8.9. Neural network modeling (line) vs. CST modeling (dot) for magnitude of S-parameter (S_{12}) with five different values of L_1.

8.6.2. *PAMG Used for Device Modeling by Driving a Physics-Based Device Simulator*

This example illustrates the application of the PAMG technique for nonlinear microwave device modeling by driving a physics-based device simulator. A metal semiconductor field effect transistor (MESFET) device is modeled using dc and bias-dependent S-parameter data generated from MINIMOS by solving the device Poison equations.

The ANN model takes four geometrical and electrical inputs (gate width, gate voltage, drain voltage, and frequency) and produces nine outputs (drain current, real and imaginary parts of S-parameters) as shown in Fig. 8.10. PAMG is implemented using OpenMP in C++ on an Intel®

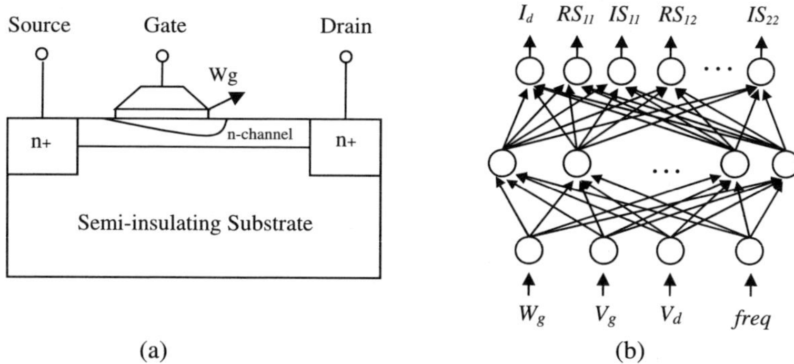

Fig. 8.10. (a) Cross-section of a MESFET structure used in MINIMOS. Wg represents gate width, (b) ANN MESFET model generated by AMG.

Table 8.2. Comparisons of the sequential and parallel data generations for the MESFET example.

No. of Data Generated	100	500	1000
Original AMG Time (min)	25.74	120.6	246.6
PAMG Time (min)	13.46	62.4	128.4
Parallel Efficiency (%)	95.5	96.5	96.0

Table 8.3. Comparisons of training time in the MESFET example. The ANN with 30 hidden neurons is trained by three sets of data to satisfy accuracy of 1% training error.

No. of Training Data	100	200	500
Original AMG Time (min)	2.44	3.46	6.89
PAMG Time (min)	1.30	1.90	3.68
Parallel Efficiency (%)	93.8	91.1	93.6

Core^{TM}2 Duo Processor (two processors). Tables 8.2 and 8.3 compare CPU time between the sequential AMG and the PAMG for the adaptive-sampling/data-generation and training respectively. Since the PAMG is designed to maximally parallelize computationally intensive processes in AMG while minimizing the sequential overhead, parallel efficiency above 90% is achieved. For more comparisons, three MESFET models are developed with different input ranges as shown in Table 8.4. It is estimated that the sequential overhead Ts, total data generation time, and total training time are 3.6, 62.4, and 18.4 minutes respectively, for developing a MESFET

Table 8.4.   Comparisons between sequential and parallel AMG
for three AMG processes of model developments.

	AMG1	AMG2	AMG3
No. of Data Generated	109	200	516
No. of Training Iterations	22043	52604	62123
Accuracy (E_{tr} in %)	0.89	0.79	0.76
CPU for Original AMG (min)	28.03	63.33	157.79
CPU for AMG (min)	15.07	34.23	84.38
Parallel Efficiency (%)	93.0	92.5	93.5

model with 0.76% training error. It is also confirmed that PAMG has high
parallel efficiency and retains the same accuracy as the sequential AMG.

8.7. Conclusions

We have provided an introduction of ANN from theory to practice including
neural network structures, the neural network modeling process, and
AMG, PAMG, and neural network applications in RF/microwave modeling.
Through this chapter, we have aimed to build a technical bridge between
neural network fundamentals and practical applications. AMG and PAMG
techniques have been introduced, showing a fast and automatic way of
driving data generators to train ANN converting human-based modeling
process into an automated CAD process.

References

Broyden, C.G. (1967). Quasi-Newton methods and their application to function
minimization, *Math. Comput.*, **21**, 368–381.

CST Microwave Studio (2010). CST AG, Bad Nauheimer Str. 19, D-64289
Darmstadt, Germany.

Devabhaktuni, V.K., Yagoub, M.C.E. and Zhang, Q.J. (2001). A robust algorithm for automatic development of neural-network models for microwave
applications, *IEEE T. Microw. Theory*, **49**, 2283–2285.

Ergezinger, S. and Thomsen, E. (1995). An accelerated learning algorithm
for multilayer perceptrons: optimization layer by layer, *IEEE T. Neural
Networ.*, **6**, 31–42.

Goldberg, D.E. (1989). *Genetic Algorithms in Search, Optimization and Machine
Learning*, Addison-Wesley, Reading: MA.

Hornik, K., Stinchcombe, M. and White, H. (1989). Multilayer feedforward
networks are universal approximators, *Neural Networks*, **2**, 359–366.

Jacobs, R.A. (1988). Increased rate of convergence through learning rate adaptation, *Neural Networks*, **1**, 295–307.

Kabir, H., Zhang, L., Yu, M., Aaen, P., Wood, J. and Zhang, Q.J. (2010). Smart modeling of microwave devices, *IEEE Microw. Mag.*, **11**, 105–118.

Kirkpatrick, S., Gelatt, C.D. and Vecchi, M.P. (1983). Optimization by simulated annealing, *Science*, **220**, 671–680.

Lengelle, R. and Denceux, T. (1996). Training MLPs layer by layer using an objective function for internal representations, *Neural Networks*, **9**, 83–97.

Luenberger, D.G. (1989). *Linear and Nonlinear Programming*, Addison-Wesley, Reading: MA.

McLoone, S. and Irwin, G.W. (1997). Fast parallel off-line training of multilayer perceptrons, *IEEE T. Neural Networ.*, **8**, 646–653.

MINIMOS 6.1 Win, Institute for Microelectronics, Technical University, Vienna, Austria.

Nakano, K.R. (1997). Partial BFGS update and efficient step-length calculation for three-layer neural networks, *Neural. Comput.*, **9**, 123–141.

Ochiai, K., Toda, N. and Usui, S. (1994). Kick-out learning algorithm to reduce the oscillation of weights, *Neural Networks*, **7**, 797–807.

Perantonis, S.J. and Karras, D.A. (1995). An efficient constrained learning algorithm with momentum acceleration, *Neural Networks*, **8**, 237–249.

Rumelhart, D.E., Hinton, G.E. and Williams, R.J. (1986). Learning internal representations by error propagation, in Rumelhart, D.E. and McClelland, J.L. (eds.), *Parallel Distributed Processing*, The MIT Press, Cambridge: MA, pp. 318–362.

Wang, G.J. and Chen, C.C. (1996). A fast multilayer neural network training algorithm based on the layer-by-layer optimizing procedures, *IEEE T. Neural Networ.*, **7**, 768–775.

Xi, C., Wang, F., Devabhaktuni, V.K. and Zhang, Q.J. (1999). Huber optimization of neural networks: a robust training method, *Int. Joint Conf. Neural Networks*, Washington: DC, pp. 1639–1642.

Zhang, Q.J. and Gupta, K.C. (2000). *Neural Networks for RF and Microwave Design*, Artech House, Norwood: MA.

Zhang, Q.J., Deo, M. and Xu, J. (2005). Neural Networks for Microwave Circuits, in Chang, K. (ed.), *Encyclopedia of RF and Microwave Engineering*, Wiley, New York: NY, p. 3390.

Zhang, L., Cao, Y., Wan, S., Kabir, H. and Zhang, Q.J. (2007). Parallel automatic model generation technique for microwave modeling, *Microwave Symposium, IEEE/MTT-S International*, Honolulu: HI, pp. 103–106.

Chapter 9

Parametric Modeling of Microwave Passive Components Using Combined Neural Network and Transfer Function

Yazi Cao, Venu-Madhav-Reddy Gongal-Reddy,
and Qi-Jun Zhang

This chapter presents a combined neural network and transfer function (neuro-transfer-function) technique for parametric modeling of microwave passive components. The chapter starts with a brief description of modeling issues in transfer-function-based neural network techniques for microwave applications. We also discuss three typical training methods to develop the neuro-transfer-function model for microwave passive components. We then introduce an advanced training method to guarantee the continuity of coefficients in transfer functions and simultaneously allow better model accuracy in challenging applications involving high-order transfer functions, wide frequency range, and large geometrical variations. Finally, parametric modeling examples using the neuro-transfer-function technique are provided.

9.1. Introduction

Artificial neural network (ANN) techniques have been recognized as a powerful tool in electromagnetic (EM)-based modeling and design optimization of microwave passive components (Burrascano *et al.*, 1999; Steer *et al.*, 2002; Rayas-Sánchez, 2004; Rizzoli *et al.*, 2004). Design optimization of microwave passive components often requires repetitive adjustments of the values of geometrical or material parameters and can be very time consuming. ANNs can learn EM responses as a function of geometrical

215

variables through an automated training process, and the trained ANNs can be used as accurate and fast models in the design optimization (Zhang *et al.*, 2003).

A pure ANN such as the widely used multilayer perceptron neural network has the ability to learn EM behavior without having to rely on the complicated internal details of passive components. The trained ANN model can be subsequently implemented in high-level circuit and system designs, allowing fast simulation and optimization (Zhang and Gupta, 2000). A further approach, named knowledge-based neural network, was introduced for improving the modeling accuracy and reliability with less hidden neurons and training data. In this knowledge-based approach, prior knowledge such as analytical expressions (Wang and Zhang, 1997; Devabhaktuni *et al.*, 2003), empirical models (Bandler *et al.*, 1999; Rayas-Sánchez and Gutiérrez-Ayala, 2006) or equivalent circuits (Cao and Wang, 2007) is incorporated into the model structure for enhancing the capability of learning and generalization of the overall models.

This chapter presents a combined neural network and transfer function (neuro-transfer-function) technique for parametric modeling of microwave passive components. This neuro-transfer-function technique can be used even if the problem-dependent prior knowledge is unavailable (Ding *et al.*, 2004). The generalized behavior of passive components versus frequency can be expressed by the widely used transfer functions. Moreover, the use of transfer functions makes it possible to conveniently represent the model by equivalent circuit elements and easily link the model with a standard circuit simulator such as SPICE (Choi and Swaminathan, 2000). In this technique, the neural network is used for mapping the geometrical variables of passive components onto the coefficients of transfer functions. We present three typical training methods to develop the neuro-transfer-function model for microwave passive components. We further introduce an advanced training method to develop the neuro-transfer-function model for covering challenging applications involving high-order transfer functions, wide frequency range, and/or large geometrical variations (Cao *et al.*, 2009). This neuro-transfer-function technique is also expanded to include bilinear function format in place of rational function format for higher-order and wider frequency-range applications. Parametric modeling examples of the use of the neuro-transfer-function technique are provided in this chapter.

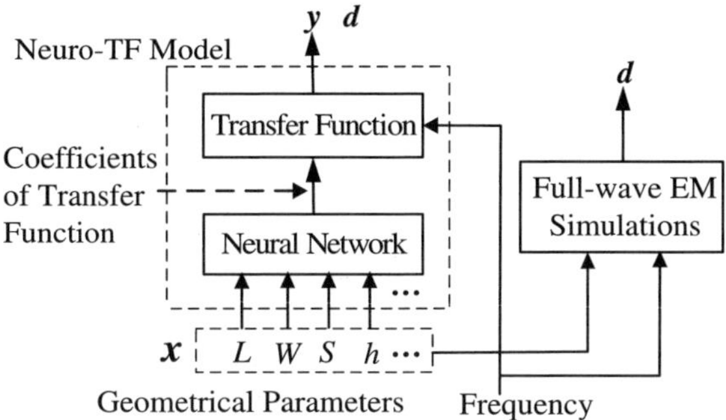

Fig. 9.1.   Structure of the neuro-transfer-function (neuro-TF) model (Cao *et al.*, 2009).

9.2. Formulation of the Neuro-Transfer-Function Model

The structure of the neuro-transfer-function model is illustrated in Fig. 9.1. It consists of a neural network and a transfer function. Let x represent the inputs of the neural network containing the geometrical variables of passive components, such as length (L), width (W), space (S) between lines, and thickness of substrates (h) in microstrip structures. Let the outputs of the neural network be coefficients in the transfer function. The popular vector fitting method (Gustavsen and Semlyen, 1999) is used to extract the coefficients from the corresponding EM responses. Here coefficients in the transfer function are used as the ANN outputs instead of poles and residues. This is because when the geometrical parameters vary, the tracking and indexing of poles and residues used for different geometries are not easy. The neural network is trained to learn the high-dimensional and nonlinear mapping between x and the coefficients. Let y be a vector representing real and imaginary parts of the transfer function. Let d be a vector representing the outputs of the EM simulations such as real and imaginary parts of S-parameters. The objective here is to adjust the neural network internal weights such that the error between y and the training data d is minimized.

We use a general type of transfer function encompassing both rational and bilinear function formats in the frequency domain as:

$$H(s) = \frac{\sum_{i=0}^{M} a_i s^i}{1 + \sum_{i=1}^{N} b_i s^i}, \tag{9.1}$$

where the s variable can be represented in two ways:

$$s = \begin{cases} j\omega & \text{for rational function format of } H(s) \\[2mm] \dfrac{1 - j\omega}{1 + j\omega} = e^{-j\theta} & \text{for bilinear function format of } H(s) \end{cases} \tag{9.2}$$

and $\theta = 2\arctan(\omega)$. In (9.1), $\{a_0 a_1 a_2 \ldots a_M b_1 b_2 \ldots b_N\}$ are real coefficients. b_0 is normalized to unity. M and N are the orders of the numerator and denominator, respectively. For simplicity of the program realization, $M = N - 1$ and $M = N$ are chosen in the rational and bilinear function formats, respectively.

If the transfer function in (9.1) is in rational function format, the neuro-transfer-function model structure is the same as that in Ding *et al.* (2004). However, ordinary power series $\{(j\omega)^0 (j\omega)^1 (j\omega)^2 \ldots (j\omega)^N\}$ in rational function format has a very large dynamic range if the order N is high or the bandwidth of frequency ω is wide (Beyene, 2001). Such large dynamic range makes the value of the transfer function sensitive to the coefficients $\{a_0 a_1 a_2 \ldots a_M b_1 b_2 \ldots b_N\}$. Consequently, it will affect the training efficiency and the final model accuracy.

Because of the above phenomenon, we introduce the bilinear function format into the neuro-transfer-function model as an expansion beyond Ding *et al.* (2004) since the magnitude of power series $\{e^0 e^{-j\theta} e^{-j2\theta} e^{-j3\theta} \ldots e^{-jN\theta}\}$ in bilinear function format is a constant and only the phase is varying regardless of the order and the frequency range. Moreover, the relationship of coefficients between these two formats is linear (Kang, 2006). It is easy to convert the coefficients between these two different formats through the Pascal matrix (Psenicka *et al.*, 2002) which only depends on the orders of transfer functions.

9.3. Typical Training Methods for Developing the Neuro-Transfer-Function Model

This section introduces three typical training methods: variable-order method, constant-order method, and the direct optimization method to develop the neuro-transfer-function model. In variable-order method, the orders of transfer functions vary with different geometries. Constant-order method is similar to variable-order method except that the orders of transfer functions are fixed to the maximum one among all geometries. In the direct optimization method, the combined neural network and transfer functions are directly optimized to approximate the EM responses.

All typical training methods begin with EM data $(\boldsymbol{x}_k, \boldsymbol{d}_k)$, where the subscript k is the index indicating the kth set of values for geometrical parameters $(k = 1, 2 \ldots n)$, and n is the number of training geometries in the geometrical parameter space. Inputs $\boldsymbol{x}$ represents the geometrical parameters excluding frequency. Frequency is an input for the rational or bilinear transfer functions. Real and imaginary parts of S-parameters generated from detailed EM simulations are represented by the symbol $\boldsymbol{d}$.

Next, the parameter extraction is repetitively carried out for each sample in the geometrical parameter space (also called each geometry for convenience) and subsequently coefficients of the transfer function are obtained by using the vector fitting method (Gustavsen and Semlyen, 1999). Using S-parameters as inputs and with an initial guess of poles, vector fitting method is used iteratively to refine the poles and obtain the coefficients in a rational transfer function. In variable-order method, the orders of transfer functions vary different geometries. The orders of transfer functions are determined by repetitively applying vector fitting starting with a low-order transfer function, and increasing the order until the error between the values of transfers functions and training data (i.e., S-parameters $\boldsymbol{d}$) drops below a threshold level. Constant-order method is similar to variable-order method except that the orders of transfer functions are fixed to the maximum one among all geometries.

Once the coefficients in transfer functions are extracted, a pure neural network is trained to learn the mapping between the geometrical variables and the extracted coefficients. In the direct optimization method, there is no separate training of the ANN to learn the extracted coefficients. Finally, the neural network using the results from the previous step as initial solutions in combination with transfer functions are refined to approximate the EM responses.

9.4. An Advanced Training Method for Developing the Neuro-Transfer-Function Model

The above typical training methods for developing the neuro-transfer-function model are important advances of parametric modeling of passive components without having to rely on the prior knowledge. However, these training methods may not be adequate to ensure the model accuracy in complicated applications involving high-order transfer functions, wide frequency range, and/or large geometrical variations. For example, numerical solutions of coefficients obtained by the parameter extraction cannot be guaranteed continuous with respect to the geometrical variables even though the EM responses vary continuously. Consequently, the ANN training will be ineffective because of the discontinuity of coefficients in the training data, and the final model obtained will not be accurate enough. This problem is especially prominent when the order of transfer function is high, the frequency range is wide, and/or the geometrical variations are large.

This section presents an advanced training method, called generalized-order method, to develop the neuro-transfer-function model covering challenging applications involving high-order transfer functions, wide frequency range, and/or large geometrical variations (Cao *et al.*, 2009). The discontinuity problem of coefficients in transfer functions with respect to geometrical variables is addressed. Minimum effective orders of transfer functions for different regions of geometrical parameter space are identified. Investigations show that varied orders for different regions will result in abrupt changes thus the discontinuity of coefficients. An order-changing module is formulated to bridge the gaps between the orders used for different regions. This module can guarantee the continuity of coefficients and simultaneously maintain the model accuracy through an ANN optimization process.

9.4.1. *Problem Analysis: Discontinuity of the Coefficients in Transfer Functions*

The major issue is the discontinuity of coefficients in transfer functions with respect to the geometrical variables. This discontinuous phenomenon will result in two problems. One is that it is hard for the ANN to learn the discontinuous mapping between the geometrical variables and the coefficients. Thus the final models obtained will not be accurate enough even after many iterations of training. The other is that even if the ANN is accurately trained to predict the discontinuous coefficients with more hidden neurons, it will result in the overlearning problem (Zhang and Gupta, 2000). Thus it will lead to a large testing error when the testing data happen to be in the discontinuous region.

When the geometrical variables of passive components have large variations, the corresponding EM responses for different geometries could be substantially different, resulting in transfer functions of different orders in order to represent the EM responses accurately. Lower orders for transfer functions are needed for some simple responses and high orders are needed for some complicated responses. Otherwise, if transfer functions for all geometries use the low orders, the model accuracy cannot be guaranteed. If transfer functions for all geometries use the high orders, it will result in the non-unique numerical solutions of the coefficients. Our investigations show that the orders of transfer functions have an important effect on the discontinuity of coefficients.

There are often two ways to define the orders of transfer functions. One way is to fix all orders to the maximum one among all geometries, i.e., constant-order method. The higher the order is, the more accurate approximation can be obtained. However, when using a high-order transfer function for a low-order problem, a typical coefficient extraction process will lead to the non-unique numerical solutions. For example in the vector fitting method, using a high-order transfer function for a low-order problem will force the over-determined matrix (Gustavsen, 2006) in the fitting equations to become highly ill-conditioned and singular. It is well known that as the condition number of the matrix becomes high, a small perturbation error can result in a quite different solution (Golub and Van Loan, 1996). In this case, as the geometrical parameters vary, the coefficients obtained from the vector fitting method will be non-unique and arbitrary, causing the discontinuity of the coefficients even though the EM responses vary continuously.

Another way to define the orders of transfer functions, which can resolve the non-uniqueness problem described above, is to use minimum effective order of the transfer function for each geometry, i.e., variable-order method. This leads to different orders of transfer functions for different geometries. However, these varied orders will result in abrupt changes of the coefficients. Table 9.1 shows an example how this discontinuous problem happens when the orders are changed, e.g., when a new pole is added. Suppose that when the geometrical parameters length (L) and width (W) of passive components switch across the boundary from $[L_1 W_1]^T$ to $[L_1 + \Delta L W_1 + \Delta W]^T$, we need to add a new pole p_2 and residue Δr_2 in order to match the corresponding EM responses accurately. When ΔL and ΔW approach to zeros, the values of $H_1(s)$ and $H_2(s)$ for these two geometries are the same. But coefficients $\{a_0\, a_1\, a_2\, b_0\, b_1\, b_2\}$ in $H_1(s)$ and $H_2(s)$ do not approach each other even if Δr_1, Δr_2 and Δp_1 become zeros. In other words, the limits of corresponding coefficients of $H_2(s)$ are quite different from those of $H_1(s)$. For example, for b_0, the value $-p_1$ in $H_1(s)$ is quite different from the limit value $p_2 p_1$ in $H_2(s)$. In this case, these coefficients in each column of Table 9.1 vary abruptly when the orders of transfer functions are changed over the geometrical variables.

Table 9.1. Discontinuity of coefficients when the orders of transfer functions are changed (e.g., adding a new pole) (Cao *et al.*, 2009).

Parameters		Transfer Functions	Corresponding coefficients					
L	W		A_0	a_1	a_2	b_0	b_1	b_2
L_1	W_1	$H_1(s) = \dfrac{r_1}{s - p_1}$	R_1	0	0	$-p_1$	1	0
$L_1 + \Delta L$	$W_1 + \Delta W$	$H_2(s) = \dfrac{r_1 + \Delta r_1}{s - (p_1 + \Delta p_1)} + \dfrac{\Delta r_2}{s - p_2}$	$-p_2(r_1 + \Delta r_1) - \Delta r_2(p_1 + \Delta p_1)$	$r_1 + \Delta r_2 + \Delta r_1$	0	$p_2(p_1 + \Delta p_1)$	$-(p_1 + p_2 + \Delta p_1)$	1
$L_1 + \Delta L$	$W_1 + \Delta W$	$\displaystyle \lim_{\substack{\Delta r_1, \Delta r_2 \to 0 \\ \Delta p_1 \to 0}} H_2(s)$	$-p_2 r_1$	r_1	0	$p_2 p_1$	$-(p_1 + p_2)$	1

9.4.2. *Generalized-Order Method for Developing the Neuro-Transfer-Function Model*

In order to guarantee the continuity of coefficients in transfer functions, this section introduces the generalized-order method for developing the neuro-transfer-function model. The idea is based on multiplying a new pole both in the numerator and denominator in transfer functions. This approach resolves the discontinuity problem of coefficients while maintaining the accuracy of transfer functions as shown in Table 9.2. When ΔL and ΔW approach to zeros (i.e., as Δr_1, Δr_2 and Δp_1 become zeros), not only the values of $H_1(s)$ and $H_2(s)$ are same, but also the coefficients $\{a_0\,a_1\,a_2\,b_0\,b_1\,b_2\}$ in $H_1(s)$ and $H_2(s)$ approach each other. For example, for b_0, the value $p_2\,p_1$ in $H_1(s)$ is the same as the limit value $p_2\,p_1$ in $H_2(s)$. In this case, these coefficients in each column of Table 9.2 are continuous when the transfer function $H_1(s)$ switches to $H_2(s)$.

This idea, while simple, requires an advanced algorithm to be used for practical problems. In reality, there will be many transfer functions $(H_1, H_2, H_3 \ldots)$ because of various regions in the high-dimensional space of geometrical parameters. How to identify the boundary for switching among different transfer functions in the geometrical parameter space, which pair of transfer functions to pick at a given switching boundary, and what values of poles to add to the transfer functions? There are no straightforward analytical answers. In this section, we introduce an advanced training algorithm that enables us to implement the idea of Table 9.2 and resolves the discontinuity problem for practical microwave modeling applications. Similar to the typical training methods, the generalized-order method also begins with EM data $(\boldsymbol{x}_k, \boldsymbol{d}_k)$, where the subscript k is the index indicating the kth geometry $(k = 1, 2 \ldots n)$, and n is the number of training samples in the geometrical parameter space.

9.4.2.1. *Parameter extraction phase*

In the first phase (i.e., the parameter extraction phase), the parameter extraction is repetitively carried out for each geometry to extract the corresponding coefficients in transfer functions. This is achieved using the vector fitting method (Gustavsen and Semlyen, 1999). Using S-parameters as inputs and with an initial guess of poles, vector fitting method is used to iteratively refine the poles and obtain the coefficients in the transfer function. The orders of transfer functions are determined by repetitively applying vector fitting starting with a low-order transfer function, and

Table 9.2. Elimination of discontinuity of coefficients by introducing a new pole (Cao *et al.*, 2009).

Parameters		Transfer Functions	Corresponding coefficients					
L	W		a_0	a_1	a_2	b_0	b_1	b_2
L_1	W_1	$H_1(s) = \dfrac{r_1(s - p_2)}{(s - p_1)(s - p_2)}$	$-p_2 r_1$	r_1	0	$p_2 p_1$	$-(p_1 + p_2)$	1
$L_1 + \Delta L$	$W_1 + \Delta W$	$H_2(s) = \dfrac{r_1 + \Delta r_1}{s - (p_1 + \Delta p_1)} + \dfrac{\Delta r_2}{s - p_2}$	$-p_2(r_1 + \Delta r_1) - \Delta r_2(p_1 + \Delta p_1)$	$r_1 + \Delta r_2 + \Delta r_1$	0	$p_2(p_1 + \Delta p_1)$	$-(p_1 + p_2 + \Delta p_1)$	1
$L_1 + \Delta L$	$W_1 + \Delta W$	$\displaystyle\lim_{\substack{\Delta r_1, \Delta r_2 \to 0 \\ \Delta p_1 \to 0}} H_2(s)$	$-p_2 r_1$	r_1	0	$p_2 p_1$	$-(p_1 + p_2)$	1

increasing the order until the error between the values of transfers functions and training data (i.e., S-parameters $\boldsymbol{d}$) drops below a threshold level E_t. The initial poles are some pairs of complex conjugate poles with imaginary parts linearly distributed over the frequency range of interest. When the order is increased by one, a new real starting pole is added. When the order is increased by two, a new pair of complex conjugate poles is added. When the order stays unchanged as the geometrical parameters vary, the same locations of starting poles are used.

Under this strategy, the minimum effective orders of transfer functions for different geometries are identified. For example, the minimum effective order of the transfer function for the kth geometry can be defined as

$$N_k = \mathop{\mathrm{Min}}_{\left\|[\mathrm{Re}(H_k(s))\,\mathrm{Im}(H_k(s))]^T - d_k\right\| \leq E_t} \{N\}, \tag{9.3}$$

where $H_k(s)$ is a N_k^{th} order transfer function in the form of (9.1) for the kth geometry. Let $\hat{\boldsymbol{C}}_{TF,k}$ represent the extracted coefficients defined as

$$\hat{\boldsymbol{C}}_{TF,k} = \begin{bmatrix} \boldsymbol{a}^{(k)} \\ \boldsymbol{b}^{(k)} \end{bmatrix} = \left[a_0^{(k)} a_1^{(k)} \cdots a_{N_k}^{(k)} b_1^{(k)} b_2^{(k)} \cdots b_{N_k}^{(k)}\right]^T, \tag{9.4}$$

where $\boldsymbol{a}^{(k)}$ is a vector defined as $\boldsymbol{a}^{(k)} = [a_0^{(k)} a_1^{(k)} a_2^{(k)} \cdots a_{N_k}^{(k)}]^T$, and $\boldsymbol{b}^{(k)}$ is a vector defined as $\boldsymbol{b}^{(k)} = [b_1^{(k)} b_2^{(k)} \cdots b_{N_k}^{(k)}]^T$. $a_i^{(k)}$ and $b_i^{(k)} (i = 1, 2 \ldots N_k)$ are the coefficients as expressed in (9.1) extracted for the kth geometry. Under the same conditions, this process is repeated to extract the coefficients $\hat{\boldsymbol{C}}_{TF,k}$ for all geometries. The maximum order among all geometries is specified as

$$N_{max} = \mathop{\mathrm{Max}}_{k \in T_r}\{N_k\}, \tag{9.5}$$

where T_r is an index set defined as $T_r = \{1, 2 \ldots n\}$, and each index in the set represents a particular geometry.

Note that the extracted coefficients obtained by the vector fitting method are in rational function format. These coefficients can be converted into those in bilinear function format through the bilinear transformation (Kang, 2006). Since this bilinear transformation is linear, the continuity properties of coefficients will not change during this transformation. At the end of this phase, the extracted coefficient pairs $(\boldsymbol{x}_k, \hat{\boldsymbol{C}}_{TF,k})$ in rational or bilinear function formats are available to accurately represent the EM responses. But the extracted coefficients $\hat{\boldsymbol{C}}_{TF,k}$ (i.e., $\boldsymbol{a}^{(k)}$ and $\boldsymbol{b}^{(k)}$, where $k = 1, 2 \ldots n$) obtained so far cannot be guaranteed to be continuous since the effective orders of transfer functions are changed over the geometrical variables.

9.4.2.2. *Order-changing phase*

In the second phase (i.e., the order-changing phase), an order-changing module is formulated to guarantee the continuity of coefficients in transfer functions and simultaneously maintain the model accuracy through an ANN optimization process. When the effective order N_k is less than the maximum order N_{max}, the gaps between the orders are bridged here. This process is performed through multiplying an extra polynomial $(1 + q_1^{(k)}s + q_2^{(k)}s^2 \cdots + q_{M_k}^{(k)}s^{M_k})$ to both the numerator and denominator in the transfer function:

$$
\begin{aligned}
H_k(s) &= \frac{\sum_{i=0}^{N_k} a_i^{(k)} s^i}{1 + \sum_{i=1}^{N_k} b_i^{(k)} s^i} \\[2mm]
&= \frac{(a_0^{(k)} + a_1^{(k)}s \cdots + a_{N_k}^{(k)}s^{N_k})(1 + q_1^{(k)}s + q_2^{(k)}s^2 \cdots + q_{M_k}^{(k)}s^{M_k})}{(1 + b_1^{(k)}s \cdots + b_{N_k}^{(k)}s^{N_k})(1 + q_1^{(k)}s + q_2^{(k)}s^2 \cdots + q_{M_k}^{(k)}s^{M_k})}, \\[2mm]
&= \frac{A_0^{(k)} + A_1^{(k)}s + A_2^{(k)}s^2 \cdots + A_{N_{max}}^{(k)} s^{N_{max}}}{1 + B_1^{(k)}s + B_2^{(k)}s^2 \cdots + B_{N_{max}}^{(k)} s^{N_{max}}} \\[2mm]
&= H(\boldsymbol{C}_{TF,k}, s)
\end{aligned}
\tag{9.6}
$$

where $\boldsymbol{C}_{TF,k}$ is defined as

$$
\boldsymbol{C}_{TF,k} = \begin{bmatrix} \boldsymbol{A}^{(k)} \\ \boldsymbol{B}^{(k)} \end{bmatrix} = \begin{bmatrix} A_0^{(k)} A_1^{(k)} \cdots A_{N_{max}}^{(k)} B_1^{(k)} B_2^{(k)} \cdots B_{N_{max}}^{(k)} \end{bmatrix}^T, \tag{9.7}
$$

$\boldsymbol{A}^{(k)}$ is a vector defined as $\boldsymbol{A}^{(k)} = [A_0^{(k)} A_1^{(k)} A_2^{(k)} \cdots A_{N_{max}}^{(k)}]^T$, and $\boldsymbol{B}^{(k)}$ is a vector defined as $\boldsymbol{B}^{(k)} = [B_1^{(k)} B_2^{(k)} \cdots B_{N_{max}}^{(k)}]^T$. $H_k(s)$ can be rewritten to $H(\boldsymbol{C}_{TF,k}, s)$ as shown in (9.6). The real and imaginary parts of the rational or bilinear function can be represented as

$$
\boldsymbol{y}(\boldsymbol{C}_{TF,k}, s) = [\text{Re}(H(\boldsymbol{C}_{TF,k}, s))\ \text{Im}(H(\boldsymbol{C}_{TF,k}, s))]^T. \tag{9.8}
$$

The order of the extra polynomial is defined as

$$
M_k = N_{max} - N_k, \tag{9.9}
$$

i.e., the difference between the effective order N_k and the maximum order N_{max}. The purpose of introducing the coefficients $\boldsymbol{A}^{(k)}$ and $\boldsymbol{B}^{(k)}$ in (9.6) is to find a way to guarantee their continuity, because the extracted values of the original coefficients $\boldsymbol{a}^{(k)}$ and $\boldsymbol{b}^{(k)}$ in (9.1) are discontinuous. The formulation works in such a way that the ANN outputs will be the coefficients $\boldsymbol{A}^{(k)}$ and $\boldsymbol{B}^{(k)}$ (instead of the extracted coefficients $\boldsymbol{a}^{(k)}$ and $\boldsymbol{b}^{(k)}$, and the transfer function to be used in the structure of our neuro-transfer-function model will be the formula (9.6) instead of the original formula (9.1). In this way, we make it possible to eliminate the discontinuity of coefficients during modeling the EM behavior over the geometrical variables. Because $\boldsymbol{A}^{(k)}$ and $\boldsymbol{B}^{(k)}$ play such an important role in interfacing the ANN with the transfer function, we call the coefficients $\boldsymbol{A}^{(k)}$ and $\boldsymbol{B}^{(k)}$ (i.e., $\boldsymbol{C}_{TF,k}$) interface coefficients in the rest of the paper.

The relationship between the coefficients of the numerator in (9.1), i.e., $\boldsymbol{a}^{(k)}$, and that in (9.6), i.e., $\boldsymbol{A}^{(k)}$, can be obtained by:

$$
\begin{bmatrix} A_0^{(k)} \\ A_1^{(k)} \\ A_2^{(k)} \\ \vdots \\ A_{N_{max}}^{(k)} \end{bmatrix} =
\begin{bmatrix}
1 & 0 & 0 & \cdots & 0 \\
q_1^{(k)} & 1 & 0 & \cdots & 0 \\
q_2^{(k)} & q_1^{(k)} & 1 & \cdots & 0 \\
\vdots & q_2^{(k)} & q_1^{(k)} & \cdots & 0 \\
\vdots & \vdots & q_2^{(k)} & \cdots & \vdots \\
\vdots & \vdots & \vdots & \cdots & \vdots \\
q_{M_k}^{(k)} & q_{M_k-1}^{(k)} & q_{M_k-2}^{(k)} & \cdots & \vdots \\
0 & q_{M_k}^{(k)} & q_{M_k-1}^{(k)} & \cdots & 0 \\
0 & 0 & q_{M_k}^{(k)} & \cdots & 1 \\
\vdots & 0 & 0 & \cdots & q_1^{(k)} \\
\vdots & \vdots & 0 & \cdots & \vdots \\
\vdots & \vdots & \vdots & \cdots & q_{M_k-1}^{(k)} \\
0 & 0 & 0 & \cdots & q_{M_k}^{(k)}
\end{bmatrix}
\begin{bmatrix} a_0^{(k)} \\ a_1^{(k)} \\ a_2^{(k)} \\ \vdots \\ a_{N_k}^{(k)} \end{bmatrix}
\tag{9.10a}
$$

228   *Yazi Cao et al.*

or

$$A^{(k)} = F(q^{(k)})a^{(k)}, \tag{9.10b}$$

where $\boldsymbol{F}$ represents the square matrix in (9.10a), and $\boldsymbol{q}^{(k)} = [q_1^{(k)} q_2^{(k)} \cdots q_{M_k}^{(k)}]^T$ are the parameters in the extra polynomial for the kth geometry. From (9.10), the coefficients $\boldsymbol{a}^{(k)}$ can be back-deduced from $\boldsymbol{A}^{(k)}$. The relationship between the coefficients of two denominators $\boldsymbol{b}^{(k)}$ and $\boldsymbol{B}^{(k)}$ are similar to (9.10) except $b_0^{(k)} = 1$ and $B_0^{(k)} = 1$. However, the parameters $\boldsymbol{q}^{(k)}$ are unknown since the newly added poles are different for different geometries. These parameters also cannot be obtained by conventional techniques such as vector fitting method due to the non-uniqueness problem.

Here an ANN optimization process is used to determine the parameters $\boldsymbol{q}^{(k)}$ as shown in Fig. 9.2. It consists of several order-changing units and a neural network. Each geometry has a unique order-changing unit. The inputs of the overall module are geometrical variables $\boldsymbol{x}_k$ (e.g., $L, W, S, h \ldots$), while the outputs are the extracted coefficients $\hat{\boldsymbol{C}}_{TF,k}$ (i.e., $\boldsymbol{a}^{(k)}$ and $\boldsymbol{b}^{(k)}$ obtained from the parameter extraction phase. The module

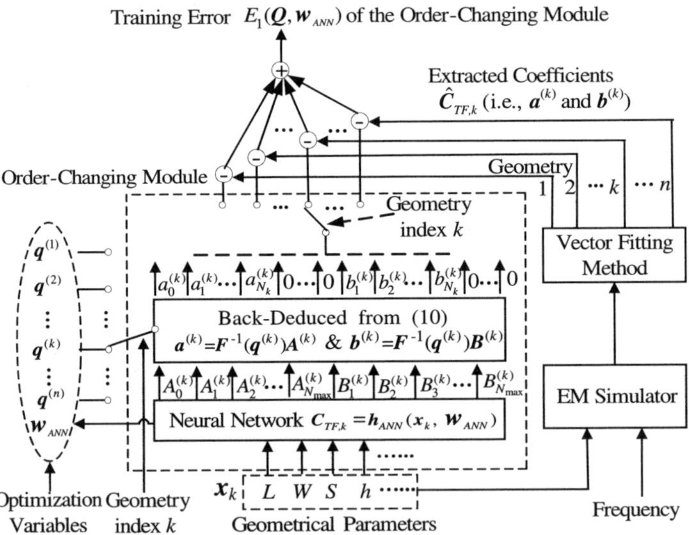

Fig. 9.2. Structure of the order-changing module, and the formulation of optimization to determine the parameters $\boldsymbol{q}^{(k)}$ and $\boldsymbol{w}_{ANN}$ in this module. Note that for different values of geometrical parameters in the training data, the module uses the common ANN weights $\boldsymbol{w}_{ANN}$ but different parameters $\boldsymbol{q}^{(k)}$ (Cao et al., 2009).

also has an additional input k for identifying the index of geometry samples. The minimum effective orders of transfer functions for different geometries are different as defined in (9.3). Note that the number of outputs of the module is a constant determined by the maximum order N_{max} among all geometries. When the effective order for a particular geometry is lower than the maximum order N_{max}, the deficient parts in the outputs are set to be zeros. The outputs of the ANN are the interface coefficients $C_{TF,k}$ (i.e., $A^{(k)}$ and $B^{(k)}$ after converting the effective order N_k to the maximum order N_{max} through the multiplication of the extra polynomial as shown in (9.6). The ANN can be represented as

$$C_{TF,k} = h_{ANN}(x_k, w_{ANN}),\qquad(9.11)$$

where w_{ANN} is a vector of the ANN internal weights in h_{ANN}. This overall optimization process is carried out by adjusting the parameters of order-changing units

$$Q = [\overbrace{q_1^{(1)} q_2^{(1)} \cdots q_{M_1}^{(1)}}^{q^{(1)}} \overbrace{q_1^{(2)} q_2^{(2)} \cdots q_{M_2}^{(2)}}^{q^{(2)}} \cdots\cdots \overbrace{q_1^{(n)} q_2^{(n)} \cdots q_{M_n}^{(n)}}^{q^{(n)}}]^T \qquad(9.12)$$

and the common weights w_{ANN} of the ANN to minimize the error function

$$E_1(Q, w_{ANN})$$
$$= \frac{1}{2n} \sum_{k=1}^{n} \left\| diag\{F^{-1}(q^{(k)}), F^{-1}(q^{(k)})\} \cdot h_{ANN}(x_k, w_{ANN}) - \hat{C}_{TF,k} \right\|^2.$$
$$(9.13)$$

This optimization process is accomplished using the quasi-Newton algorithm in NeuroModelerPlus. This optimization simultaneously trains the ANN and processes order changing. By formulating $q^{(1)}$, $q^{(2)}$,..., and $q^{(n)}$ as optimization variables, our method allows the order-changing between transfer functions for different geometries to be automatically sorted out. The optimization also automatically determines the values of poles which are added to transfer functions in order to achieve order changing across the boundaries of geometrical regions. At the end of this phase, since all geometry samples are involved in a batch-mode optimization and the parameters $q^{(k)}$ are constrained by the continuous behavior of the ANN, the interface coefficients $C_{TF,k}$ (i.e., $A^{(k)}$ and $B^{(k)}$ of (9.7) obtained by the order-changing module can be guaranteed to be continuous while maintaining the approximate accuracy of transfer functions.

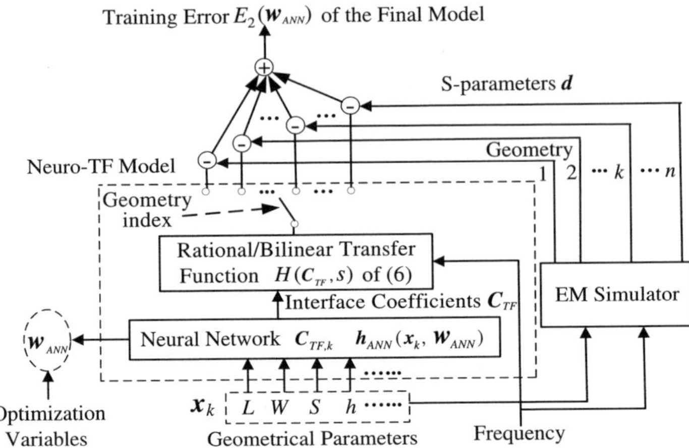

Fig. 9.3.   Structure of the neuro-transfer-function (neuro-TF) model, and the formulation of refinement training to finalize the ANN weights w_{ANN} (Cao *et al.*, 2009).

9.4.2.3. *Model refinement and testing phase*

In the third phase (i.e., the model refinement phase), a refinement optimization is performed to refine the final outputs of the overall model. The structure of the model refinement phase is shown in Fig. 9.3. It consists of the rational or bilinear function of (9.6) and the ANN obtained from the order-changing phase as initial solutions. This optimization process is carried out through adjusting the weights w_{ANN} of the ANN to minimize the error function

$$E_2(w_{ANN}) = \frac{1}{2n} \sum_{k=1}^{n} \sum_{l=1}^{N_s} \left\| y(h_{ANN}(x_k, w_{ANN}), s_l) - d_{k,l} \right\|^2, \qquad (9.14)$$

where N_s is the number of frequency points. y as defined in (9.8) represent the outputs of the overall model as a function of frequency point s_l, as well as x_k, and w_{ANN} which are expressed through the ANN. Note that the frequency point $s_l(l = 1, 2 \ldots N_s)$ should be normalized in the parameter extraction phase. At the end of this phase, a refined neuro-transfer-function model is obtained for independent quality assessment. Although the whole training process involved poles, coefficients, and modules, the final model is simple, consisting of the trained ANN and the rational or bilinear transfer function of (9.6).

In the final phase (i.e., the testing phase), an independent set of testing data which are never used during the training is utilized to test the quality of the trained neuro-transfer-function model. If the testing error is acceptable, this model is ready to be used for high-level applications. Otherwise, the model refinement phase will be repeated using incremental EM data. A flowchart demonstrating the neuro-transfer-function model development process is shown in Fig. 9.4. In this flow, E_{oc}, E_{train}, and E_{test} are defined as user-specified error criterions in the order-changing phase, the model refinement phase, and the model testing phase, respectively.

9.4.3. *Discussion*

Notice that the extracted coefficients $\hat{C}_{TF,k}$ (i.e., $a^{(k)}$ and $b^{(k)}$ in (9.4)) obtained by the parameter extraction are discontinuous since the effective orders of transfer functions are changed over the geometrical variables. Had we used the direct parameter extraction also to compute the interface coefficients $C_{TF,k}$ (i.e., $A^{(k)}$ and $B^{(k)}$ in (9.7), the coefficients obtained would be non-unique and discontinuous because of the use of high-order transfer functions for low-order problems. In contrast, by using the generalized-order method, the interface coefficients $C_{TF,k}$ (i.e., $A^{(k)}$ and $B^{(k)}$ in our final model are guaranteed to be continuous, while maintaining the model accuracy.

The generalized-order method in bilinear function format can be used for both low- and high-order cases of modeling. For the case of relatively low-order transfer functions, e.g., $N_{max} \leq 14$, the generalized-order method in rational function format, which is simpler, can also be used.

For modeling multiport components requiring multiple transfer functions (e.g., three transfer functions to represent S_{11}, S_{12} and S_{22}) simultaneously, the generalized-order method works in the same way as described except that the multiple transfer functions will share a common denominator.

More specifically, we will have a common set of denominator coefficients $b^{(k)}$ and multiple sets (e.g., three sets) of numerator coefficients $a^{(k)}$ in the parameter extraction phase. Similarly there will be a common set of interface denominator coefficients $B^{(k)}$, and multiple sets (e.g., three sets) of interface numerator coefficients $A^{(k)}$ in the order-changing phase. The optimization variables in the order-changing phase, i.e., $q^{(k)}$, $k = 1, 2, \ldots, n$ will be common for the multiple transfer functions.

In addition to the orders of transfer functions as discussed above, our investigations show that some other factors may also have effects on the continuity of the coefficients, including: (i) mesh density in EM simulations,

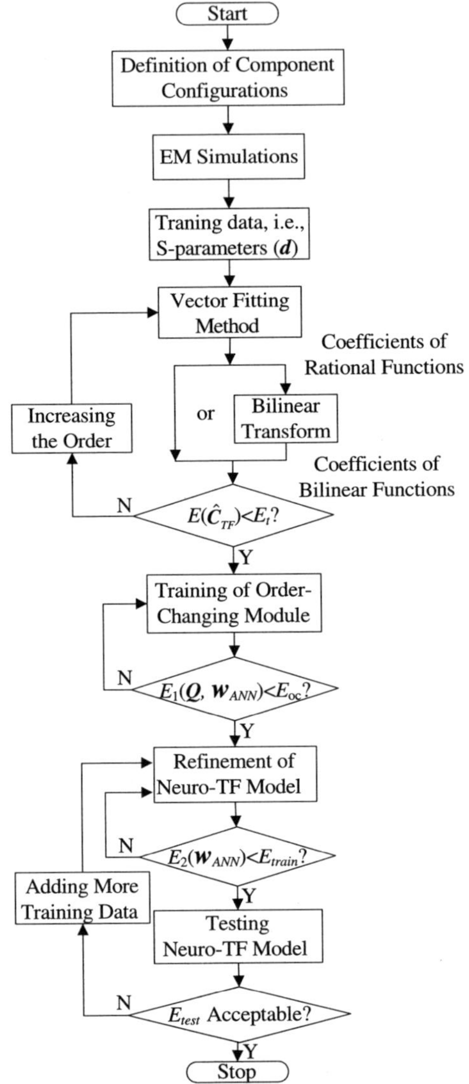

Fig. 9.4. Flowchart of the generalized-order method for developing the neuro-transfer-function (neuro-TF) model. $E(\hat{C}_{TF})$ represents the error between the transfer function of (9.1) with the extracted coefficients $\hat{C}_{TF}$ and the training data, i.e., S-parameters d. $E_1(Q, w_{ANN})$ and $E_2(w_{ANN})$ are defined in (9.13) and (9.14), respectively (Cao *et al.*, 2009).

(ii) locations of initial poles in vector fitting method, (iii) the number of iterations in vector fitting method, and (iv) frequency sampling strategy. In order to minimize the discontinuity effect due to these factors, the following guidelines can be used: adequate mesh density in EM simulations suitable for different values of the geometrical variables, fixed locations of initial poles for a given order in vector fitting method, use of same iterations in the vector fitting method, and uniform sampling frequency steps. Under the above guidelines, the continuity of coefficients will depend on one main factor: the orders of transfer functions.

9.5. Application Examples

9.5.1. *Illustration of the Discontinuity Problem Using the Patch Antenna Model*

The structure of a slotted patch antenna (Soliman *et al.*, 2004) is illustrated in Fig. 9.5. It has four input geometrical variables: length (L) and width (W) of the main radiation board, space (s) and length (l) of slots. The length and width of feed lines are fixed to 30 mm and 6 mm, respectively. The substrate parameters are as follows: relative permittivity of substrate $\varepsilon_r = 2.17$, thickness of substrate $h = 0.508$ mm and conductor thickness $T = 0.017$ mm. EM-field simulations are accurate but time consuming when the values of geometrical parameters vary repetitively. Here we apply different training methods to model EM responses versus the geometrical parameters and frequency.

For convenient illustration of the continuity of coefficients, only two geometrical parameters $x = [LW]^T$ are used as the input variables which vary synchronously and continuously. The model has two outputs, i.e., $y = [RS_{11}\ IS_{11}]^T$, which are real and imaginary parts of S_{11}. The training

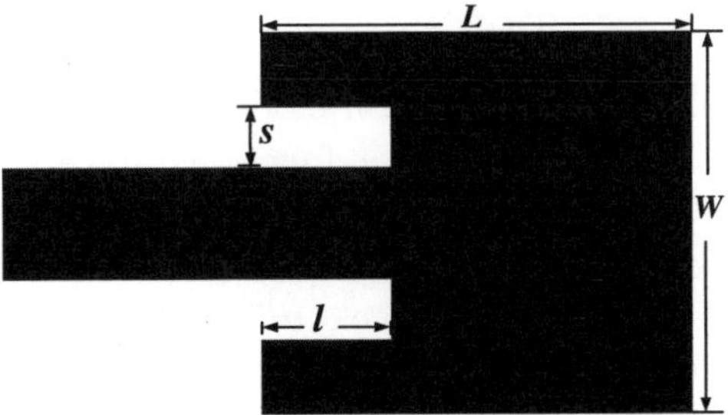

Fig. 9.5.   Structure of a slotted patch antenna (Cao *et al.*, 2009).

Table 9.3.   Definition of training and testing data for patch antenna example (Cao *et al.*, 2009).

Parameters	Training Data (78 sets)			Testing Data (77 sets)		
	Min	Max	Step	Min	Max	Step
L (mm)	16.5	55	0.5	16.75	54.75	0.5
W (mm)	26.5	65	0.5	26.75	64.75	0.5

and testing data generated from detailed EM simulator ADS Momentum (2008) are defined in Table 9.3. Space (s) and length (l) of the slots are fixed to 10 mm and 5 mm, respectively. The frequency range is from DC to 4 GHz with a step size of 25 MHz.

The 78 training sets of S_{11} generated from EM simulations of 78 sets of $[LW]^T$ values are depicted in Fig. 9.6. These training sets will be used as the inputs in the parameter extraction phase. As shown in this figure, the shapes of S-parameter curves vary continuously with respect to the geometrical variables.

The generalized-order method for developing the neuro-transfer-function model is applied in this example. In the parameter extraction phase, the effective orders $N_k(k = 1, 2 \ldots 78)$ of transfer functions vary from 7 to 16 over the 78 sets of geometrical variables. The maximum order N_{max} is 16. Here we use bilinear function format since the ordinary power series $\{(j\omega)^0 (j\omega)^1 (j\omega)^2 \cdots (j\omega)^{16}\}$ in rational function format have a very large dynamic range. The algorithm of order-changing phase as described in Section 9.4.2 is then applied to train the order-changing module using the extracted coefficients obtained from the parameter extraction phase. The average error $E_1(\boldsymbol{Q}, \boldsymbol{w}_{ANN})$ as defined in (9.13) is reduced to 0.413%. Next, a refinement phase using above results as initial solutions is performed to improve the model accuracy to 0.211% of average training error $E_2(\boldsymbol{w}_{ANN})$ as defined in (9.14).

For comparison purpose, we also use three typical training methods to develop the neuro-transfer-function model in this example. In variable-order method, the orders of transfer functions are always the minimum effective ones which vary between 7 and 16 over different geometries. The parameter extraction process is firstly performed to extract the corresponding coefficients $\hat{\boldsymbol{C}}_{TF,k}$ (i.e., $\boldsymbol{a}^{(k)}$ and $\boldsymbol{b}^{(k)}$. Next, a four-layer neural network, which has better generalization capability than the three-layer neural network in this example (Tamura and Tateishi, 1997), tries to learn the mapping between the geometrical variables and the extracted

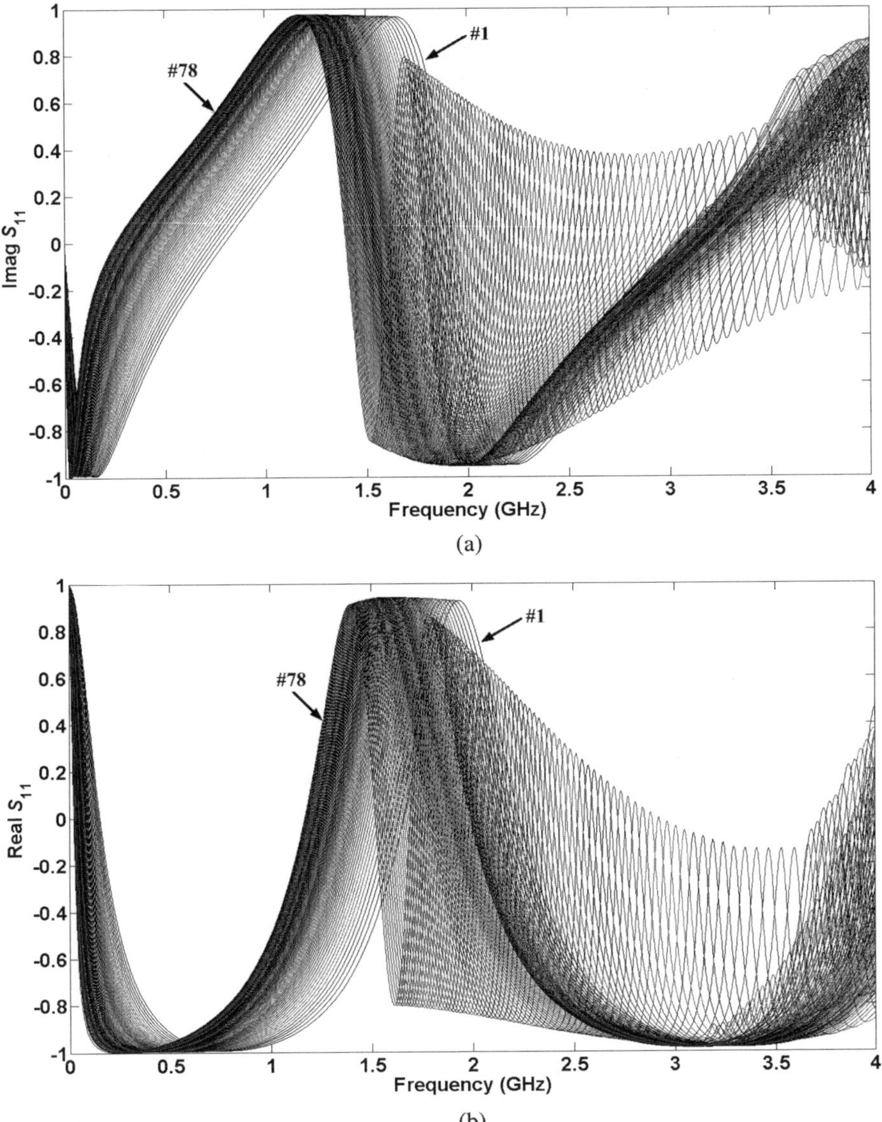

Fig. 9.6. Real part (a) and imaginary part (b) of S_{11} under continuously varied geometrical parameters for the patch antenna example. There are 78 curves in each figure corresponding to 78 different sets of geometrical variables. As shown in this figure, the shapes of S_{11} curves vary continuously with respect to the geometrical variables (Cao *et al.*, 2009).

coefficients $\hat{C}_{TF,k}$. At last, the neural network, using the results from the above phase as initial solutions in combination with bilinear functions, is trained to approximate the EM responses. Constant-order method is similar to the variable-order method except that where the orders of transfer functions are fixed to the maximum one among all geometries, i.e., 16. In the direct optimization method, the combined four-layer neural network and transfer functions are optimized to approximate the EM responses directly.

Comparisons among the results from these training methods are included in Table 9.4. The training time between various methods is similar except that there is an incremental time due to the order-changing phase in the generalized-order method. In constant-order method, we trained the model with two different four-layer ANN structures, i.e., 2-20-30-33 (2 neurons in input layer, 20 hidden neurons in second layer, 30 hidden neurons in third layer, and 33 neurons in output layer) and 2-30-40-33.

Note that the training of all ANN structures shown in this chapter is done with optimized selection of the number of hidden neurons. With less hidden neurons (the same number of hidden neurons as in the generalized-order method), it is difficult for the ANN in the constant-order method to directly learn the discontinuous mapping between the extracted coefficients and the geometrical variables. As a consequence, it is hard to reduce the training and testing errors even though the refinement phase is used. With more hidden neurons, the training error is reduced, but the testing error is still relatively large since several testing data happen to be in the discontinuous region. A similar situation happens with the

Table 9.4. Comparison of various methods for patch antenna example (Cao *et al.*, 2009).

Method	ANN Structure	Coeff. Continuity?	Average Training Error	Average Testing Error
Constant-Order Method	2-20-30-33	No	2.23%	3.06%
	2-30-40-33	No	0.27%	0.89%
Variable-Order Method	2-20-30-33	No	0.85%	2.46%
	2-30-40-33	No	0.36%	1.01%
Direct Optimization Method	2-30-40-33	Yes	38.8%	38.85%
Generalized-Order Method	2-20-30-33	Yes	0.21%	0.30%

variable-order method as well. The direct optimization method cannot train the model accurately in this complicated problem because of the lack of prior knowledge of coefficients. In contrast, the generalized-order method not only obtains the smallest training error but also the smallest testing error even with less hidden neurons. This is due to the fact that there are parameters Q, working with the ANN, to learn the coefficients in the order-changing phase. Also, the training error and testing error are consistent due to the assurance of the continuity of coefficients by the order-changing module.

Figure 9.7 depicts the continuity of coefficients in bilinear functions obtained by the generalized-order method, in comparison with variable-order method and constant-order method, for this patch antenna example. In constant-order method, where the orders of transfer functions are fixed to the maximum one among all geometries, i.e., 16, the extracted coefficients $\hat{C}_{TF,k}$ (i.e., $a^{(k)}$ and $b^{(k)}$) are arbitrary and discontinuous as shown in Fig. 9.7. This is because of the use of a high-order transfer function (e.g., order $= 16$) for low-order problems (e.g., order $= 7$), which leads to the highly ill-conditioned problem, thus the non-unique solutions of coefficients. In variable-order method, the orders of transfer functions are always the minimum effective ones varying between 7 and 16 over different geometries. In variable-order method, the extracted coefficients $\hat{C}_{TF,k}$ (i.e., $a^{(k)}$ and $b^{(k)}$) vary abruptly (i.e., are discontinuous) since the effective orders are changed over the geometrical variables. On the other hand, although the extracted coefficients $\hat{C}_{TF,k}$ (i.e., $a^{(k)}$ and $b^{(k)}$) from the parameter extraction phase are discontinuous, the interface coefficients $C_{TF,k}$ (i.e., $A^{(k)}$ and $B^{(k)}$) obtained by the generalized-order method vary continuously and smoothly over the geometrical variables since the order-changing module eliminates the discontinuity of coefficients.

9.5.2. *Parametric Modeling of a Bandstop Microstrip Filter*

In this example, different training methods are applied to develop the parametric neuro-transfer-function model of bandstop microstrip filters with quarter-wave resonant open stubs (Bakr *et al.*, 2000), as illustrated in Fig. 9.8, where L_1 and L_2 are the open stub lengths, W_1 and W_2 are the width, and L_0 is the length of the middle connected line. An alumina substrate with thickness $h = 0.635\,\text{mm}$ (25 mils), width $W_0 = 0.635\,\text{mm}$, dielectric constant $\varepsilon_r = 9.4$, and conductor thickness $T = 0.035\,\text{mm}$ is used for a $50\,\Omega$ feed line.

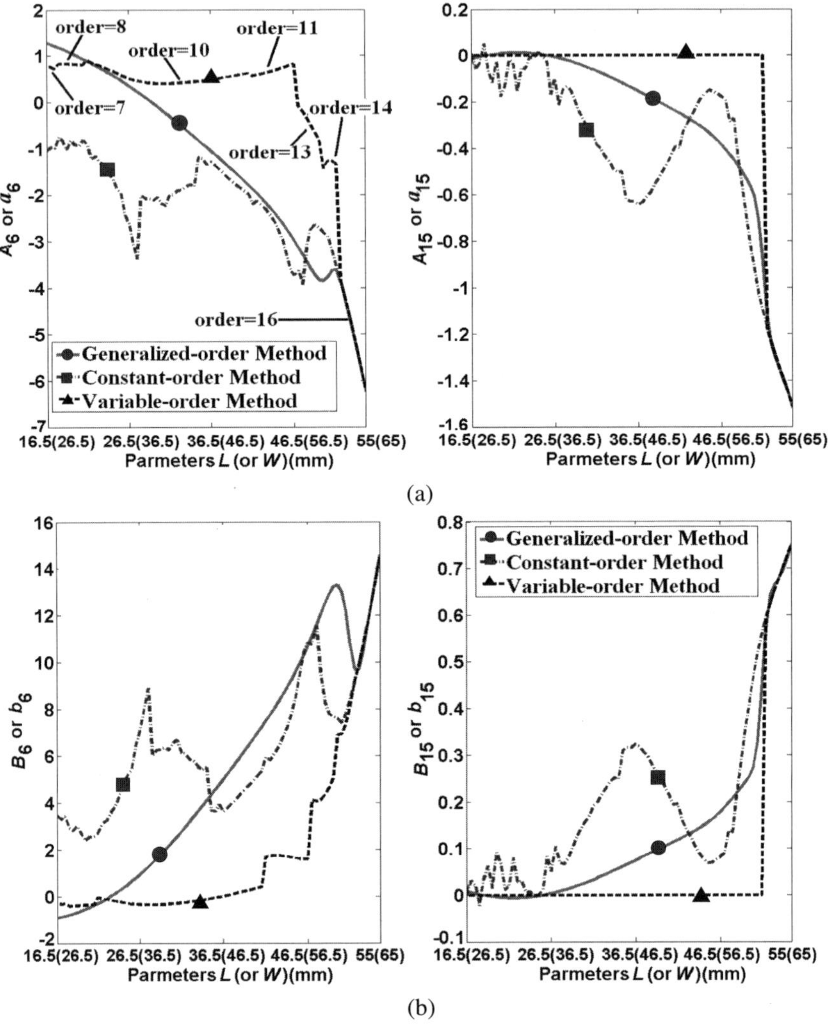

Fig. 9.7. Illustration of the continuity of coefficients versus geometrical variables L or W for the patch antenna example. The coefficients shown here are the interface coefficients $(A_6^{(k)}, A_{15}^{(k)}, B_6^{(k)},$ and $B_{15}^{(k)})$ obtained by the generalized-order method and the extracted coefficients $(a_6^{(k)}, a_{15}^{(k)}, b_6^{(k)},$ and $b_{15}^{(k)})$ obtained by constant-order and variable-order methods. Different k corresponds to different geometry, e.g., $k = 1$ means L (or W) $= 16.5$ mm (26.5 mm), and $k = 10$ means L (or W) $= 26.5$ mm (36.5 mm). Notice that in variable-order method the orders of transfer functions are changed over the geometrical variables L or W. The orders at different segments of the a_6 curve are indicated in the top-left subfigure. As shown in this figure, the coefficients obtained by constant-order and variable-order methods are discontinuous over the geometrical variables, while the coefficients obtained by the generalized-order method are continuous (Cao *et al.*, 2009).

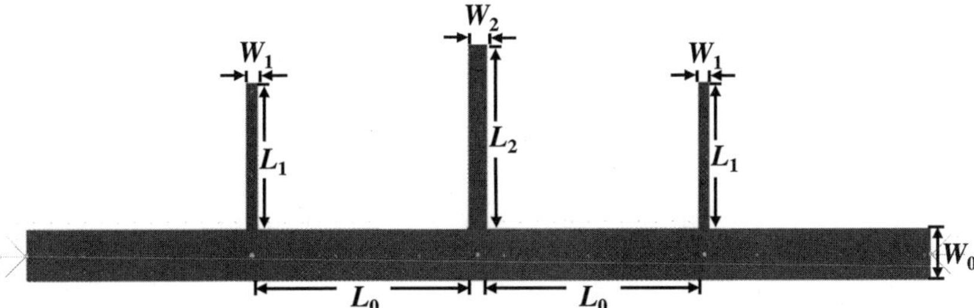

Fig. 9.8. Structure of a bandstop microstrip filter (Cao *et al.*, 2009).

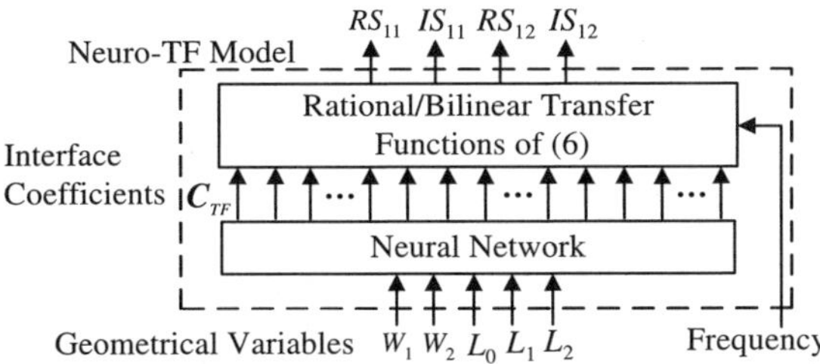

Fig. 9.9. Structure of the parametric neuro-tranfer-function (neuro-TF) model for bandstop microstrip filters (Cao *et al.*, 2009).

Table 9.5. Definition of training and testing data for bandstop filter example (Cao *et al.*, 2009).

Geometrical Parameters	Training Data (450 sets)			Testing Data (150 sets)		
	Min	Max	Step	Min	Max	Step
W_1 (mm)	0.1	0.35	0.05	0.13	0.33	0.05
W_2 (mm)	0.1	0.35	0.05	0.13	0.33	0.05
L_0 (mm)	1.27	3.3	0.13	1.35	3.25	0.13
L_1 (mm)	1.27	3.3	0.13	1.35	3.25	0.13
L_2 (mm)	1.27	3.3	0.13	1.35	3.25	0.13

This model has five input geometrical variables, i.e., $\boldsymbol{x} = [W_1 \, W_2 \, L_0 \, L_1 \, L_2]^T$, plus frequency as an additional input of transfer functions, as illustrated in Fig. 9.9. The model has four outputs, i.e., $\boldsymbol{y} = [RS_{11} \, IS_{11} \, RS_{12} \, IS_{12}]^T$, which are real and imaginary parts of S_{11} and S_{12}. The training and testing data generated from detailed EM simulator ADS Momentum are defined in Table 9.5, with data samples determined by the design of experiment method (Schmidt and Launsby, 1992). In order to represent S_{11} and S_{12} simultaneously, the model uses two transfer functions sharing a common denominator.

The generalized-order method for developing the neuro-transfer-function model is applied to four different cases, i.e., Cases 1, 2, and 3 in rational function format, and Case 4 in bilinear function format. In Case 1 (narrower frequency range and less variables), only $[W_1\ W_2]^T$ in Table 9.8 are set as the variables. L_0, L_1 and L_2 are all fixed to 2.54 mm. The frequency range is from 9 GHz to 13 GHz with a step size of 25 MHz, and the minimum effective orders N_k of rational functions vary slightly from 7 to 8, i.e., the maximum order $N_{max} = 8$. Case 2 (narrower frequency range and more variables) is similar to Case 1 except that five input parameters in Table 9.8 are all set as input variables, and the effective orders of rational functions vary from 5 to 10, i.e., $N_{max} = 10$. Case 3 (wider frequency range and more variables) is similar to Case 2 except that the frequency range is enlarged. The enlarged frequency range is from 6 GHz to 15 GHz, and the effective orders of rational functions vary from 6 to 15, i.e., $N_{max} = 15$. Case 4 (wider frequency range and more variables) is similar to Case 3 except that the bilinear function format instead of the rational function format is used. The algorithm of the order-changing phase as described in Section 9.4.2 is then applied to train the order-changing module using the extracted coefficients obtained from the parameter extraction phase. The average errors $E_1(\boldsymbol{Q},\ \boldsymbol{w}_{ANN})$ as defined in (9.13) are reduced to 0.527%, 0.647%, 0.589%, and 0.623% for the four cases, respectively. Next, a refinement phase as described in Section 9.4.2 using above results as initial solutions is performed to refine the overall models. The average errors $E_2(\boldsymbol{w}_{ANN})$ as defined in (9.14) are reduced to 0.633%, 0.686%, 4.927%, and 1.029% for the four cases, respectively.

For comparison purpose, we also apply variable-order method, constant-order method, and the direct optimization method to develop the neuro-transfer-function model for these four cases. Tables 9.6–9.9 compare these various methods in terms of the size of ANN structures, coefficients' continuity, and average training and testing error. As shown in Table 9.6 for Case 1, since the effective orders are almost the same over geometrical variables, there is no obvious discontinuity in transfer function coefficients. Consequently, the constant-order and variable-order methods become special cases of the generalized-order method. All methods except the direct optimization method obtain comparatively small training and testing errors. In Case 2, more geometrical variables make the orders of rational

Table 9.6. Comparison of various methods for bandstop filter example for Case 1 (Cao *et al.*, 2009).

	Method	ANN Structure	Coeff. Continuity?	Average Training Error	Average Testing Error
Rational Function (narrower freq. range and less variables)	Constant-Order Method	2-20-20-24	No	0.667%	0.694%
	Variable-Order Method	2-20-20-24	No	0.689%	1.091%
	Direct Optimization Method	2-20-20-24	Yes	7.23%	7.46%
	Generalized-Order Method	2-20-20-24	Yes	0.633%	0.708%

Table 9.7. Comparison of various methods for bandstop filter example for Case 2 (Cao *et al.*, 2009).

	Method	ANN Structure	Coeff. Continuity?	Average Training Error	Average Testing Error
Rational Function (narrower freq. range and more variables)	Constant-Order Method	5-30-40-30	No	7.521%	16.65%
	Variable-Order Method	5-30-40-30	No	10.04%	12.22%
	Direct Optimization Method	5-30-40-30	Yes	45.92%	46.13%
	Generalized-Order Method	5-20-30-30	Yes	0.686%	0.778%

functions vary more significantly over the geometrical variables and the coefficients obtained by vector fitting method have obvious discontinuity. As shown in Table 9.7, the model computed by the generalized-order method is much better than those by constant-order and variable-order methods, and the direct optimization method. As shown in Table 9.8 for Case 3, i.e., the high-order and wide frequency-range case, due to the large dynamic range of ordinary power series (i.e., the high order and wide frequency range in $\{(j\omega)^0 (j\omega)^1 (j\omega)^2 \ldots (j\omega)^{15}\}$) in rational functions, all methods in rational function format fail to work well. When the bilinear function format is used, all the methods lead to better accuracy than those in rational function format as shown in Table 9.9 for Case 4. However, the testing errors of constant-order and variable-order methods are still not small enough

Table 9.8.   Comparison of various methods for bandstop filter example for Case 3 (Cao *et al.*, 2009).

Method		ANN Structure	Coeff. Continuity?	Average Training Error	Average Testing Error
Rational Function (wider freq. range and more variables)	Constant-Order Method	5-30-50-45	No	6.822%	8.418%
	Variable-Order Method	5-30-50-45	No	12.41%	15.25%
	Direct Optimization Method	5-30-40-45	Yes	35.52%	36.16%
	Generalized-Order Method	5-30-40-45	Yes	4.927%	5.991%

Table 9.9.   Comparison of various methods for bandstop filter example for Case 4 (Cao *et al.*, 2009).

Method		ANN Structure	Coeff. Continuity?	Average Training Error	Average Testing Error
Bilinear Function (wider freq. range and more variables)	Constant-Order Method	5-30-50-47	No	3.162%	5.594%
	Variable-Order Method	5-30-50-47	No	2.897%	4.013%
	Direct Optimization Method	5-30-50-47	Yes	34.48%	36.61%
	Generalized-Order Method	5-30-40-47	Yes	1.029%	1.059%

because of the discontinuity of coefficients. The direct optimization method cannot train the model accurately in these complicated problems because of the lack of prior knowledge of coefficients. In contrast, the generalized-order method in bilinear function format provides not only the smallest training error but also the smallest testing error among these methods.

Figure 9.10 depicts the outputs of the model developed by the generalized-order method for three different bandstop filter geometries #1, #2, and #3 in the testing data of Case 4, and its comparison with EM data, and models developed by constant-order and variable-order methods. The geometrical parameters for the three bandstop filters are as follows: (#1) $W_1 = 0.127\,\text{mm}$, $W_2 = 0.33\,\text{mm}$, $L_0 = 2.11\,\text{mm}$, $L_1 = 3\,\text{mm}$, and $L_2 = 3.12\,\text{mm}$, (#2) $W_1 = 0.127\,\text{mm}$, $W_2 = 0.254\,\text{mm}$, $L_0 = 3.25\,\text{mm}$,

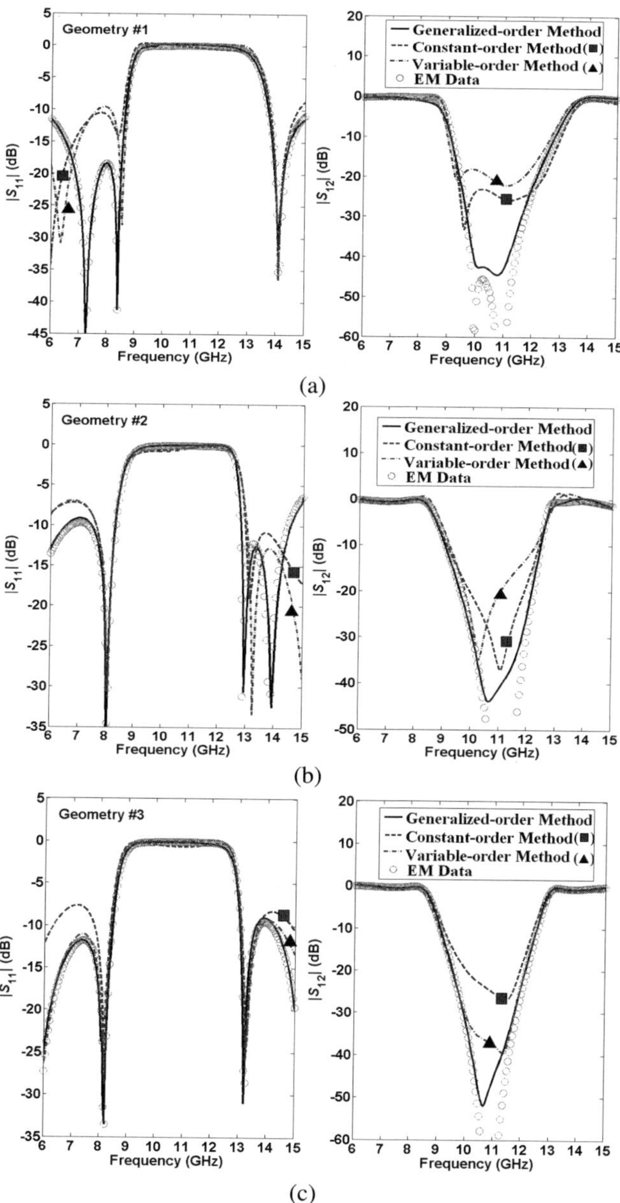

Fig. 9.10. Comparison of the magnitude of S_{11} and S_{12} of models developed by the generalized-order, constant-order, and variable-order methods, and EM data for different geometries #1 (a), #2 (b), and #3 (c) of Case 4 for the bandstop filter example (Cao *et al.*, 2009).

$L_1 = 3.12\,\text{mm}$, and $L_2 = 2.74\,\text{mm}$, (#3) $W_1 = 0.127\,\text{mm}$, $W_2 = 0.33\,\text{mm}$, $L_0 = 2.87\,\text{mm}$, $L_1 = 3.12\,\text{mm}$, and $L_2 = 2.87\,\text{mm}$. The neuro-transfer- function model developed by the generalized-order method can predict S_{11} and S_{12} accurately even though the values of geometrical variables are different from those in training.

9.6. Summary

In this chapter, a neuro-transfer-function technique for parametric modeling of microwave passive components has been described. This technique further enhances the flexibilities of the neural-based schemes in passive components modeling and optimization including EM effects. The parametric model developed by the neuro-transfer-function technique is general, and can be applied to model passive component even if accurate empirical models are unavailable. It has a great potential to develop computer-based approaches for automated generation of accurate and fast models for passive components over geometrical variables.

References

ADS Momentum (2008). Agilent Technologies, Santa Rosa: CA.

Bakr, M.H., Bandler, J.W., Ismail, M.A., Rayas-Sánchez, J.E. and Zhang, Q.J. (2000). Neural space-mapping optimization for EM-based design, *IEEE T. Microw. Theory*, **48**, 2307–2315.

Bandler, J.W., Ismail, M.A., Rayas-Sánchez, J.E. and Zhang, Q.J. (1999). Neuromodeling of microwave circuits exploiting space-mapping technology, *IEEE T. Microw. Theory*, **47**, 2417–2427.

Beyene, W.T. (2001). Improving time-domain measurements with a network analyzer using a robust rational interpolation technique, *IEEE T. Microw. Theory*, **49**, 500–508.

Burrascano, P., Fiori, S. and Mongiardo, M. (1999). A review of artificial neural networks applications in microwave computer-aided design, *Int. J. RF Microw. CE*, **9**, 158–174.

Cao, Y. and Wang, G. (2007). A wideband and scalable model of spiral inductors using space-mapping neural network, *IEEE T. Microw. Theory*, **55**, 2473–2480.

Cao, Y., Wang, G. and Zhang, Q.J. (2009). A new training approach for parametric modeling of microwave passive components using combined neural networks and transfer functions, *IEEE T. Microw. Theory*, **57**, 2727–2742.

Choi, K.L. and Swaminathan, M. (2000). Development of model libraries for embedded passives using network synthesism *IEEE T. Circuits Sys.*, **47**, 249–260.

Devabhaktuni, V.K., Chattaraj, B., Yagoub, M.C.E. and Zhang, Q.J. (2003). Advanced microwave modeling framework exploiting automatic model generation, knowledge neural networks, and space mapping, *IEEE T. Microw. Theory*, **51**, 1822–1833.

Ding, X., Devabhaktuni, V.K., Chattaraj, B., Yagoub, M.C.E., Deo, M., Xu, J. and Zhang, Q.J. (2004). Neural-network approaches to electromagnetic-based modeling of passive components and their applications to high-frequency and high-speed nonlinear circuit optimization, *IEEE T. Microw. Theory*, **52**, 436–449.

Gustavsen, B. and Semlyen, A. (1999). Rational approximation of frequency domain responses by vector fitting, *IEEE T. Power Deliver.*, **14**, 1052–1061.

Gustavsen, B. (2006). Relaxed vector fitting algorithm for rational approximation of frequency domain responses, *IEEE Signal Propagation on Interconnects Workshop*, Berlin, Germany, pp. 97–100.

Golub, G.H. and Van Loan, C.F. (1996). *Matrix Computations*, (3rd ed.), The Johns Hopkins University Press, London.

Kang, Z. (2006). An analytical pascal matrix transform for s-to-z domain transfer functions, *IEEE Signal Proc. Let.*, **13**, 597–600.

NeuroModelerPlus, Department of Electronics, Carleton University, 1125 Colonel By Drive, Ottawa, Ontario, K1S 5B6, Canada.

Psenicka, B., García-Ugalde, F. and Herrera-Camacho, A. (2002). The bilinear Z transform by pascal matrix and its application in the design of digital filters, *IEEE Signal Proc. Let.*, **11**, 368–370.

Rayas-Sánchez, J.E. (2004). EM-based optimization of microwave circuits using artificial neural networks: The state-of-the-art, *IEEE T. Microw. Theory*, **52**, 420–435.

Rayas-Sánchez, J.E. and Gutiérrez-Ayala, V. (2006). EM-based monte carlo analysis and yield prediction of microwave circuits using linear-input neural-output space mapping, *IEEE T. Microw. Theory*, **54**, 4528–4537.

Rizzoli, V., Costanzo, A., Masotti, D., Lipparini, A. and Mastri, F. (2004). Computer-aided optimization of nonlinear microwave circuits with the aid of electromagnetic simulation, *IEEE T. Microw. Theory*, **52**, 362–377.

Schmidt, S.R. and Launsby, R.G. (1992). *Understanding Industrial Designed Experiments*, Air Force Academy, Colorado Springs: CO.

Soliman, E.A., Bakr, M.H. and Nikolova, N.K. (2004). Neural networks-method of moments (NN-MoM) for the efficient filling of the coupling matrix, *IEEE T. Antenn. Propag.*, **52**, 1521–1529.

Steer, M.B., Bandler, J.W. and Snowden, C.M. (2002). Computer-aided design of RF and microwave circuits and systems, *IEEE T. Microw. Theory*, **50**, 996–1005.

Tamura, S. and Tateishi, M. (1997). Capabilities of a four-layered feedforward neural network: Four layer versus three, *IEEE T. Neural Networ.*, **8**, 251–255.

Wang, F. and Zhang, Q.J. (1997). Knowledge-based neural models for microwave design, *IEEE T. Microw. Theory*, **45**, 2333–2343.

Zhang, Q.J. and Gupta, K.C. (2000). *Neural Networks for RF and Microwave Design*, Artech House, Norwood: MA.

Zhang, Q.J., Gupta, K.C. and Devabhaktuni, V.K. (2003). Artificial neural networks for RF and microwave design — from theory to practice, *IEEE T. Microw. Theory*, **51**, 1339–1350.

Chapter 10

Parametric Sensitivity Macromodels for Gradient-Based Optimization

Krishnan Chemmangat, Francesco Ferranti,
Tom Dhaene, and Luc Knockaert

10.1. Introduction

A typical design process of high-speed microwave systems consists of different design activities, such as design optimization, design space exploration, and sensitivity analysis. The final aim of a design process is to obtain the optimal values of the design variables such as the layout and substrate features for which the system response satisfies the design specifications. The design process is usually carried out through electromagnetic (EM) simulations. Optimal values of the design variables are often determined using optimization algorithms (optimizers) which drive the EM simulator to compute the system responses and corresponding sensitivities in consecutive optimization iterations. Unfortunately, multiple EM simulations are often computationally expensive. This makes the overall design process an engineering challenge in terms of the computational burden.

The use of parametric macromodels for design space exploration, design optimization, and sensitivity analysis of complex systems is becoming increasingly important because of their efficiency in terms of computational resources with respect to EM simulators. It is important that these parametric macromodels can accurately calculate the parametric sensitivity responses over the entire design space of interest, to use this information in gradient-based design optimization. Efficient and accurate parametric sensitivity information is required by optimizers which employ state-of-the-art gradient-based techniques to calculate the optimal values of design parameters.

One of the most common approaches to calculate local sensitivities (only around one nominal point in the design space) is the adjoint variable method. The main attractiveness of this approach is that sensitivity information can be obtained from at most two systems analyses regardless of the number of designable parameters (Dounavis *et al.*, 2001; Nikolova *et al.*, 2004; Li *et al.*, 2007). However, these methods involve the calculation of the system matrix derivatives with respect to each of the design parameters, which is mostly done by finite difference approximations. Also, these methods can only compute local sensitivities.

In recent years, there has been ongoing research on interpolation-based parametric macromodeling (Deschrijver and Dhaene, 2008; Ferranti *et al.*, 2009, 2010a, 2010b; Triverio *et al.*, 2010a, 2010b). These methods interpolate at different levels a set of initial univariate macromodels, called *root macromodels*. A *root macromodel* defines the frequency response of a system as a function of frequency in a rational form. The coefficients of a *root macromodel* are found from the frequency response data using one of the most popular rational approximation techniques known as vector fitting (VF) (Gustavsen and Semlyen, 1999).

In Deschrijver and Dhaene (2008) and Ferranti *et al.* (2009, 2010a), a parametric macromodel is built by interpolating a set of *root macromodels* at an input–output level, while in Ferranti *et al.* (2010b) and Triverio *et al.* (2010a, 2010b) the interpolation process is applied to the internal state-space matrices of the *root macromodels*, therefore at a deeper level than in the transfer function-based interpolation approaches (Deschrijver and Dhaene, 2008; Ferranti *et al.*, 2009, 2010a). The methods (Ferranti *et al.*, 2010b; Triverio *et al.* 2010a, 2010b) allow to parameterize both poles and residues, hence their modeling capability is increased with respect to Deschrijver and Dhaene (2008) and Ferranti *et al.* (2009, 2010a), where only residues are parameterized.

In this chapter, we present a gradient-based minimax optimization of an EM system with the help of parameterized sensitivity macromodels (Chemmangat *et al.*, 2012). As in Ferranti *et al.* (2010b) and Triverio *et al.* (2010a, 2010b) an interpolation process on the internal state-space matrices of the *root macromodels* is performed to build the parametric macromodel. However, in these works the focus is on parametric macromodeling which ensures stability and passivity over the design space of interest. Since the main concern of this chapter is to model the parametric behavior as well as the parametric sensitivities of the frequency responses of EM systems for an efficient design optimization, preserving system properties is not strictly necessary. This allows the use of more powerful interpolation schemes which are at least continuously differentiable so that the system response and corresponding sensitivities are accurately modeled. The parametric sensitivity macromodels avoid the use of finite difference approximations in the optimization process. Also, in Ferranti *et al.* (2010b) and Triverio *et al.* (2010a, 2010b), computationally expensive linear matrix inequalities (LMI) are solved to guarantee the preservation of passivity, which can be avoided in this method. Parametric macromodels along with the corresponding parametric sensitivities are used in two pertinent numerical examples, which confirm the applicability of the technique to the efficient optimization process of EM systems. The importance of parameterized sensitivity information to speed up the design optimization process is shown in these numerical examples.

This chapter is organized as follows. A detailed description of the generation of frequency-dependent rational models (*root macromodels*) is presented in Section 10.2. In Section 10.3, the parameterization of these *root macromodels* as a function of the design variables is explained with several possible interpolation schemes. Section 10.4 discusses the analytical calculation of sensitivities of the parametric macromodels with respect to the design variables. Section 10.5 describes the gradient-based minimax optimization along with typical design constrains and the formulation of a general cost function for the optimization. Two pertinent numerical examples are presented in Section 10.6 which validate the method. Section 10.7 concludes the chapter.

10.2. Generation of *Root Macromodels*

Parametric sensitivity macromodeling starts from a set of data samples $\{(s, \vec{g})_k, \mathbf{H}(s, \vec{g})_k\}_{k=1}^{K_{tot}}$, which depend on the complex frequency $s = j\omega$ and several design variables $\vec{g} = (g^{(n)})_{n=1}^{N}$, such as layout features or substrate parameters. The first step of the whole process is to build frequency-dependent rational models which are called the *root macromodels* for each design space point. The important aspects of the generation of these *root macromodels* is explained in this section.

10.2.1. *Vector Fitting*

Vector fitting is an efficient and robust technique to build frequency-dependent rational function approximation of system responses that are defined by admittance ($\mathbf{Y}$), impedance ($\mathbf{Z}$), and scattering ($\mathbf{S}$)

representations (Deschrijver and Dhaene, 2008; Gustavsen, 2006; Gustavsen and Semlyen, 1999). Using VF, each *root macromodel* is represented in the form:

$$\mathbf{R}_{\vec{g}_k}(s) = \sum_{n=1}^{N_P} \frac{\mathbf{c}_n^{\vec{g}_k}}{s - a_n^{\vec{g}_k}} + \mathbf{d}^{\vec{g}_k}, \tag{10.1}$$

where $a_n^{\vec{g}_k}$, $\mathbf{c}_n^{\vec{g}_k}$, and $\mathbf{d}^{\vec{g}_k}$ represent poles, residues and feed forward terms, respectively at the design point $\vec{g}_k = (g_{k_1}^{(1)}, \ldots, g_{k_N}^{(N)})$. The idea of VF is to recast the nonlinear problem (10.1) into a linear problem by introducing a set of starting poles b_n, which are chosen by a rule presented in Gustavsen and Semlyen (1999) and an unknown function $\sigma(s)$ such that:

$$\begin{bmatrix} \sigma(s)\mathbf{H}_{\vec{g}_k}(s) \\ \sigma(s) \end{bmatrix} = \begin{bmatrix} \sum\limits_{n=1}^{N_P} \dfrac{\mathbf{c}_n^{\vec{g}_k}}{s - b_n} + \mathbf{d}^{\vec{g}_k} \\ \sum\limits_{n=1}^{N_P} \dfrac{\tilde{\mathbf{c}}_n^{\vec{g}_k}}{s - b_n} + 1 \end{bmatrix}. \tag{10.2}$$

From (10.2) we have:

$$\sum_{n=1}^{N_P} \frac{\mathbf{c}_n^{\vec{g}_k}}{s - b_n} + \mathbf{d}^{\vec{g}_k} = \left[\sum_{n=1}^{N_P} \frac{\tilde{\mathbf{c}}_n^{\vec{g}_k}}{s - b_n} + 1 \right] \mathbf{H}_{\vec{g}_k}(s). \tag{10.3}$$

The unknowns $\mathbf{c}_n^{\vec{g}_k}$, $\tilde{\mathbf{c}}_n^{\vec{g}_k}$, and $\mathbf{d}^{\vec{g}_k}$ in (10.3) are found by solving an overdetermined linear problem over several frequency samples iteratively (Gustavsen and Semlyen, 1999). If unstable poles are generated during an iteration, a pole-flipping scheme is used to enforce stability (Gustavsen and Semlyen, 1999). The passivity assessment and enforcement can be accomplished using the standard techniques (Gustavsen, 2008, 2010; Gustavsen and Semlyen, 2008, 2009). This initial step results in a set of stable and passive frequency-dependent rational models, called *root macromodels*.

10.2.2. *The Estimation and Validation Design Space Grids*

In this chapter, the design space contains all design variables such as layout features represented by $\vec{g}$, and the frequency is not considered as a part of the design space. Two design space grids are used in the modeling process: an estimation grid and a validation grid. The estimation grid is utilized to build the parametric macromodels. The second grid is utilized to assess

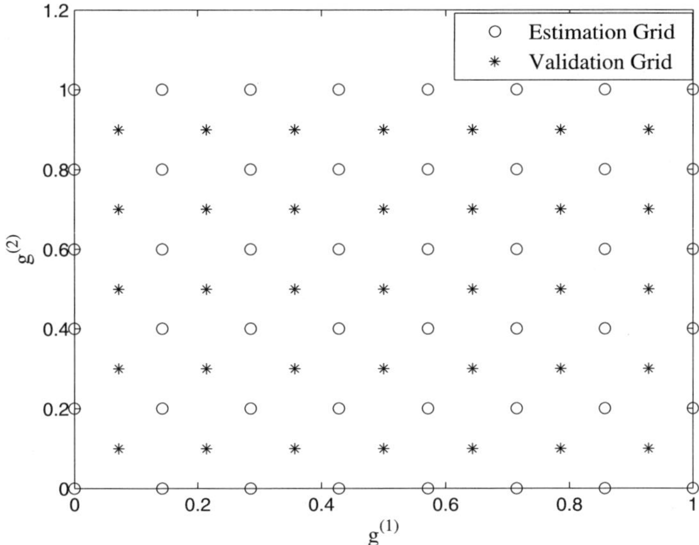

Fig. 10.1.   Estimation and validation grids for a general two parameter design space (Chemmangat *et al.*, 2012).

the capability of parametric macromodels of describing the system under study in a set of points of the design space previously not used for their construction. To clarify the use of these two design space grids, we show in Fig. 10.1 a possible estimation and validation design space grid in the case of two design parameters $\vec{g} = (g^{(1)}, g^{(2)})$. A *root macromodel* is built for each estimation grid point in the design space. This set of *root macromodels* is interpolated, which will be explained in Section 10.3, to build a parametric macromodel that is evaluated and compared with original data related to the validation design space points.

10.2.3. *Definition of the Error Measure*

Defining a proper error measure to assess the accuracy of the parametric macromodel at the validation points is a critical task in a modeling process. Let us define the absolute error

$$Err(\vec{g}) = \max_{i,j,k}(|(R_{i,j}(s_k, \vec{g}) - H_{i,j}(s_k, \vec{g})|)$$

$$i = 1, \ldots, P_{in}, \quad j = 1, \ldots, P_{out}, \quad k = 1, \ldots, N_s, \tag{10.4}$$

where P_{in} and P_{out} are the number of inputs and outputs of the system, respectively, and N_s is equal to the number of frequency samples. The worst case absolute error over the validation grid is chosen to assess the accuracy and the quality of parametric macromodels:

$$\vec{g}_{max} = \underset{\vec{g}}{\mathrm{argmax}}[Err(\vec{g})], \quad \vec{g} \in validation\ grid \tag{10.5}$$

$$Err_{max} = Err(\vec{g}_{max}). \tag{10.6}$$

10.3. Parametric Macromodeling

Each *root macromodel* $\mathbf{R}_{\vec{g}_k}(s)$, corresponding to a specific design space point $\vec{g}_k = (g_{k_1}^{(1)}, ..., g_{k_N}^{(N)})$, is converted from the pole-residue form (10.1) into a state-space form as given in Gustavsen and Semlyen (2009):

$$\mathbf{R}_{\vec{g}_k}(s) = \mathbf{C}_{\vec{g}_k}(s\mathbf{I} - \mathbf{A}_{\vec{g}_k})^{-1}\mathbf{B}_{\vec{g}_k} + \mathbf{D}_{\vec{g}_k}. \tag{10.7}$$

Then, this set of state-space matrices $\mathbf{A}_{\vec{g}_k}, \mathbf{B}_{\vec{g}_k}, \mathbf{C}_{\vec{g}_k}, \mathbf{D}_{\vec{g}_k}$ is interpolated entry-by-entry and the multivariate models $\mathbf{A}(\vec{g}), \mathbf{B}(\vec{g}), \mathbf{C}(\vec{g}), \mathbf{D}(\vec{g})$ are built using multivariate interpolation schemes to generate a parametric macromodel $\mathbf{R}(s, \vec{g})$ for the entire design space (Ferranti *et al.*, 2010b; Triverio *et al.*, 2010a, 2010b):

$$\mathbf{R}(s, \vec{g}) = \mathbf{C}(\vec{g})(s\mathbf{I} - \mathbf{A}(\vec{g}))^{-1}\mathbf{B}(\vec{g}) + \mathbf{D}(\vec{g}). \tag{10.8}$$

The computationally cheap piecewise linear interpolation cannot be used to generate parametric sensitivity macromodels, since it is not continuously differentiable. A proper choice of interpolation schemes which are at least continuously differentiable is necessary. In this chapter, three interpolation methods are investigated, namely the cubic spline (CS) interpolation, the piecewise cubic Hermite interpolation (PCHIP) and the shape preserving C2 cubic spline interpolation (SPC2). They are briefly described in what follows.

10.3.1. *Cubic Spline Interpolation*

Given some data samples $(x_i, y_i)_{i=1}^n$, the CS interpolation method builds a cubic polynomial for each interval of the dataset $x_i \leq x \leq x_{i+1}, i = 1, \ldots, n$:

$$s_i(x) = a_i(x - x_i)^3 + b_i(x - x_i)^2 + c_i(x - x_i) + d_i. \tag{10.9}$$

The coefficients of the cubic polynomials are obtained by imposing the first-and-second order derivative continuity at each data point along with a *not-a-knot* end condition, and then solving a tridiagonal linear system (de Boor, 2001). Once these coefficients are computed, the derivatives of the overall spline interpolation function can be analytically calculated in terms of its coefficients a_i, b_i, and c_i for $x_i \le x < x_{i+1}$, $i = 1, 2, ..., n - 1$. If the data under interpolation is in a matrix form, each entry of the matrices is independently interpolated.

The univariate CS interpolation can be extended to higher dimensions by means of a tensor product implementation (de Boor, 2001).

10.3.2. *Piecewise Cubic Hermite Interpolation*

The PCHIP method is a monotonic shape preserving interpolation scheme. As in the CS interpolation, each data interval is modeled by a cubic polynomial similar to (10.9):

$$p_i(x) = f_i H_1(x) + f_{i+1} H_2(x) + d_i H_3(x) + d_{i+1} H_4(x), \qquad (10.10)$$

where $d_j = \frac{dp(x_j)}{dx}$, $j = i, i + 1$, and the $H_k(x)$ are the usual cubic Hermite basis functions for the interval $x_i \le x < x_{i+1}$, $i = 1, 2, \ldots, n - 1$: $H_1(x) = \phi((x_{i+1} - x)/h_i)$, $H_2(x) = \phi((x - x_i)/h_i)$, $H_3(x) = -h_i \psi((x_{i+1} - x)/h_i)$, $H_4(x) = h_i \psi((x - x_i)/h_i)$, where $h_i = x_{i+1} - x_i$, $\phi(t) = 3t^2 - 2t^3$, $\psi(t) = t^3 - t^2$. The first-order derivative d_i at each data point x_i is calculated such that the local monotonicity is preserved (Fritsch and Carlson, 1980). An extension to higher dimension can be performed by a tensor product implementation (de Boor, 2001). The calculation of derivatives is done in the same way as in the CS interpolation case. This interpolation scheme works better for non-smooth datasets, wherein the CS scheme could result in overshoots or oscillatory behavior of the derivatives. However, the PCHIP method is only continuous in first derivatives, which affects the smoothness of the derivatives (Fritsch and Carlson, 1980).

10.3.3. *Shape Preserving C2 Cubic Spline Interpolation*

The SPC2 interpolation is a monotonicity-preserving interpolation scheme similar to PCHIP. However, in contrast to the PCHIP method which is only continuous in first derivative, the SPC2 method is a second-order derivative continuous interpolation scheme. The idea here is to add two extra break points in each subinterval of the data, such that enough degrees of freedom are generated to construct a cubic spline interpolant, which is globally second-order derivative continuous (Pruess, 1993). Since the monotonicity of the data is preserved, this scheme works better with respect to the CS method for non-smooth data sets. Similar to the CS and the PCHIP interpolation schemes, a multivariate SPC2 interpolation is performed using a tensor product implementation (de Boor, 2001).

10.4. Parametric Sensitivity Macromodels

Once the parametric macromodel $\mathbf{R}(s, \vec{g})$ is built, the corresponding sensitivities can be computed by differentiating (10.8) with respect to the design parameters $\vec{g}$:

$$\frac{\partial}{\partial \vec{g}} \mathbf{R}(s, \vec{g}) = \frac{\partial \mathbf{C}(\vec{g})}{\partial \vec{g}} (s\mathbf{I} - \mathbf{A}(\vec{g}))^{-1} \mathbf{B}(\vec{g})$$

$$+ \mathbf{C}(\vec{g})(s\mathbf{I} - \mathbf{A}(\vec{g}))^{-1} \frac{\partial \mathbf{A}(\vec{g})}{\partial \vec{g}} (s\mathbf{I} - \mathbf{A}(\vec{g}))^{-1} \mathbf{B}(\vec{g})$$

$$+ \mathbf{C}(\vec{g})(s\mathbf{I} - \mathbf{A}(\vec{g}))^{-1} \frac{\partial \mathbf{B}(\vec{g})}{\partial \vec{g}} + \frac{\partial \mathbf{D}(\vec{g})}{\partial \vec{g}}. \qquad (10.11)$$

In (10.11), $\frac{\partial}{\partial \vec{g}} \mathbf{R}(s, \vec{g})$ is based on the parameterized state-space matrices $\mathbf{A}(\vec{g}), \mathbf{B}(\vec{g}), \mathbf{C}(\vec{g}), \mathbf{D}(\vec{g})$ and the corresponding derivatives, which are computed efficiently and analytically using the interpolation methods described in Section 10.3.

10.5. Gradient-Based Minimax Optimization

Parametric sensitivity macromodels can be used in the optimization process of EM systems. Considering microwave filters, a typical optimization process begins by defining passband and stopband specifications in terms of the frequency responses, which are reformulated in the form of a cost function $F_i(\vec{g})$, to be minimized at optimization frequency samples s_i, $i = 1, 2, \ldots N_s$:

$$F_i(\vec{g}) = R_{\mathrm{L}}^i - R(s_i, \vec{g}) \quad \text{or} \quad R(s_i, \vec{g}) - R_{\mathrm{U}}^i, \qquad i = 1, 2, \ldots, N_s. \quad (10.12)$$

In (10.12), R_{L}^i and R_{U}^i represent lower and upper frequency response thresholds, respectively, at frequency samples s_i spread over the frequency range of interest. A negative cost indicates that the corresponding specification is satisfied, while a positive cost denotes that the specification is violated. The minimization (10.12) can be performed by several state-of-the-art optimization algorithms. In this chapter, we use a minimax optimization algorithm (Du and Pardalos, 1995) which uses the cost function (10.12) and its gradients with respect to design parameter $\vec{g}$ giving the optimum design

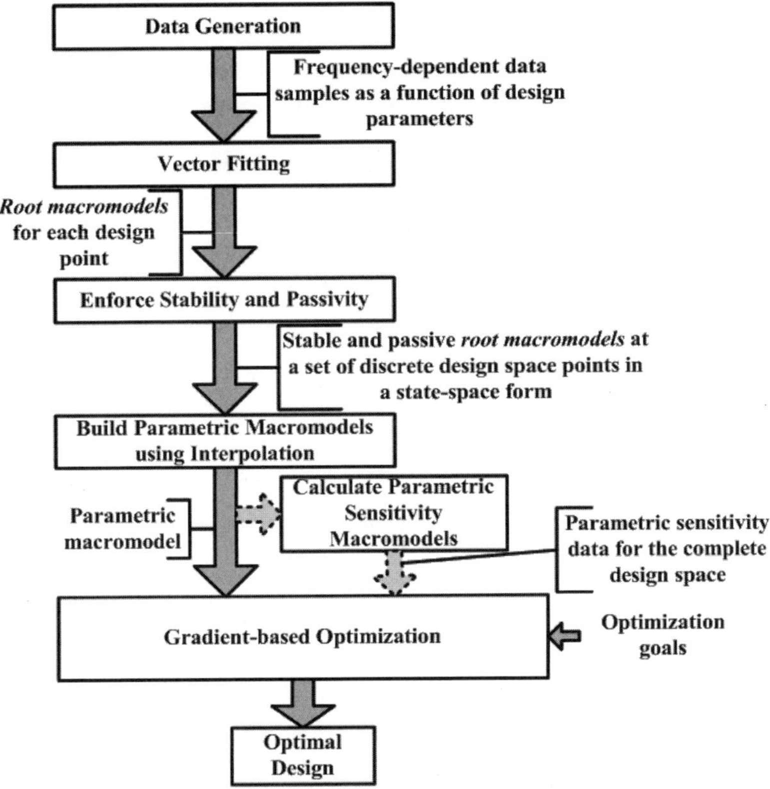

Fig. 10.2.   Complete optimization process flowchart (Chemmangat *et al.*, 2012).

parameters $\vec{g}^*$ as,

$$\vec{g}^* = \operatorname*{argmin}_{\vec{g}} \{ \max_i [F_i(\vec{g})] \}. \tag{10.13}$$

The complete optimization process starting from the parametric macro-modeling technique is depicted in Fig. 10.2.

10.6. Numerical Examples

10.6.1. *Double Folded Stub Microwave Filter*

A double folded stub (DFS) microwave filter on a substrate with relative permittivity $\epsilon_r = 9.9$ and a thickness of 0.127 mm is modeled in this example. The layout of this DFS filter is shown in Fig. 10.3. The spacing

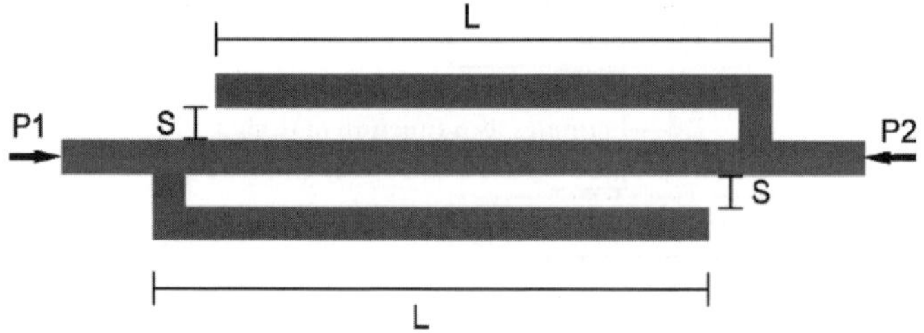

Fig. 10.3.   Layout of the DFS band-stop filter (Chemmangat *et al.*, 2012).

Table 10.1.   Design parameters of the DFS band-stop filter (Chemmangat *et al.*, 2012).

Parameter	Min	Max
Frequency (*freq*)	5 GHz	20 GHz
Spacing (S)	0.1 mm	0.25 mm
Length (L)	2.0 mm	3.0 mm

S between the stubs and the length L of the stub are chosen as design variables in addition to frequency. Their corresponding ranges are shown in Table 10.1. The design specifications of this band-stop filter are given in terms of the scattering parameter, similarly to Ureel and De Zutter (1997),

$$|S_{21}| \geq -3\,\text{dB} \quad \text{for} \quad \textit{freq} \leq 9\,\text{GHz} \quad \text{or} \quad \textit{freq} \geq 17\,\text{GHz} \tag{10.14a}$$

$$|S_{21}| \leq -30\,\text{dB} \quad \text{for} \quad 12\,\text{GHz} \leq \textit{freq} \leq 14\,\text{GHz}. \tag{10.14b}$$

From the design specifications (10.14), a cost function (10.12) is formulated in terms of S_{21} and $\vec{g} = (S, L)$. The scattering matrix $\mathbf{S}(s, S, L)$ has been computed using the ADS Momentum (2009) software. The number of frequency samples were chosen to be equal to 31. The estimation and validation grid points for the design variables are shown in Fig. 10.4. The average simulation time for each design space point (S, L) has been found to be equal to, $T_{SimAvg} = 32.87\,\text{sec}$. A set of stable and passive *root macromodels* has been built for all design space points in the estimation grid of Fig. 10.4 by means of VF with a fixed number of poles $N_P = 18$, selected using an error-based bottom-up approach. Each *root macromodel* has been converted into a state-space form and the set of state-space matrices has been interpolated using the CS, PCHIP, and SPC2 methods. The maximum absolute error (10.6) for the parametric macromodel over the validation grid

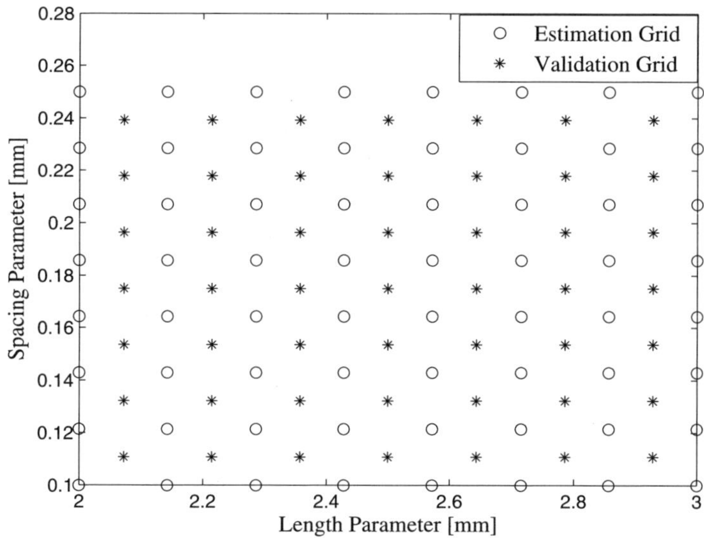

Fig. 10.4.   Estimation and validation grids for the parametric macromodeling of the DFS Filter (Chemmangat *et al.*, 2012).

of Fig. 10.4 is -58.45 dB, -50.23 dB, and -50.23 dB, respectively, using the CS, PCHIP, and SPC2 interpolation schemes. The CS interpolation scheme gives the minimum worst-case error (10.6) for this specific example and hence it has been used in the optimization process to generate the parametric macromodel and corresponding sensitivities. Figure 10.5 shows the parametric behavior of the magnitude of S_{21} as a function of frequency for five different values of S, and $L = 2.5$ mm. In Fig. 10.5, the darkest curve of S_{21} corresponds to the largest value of S. Similarly, Fig. 10.6 shows the magnitude of S_{21} for five different values of L, with $S = 0.175$ mm.

The cost function (10.12) and its gradients calculated using (10.11), have been supplied to the minimax optimization algorithm (10.13), resulting in the optimum design parameter values S^* and L^*. To show the advantage of supplying parametric sensitivity information to the optimizer to speed up the optimization process, two cases have been considered:

- Case 1: No sensitivity information is supplied to the minimax algorithm and the algorithm estimates it with the help of finite difference approximation computed using the parametric macromodel.
- Case 2: The sensitivity information is calculated from (10.11) and supplied to the minimax algorithm.

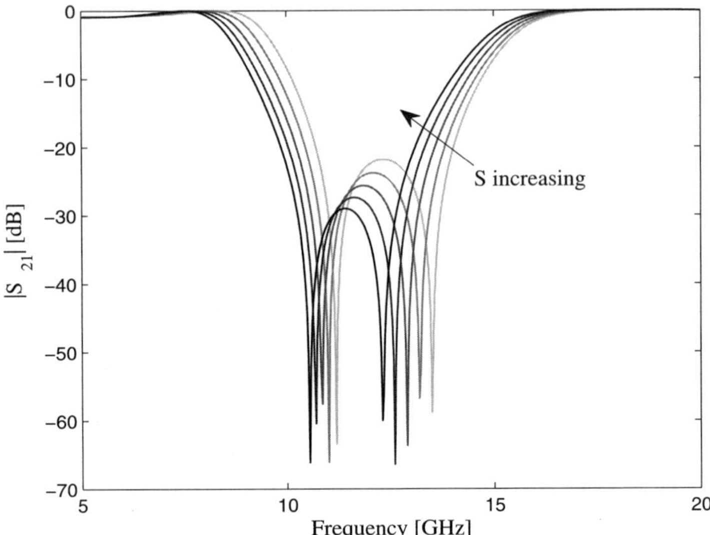

Fig. 10.5.   DFS filter: Magnitude of S_{21} as a function of frequency for five values of S and $L = 2.5\,\text{mm}$ (Chemmangat *et al.*, 2012).

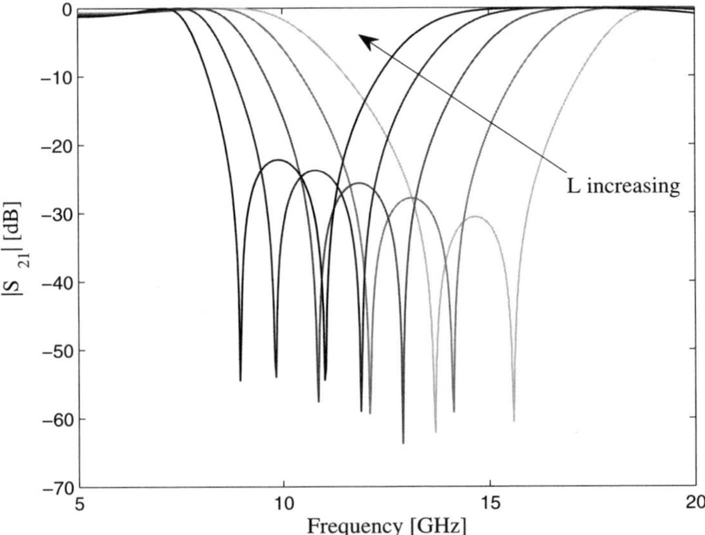

Fig. 10.6.   DFS filter: Magnitude of S_{21} as a function of frequency for five values of L and $S = 0.175\,\text{mm}$ (Chemmangat *et al.*, 2012).

Table 10.2.   DFS filter: Optimization using parametric macromodel and ADS Momentum software (Chemmangat *et al.*, 2012).

Method	(S^0, L^0) [mm]	(S^*, L^*) [mm]	Optimization Time [sec]	Cost Function at (S^*, L^*) [dB][†]
Case 1	(0.1500, 2.6364)	(0.2408, 2.1802)	19.88	−1.67
Case 2	(0.1500, 2.6364)	(0.2408, 2.1802)	1.23	−1.67
ADS Momentum	(0.1500, 2.6364)	(0.2303, 2.1580)	5115.00	0.00

[†]The gradient-based minimax optimization routine of ADS Momentum software uses the minimax L1 error function which cannot take a value less than zero.

In addition to that, in order to show the advantage of using a parametric macromodel, the same optimization problem has been performed using the gradient-based minimax optimization routine in the ADS Momentum software. Table 10.2 compares these three optimization approaches in terms of optimization time for a particular optimization case. Table 10.2 shows the relevant speed-up in the optimization process obtained using the parametric macromodel. We note that the generation of the parametric macromodel requires some initial ADS Momentum simulations and therefore an initial computational effort of 3714.31 sec (for the estimation and validation design space points in Fig. 10.4). However, once the parametric macromodel is generated and validated, it acts as an accurate and efficient surrogate of the original system and can be used for multiple design optimization scenarios, for instance, changing filter specifications. Figure 10.7 shows the magnitude of S_{21} for the optimization case of Table 10.2. The actual data generated by the ADS Momentum software at the optimum design space point (S^*, L^*) obtained using the parametric macromodel is shown by asterisks in Fig. 10.7. As seen, this is in good agreement with the parametric macromodel response. Figure 10.8 shows the solution obtained using the gradient-based minimax optimization routine of ADS Momentum software. As clearly seen in Figs. 10.7–10.8, the optimal solutions satisfy all the requirements (10.14), which are shown by thin solid black lines.

The convergence time taken by Case 1 and 2 are 13.76 sec and 1.73 sec, respectively. The trajectory of the optimum design space point (S^*, L^*) during optimization for Case 1 is shown in Fig. 10.9 with different points showing the output of some particular iterations along with the elapsed time. A similar curve is plotted in Fig. 10.10 for Case 2. Comparing Figs. 10.9–10.10, it is seen that the time for convergence of Case 2 is considerably less compared with Case 1.

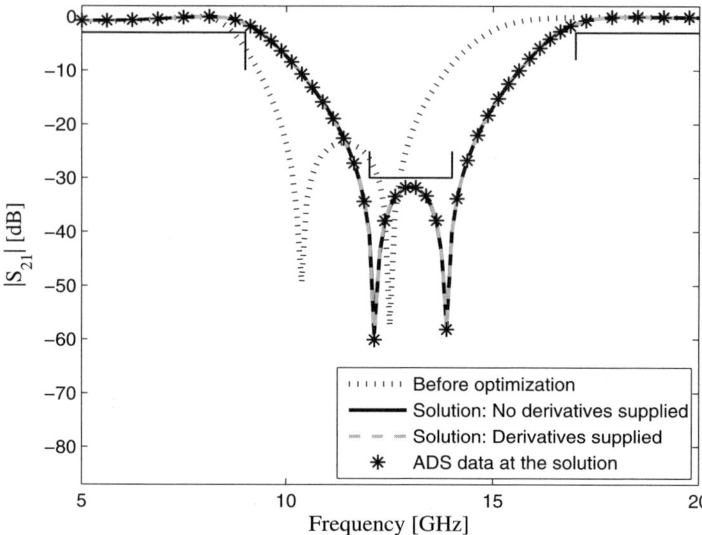

Fig. 10.7. DFS filter: Magnitude of S_{21} before and after optimization (Chemmangat *et al.*, 2012).

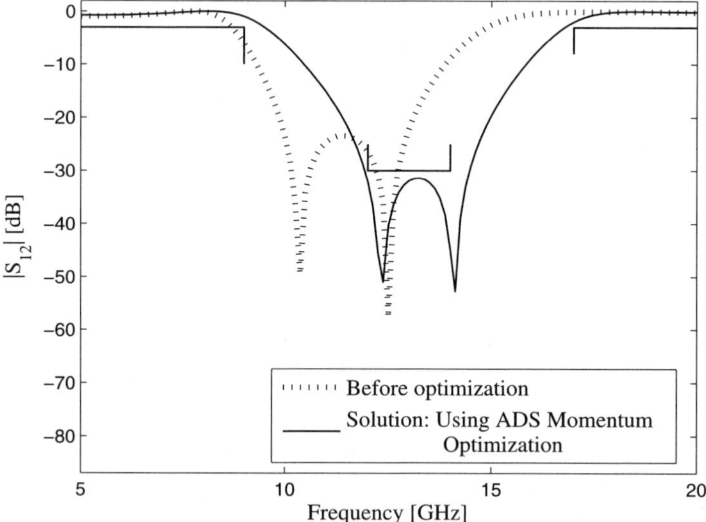

Fig. 10.8. DFS filter: Magnitude of S_{21} before and after optimization (Chemmangat *et al.*, 2012).

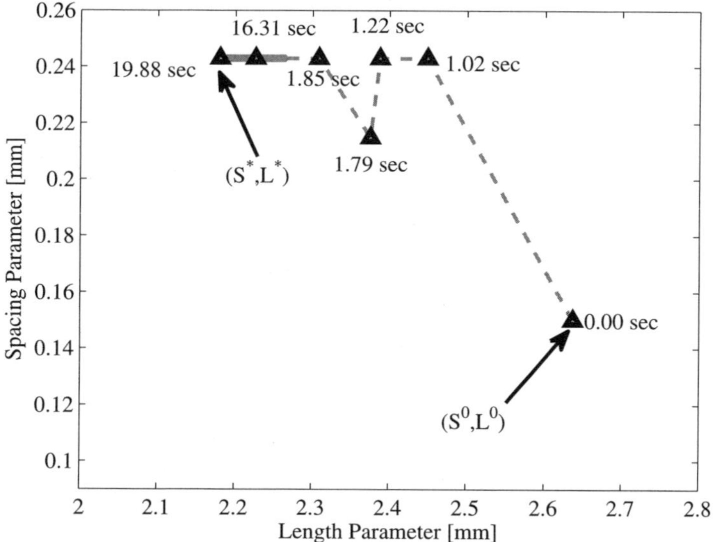

Fig. 10.9.   DFS filter: The trajectory of the optimal design space point (S^*, L^*) during optimization for Case 1 (Chemmangat *et al.*, 2012).

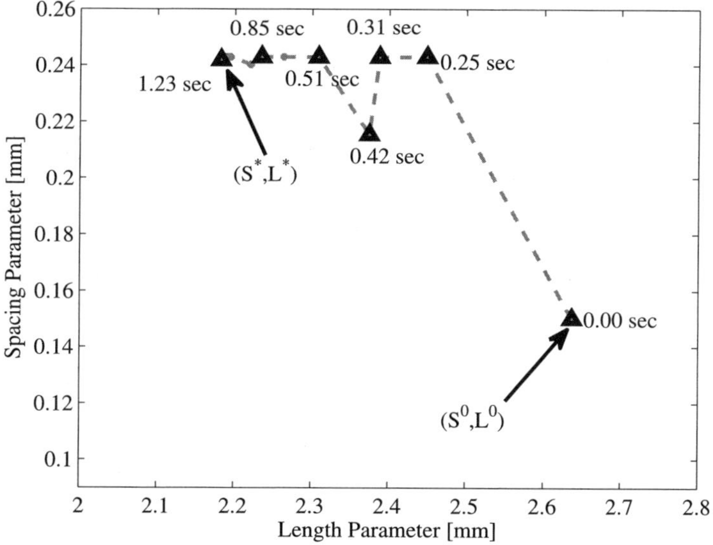

Fig. 10.10.   DFS filter: The trajectory of the optimal design space point (S^*, L^*) during optimization for Case 2 (Chemmangat *et al.*, 2012).

Table 10.3.   DFS filter: Comparison between Cases 1 and 2 (Chemmangat *et al.*, 2012).

Method	Number of Cost Function Evaluations			Optimization Time [sec]		
	Max	Mean	STD	Max	Mean	STD
Case 1	20004	861.33	2360.10	135.30	6.05	16.31
Case 2	1601	50.56	138.82	41.10	1.43	3.62

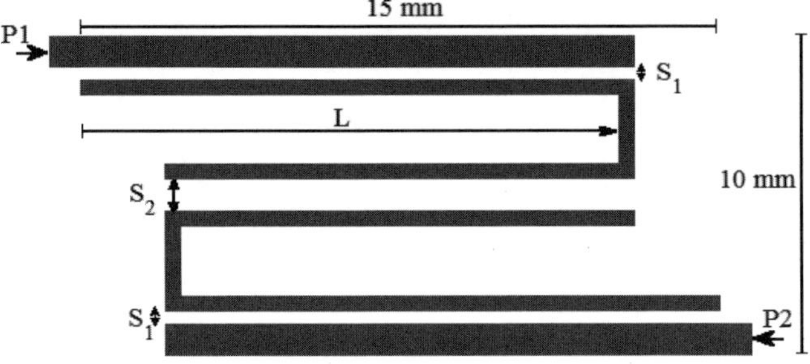

Fig. 10.11.   Layout of the hairpin bandpass filter (Chemmangat *et al.*, 2012).

Table 10.3 shows the comparison of Case 1 and 2 for some important optimization measures which are related to 200 optimization trials, starting from different initial design points spread over the design space. Table 10.3 confirms that there is a considerable reduction in the number of cost function evaluations and the optimization time if derivatives are supplied (Case 2).

10.6.2. *Hairpin Bandpass Microwave Filter*

In this example, a hairpin bandpass filter with the layout shown in Fig. 10.11 is modeled (Ureel and De Zutter, 1997). The relative permittivity of the substrate is $\epsilon_r = 9.9$, while its thickness is equal to $0.635\,\text{mm}$. The specifications for the bandpass filter are given in terms of the scattering parameters S_{21} and S_{11}:

$$|S_{21}| > -2.5\,\text{dB} \quad \text{for} \quad 2.4\,\text{GHz} < freq < 2.5\,\text{GHz} \tag{10.15a}$$

$$|S_{11}| < -7\,\text{dB} \quad \text{for} \quad 2.4\,\text{GHz} < freq < 2.5\,\text{GHz} \tag{10.15b}$$

$$|S_{21}| < -40\,\text{dB} \quad \text{for} \quad freq < 1.7\,\text{GHz} \tag{10.15c}$$

$$|S_{21}| < -25\,\text{dB} \quad \text{for} \quad 3.0\,\text{GHz} < freq. \tag{10.15d}$$

As shown in Fig. 10.11, three design parameters have been chosen for the design process namely, the spacing between the port lines and the filter

Table 10.4.   Design parameters of the hairpin bandpass filter (Chemmangat *et al.*, 2012).

Parameter	Min	Max
Frequency (*freq*)	1.5 GHz	3.5 GHz
Length (L)	12.0 mm	12.5 mm
Spacing (S_1)	0.27 mm	0.32 mm
Spacing (S_2)	0.67 mm	0.72 mm

lines S_1, the spacing between the two filter lines S_2, and the overlap length L in addition to frequency. The ranges of the different design variables are shown in Table 10.4. The scattering matrix $\mathbf{S}(s, S, L)$ has been computed using the ADS Momentum (2009) software. The number of frequency samples were chosen to be equal to 41. As in the first example, two design space grids are used in the modeling process. The average simulation time for each design space point (L, S_1, S_2) has been found to be equal to $T_{SimAvg} = 34.30$ sec. A set of stable and passive *root macromodels* has been built for the estimation grid of $6 \times 4 \times 4$ $(L \times S_1 \times S_2)$ samples by means of VF with a fixed number of poles, $N_P = 18$, selected using an error-based bottom-up approach. Each *root macromodel* has been converted into a state-space form and the set of state-space matrices has been interpolated by the CS, PCHIP, and SPC2 methods. The maximum absolute error (10.6) of the models over the validation grid of $5 \times 3 \times 3$ $(L \times S_1 \times S_2)$ samples is equal to -42.57 dB, -38.27 dB, and -38.27 dB for the CS, PCHIP, and SPC2 methods, respectively. The CS technique has been used in the optimization process of this example, since it gives the best accuracy. Figure 10.12 shows the parametric behavior of the magnitude of S_{21} as a function of frequency for five different values of L and $S_1 = 0.295$ mm, $S_2 = 0.695$ mm. Figure 10.13 shows the magnitude of S_{21} when the parameter S_1 changes.

The cost function (10.12) and its gradients calculated using (10.11), have been supplied to the minimax optimization algorithm (10.13), resulting in the optimum design parameter values L^*, S_1^*, and S_2^*. To show the improved convergence of the optimization when derivatives are supplied, two cases are considered as in the previous example. In addition to that, in order to show the advantage of using a parametric macromodel, the same optimization problem has been performed using the gradient-based minimax optimization routine in the ADS Momentum software. Table 10.5 compares these three optimization approaches in terms of optimization time for a particular optimization case. Table 10.5 shows the relevant speed-up in the optimization process obtained using the parametric

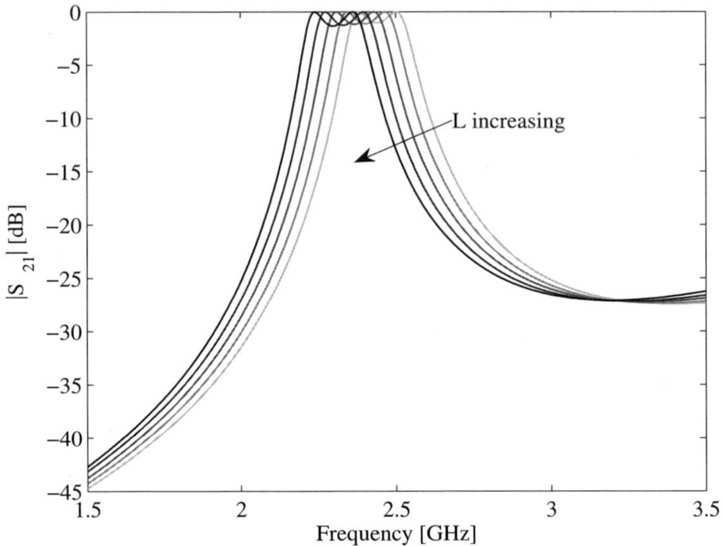

Fig. 10.12. Hairpin filter: Magnitude of S_{21} as a function of frequency for five values of L with $(S_1, S_2) = (0.295, 0.695)\,\mathrm{mm}$ (Chemmangat *et al.*, 2012).

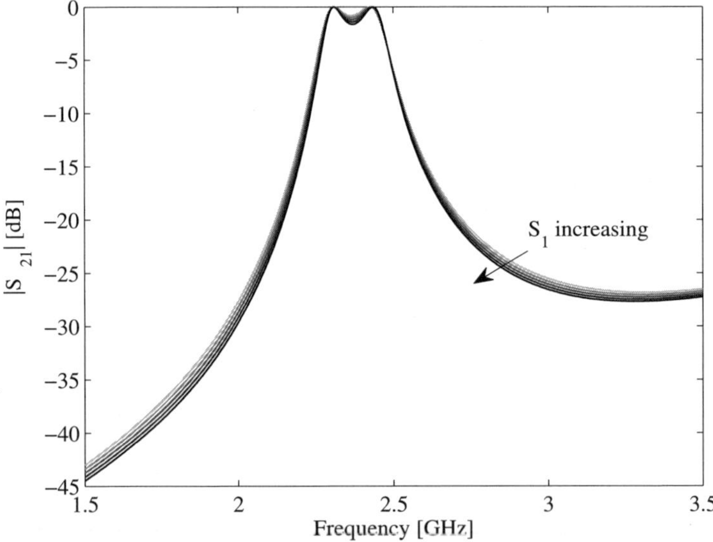

Fig. 10.13. Hairpin filter: Magnitude of S_{21} as a function of frequency for five values of S_1 with $(L, S_2) = (12.25, 0.695)\,\mathrm{mm}$ (Chemmangat *et al.*, 2012).

Table 10.5.   Hairpin filter optimization using parametric macromodel and ADS Momentum software (Chemmangat *et al.*, 2012).

Method	(L^0, S_1^0, S_2^0) [mm]	(L^*, S_1^*, S_2^*) [mm]	Optimization Time [sec]	Cost Function at (L^*, S_1^*, S_2^*) [dB][†]
Case 1	(12.2778, 0.2700, 0.6867)	(12.0568, 0.2984, 0.7200)	13.46	−0.76
Case 2	(12.2778, 0.2700, 0.6867)	(12.0568, 0.2984, 0.7200)	0.79	−0.76
ADS Momentum	(12.2778, 0.2700, 0.6867)	(12.0036, 0.2700, 0.6826)	1251.00	0.00

[†]The gradient-based minimax optimization routine of ADS Momentum software uses the minimax L1 error function which cannot take a value less than zero.

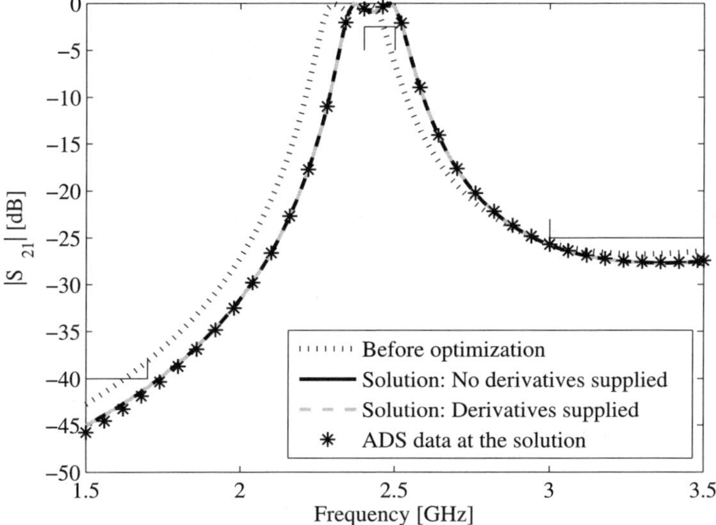

Fig. 10.14.   Hairpin filter: Magnitude of S_{21} before and after optimization (Chemmangat *et al.*, 2012).

macromodel. As mentioned in the previous example, the generation of the parametric macromodel requires an initial ADS Momentum simulation cost of 4836.30 sec. However, once the parametric macromodel is generated and validated, it acts as an accurate and efficient surrogate of the original system and can be used for multiple design optimization scenarios, for instance, changing filter specifications. Figure 10.14 shows the magnitude of S_{21} for the optimization case of Table 10.5. The actual data generated by the

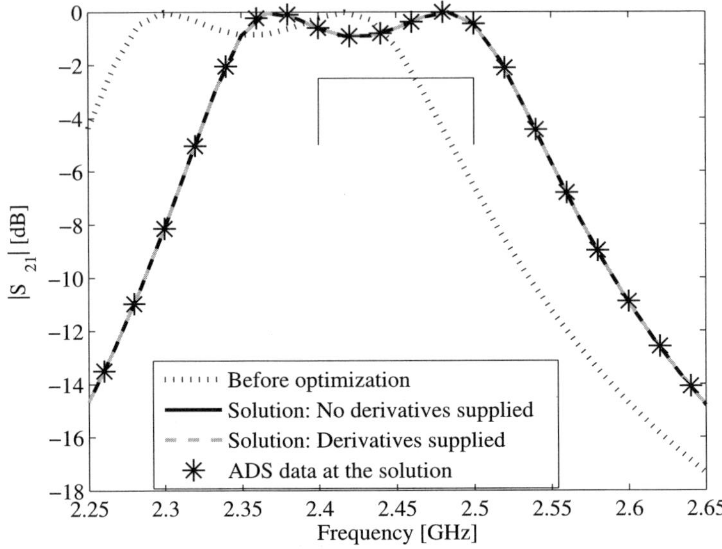

Fig. 10.15.   Hairpin filter: A zoomed in view of Fig. 10.14 (Chemmangat *et al.*, 2012).

ADS Momentum software at the optimum design space point (L^*, S_1^*, S_2^*) obtained using the parametric macromodel is shown by asterisks in Fig. 10.14. As seen, this is in agreement with the parametric macromodel response. The requirements (10.15) are shown by the thin black solid lines. A magnified view of the passband is shown in Fig. 10.15 for clarity. Similar results are given for the magnitude of S_{11} in Fig. 10.16. As clearly seen, all the requirements are met for the optimal design point. Figures 10.17–10.18 shows similar results for the solution obtained using the gradient-based minimax optimization routine of ADS Momentum software. Here also, all the design specifications (10.15) are met. Some important measures of the optimization process related to 1,000 trial runs are shown in Table 10.6, which clearly shows the better convergence properties of Case 2.

Another important thing to be noted here is that, with the increase in the number of design parameters the initial number of EM simulations increases considerably due to the curse of dimensionality. Adaptive sampling schemes, which take into account influence of the design parameters on the system behavior, could be used to properly sample the design space prior to the parametric macromodeling and help resolve this issue. For instance, in the second example, from Figs. 10.12 and 10.13 it is seen that the overlap length L of the hairpin filter is more influential than the

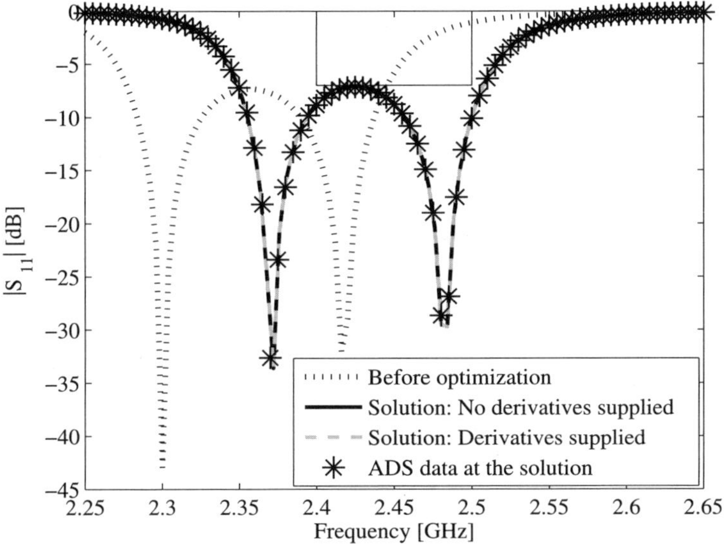

Fig. 10.16. Hairpin filter: Magnitude of S_{11} before and after optimization (Chemmangat *et al.*, 2012).

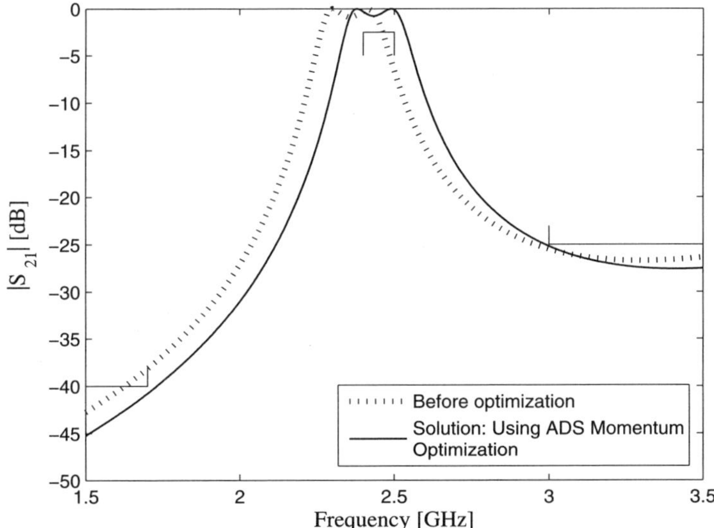

Fig. 10.17. Hairpin filter: Magnitude of S_{21} before and after optimization (Chemmangat *et al.*, 2012).

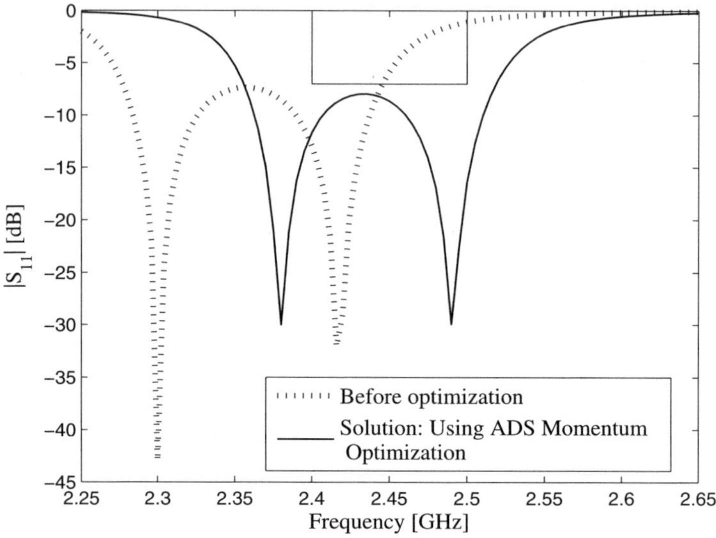

Fig. 10.18.   Hairpin filter: Magnitude of S_{11} before and after optimization (Chemmangat *et al.*, 2012).

Table 10.6.   Hairpin filter: Comparison between the Cases 1 and 2 (Chemmangat *et al.*, 2012).

Method	Number of Cost Function Evaluations			Optimization Time [sec]		
	Max	Mean	STD	Max	Mean	STD
Case 1	53183	114.39	1689.71	565.14	1.29	17.95
Case 2	99	13.73	5.79	4.69	0.76	0.28

spacing S_1, which allows one to sample along S_1 more sparsely using a wise adaptive sampling scheme, thereby reducing the number of initial EM simulations needed for the construction of parametric macromodels and the corresponding computational effort.

10.7. Conclusion

In this chapter, the design optimization of high-speed microwave systems using parametric sensitivity macromodels has been discussed. Providing parametric sensitivity information to the gradient-based optimizers can bring down the convergence time of the optimization. This is true in cases where the gradient of the cost function is difficult to model using finite difference approximations, which affects the efficiently of the optimizers. In addition to that, parametric macromodels discussed here can act as an efficient replacement model for the actual EM simulator. Therefore, these parametric sensitivity macromodels can be used in design activities where multiple EM simulations are needed (e.g., design-space exploration and sensitivity analysis) and its use is not just limited to design optimization. The use of parametric macromodels combined with adaptive sampling schemes to select the optimal samples to build these models is a powerful tool for the overall design process of high-speed microwave systems. It should be noted however that the interpolation of state-space matrices to build parametrized state-space models is still a challenging issue, since finding the best state-space representation to use in the interpolation process of state-space matrices is still an open problem and ongoing area for research.

Acknowledgment

This research was supported by the Research Foundation Flanders (FWO).

References

ADS Momentum (2009). Momentum EEsof EDA, Agilent Technologies, Santa Rosa: CA.

Chemmangat, K., Ferranti, F., Dhaene, T. and Knockaert, L. (2012). Gradient-based optimization using parametric sensitivity macromodels, *International Journal of Numerical Modelling: Electronic Networks, Devices and Fields*, **25**, 347–361. [Online] Available at: http://dx.doi.org/10.1002/jnm.839. Date accessed 16 June 2011.

de Boor, C. (2001). *A Practical Guide to Splines*, Springer-Verlag, New York: NY.

Deschrijver, D. and Dhaene, T. (2008). Stability and passivity enforcement of parametric macromodels in time and frequency domain, *IEEE T. Microw. Theory*, **56**, 2435–2441.

Dounavis, A., Achar, R. and Nakhla, M. (2001). Efficient sensitivity analysis of lossy multiconductor transmission lines with nonlinear terminations, *IEEE T. Microw. Theory*, **49**, 2292–2299.

Du, D.Z. and Pardalos, P.M. (1995). *Minimax and Applications*, Kluwer Academic Publishers, Dordrecht.

Ferranti, F., Knockaert, L. and Dhaene, T. (2009). Parameterized S-parameter-based macromodeling with guaranteed passivity, *IEEE Microw. Wirel. Co.*, **19**, 608–610.

Ferranti, F., Knockaert, L. and Dhaene, T. (2010a). Guaranteed passive parameterized admittance-based macromodeling, *IEEE T. Adv. Packaging*, **33**, 623–629.

Ferranti, F., Knockaert, L., Dhaene, T. and Antonini, G. (2010b). Passivity-preserving parametric macromodeling for highly dynamic tabulated data based on Lur'e equations, *IEEE T. Microw. Theory*, **58**, 3688–3696.

Fritsch, F.N. and Carlson, R.E. (1980). Monotone piecewise cubic interpolation, *SIAM J. Numer. Anal.*, **17**, 238–246.

Gustavsen, B. (2006). Improving the pole relocating properties of vector fitting, *IEEE T. Power Deliv.* **21**, 1587–1592.

Gustavsen, B. (2008). Fast passivity enforcement for pole-residue models by perturbation of residue matrix eigenvalues, *IEEE T. Power Deliv.* **23**, 2278–2285.

Gustavsen, B. (2010). Fast passivity enforcement for S-parameter models by perturbation of residue matrix eigenvalues, *IEEE T. Adv. Packaging*, **33**, 257–265.

Gustavsen, B. and Semlyen, A. (1999). Rational approximation of frequency domain responses by vector fitting, *IEEE T. Power Deliv.*, **14**, 1052–1061.

Gustavsen, B. and Semlyen, A. (2008). Fast passivity assessment for S-parameter rational models via a half-size test matrix, *IEEE T. Microw. Theory*, **56**, 2701–2708.

Gustavsen, B. and Semlyen, A. (2009). A half-size singularity test matrix for fast and reliable passivity assessment of rational models, *IEEE T. Power Deliv.*, **24**, 345–351.

Li, D., Zhu, J., Nikolova, N.K., Bakr, M.H. and Bandler, J. (2007). Electromagnetic optimisation using sensitivity analysis in the frequency domain, *IET Microw. Antenna. P.*, **1**, 852–859.

Nikolova, N.K., Bandler, J.W. and Bakr, M.H. (2004). Adjoint techniques for sensitivity analysis in high-frequency structure CAD, *IEEE T. Microw. Theory*, **52**, 403–419.

Pruess, S. (1993). Shape preserving C2 cubic spline interpolation, *IMA J. Numer. Anal.*, **13**, 493–507.

Triverio, P., Nakhla, M. and Grivet-Talocia, S. (2010a). Passive parametric macromodeling from sampled frequency data, *2010 IEEE 14th Workshop on Signal Propagation on Interconnects*, Hildesheim, Germany, pp. 117–120.

Triverio, P., Nakhla, M. and Grivet-Talocia, S. (2010b). Passive parametric modeling of interconnects and packaging components from sampled impedance, admittance or scattering data, *3rd Electronic System-Integration Technology Conference*, Berlin, Germany, pp. 1–6.

Ureel, J. and De Zutter, D. (1997). Gradient-based minimax optimization of planar microstrip structures with the use of electromagnetic simulations, *Int. J. Microw. Mill.*, **7**, 29–36.

Chapter 11

Neural Space Mapping Methods for Electromagnetics-Based Yield Estimation

José E. Rayas-Sánchez

Efficient neural space mapping methods for highly accurate electromagnetics (EM)-based design optimization, statistical analysis, and yield estimation are described and contrasted in this chapter. Statistical analysis of radio frequency (RF) and microwave circuits is crucial for manufacturability, typically requiring a massive amount of high-fidelity simulations. This chapter describes several methodologies for accurate and inexpensive yield estimations after an efficient design optimization process based on space mapping. Comparisons are presented between different strategies for developing neural models that implement suitable mappings. First the process for developing an input space mapping approach to enhance available coarse models for efficient statistical analysis is described. This is realized in a linear fashion, using a Broyden-based approach, as well as in a nonlinear manner using neural networks. The corresponding neural space mapping is trained using reduced sets of full-wave EM data. It is illustrated how these input space mapping techniques can be employed to perform efficient yield estimations. Finally, the combination of linear input mappings with neural output mappings to develop more accurate and equally efficient surrogate models is described. This last strategy allows the transformation of conventional equivalent circuit models into accurate vehicles for EM-based statistical analysis and design. The design optimization and yield prediction of synthetic problems as well as practical microstrip circuits, using commercially available computer-aided design (CAD) tools, illustrate the methodologies.

11.1. Introduction

The real performance of practical electronic systems is subject to manufacturing tolerances and uncertainties. Statistical analysis and yield estimations of electronic circuits are crucially important for manufacturability-driven design. Reliable yield prediction typically requires massive amounts of accurate simulations to cover the entire statistics of possible outcomes of a given manufacturing process. In the case of RF and microwave circuits, such accurate simulations are usually provided by full-wave field solvers. Performing Monte Carlo yield analysis by directly using high-fidelity full-wave EM simulations is not feasible for most practical problems.

The incorporation of field solvers in statistical analysis and design has been addressed in several innovative methodologies. EM-based yield optimization was proposed by Bandler *et al.* (1993, 1994), by using multidimensional quadratic models. Other approaches based on principal component analysis combined with quadratic interpolation aim at avoiding redundant Monte Carlo simulations by using techniques such as Latin hypercube sampling (Swidzinski and Chang, 2000).

Artificial neural networks (ANNs) have also been employed to perform efficient EM-based statistical analysis and design. The most basic approach consists of training an ANN over a certain region of interest and then applying Monte Carlo analysis and yield optimization techniques to the computationally inexpensive but accurate neuromodel, as demonstrated by Zaabab *et al.* (1995), Burrascano *et al.* (1998, 1999), Creech and Zurada (1999), and Zhang and Gupta (2000).

The efficiency of neural model development can be significantly enhanced by incorporating prior knowledge in the ANN training process (Rayas-Sánchez, 2004). One of the most effective ways to perform this is by following a space mapping (SM) approach, which essentially consists of combining the original high-fidelity model with a coarse circuital model (Bandler *et al.*, 2004a; Koziel *et al.*, 2008). Several methodologies exploiting space mapping and neural networks have been proposed to realize efficient statistical analysis and yield estimations. The use of input space mapped neural models in the frequency-domain was demonstrated by Bandler *et al.* (2002) for efficient EM-based statistical analysis and yield optimization. In the work by Rayas-Sánchez *et al.* (2006a, 2006b), an output neural mapping modeling process is implemented on a linearly input mapped coarse model, in which not only the responses but the design parameters and independent variable are used as inputs to the output neural network.

This chapter describes advanced techniques for accurate and inexpensive yield estimations after an efficient design optimization process based on conventional input space mapping. Comparisons are presented between different strategies for developing neural models that implement suitable mappings.

First it is described how to develop an input space mapping approach to enhance available coarse models for efficient statistical analysis. This is realized in a linear fashion, using a Broyden-based approach, as well as in a nonlinear manner using ANNs. The corresponding neural space mapping is trained using reduced sets of full-wave EM data. It is illustrated how these input space mapping techniques can be employed to perform efficient yield estimations.

Finally, the combination of linear input mappings with neural output mappings to develop more accurate and equally efficient surrogate models is described. This last strategy allows the transformation of conventional equivalent circuit models into accurate vehicles for EM-based statistical analysis and design.

The design optimization and yield prediction of synthetic problems as well as practical microstrip circuits, using commercially available CAD tools, illustrate the methodologies.

11.2. Basic Assumptions and Fundamental Terminology

The algorithmic methodologies described in this chapter assume that there are at least two models available for the device whose statistical behavior is required: a fine model and a coarse model. The fine model is regarded as very accurate but computationally expensive. It usually consists of high-resolution full-wave EM simulations. By contrast, the coarse model is a low-fidelity and a computationally cheap representation of the physical device: it is less accurate but computationally much more efficient than the fine model. It is usually implemented by an equivalent circuit model combined with some empirical equations. It is assumed that the coarse model can be intensively evaluated with no significant cost.

11.2.1. *The Fine Model*

Let $\boldsymbol{R}_\mathrm{f} \in \Re^p$ denote a fine model response vector, sampled at p independent-variable points. Response $\boldsymbol{R}_\mathrm{f}$ can be at any domain (frequency-domain, time-domain, or static). In general, we consider that $\boldsymbol{R}_\mathrm{f}$ depends on three vectors: the selected design variables, $\boldsymbol{x}_\mathrm{f} \in \Re^n$, some pre-assigned model parameters, $z \in \Re^m$, and the fine model independent variable according to the kind of simulation (frequency, time, etc.), $\psi_f \in \Re$. Selecting suitable design variables x for a given problem usually requires engineering expertise, while most of the pre-assigned parameters z are usually determined by manufacturing constraints as well as by engineering knowledge about the influence of the physical parameters on the circuit responses.

From the design optimization perspective, the fine model can be treated as a multidimensional (not necessarily continuous) vector function, $\boldsymbol{R}_f(\boldsymbol{x}_f) : X_f \to \Re^p$ whose domain is $X_f \subseteq \Re^n$. As mentioned before, evaluating $\boldsymbol{R}_f(\boldsymbol{x}_f)$ is computationally expensive. Direct optimization of the fine model using classical optimization methods is prohibitive given its typically high computational cost. It requires solving

$$\boldsymbol{x}_f^* = \arg \min_{\boldsymbol{x}_f \in X_f} U(\boldsymbol{R}_f(\boldsymbol{x}_f)), \qquad (11.1)$$

where $U : \Re^n \to \Re$ is the objective function expressed in terms of the design specifications, and vector $\boldsymbol{x}_f^*$ contains the optimal fine model design.

Performing statistical analysis by directly using the fine model is not feasible.

11.2.2. *The Coarse Model*

The coarse model is denoted as $\boldsymbol{R}_c \in \Re^p$ (it is in the same domain as $\boldsymbol{R}_f$). Similarly, $\boldsymbol{R}_c$ depends on three vectors: the selected design variables, $\boldsymbol{x}_c \in \Re^n$, some pre-assigned parameters, $z \in \Re^m$, and the coarse model independent variable, ψ_c. In most cases it is assumed that $\psi_c = \psi_f = \psi$, however, some shifting in the independent variable can be imposed between the two models. The coarse model is also treated as a multidimensional vector function, $\boldsymbol{R}_c(\boldsymbol{x}_c) : X_c \to \Re^p$ whose domain is $X_c \subseteq \Re^n$.

Given its computational efficiency, the coarse model can be directly optimized using classical optimization methods,

$$\boldsymbol{x}_c^* = \arg \min_{\boldsymbol{x}_c \in X_c} U(\boldsymbol{R}_c(\boldsymbol{x}_c)), \qquad (11.2)$$

where vector $\boldsymbol{x}_c^*$ contains the optimal coarse model design, and $\boldsymbol{R}_c^* = \boldsymbol{R}_c(\boldsymbol{x}_c^*)$ is the corresponding optimal coarse model response.

11.3. Input Space Mapping

The input space mapping approach to design optimization is described in this section. First, a Broyden-based linear input space mapping technique is described. In this technique, a simple linear transformation locally approximates the existing mapping from the fine model design space to the coarse model design space. Second, a neural input space mapping approach to design optimization is described, where the corresponding mapping can be arbitrary nonlinear. It will be later shown how, after applying these techniques, a simple enhanced coarse model is available to perform inexpensive yield estimations.

11.3.1. *Linear Input Space Mapping*

The concept of linear input space mapping is illustrated in Fig. 11.1. This approach aims at developing a simple linear mapping function P to adjust the coarse model design parameters (x_c) with respect to the actual physical design parameters (x_f), such that the coarse model response is as close as possible to the fine model response, as illustrated in Fig. 11.1. An iterative process can be applied to gradually develop the mapping function P, as in the original aggressive space mapping formulation (Bandler *et al.*, 1995b).

The linear mapping function P in Fig. 11.1 can be found through an algorithmic design process that simultaneously reveals a space mapped solution, x_f^{SM}, that is a fine model design that makes the fine model response $R_f(x_f^{SM}, \psi)$ close enough (from a practical engineering perspective) to a target response, which is given by the optimal coarse model response, i.e. $R_f(x_f^{SM}, \psi) \approx R_c(x_c^*, \psi)$. Ideally, the resultant input mapped coarse model, $R_c(P(x_f), \psi)$, could be used as an inexpensive vehicle to estimate the fine model yield around x_f^{SM}.

A linear input SM optimization algorithm can be formulated by finding an approximate root of the following system of nonlinear equations:

$$f(x_f) = P(x_f) - x_c^*, \tag{11.3}$$

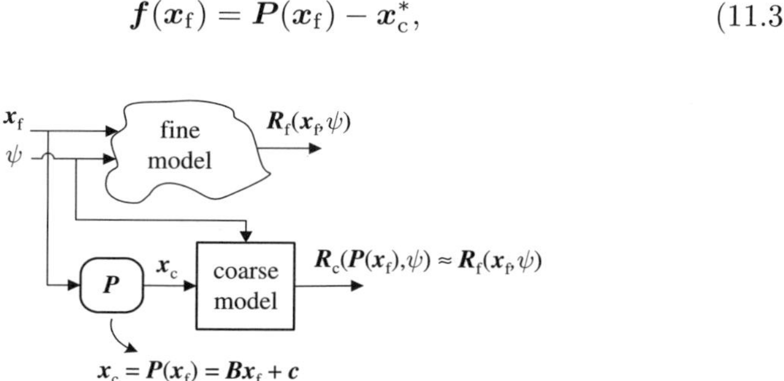

Fig. 11.1.   Linear input space mapping concept. A simple linear transformation aims at aligning both models within some region of design parameters.

where the multidimensional vector function $\boldsymbol{x}_c = \boldsymbol{P}(\boldsymbol{x}_f)$ is evaluated through a local alignment of the two models at the current iterate

$$\boldsymbol{P}(\boldsymbol{x}_f) = \arg \min_{\boldsymbol{x}_c} \|\boldsymbol{e}_1^T \;\cdots\; \boldsymbol{e}_p^T\|_2^2, \tag{11.4a}$$

where p is the number of points (samples) of the independent variable and the jth parameter extraction error vector is given by

$$\boldsymbol{e}_j(\boldsymbol{x}_f) = \boldsymbol{R}_{\text{fs}}(\boldsymbol{x}_f, \psi_j) - \boldsymbol{R}_{\text{cs}}(\boldsymbol{x}_c, \psi_j), \quad j = 1 \ldots p. \tag{11.4b}$$

Aligning not only the optimizable responses $\boldsymbol{R}_c$ and $\boldsymbol{R}_f$ but the complete set of characterizing responses available in the two models, denoted by $\boldsymbol{R}_{\text{cs}}$ and $\boldsymbol{R}_{\text{fs}}$, respectively, is an effective practicality to avoid poor local minima when solving the parameter extraction problem in (11.4), as described in Bandler *et al.* (2003).

11.3.1.1. *Algorithm for Broyden-based input space mapping*

The employed algorithmic design process exploiting linear input space mapping is shown in Fig. 11.2. It consists of an enhanced Broyden-based "direct" SM algorithm (Rayas-Sánchez *et al.*, 2006b). It aims at solving (11.3) by

Begin

   find $\boldsymbol{x}_c^*$, define $\boldsymbol{x}_f^{\min}$ and $\boldsymbol{x}_f^{\max}$

   $i = 1,\, \boldsymbol{x}_f^{(i)} = \boldsymbol{x}_c^*,\, \boldsymbol{B}^{(i)} = \boldsymbol{I},\, \alpha = 0.9$

   repeat until *StoppingCriteria*

      $\boldsymbol{x}_c^{(i)} = \arg \min_{\boldsymbol{x}_c} \boldsymbol{R}_f(\boldsymbol{x}_f^{(i)}) - \boldsymbol{R}_c(\boldsymbol{x}_c)$

      $\boldsymbol{f}^{(i)} = \boldsymbol{x}_c^{(i)} - \boldsymbol{x}_c^*$

      solve $\boldsymbol{B}^{(i)}\boldsymbol{h}^{(i)} = -\boldsymbol{f}^{(i)}$ for $\boldsymbol{h}^{(i)}$

      $\boldsymbol{x}_f^{\text{test}} = \boldsymbol{x}_f^{(i)} + \boldsymbol{h}^{(i)}$

      while $\boldsymbol{x}_f^{\text{test}} \prec \boldsymbol{x}_f^{\min} \vee \boldsymbol{x}_f^{\text{test}} \succ \boldsymbol{x}_f^{\max}$

         $\boldsymbol{h}^{(i)} = \alpha \boldsymbol{h}^{(i)}$

         $\boldsymbol{x}_f^{\text{test}} = \boldsymbol{x}_f^{(i)} + \boldsymbol{h}^{(i)}$

      end

      $\boldsymbol{x}_f^{(i+1)} = \boldsymbol{x}_f^{\text{test}}$

      $\boldsymbol{B}^{(i+1)} = \boldsymbol{B}^{(i)} + \dfrac{\boldsymbol{f}^{(i)}\boldsymbol{h}^{(i)T}}{\boldsymbol{h}^{(i)T}\boldsymbol{h}^{(i)}}$ (Broyden update)

      $i = i + 1$

   end

end

Fig. 11.2. Pseudo-code implementing an algorithm for linear input space mapping.

using the classical Broyden method for solving systems of nonlinear equations, where matrix $\boldsymbol{B}$ is an approximation of the Jacobian of $\boldsymbol{f}$ with respect to $\boldsymbol{x}_\mathrm{f}$ at the current iterate. It is initialized by the identity matrix and updated by using

$$\boldsymbol{B}^{(i+1)} = \boldsymbol{B}^{(i)} + \frac{\boldsymbol{f}^{(i)}\boldsymbol{h}^{(i)T}}{\boldsymbol{h}^{(i)T}\boldsymbol{h}^{(i)}}, \tag{11.5}$$

where $\boldsymbol{h}^{(i)}$ is the corresponding quasi-Newton step at the ith iteration.

The algorithm in Fig. 11.2 avoids instabilities (Rayas-Sánchez *et al.*, 2005) by constraining the fine model parameters within user-defined limits, $\boldsymbol{x}_\mathrm{f}^{\min}$ and $\boldsymbol{x}_\mathrm{f}^{\max}$. When the next candidate, $\boldsymbol{x}_\mathrm{f}^{(\mathrm{test})}$, falls outside these limits, the step size is decreased in the same quasi-Newton direction. This simple improvement has been confirmed extremely effective, and avoids the need of extra fine model evaluations, as in the case of trust region methodologies. Implementation in Fig. 11.2 uses a shrinking factor $\alpha = 0.9$, but other values of α have been tested, yielding similar results.

When exiting the algorithm in Fig. 11.2, we have not only the space mapped solution $\boldsymbol{x}_\mathrm{f}^{\mathrm{SM}} = \boldsymbol{x}_\mathrm{f}^{(i)}$ available, but also the linear input mapping function between the two models,

$$\boldsymbol{P}(\boldsymbol{x}_\mathrm{f}) = \boldsymbol{B}\boldsymbol{x}_\mathrm{f} + \boldsymbol{c} \tag{11.6}$$

where the mapping parameters are $\boldsymbol{B} = \boldsymbol{B}^{(i)}$ and $\boldsymbol{c} = \boldsymbol{x}_\mathrm{c}^* - \boldsymbol{B}\boldsymbol{x}_\mathrm{f}^{\mathrm{SM}}$.

11.3.1.2. *Termination criteria*

The algorithm in Fig. 11.2 terminates when: the relative change in the fine model design parameters is small enough; or the relative difference between the extracted coarse model parameters and the optimal coarse model design is small enough; or the relative difference between the maximum absolute error of the fine model response with respect to the target response is small enough; or a maximum number of iterations is reached,

$$\|\boldsymbol{x}_\mathrm{f}^{(i+1)} - \boldsymbol{x}_\mathrm{f}^{(i)}\|_2 \leq \varepsilon_1(\varepsilon_1 + \|\boldsymbol{x}_\mathrm{f}^{(i)}\|_2) \vee \ldots \tag{11.7a}$$

$$\|\boldsymbol{x}_\mathrm{c}^{(i)} - \boldsymbol{x}_\mathrm{c}^*\|_2 \leq \varepsilon_2(\varepsilon_2 + \|\boldsymbol{x}_\mathrm{c}^*\|_2) \vee \ldots \tag{11.7b}$$

$$\|\boldsymbol{R}_\mathrm{f}(\boldsymbol{x}_\mathrm{f}^{(i)}) - \boldsymbol{R}_\mathrm{c}(\boldsymbol{x}_\mathrm{c}^*)\|_\infty \leq \varepsilon_3(\varepsilon_3 + \|\boldsymbol{R}_\mathrm{c}(\boldsymbol{x}_\mathrm{c}^*)\|_\infty) \vee \ldots \tag{11.7c}$$

$$i > 4n. \tag{11.7d}$$

A typical implementation uses $\varepsilon_1 = \varepsilon_2 = 10^{-3}$ and $\varepsilon_3 = 10^{-2}$.

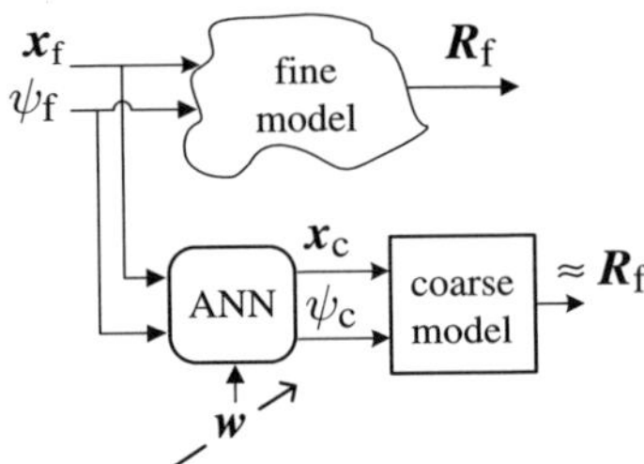

Fig. 11.3.   Neural input space mapping concept. An artificial neural network is trained to learn the mapping between the design spaces of both models.

11.3.2. *Neural Input Space Mapping*

The mathematical relationship that maps the fine model design space to the coarse model design space can also be implemented in a nonlinear fashion, aiming at producing an enhanced input mapped coarse model able to approximate the fine model with higher accuracy over a larger region of design parameters. This is the main reason for using neural input space mapping (NSM), which exploits the SM-based neuromodeling techniques described in Bandler *et al.* (1999). The NSM concept is illustrated in Fig. 11.3, where a neural network implements the input mapping. In this approach, the ANN can map not only the design parameters but also the independent variable, providing a very powerful interpolation capability to the resultant input mapped coarse model.

NSM requires a reduced set of upfront learning base points. From the neural modeling perspective, the coarse model is used not only as a source of knowledge that reduces the amount of learning data and improves the generalization performance of the SM-based neuromodel, but also as a means to select the initial learning base points through sensitivity analysis (Bakr *et al.*, 2000).

11.3.2.1. *Algorithm for neural input space mapping*

Figure 11.4 shows a flow diagram of NSM optimization. It starts by directly optimizing the coarse model to find its optimal design x_c^* that yields the desired response.

Next, $2n$ additional points are selected following an n-dimensional star-distribution (Biernacki *et al.*, 1989) centered at x_c^*, as shown in Fig. 11.5, where n is the number of design parameters. The percentage of deviation from x_c^* for each design parameter is determined according to the coarse

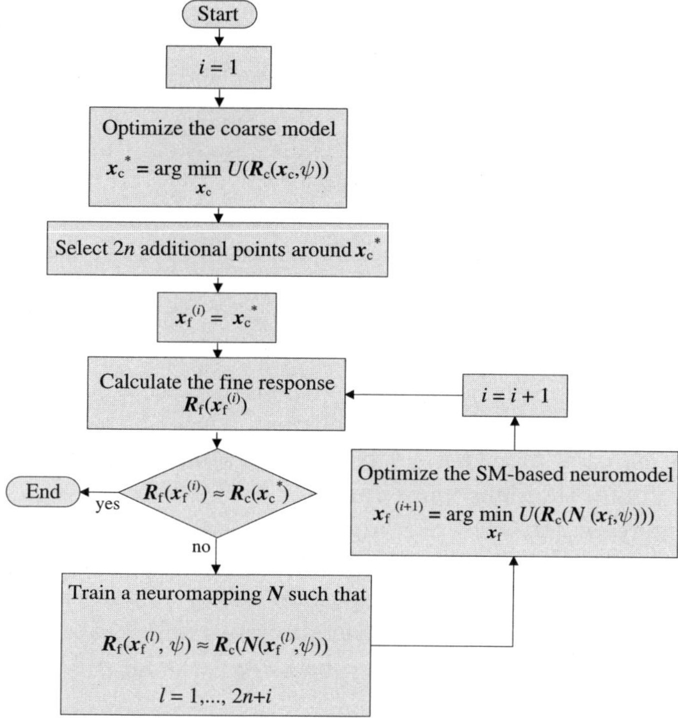

Fig. 11.4. Flow diagram for neural input space mapping optimization. The input–output relationship of the neural network is represented by multidimensional vector function N.

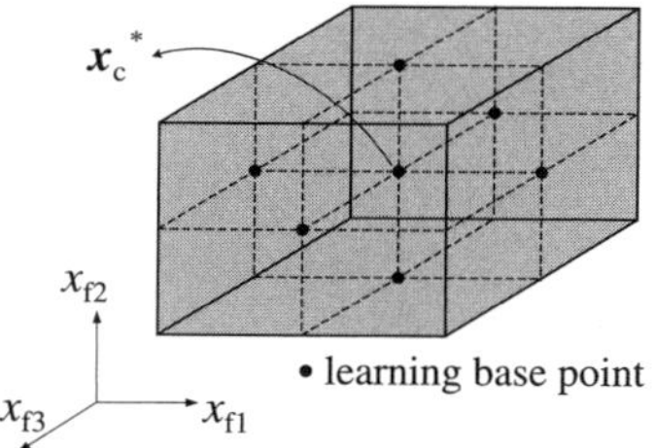

Fig. 11.5. Star-distribution of learning base points to train the initial neural input space mapping.

model sensitivities: the larger the sensitivity of the coarse model response with respect to a certain parameter, the smaller the percentage of variation of that parameter. The underlying assumption here is that the coarse model sensitivities are similar to those of the fine model, which usually happens when the coarse model is a physics-based representation of the original structure.

The fine model response $\boldsymbol{R}_\mathrm{f}$ at the optimal coarse model solution $\boldsymbol{x}_\mathrm{c}^*$ is then calculated (see Fig. 11.4). If $\boldsymbol{R}_\mathrm{f}$ is approximately equal to the desired response, the algorithm ends, otherwise an SM-based neuromodel over the initial $2n + 1$ fine model points is trained.

Once an SM-based neuromodel with small learning errors is available, it is used as an enhanced coarse model, optimizing its parameters to generate the desired response. The solution to this optimization problem becomes the next point in the fine model parameter space, and it is included in the learning set (see Fig. 11.4).

The fine model response at the new point is calculated and compared with the desired response. If it is still different, a new SM-based neuromodel is trained over the extended set of learning samples and the algorithm continues. If not, the algorithm terminates.

It is interesting to notice that input neural space mapping optimization does not require troublesome parameter extraction to predict the next iterate. Additionally, the SM-based neuromodels are developed without using testing points, since their generalization performance is controlled by gradually increasing the ANN complexity, starting with a two-layer perceptron (Bakr *et al.*, 2000).

11.3.2.2. *Training the input neural mapping*

As seen in Fig. 11.4, at the ith iteration of the neural input space mapping optimization algorithm, the neuromapping $\boldsymbol{N}$ is trained such that the coarse model using that mapping approximates the fine model at all the learning base points. This is realized by solving the optimization problem

$$\boldsymbol{w}^* = \arg\min_{\boldsymbol{w}} \| [\cdots \ \boldsymbol{e}_k^\mathrm{T} \ \cdots]^\mathrm{T} \|, \tag{11.8}$$

with

$$\boldsymbol{e}_k = \boldsymbol{R}_\mathrm{f}(\boldsymbol{x}_\mathrm{f}^{(l)}, \psi_{\mathrm{f}j}) - \boldsymbol{R}_\mathrm{c}(\boldsymbol{x}_{\mathrm{c}_j}^{(l)}, \psi_{\mathrm{c}_j}) \tag{11.9}$$

$$\begin{bmatrix} \boldsymbol{x}_{\mathrm{c}_j}^{(l)} \\ \psi_{\mathrm{c}_j} \end{bmatrix} = \boldsymbol{N}(\boldsymbol{x}_\mathrm{f}^{(l)}, \psi_{\mathrm{f}_j}, \boldsymbol{w}) \tag{11.10}$$

$$j = 1, \ldots, p \tag{11.11a}$$

$$l = 1, \ldots, 2n + i \qquad (11.11\text{b})$$

$$k = j + p(l - 1), \qquad (11.11\text{c})$$

where $2n + i$ is the number of training base points for the input design parameters and p is the number of independent variable points. It is seen that the total number of learning samples at the ith iteration is $(2n + i)p$.

The input–output relationship of the ANN (11.10) implements the mapping at the ith iteration. Vector $\boldsymbol{w}$ contains the internal free parameters (weights, bias, etc.) of the neural network. The most widely used paradigm to implement $\boldsymbol{N}$ is a three-layer perceptron (Rayas-Sánchez, 2004). The complexity of $\boldsymbol{N}$ (the number of hidden neurons) is gradually increased according to the learning error, starting with a linear mapping (three-layer perceptron with 0 hidden neurons) (Bakr *et al.*, 2000).

It has been demonstrated (Gutiérrez-Ayala and Rayas-Sánchez, 2010) that a better way to control the generalization performance of the neural input mapping $\boldsymbol{N}$ is by using a two-layer perceptron with optimized nonlinearity. In this case, further advantage is taken of the coarse model implicit knowledge, exploiting the fact that typically the resultant input mapping between both models (coarse and fine models) is not too complex, such that after initiating with a linear system, if the problem requires it, the algorithm switches to a 2LP with nonlinear activation functions to implement the mapping needed. Control of the nonlinearity is performed during all the training process at each iteration of the algorithm (Gutiérrez-Ayala and Rayas-Sánchez, 2010).

11.3.2.3. *Example of neural input space mapping for design optimization*

Consider the microstrip notch filter with mitered bends and open stubs illustrated in Fig. 11.6 (Rayas-Sánchez *et al.*, 2006b). L_o is the open stubs length, L_c is the length of the coupled lines, and S_g is the separation gap. The width W_{50} is the same for all the sections as well as for the input and output lines, of length L_p. A substrate with thickness H and relative dielectric constant ε_r is used.

The design parameters are $\boldsymbol{x}_\text{f} = [L_\text{c} \; L_\text{o} \; S_\text{g}]^\text{T}$ (mil). The pre-assigned parameter values are $H = 10\,\text{mil}$, $W_{50} = 31\,\text{mil}$ and $\varepsilon_\text{r} = 2.2$ (RT Duroid 5880, with loss tangent $\tan \delta = 0.0009$). Lossless metals are considered.

The design specifications are: $|S_{21}| \leq 0.05$ for frequencies between $13.19\,\text{GHz}$ and $13.21\,\text{GHz}$, and $|S_{21}| \geq 0.95$ for frequencies below $13\,\text{GHz}$ and above $13.4\,\text{GHz}$.

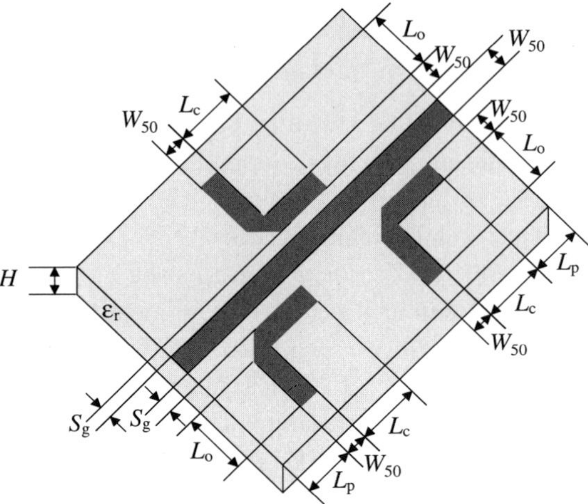

Fig. 11.6.   Microstrip notch filter with mitered bends and open stubs.

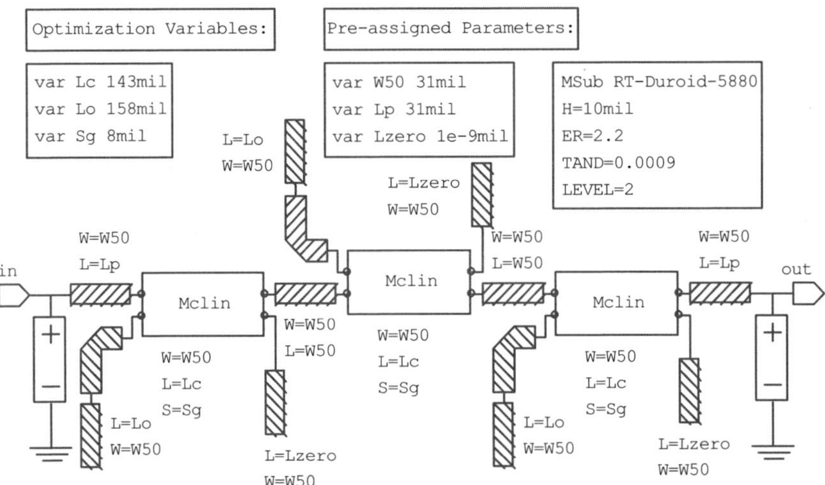

Fig. 11.7.   Coarse model of the microstrip notch filter with mitered bends and open
stubs, implemented in APLAC.

The coarse model is illustrated in Fig. 11.7. It consists of an equivalent distributed circuit implemented in APLAC (2005) using its built-in
microstrip circuit models available for lines, open circuited lines, coupled
lines, and mitered bends.

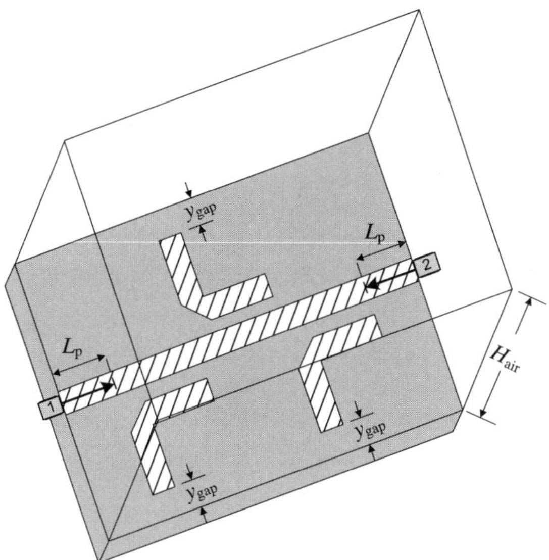

Fig. 11.8.   Fine model of the microstrip notch filter with mitered bends and open stubs, implemented in Sonnet (2009).

The fine model is implemented in the full-wave EM simulator Sonnet (2009), whose simulation box is illustrated in Fig. 11.8. It uses a very high-resolution grid (cell size of $0.5\,\mathrm{mil} \times 0.5\,\mathrm{mil}$) with $H_{\mathrm{air}} = 60\,\mathrm{mil}$, $L_{\mathrm{p}} = \frac{1}{2}(L_{\mathrm{o}} + L_{\mathrm{c}})$, and $y_{\mathrm{gap}} = L_{\mathrm{o}}$. By using these values of y_{gap} and H_{air}, the effects of unwanted EM interaction and potential resonances with Sonnet's box are disabled.

After solving (11.2) for this example using the sequential quadratic programming (SQP) method available in Matlab, the optimal coarse model solution is $\boldsymbol{x}_{\mathrm{c}}^* = [143\ 158\ 8]^{\mathrm{T}}$ (mil). To generate the learning base points, six in this case, a 3% of deviation for L_{c} and L_{o} is used, while a 50% for S_{g} is applied, with respect to their values at $\boldsymbol{x}_{\mathrm{c}}^*$. The coarse and fine model responses at $\boldsymbol{x}_{\mathrm{c}}^*$ are shown in Fig. 11.9.

To contrast both formulations previously described for neural input space mapping, this example is solved using three-layer perceptrons with gradual increasing complexity (Bakr *et al.*, 2000), as well as two-layer perceptrons with optimized nonlinearity (Gutiérrez-Ayala and Rayas-Sánchez, 2010). In the later case, after only one additional evaluation of the fine model, the SM solution $\boldsymbol{x}_{\mathrm{f}}^{\mathrm{SM}} = [143\ 159.5\ 6.5]^{\mathrm{T}}$ (mil) is

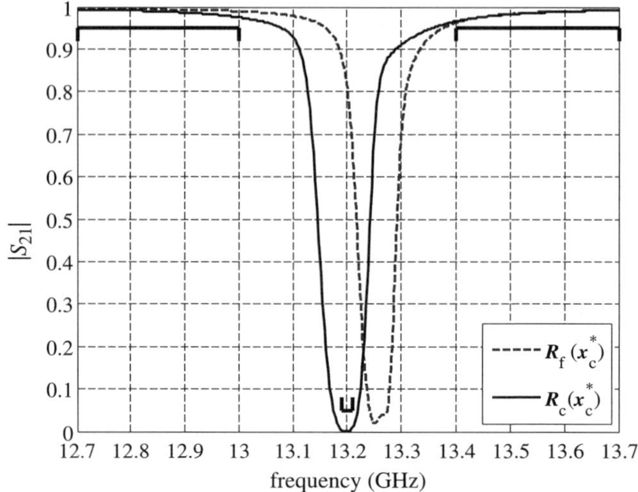

Fig. 11.9. Coarse and fine model responses at the optimal coarse model design of the microstrip notch filter.

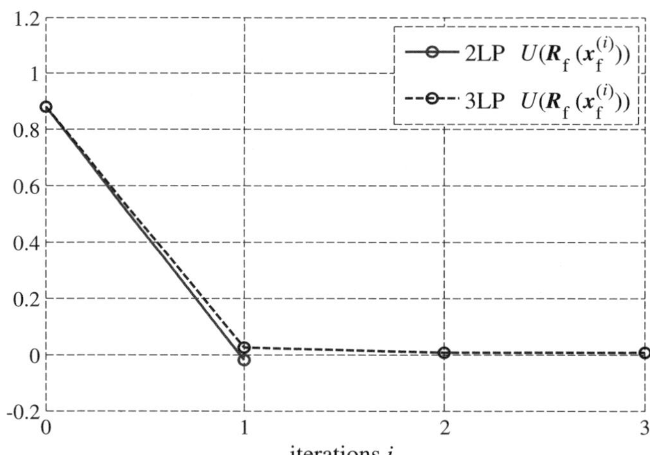

Fig. 11.10. Fine model objective function values at each neural input space mapping iteration for the microstrip notch filter.

found, while the former approach requires three additional fine model simulations.

Figure 11.10 illustrates the evolution of the fine model objective function values for each algorithm.

Table 11.1. Design parameters and fine model objective function values at each iteration of neural input space mapping for the microstrip notch filter.

Iteration	Using 3LP		Using 2LP	
	$x_{\mathrm{f}}^{(i)}$	$U(i)$	$x_{\mathrm{f}}^{(i)}$	$U(i)$
0	$[143\ 158\ 8]^{\mathrm{T}}$	0.8807	$[143\ 158\ 8]^{\mathrm{T}}$	0.8807
1	$[142.5\ 160\ 9.5]^{\mathrm{T}}$	0.02354	$[143\ 159.5\ 6.5]^{\mathrm{T}}$	-0.0215
2	$[143.5\ 159\ 9]^{\mathrm{T}}$	0.00580		
3	$[143.5\ 159\ 9]^{\mathrm{T}}$	0.00580		

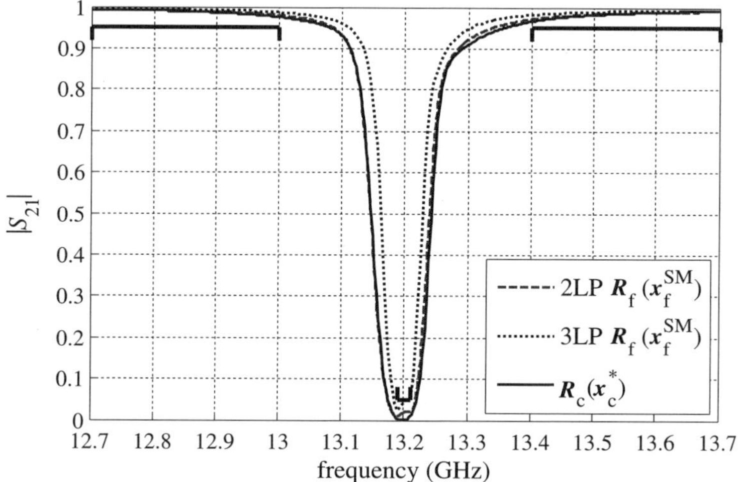

Fig. 11.11. Final results after applying neural input space mapping optimization to the microstrip notch filter, using three-layer perceptrons with gradual increasing complexity and two-layer perceptrons with optimized nonlinearity.

Table 11.1 shows the fine model design parameter values at each iteration of the algorithms, as well as the corresponding fine model objective function values. It is seen that formulation using two-layer perceptrons yields a better final objective function value in less fine model evaluations.

Finally, the target response and fine model responses evaluated at the SM solutions found are shown in Fig. 11.11. It is seen that the space mapped response found $R_{\mathrm{f}}(x_{\mathrm{f}}^{\mathrm{SM}}, \psi)$, by using two-layer perceptrons with optimized nonlinearity, is significantly better than that found by using three-layer perceptrons with gradual increasing complexity.

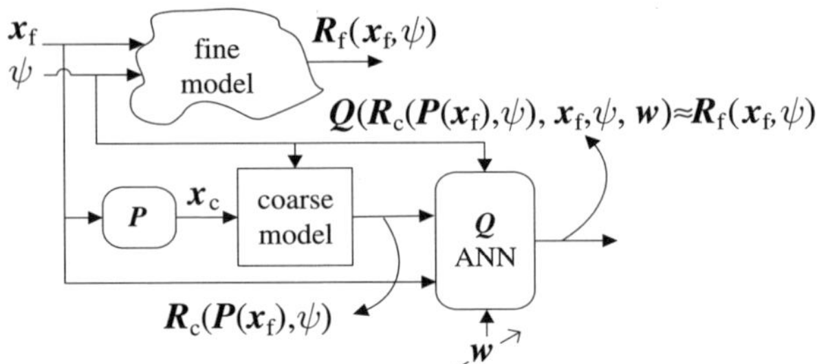

Fig. 11.12. Linear-input neural-output space mapping concept. The output neural network Q eliminates any possible residual that might exist between the input mapped coarse model and the fine model.

11.4. Linear-Input Neural-Output Space Mapping

An algorithmic method for efficient and highly accurate EM-based statistical analysis and yield prediction is described in this section. This method is based on a linear-input neural-output space mapping strategy (Rayas-Sánchez *et al.*, 2006a, 2006b), whose main concept is illustrated in Fig. 11.12.

In this approach (see Fig. 11.12), the linear mapping function P is found first through the algorithmic design process described in Section 11.3.1. Ideally, the resultant input mapped coarse model, $R_\mathrm{c}(P(x_\mathrm{f}^\mathrm{SM}), \psi)$, could be used as an inexpensive vehicle to estimate the fine model yield around x_f^SM. In many practical cases, however, $R_\mathrm{c}(P(x_\mathrm{f}^\mathrm{SM}), \psi)$ slightly deviates from $R_\mathrm{f}(x_\mathrm{f}^\mathrm{SM}, \psi)$ due to an inherent residual that cannot be eliminated through an input mapping P, regardless of the complexity of P. In those cases, any yield prediction using $R_\mathrm{c}(P(x_\mathrm{f}^\mathrm{SM}), \psi)$ is unreliable, especially when the design specifications are tight. To overcome this problem, an output neural model Q whose internal free-parameters are in vector w is trained around x_f^SM to eliminate the typical residual between $R_\mathrm{f}(x_\mathrm{f}^\mathrm{SM}, \psi)$ and $R_\mathrm{c}(P(x_\mathrm{f}^\mathrm{SM}), \psi)$, as illustrated in Fig. 11.12.

The linear-input neural-output space mapping method consists of three sub-processes: (i) find a target response $R_\mathrm{c}(x_\mathrm{c}^*, \psi)$ by optimizing the coarse model (solving (11.2) by classical optimization methods), (ii) find x_f^SM and P through a linear input space mapping optimization, following the procedure described in Section 11.3.1, and (iii) develop a suitable neural network Q valid in the tolerance region of interest around x_f^SM. Once Q is available, the linear-input neural-output mapped coarse model can be used to perform inexpensive and highly accurate statistical analysis and yield calculations.

11.4.1. *Training the Output Neural Mapping*

The output neural network $\boldsymbol{Q}$ is developed in a user-defined tolerance region around $\boldsymbol{x}_{\mathrm{f}}^{\mathrm{SM}}$. If properly trained, $\boldsymbol{Q}$ eliminates the residual between the fine model $\boldsymbol{R}_{\mathrm{f}}(\boldsymbol{x}_{\mathrm{f}}^{\mathrm{SM}}, \psi)$ and the input mapped coarse model $\boldsymbol{R}_{\mathrm{c}}(\boldsymbol{P}(\boldsymbol{x}_{\mathrm{f}}^{\mathrm{SM}}), \psi)$.

Training the output neural mapping $\boldsymbol{Q}$ aims at finding an optimal vector of weighting factors, $\boldsymbol{w}^{*}$, such that

$$\boldsymbol{Q}(\boldsymbol{R}_{\mathrm{c}}(\boldsymbol{B}\boldsymbol{x}_{\mathrm{f}} + \boldsymbol{c}, \psi), \boldsymbol{x}_{\mathrm{f}}, \psi, \boldsymbol{w}^{*}) = \boldsymbol{R}_{\mathrm{f}}(\boldsymbol{x}_{\mathrm{f}}, \psi) \qquad (11.12)$$

for all $\boldsymbol{x}_{\mathrm{f}}$ and ψ in the training region. It is interesting to notice that in this approach, the independent variable ψ is not altered for the coarse model by the input linear mapping. In contrast to other output space mapping techniques (Bandler *et al.*, 2004b; Koziel *et al.*, 2005), this approach uses as inputs for $\boldsymbol{Q}$ not only the mapped coarse model responses but also the fine model design parameters and the independent variable (see Fig. 11.12), which gives to $\boldsymbol{Q}$ an extremely powerful interpolating capability.

The problem of training neural network $\boldsymbol{Q}$ is formulated as

$$\boldsymbol{w}^{*} = \arg\min_{\boldsymbol{w}} \|\boldsymbol{E}_{\mathrm{L}}(\boldsymbol{w})\|_{\mathrm{F}}, \qquad (11.13)$$

where $\|\cdot\|_{\mathrm{F}}$ denotes the Frobenius norm of a learning error matrix $\boldsymbol{E}_{\mathrm{L}} \in \Re^{mL \times p}$ for L learning base points, p samples for ψ, and m number of characterizing responses in the circuit. For instance, if the responses of interest are the real and imaginary parts of S_{11} and S_{21}, then $m = 4$. This learning error matrix is given by

$$\boldsymbol{E}_{\mathrm{L}}(\boldsymbol{w}) = \boldsymbol{Q}_{\mathrm{L}}(\boldsymbol{w}) - \boldsymbol{R}_{\mathrm{FL}}. \qquad (11.14)$$

The matrix storing all the fine model learning responses is $\boldsymbol{R}_{\mathrm{FL}} \in \Re^{mL \times p}$. This matrix can be organized as follows

$$\boldsymbol{R}_{\mathrm{FL}} = [\boldsymbol{R}_{\mathrm{fL}}^{(1)} \ \ldots \ \boldsymbol{R}_{\mathrm{fL}}^{(L)}]^{\mathrm{T}}, \qquad (11.15)$$

where sub-matrix $\boldsymbol{R}_{\mathrm{fL}}^{(k)} \in \Re^{m \times p}$ has the fine model characterizing responses at the kth learning base point $\boldsymbol{x}_{\mathrm{f}}(^{k})$

$$\boldsymbol{R}_{\mathrm{fL}}^{(k)} = \begin{bmatrix} \boldsymbol{R}_{\mathrm{fs}}(\boldsymbol{x}_{\mathrm{f}}^{(k)}, \psi_1)^{T} \\ \vdots \\ \boldsymbol{R}_{\mathrm{fs}}(\boldsymbol{x}_{\mathrm{f}}^{(k)}, \psi_p)^{T} \end{bmatrix}^{T} . \qquad (11.16)$$

Similarly, $\boldsymbol{Q}_{\mathrm{L}} \in \Re^{mL \times p}$ contains all the outputs of the ANN in the set of learning base points

$$\boldsymbol{Q}_{\mathrm{L}}(\boldsymbol{w}) = [\boldsymbol{q}^{(1)} \ \cdots \ \boldsymbol{q}^{(L)}]^T \tag{11.17}$$

and $\boldsymbol{q}^{(k)} \in \Re^{m \times p}$ has the outputs of the ANN at the kth learning base point $\boldsymbol{x}_{\mathrm{f}}^{(k)}$, as follows:

$$\boldsymbol{q}^{(k)} = \begin{bmatrix} \boldsymbol{Q}(\boldsymbol{R}_{\mathrm{cs}}(\boldsymbol{B}\boldsymbol{x}_{\mathrm{f}}^{(k)} + \boldsymbol{c}, \psi_1), \boldsymbol{x}_{\mathrm{f}}^{(k)}, \psi_1, \boldsymbol{w})^T \\ \vdots \\ \boldsymbol{Q}(\boldsymbol{R}_{\mathrm{cs}}(\boldsymbol{B}\boldsymbol{x}_{\mathrm{f}}^{(k)} + \boldsymbol{c}, \psi_p), \boldsymbol{x}_{\mathrm{f}}^{(k)}, \psi_p, \boldsymbol{w})^T \end{bmatrix}^T . \tag{11.18}$$

A number T of testing base points not seen during training is used to control the generalization performance while solving (11.13). Similarly, the corresponding matrices of ANN outputs and fine model responses at the T testing base points are denoted by $\boldsymbol{Q}_{\mathrm{T}}, \boldsymbol{R}_{\mathrm{FT}} \in \Re^{mT \times p}$.

The complete learning and testing errors before developing the output neuromapping are calculated by the following Frobenius norms:

$$\varepsilon_{\mathrm{L}} = \|\boldsymbol{R}_{\mathrm{CL}} - \boldsymbol{R}_{\mathrm{FL}}\|_{\mathrm{F}} \tag{11.19a}$$

$$\varepsilon_{\mathrm{T}} = \|\boldsymbol{R}_{\mathrm{CT}} - \boldsymbol{R}_{\mathrm{FT}}\|_{\mathrm{F}}, \tag{11.19b}$$

where $\boldsymbol{R}_{\mathrm{CL}} \in \Re^{mL \times p}$ and $\boldsymbol{R}_{\mathrm{CT}} \in \Re^{mT \times p}$ store all the input mapped coarse model responses evaluated in the learning and testing sets, respectively.

It is interesting to notice that the output neural network $\boldsymbol{Q}$ has only m outputs and $m + n + 1$ inputs, such that its complexity is low and not affected by the number of samples used for the independent variable ψ. A classical three-layer perceptron (3LP) with h hidden neurons can be used as an ANN paradigm for $\boldsymbol{Q}$.

An algorithm for training the output neuromapping $\boldsymbol{Q}$ is in Fig. 11.13 (Rayas-Sánchez *et al.*, 2006b). First, all the learning and testing base points around $\boldsymbol{x}_{\mathrm{f}}^{\mathrm{SM}}$ are generated, as described in the following section, as well as the corresponding responses. Next, a three-layer perceptron with m hidden neurons is trained, and the corresponding training and testing errors are calculated. The complexity of $\boldsymbol{Q}$ (the number of hidden neurons in the 3LP) is iteratively increased until the generalization performance starts to deteriorate, which happens when the current testing error is larger than the previous one and the current learning error is smaller than the current testing error (Haykin, 1999). The neural mapping training problem in (11.13), used in algorithm shown in Fig. 11.13, can be efficiently solved as a nonlinear

$$
\begin{aligned}
&\textbf{Begin} \\
&\quad \text{Generate } R_{\text{CL}}, R_{\text{CT}}, R_{\text{FL}} \text{ and } R_{\text{FT}} \\
&\quad \varepsilon_{\text{L}}^{\text{old}} = \| R_{\text{CL}} - R_{\text{FL}} \|_{\text{F}} \ , \ \varepsilon_{\text{T}}^{\text{old}} = \| R_{\text{CT}} - R_{\text{FT}} \|_{\text{F}} \\
&\quad h = m, \ i = 1 \\
&\quad w^{(i)} = \arg \min_{w} \| E_{\text{L}}(w) \|_{\text{F}} \\
&\quad \varepsilon_{\text{L}} = \| Q_{\text{L}}(w^{(i)}) - R_{\text{FL}} \|_{\text{F}} \\
&\quad \varepsilon_{\text{T}} = \| Q_{\text{T}}(w^{(i)}) - R_{\text{FT}} \|_{\text{F}} \\
&\quad \textbf{while } \ \varepsilon_{\text{T}}^{\text{old}} \geq \varepsilon_{\text{T}} \ \lor \ \varepsilon_{\text{L}} \geq \varepsilon_{\text{T}} \\
&\qquad \varepsilon_{\text{T}}^{\text{old}} = \varepsilon_{\text{T}}, \ \varepsilon_{\text{L}}^{\text{old}} = \varepsilon_{\text{L}}, \ i = i+1, \ h = h+1 \\
&\qquad w^{(i)} = \arg \min_{w} \| E_{\text{L}}(w) \|_{F} \\
&\qquad \varepsilon_{\text{L}} = \| Q_{\text{L}}(w^{(i)}) - R_{\text{FL}} \|_{\text{F}} \\
&\qquad \varepsilon_{\text{T}} = \| Q_{\text{T}}(w^{(i)}) - R_{\text{FT}} \|_{\text{F}} \\
&\quad \textbf{end} \\
&\quad w^{*} = w^{(i-1)} \\
&\textbf{end}
\end{aligned}
$$

Fig. 11.13. Pseudo-code implementing an algorithm for training the output neuromapping Q between the fine model and the input mapped coarse model.

curve-fitting problem in the least-squares sense using the Levenberg–Marquardt method available in the Matlab Optimization Toolbox.

11.4.2. *Generation of Learning and Testing Base Points*

The number of learning and testing base points should be small to keep a reduced amount of fine model simulations. This can be effectively achieved by using $L = 2n + 1$ in a star-distribution (Biernacki *et al.*, 1989) around x_{f}^{SM}, and $T = 2n$ in a rotated star-distribution, as illustrated in Fig. 11.14 for the case of a problem with three design variables ($n = 3$).

The learning and testing base points are stored in matrices $X_{\text{L}} \in \Re^{n \times L}$ and $X_T \in \Re^{n \times T}$, respectively, with the following organization

$$X_{\text{L}} = [x_l^{(1)} \ \ldots \ x_l^{(2n+1)}] \tag{11.20}$$

$$X_{\text{T}} = [x_t^{(1)} \ \ldots \ x_t^{(2n)}], \tag{11.21}$$

where $x_l^{(k)}$ and $x_t^{(k)}$ denote the kth learning and testing base point, respectively. An algorithm for generating the learning base points with a star-distribution is in Fig. 11.15, where vector $\tau \in \Re^n$ contains the fractional tolerances for each of the design parameters. Similarly, an algorithm for automatic generation of the rotated star-distribution of testing base points is illustrated in Fig. 11.16.

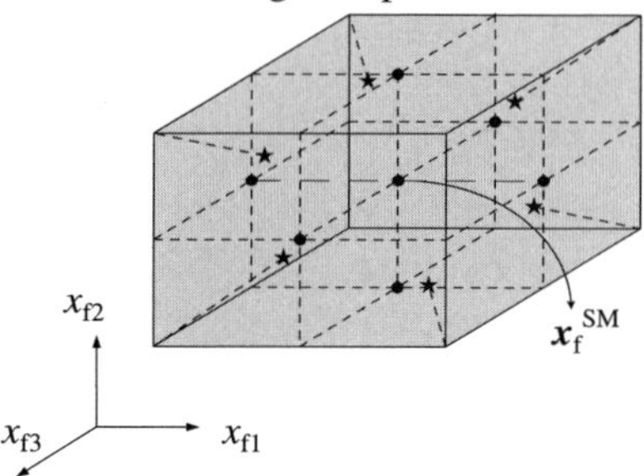

Fig. 11.14. Distribution of learning and testing base points for linear-input neural-output space mapping. Learning base points follow a star-distribution $(2n + 1)$, while testing base points follow a rotated star-distribution $(2n)$.

```
Begin
      x_l^(1) = x_f^SM ; select tolerances τ ∈ ℜ^n
      for i = 1 to n
            x_l^(2i) = x_f^SM ; x_l^(2i+1) = x_f^SM
            x_{l,i}^(2i) = x_{fi}^SM (1 + τ_i)
            x_{l,i}^(2i+1) = x_{fi}^SM (1 − τ_i)
      end
      X_L = [x_l^(1)  x_l^(2) ... x_l^(2n+1)]   (X_L ∈ ℜ^{n×2n+1})
end
```

Fig. 11.15. Pseudo-code for generating learning base points around x_f^{SM} following a star-distribution.

```
Begin
      Read matrix X_L
      S = ones (n,n)
      for i = 2 to n
            S_{i-1,i} = -1
      end
      S = (1/√2) S
      for i = 2 to 2n+1
            x_t^(i-1) = S(x_l^(i) − x_f^SM) + x_f^SM
      end
      X_T = [x_t^(1)  x_t^(2) ... x_t^(2n)]   (X_T ∈ ℜ^{n×2n})
end
```

Fig. 11.16. Pseudo-code for generating testing base points around x_f^{SM} following a rotated star-distribution.

11.5. Yield Estimation by Neural Space Mapping Approaches

Efficient procedures to realize EM-based statistical studies for microwave structures are described in this section. First, the implications of estimating the yield by directly using a fine model are indicated. Second, the neural input SM approach to statistical analysis is reviewed. Finally, the linear-input neural-output SM technique is addressed.

11.5.1. *A Formulation to Statistical Analysis with the Fine Model*

A statistical approach to analysis and design requires taking into account that the parameter values of the device to be manufactured fluctuate around their nominal values according to their statistical distributions and tolerances. It is considered that the design parameters of the kth manufactured device, outcome $x_{\mathrm{f}}^{(k)}$, are spread around the nominal space mapped solution $x_{\mathrm{f}}^{\mathrm{SM}}$ according to their tolerances and their functions of statistical distribution. These parameters can be represented as

$$x_{\mathrm{f}}^{(k)} = x_{\mathrm{f}}^{\mathrm{SM}} + \Delta x_{\mathrm{f}}^{(k)}, \quad k = 1, 2, \ldots, N \qquad (11.22\mathrm{a,\ b})$$

where N is the total number of outcomes used for statistical analysis, and $\Delta x_{\mathrm{f}}^{(k)}$ represents a random variation for the kth outcome. The perturbations in $\Delta x_{\mathrm{f}}^{(k)}$ follow some statistical distribution, typically uniform or normal (Gaussian).

A fine model acceptance index a_{f} is associated with each outcome $x_{\mathrm{f}}^{(k)}$. This acceptance index can be defined by

$$a_{\mathrm{f}}(x_{\mathrm{f}}^{(k)}) = \begin{cases} 1, & \text{if } U(\boldsymbol{R}_{\mathrm{f}}(x_{\mathrm{f}}^{(k)}, \psi)) \leq 0 \\ 0, & \text{if } U(\boldsymbol{R}_{\mathrm{f}}(x_{\mathrm{f}}^{(k)}, \psi)) > 0 \end{cases} \qquad (11.23)$$

with a minimax objective function U, as in (11.1). It is seen that $a_{\mathrm{f}} = 1$ if the corresponding outcome satisfies the specifications, otherwise $a_{\mathrm{f}} = 0$. If N is sufficiently large for statistical significance, we can approximate the fine model yield Y_{f} at the nominal space mapped solution $x_{\mathrm{f}}^{\mathrm{SM}}$ by using

$$Y_{\mathrm{f}}(x_{\mathrm{f}}^{\mathrm{SM}}) \approx \frac{1}{N} \sum_{k=1}^{N} a_{\mathrm{f}}(x_{\mathrm{f}}^{(k)}). \qquad (11.24)$$

Given the high computational cost of the fine model, evaluating (11.24) is extremely expensive, and hence unfeasible for most practical problems.

11.5.2. *Statistical Analysis Using a Neural Input Mapped Coarse Model*

Here it is assumed that a neural input mapped coarse model is already available, obtained by following the procedures described in Section 11.3.2. If properly developed, this model should satisfy

$$\boldsymbol{R}_{\mathrm{f}}(\boldsymbol{x}_{\mathrm{f}}, \psi_{\mathrm{f}}) \approx \boldsymbol{R}_{\mathrm{c}}(\boldsymbol{x}_{\mathrm{c}}, \psi_{\mathrm{c}}), \tag{11.25}$$

with an input neural mapping

$$\begin{bmatrix} \boldsymbol{x}_{\mathrm{c}} \\ \psi_{\mathrm{c}} \end{bmatrix} = \boldsymbol{N}(\boldsymbol{x}_{\mathrm{f}}, \psi_{\mathrm{f}}, \boldsymbol{w}^{*}) \tag{11.26}$$

for all $\boldsymbol{x}_{\mathrm{f}}$ and ψ_{f} in the training region. It is seen that the most general type of input neural mapping $\boldsymbol{N}$ is being considered, since (11.26) maps not only the design parameters but also the independent variable.

It is also assumed that there are r optimizable responses for the problem in hand. For instance, the magnitude of S_{11} and S_{21} at each simulated frequency $(r = 2)$. The Jacobian of the fine model responses with respect to the fine model parameters, $\boldsymbol{J}_{\mathrm{f}} \in \Re^{r \times n}$, is given by

$$\boldsymbol{J}_{\mathrm{f}} = \begin{bmatrix} \dfrac{\partial R_{\mathrm{f}}^{1}}{\partial x_{\mathrm{f}1}} & \cdots & \dfrac{\partial R_{\mathrm{f}}^{1}}{\partial x_{\mathrm{f}n}} \\ \vdots & \ddots & \vdots \\ \dfrac{\partial R_{\mathrm{f}}^{r}}{\partial x_{\mathrm{f}1}} & \cdots & \dfrac{\partial R_{\mathrm{f}}^{r}}{\partial x_{\mathrm{f}n}} \end{bmatrix} \tag{11.27}$$

On the other hand, the Jacobian of the coarse model responses with respect to the coarse model parameters and mapped independent variable value, denoted by $\boldsymbol{J}_{\mathrm{c}} \in \Re^{r \times (n+1)}$, is given by

$$\boldsymbol{J}_{\mathrm{c}} = \begin{bmatrix} \dfrac{\partial R_{\mathrm{c}}^{1}}{\partial x_{\mathrm{c}1}} & \cdots & \dfrac{\partial R_{\mathrm{c}}^{1}}{\partial x_{\mathrm{c}n}} & \dfrac{\partial R_{\mathrm{c}}^{1}}{\partial \psi_{\mathrm{c}}} \\ \vdots & \ddots & \vdots & \vdots \\ \dfrac{\partial R_{\mathrm{c}}^{r}}{\partial x_{\mathrm{c}1}} & \cdots & \dfrac{\partial R_{\mathrm{c}}^{r}}{\partial x_{\mathrm{c}n}} & \dfrac{\partial R_{\mathrm{c}}^{r}}{\partial \psi_{\mathrm{c}}} \end{bmatrix}, \tag{11.28}$$

while the Jacobian of the neural mapping with respect to the fine model parameters, denoted by $\boldsymbol{J}_{\mathrm{N}} \in \Re^{(n+1)\times n}$, is given by

$$\boldsymbol{J}_{\mathrm{N}} = \begin{bmatrix} \dfrac{\partial x_{\mathrm{c}1}}{\partial x_{\mathrm{f}1}} & \cdots & \dfrac{\partial x_{\mathrm{c}1}}{\partial x_{\mathrm{f}n}} \\[2ex] \vdots & \ddots & \vdots \\[2ex] \dfrac{\partial x_{\mathrm{c}n}}{\partial x_{\mathrm{f}1}} & \cdots & \dfrac{\partial x_{\mathrm{c}n}}{\partial x_{\mathrm{f}n}} \\[2ex] \dfrac{\partial \psi_{\mathrm{c}}}{\partial x_{\mathrm{f}1}} & \cdots & \dfrac{\partial \psi_{\mathrm{c}}}{\partial x_{\mathrm{f}n}} \end{bmatrix}. \tag{11.29}$$

It is seen from (11.27)–(11.29) that the sensitivities of the fine model responses can be approximated using (Bandler *et al.*, 2002)

$$\boldsymbol{J}_{\mathrm{f}} \approx \boldsymbol{J}_{\mathrm{c}}\,\boldsymbol{J}_{\mathrm{N}}. \tag{11.30}$$

The accuracy of the approximation of the fine model sensitivities using (11.30) depends on how well the neural input mapped coarse model reproduces the behavior of the fine model in the training region, i.e., it depends on the accuracy of the approximation (11.25)–(11.26).

The exact Jacobian of the neural network can be obtained in closed form for many neural network paradigms, as shown by Bandler *et al.* (2002) for three-layer perceptrons.

Approximation (11.30) is also valid for the case where the input mapping is implemented by a linear transformation of the design parameters only, as expressed in (11.6) (Bakr *et al.*, 1999). However, using a neural network in (11.30) yields a more accurate approximation over a larger region since the mapping is nonlinear and sensitive to the independent variable, which has proved to be very useful when dealing with coarse models based on quasi-static approximations.

11.5.2.1. *Example of neural input space mapping for statistical analysis*

Consider a high-temperature superconducting (HTS) quarter-wave parallel coupled-line microstrip filter (Bandler *et al.*, 1995a). The physical structure of this HTS filter is illustrated in Fig. 11.17. L_1, L_2, and L_3 are the lengths of the parallel coupled-line sections and S_1, S_2, and S_3 are the separations between lines. A width $W = 7\,\mathrm{mil}$ is used for all the lines, including the input and output lines of length $L_0 = 50\,\mathrm{mil}$. A lanthanum aluminate substrate with thickness $H = 20\,\mathrm{mil}$, dielectric constant $\varepsilon_{\mathrm{r}} = 23.425$, loss

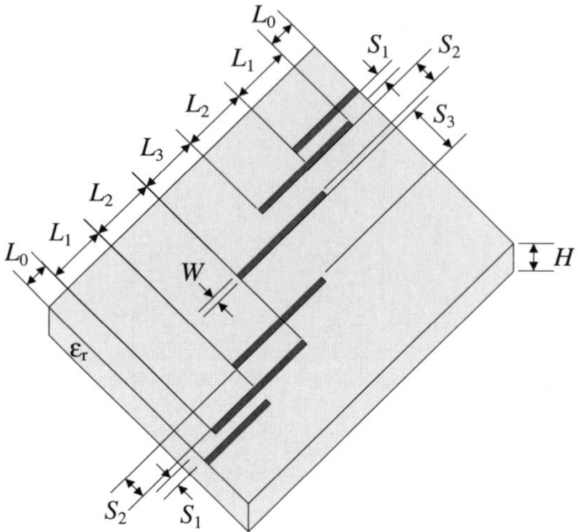

Fig. 11.17.   Microstrip notch filter with mitered bends and open stubs.

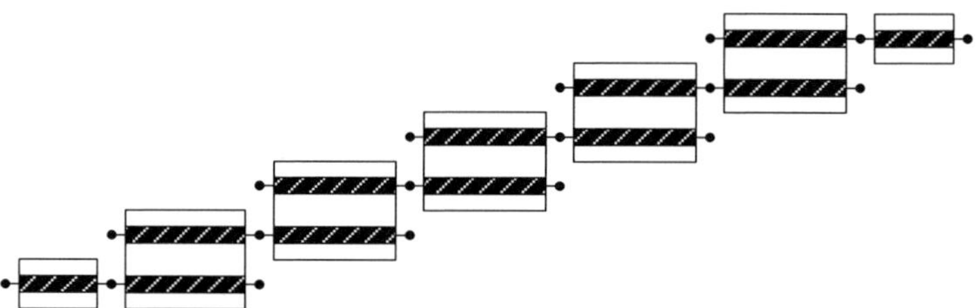

Fig. 11.18.   Schematic representation of the coarse model for the HTS microstrip filter.

tangent $\tan\delta = 3 \times 10^{-5}$ is used. The metalization is considered lossless. The design parameters are $x_{\mathrm{f}} = [L_1 \ L_2 \ L_3 \ S_1 \ S_2 \ S_3]^{\mathrm{T}}$.

The fine model of this filter is implemented in Sonnet using a high-resolution grid with a $1\,\mathrm{mil} \times 1\,\mathrm{mil}$ cell size. The coarse model is implemented in the high-frequency circuit simulator OSA90/hope^{TM} using its built-in components for microstrip lines, two-coupled microstrip lines, and microstrip opens. A schematic representation of the coarse model is illustrated in Fig. 11.18.

The design specifications are: $|S_{21}| \geq 0.95$ for $4.058\,\mathrm{GHz} \leq f \leq 4.008$ GHz (passband), and $|S_{21}| \leq 0.05$ for $f \leq 3.967\,\mathrm{GHz}$ and $f \geq 4.099\,\mathrm{GHz}$ (stopband).

A neural input space mapped coarse model is first developed (see Section 11.3.2). The neural input mapping for this particular problem is

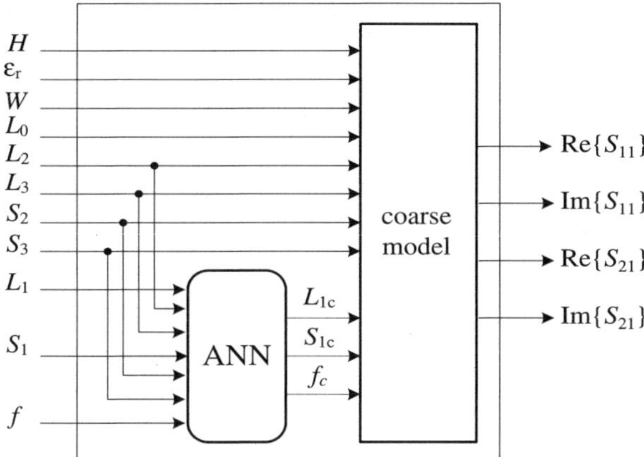

Fig. 11.19.   Neural input space mapped coarse model for the HTS microstrip filter.

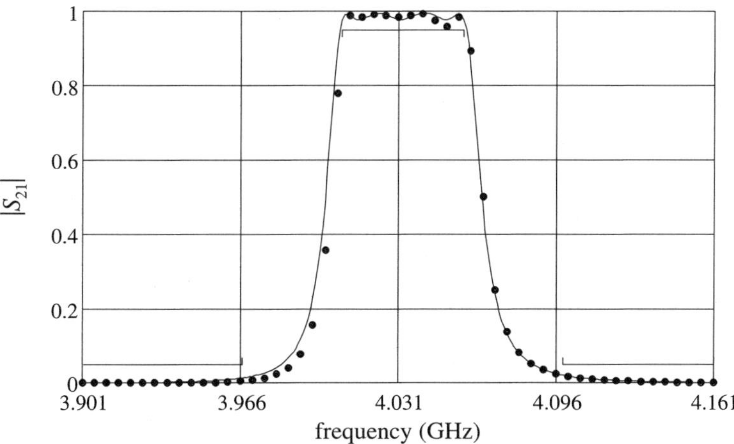

Fig. 11.20.   Fine model response $(-)$ and neural input mapped coarse model response ($\bullet$) at the space mapped solution $\boldsymbol{x}_{\mathrm{f}}^{\mathrm{SM}}$.

illustrated in Fig. 11.19, which consists of a neural network with seven hidden neurons, mapping only L_1, S_1 and the frequency (3LP:7-7-3). L_{1c} and S_{1c} in Fig. 11.9 denote the corresponding two physical dimensions as used by the coarse model, i.e., after being transformed by the mapping.

The space mapped solution found is $\boldsymbol{x}_{\mathrm{f}}^{\mathrm{SM}} = [185.79\ 194.23\ 184.91\ 21.05\ 82.31\ 89.32]^T$ (mils). Figure 11.20 shows excellent agreement between the neural input mapped coarse model response and the fine model response at $\boldsymbol{x}_{\mathrm{f}}^{\mathrm{SM}}$.

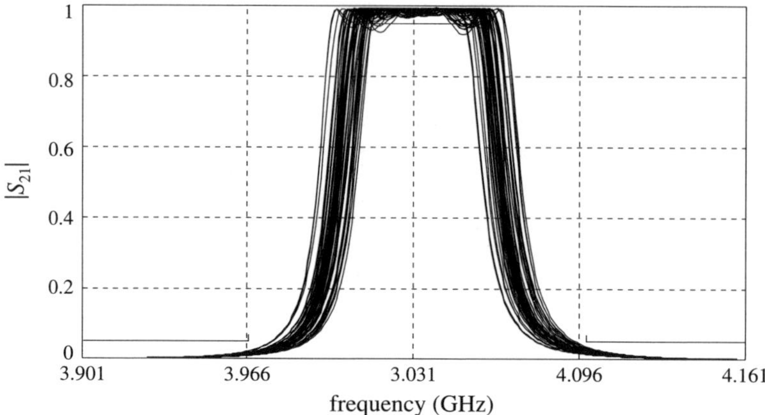

Fig. 11.21. Monte Carlo analysis using the neural input mapped coarse model of the HTS filter around x_f^SM. Estimated yield is 18.4%.

Statistical analysis is realized considering 0.2% of variation for the dielectric constant and for the loss tangent, as well as 75 microns of variation for the physical dimensions, with uniform statistical distributions.

Monte Carlo analysis and yield estimation using the inexpensive neural input mapped coarse model shown in Fig. 11.19 is realized around x_f^SM with 500 outcomes. The responses for 50 of those outcomes are shown in Fig. 11.21. The estimated yield at x_f^SM is 18.4%, which is reasonable considering the high sensitivity of this microstrip circuit.

Notice from Fig. 11.17 that it is implicitly assumed that the structure of the HTS filter has vertical and horizontal physical symmetry. Hence, the corresponding enhanced coarse model in Fig. 11.19 assumes that the random variations in the physical design parameters due to the tolerances are symmetric, which is unlikely to happen in practice. To perform a more realistic statistical analysis of the HTS filter, it should be considered that all the lengths and separations in the structure are asymmetric, as shown in Fig. 11.22.

Developing a new neural input mapping for this asymmetric structure would be much more expensive than the original one, since the dimensionality of the problem becomes very large, and many additional fine model training base points would be needed. To avoid this, the neuromapping obtained from symmetrical training data can be reused to build a first-order approximation of the fine model with asymmetric design parameter values, taking advantage of the inherent asymmetry of the coarse model. The corresponding strategy is illustrated in Fig. 11.23.

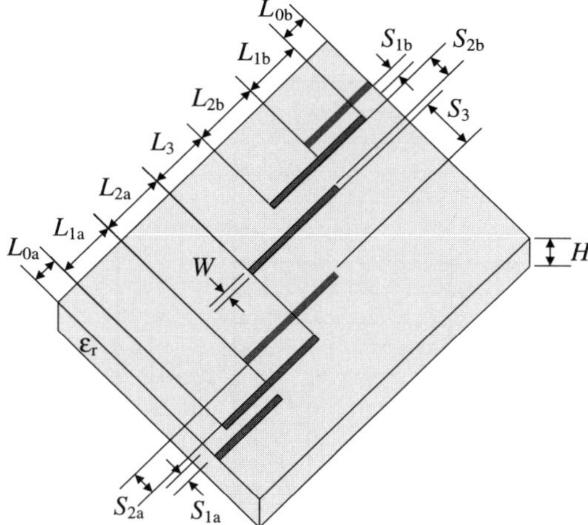

Fig. 11.22. Microstrip notch filter with mitered bends and open stubs, considering asymmetric lengths and separations.

Assigning a separate neuromapping to each couple of L_1 and S_1 dimensions (see Figs. 1.22 and 1.23) makes physical sense, since the electromagnetic interaction between the microstrip lines in either the lower-left or upper-right parts of the structure is much larger than that between the left-right or lower-upper microstrip lines. The remaining asymmetries are naturally taken by the coarse model.

A Monte Carlo yield analysis is performed now for the asymmetric neural input space mapped coarse model of the HTS filter around x_f^{SM} with 500 outcomes, obtaining a yield approximation of only 14% (Bandler *et al.*, 2002). The corresponding responses for 50 of those outcomes are shown in Fig. 11.24.

By performing Monte Carlo yield optimization of this asymmetric enhanced coarse model, the yield is increased to 68.8% estimated with 500 outcomes; 50 of those outcomes are illustrated in Fig. 11.25.

Finally, a comparison between the fine model response and the neural input SM coarse model response at the optimal yield solution is shown in Fig. 11.26, confirming an excellent agreement between both models.

11.5.3. *Statistical Analysis Using Linear-Input Neural-Output Mapped Coarse Models*

In this section, an alternative approach to statistical analysis is described. It incorporates a nonlinear output mapping suitable for highly accurate

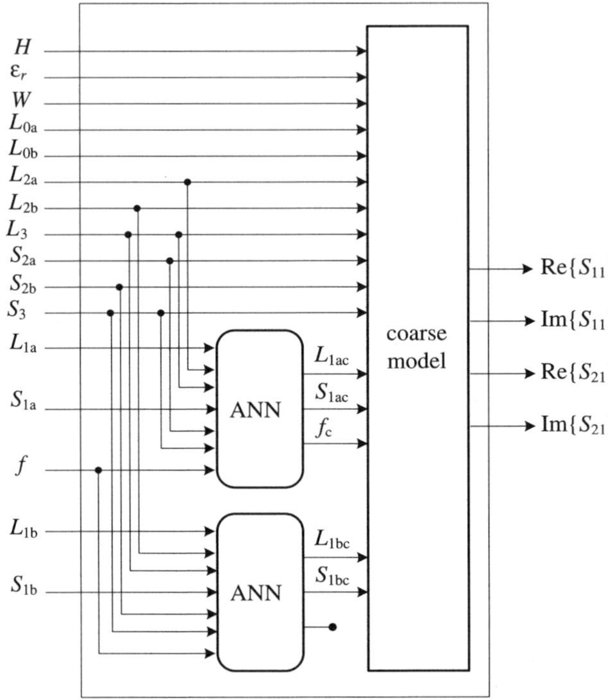

Fig. 11.23. Neural input space mapped coarse model for the HTS filter considering asymmetric random variations due to tolerances for all physical lengths and separations.

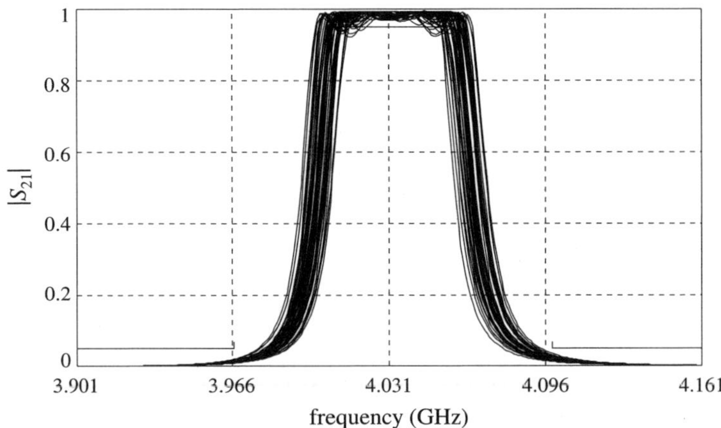

Fig. 11.24. Monte Carlo analysis using the neural input mapped coarse model of the HTS filter around $\boldsymbol{x}_f^{\mathrm{SM}}$, considering asymmetric physical lengths and separations. Estimated yield is 14%.

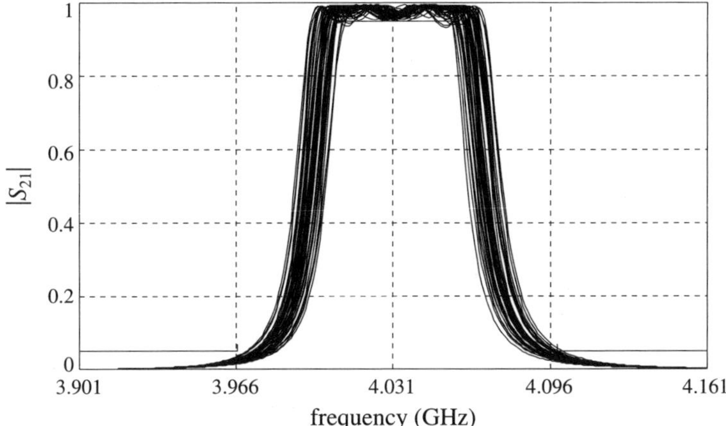

Fig. 11.25.   Monte Carlo analysis using the neural input mapped coarse model of the HTS filter around the optimal yield design, considering asymmetric physical lengths and separations. Estimated yield is increased to 68.8%.

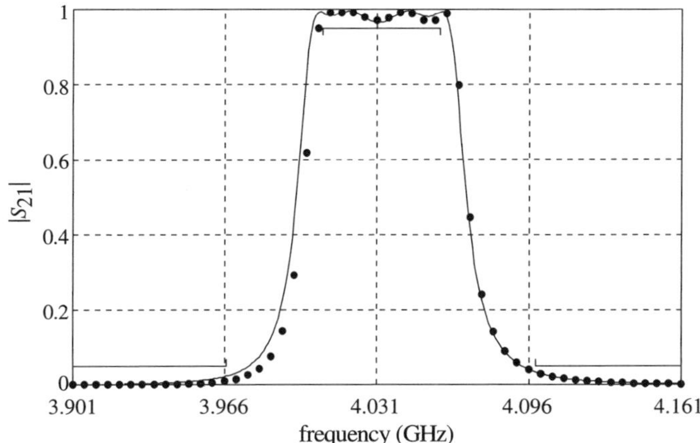

Fig. 11.26.   Fine model response ($-$) and neural input mapped coarse model response ($\bullet$) at the optimal yield solution.

yield estimations, especially useful when the design specifications are hardly satisfied at the nominal design. To illustrate how a simple linear input space mapping technique can lead to insufficiently accurate yield predictions, the statistical analysis using the linear input mapped coarse model is also addressed. It is demonstrated how this technique can be significantly enhanced by adding a neural output mapping (Rayas–Sánchez *et al.*, 2006b).

11.5.3.1. *Yield estimation using linear input space mapping*

Yield prediction using linear input mapped coarse models is formulated similarly to the formulation in Section 11.5.1, as follows.

An acceptance index a_{li} is associated to each random outcome $\boldsymbol{x}_{\mathrm{f}}^{(k)}$ spread around the space mapped solution $\boldsymbol{x}_{\mathrm{f}}^{\mathrm{SM}}$ (11.22),

$$
a_{\mathrm{li}}(\boldsymbol{x}_{\mathrm{f}}^{(k)}) = \begin{cases} 1, & \text{if } U(\boldsymbol{R}_{\mathrm{c}}(\boldsymbol{B}\boldsymbol{x}_{\mathrm{f}}^{(k)} + \boldsymbol{c}, \psi)) \leq 0 \\ 0, & \text{if } U(\boldsymbol{R}_{\mathrm{c}}(\boldsymbol{B}\boldsymbol{x}_{\mathrm{f}}^{(k)} + \boldsymbol{c}, \psi)) > 0 \end{cases} \tag{11.31}
$$

with a minimax objective function U, as in (11.1), and the mapping function defined as in (11.6). The corresponding linear input space mapped yield Y_{li} at the nominal design $\boldsymbol{x}_{\mathrm{f}}^{\mathrm{SM}}$ can be approximated by

$$
Y_{\mathrm{li}}(\boldsymbol{x}_{\mathrm{f}}^{\mathrm{SM}}) \approx \frac{1}{N} \sum_{k=1}^{N} a_{\mathrm{li}}(\boldsymbol{x}_{\mathrm{f}}^{(k)}). \tag{11.32}
$$

It is seen that evaluating (11.32) is computationally inexpensive.

11.5.3.2. *Yield estimation using linear-input neural-output space mapping*

Yield prediction using a linear-input neural-output mapped coarse model is formulated as follows. The corresponding acceptance index a_{lino} associated to each random outcome $\boldsymbol{x}_{\mathrm{f}}^{(k)}$ is defined as

$$
a_{\mathrm{lino}}(\boldsymbol{x}_{\mathrm{f}}^{(k)}) = \begin{cases} 1, & \text{if } U(\boldsymbol{Q}(\boldsymbol{R}_{\boldsymbol{c}}(\boldsymbol{B}\boldsymbol{x}_{\mathrm{f}}^{(k)} + \boldsymbol{c}, \psi), \boldsymbol{x}_{\mathrm{f}}^{(k)}, \psi)) \leq 0 \\ 0, & \text{if } U(\boldsymbol{Q}(\boldsymbol{R}_{\boldsymbol{c}}(\boldsymbol{B}\boldsymbol{x}_{\mathrm{f}}^{(k)} + \boldsymbol{c}, \psi), \boldsymbol{x}_{\mathrm{f}}^{(k)}, \psi)) > 0 \end{cases}, \tag{11.33}
$$

with the same objective function U, and the input–output mapping function defined as in (11.12). For simplicity, the optimal weighting factors $\boldsymbol{w}^{*}$ of the output neuromapping are omitted in (11.33), assuming that the neural network is already trained as described in Section 11.4.1. The corresponding linear input space mapped yield Y_{li} at the nominal design $\boldsymbol{x}_{\mathrm{f}}^{\mathrm{SM}}$ can be approximated by

$$
Y_{\mathrm{lino}}(\boldsymbol{x}_{\mathrm{f}}^{\mathrm{SM}}) \approx \frac{1}{N} \sum_{k=1}^{N} a_{\mathrm{lino}}(\boldsymbol{x}_{\mathrm{f}}^{(k)}). \tag{11.34}
$$

It is seen that evaluating (11.34) is also computationally inexpensive, however, it produces much more accurate results than (11.32).

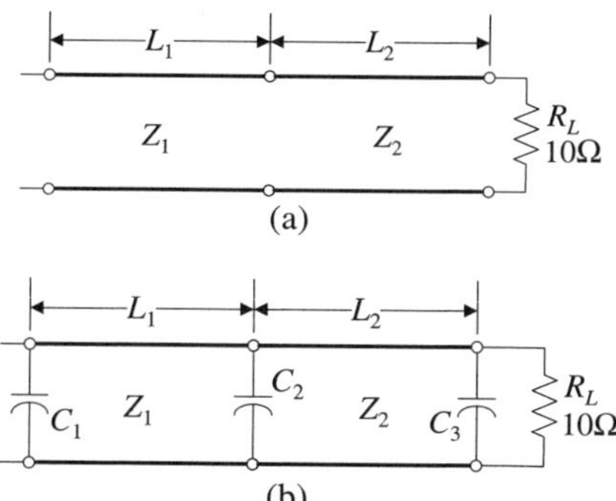

Fig. 11.27.   10:1 two-section capacitively loaded impedance transformer: (a) coarse model, (b) "fine" model.

11.5.3.3. *Example: capacitively loaded impedance transformer*

Consider the classical test problem of a capacitively-loaded 10:1 two-section impedance transformer (Bandler *et al.*, 2003). The coarse and "fine" models are shown in Fig. 11.27. The coarse model consists of ideal transmission lines, while the "fine" model consists of capacitively loaded ideal transmission lines, with $C_1 = C_2 = C_3 = 10\,\mathrm{pF}$. In this synthetic problem, the fine model is in fact computationally inexpensive. This allows testing the accuracy of approximations (11.32) and (11.34), since the actual calculation of the fine model yield for this problem is feasible.

The design specifications are: $|S_{11}| \leq 0.5$ for $0.5\,\mathrm{GHz} \leq f \leq 1.5\,\mathrm{GHz}$. The electrical lengths of the two transmission lines at 1 GHz are selected as design parameters, $\boldsymbol{x}_\mathrm{f} = [L_1\ L_2]^\mathrm{T}$ (degrees). Their characteristic impedances are kept fixed at the following values: $Z_1 = 2.23615\,\Omega$, $Z_2 = 4.47230\,\Omega$. Both models were implemented in Matlab. The optimal coarse model solution is $\boldsymbol{x}_\mathrm{c}^* = [90\ 90]^\mathrm{T}$ (degrees). The coarse and fine model responses at $\boldsymbol{x}_\mathrm{c}^*$ are shown in Fig. 11.28. Results after applying the constrained Broyden-based input space mapping algorithm described in Section 1.3.1.1 are illustrated in Fig. 11.29. The space mapped solution $\boldsymbol{x}_\mathrm{f}^\mathrm{SM} = [74.1408\ 79.6417]^\mathrm{T}$ (degrees) is found after only three fine model evaluations (three frequency sweeps).

Using tolerances of $\pm 10\%$ for the design parameters around $\boldsymbol{x}_\mathrm{f}^\mathrm{SM}$, an output neuromapping is developed for this problem. Since $n = 2$, only five learning and four testing base points are used. Figure 11.30 shows the performance of the output neural mapping training algorithm described in Section 1.4.1. Learning and testing error metrics ε_L and ε_T, used in Fig. 11.30, are defined as in (11.19). It is seen that the best performance is

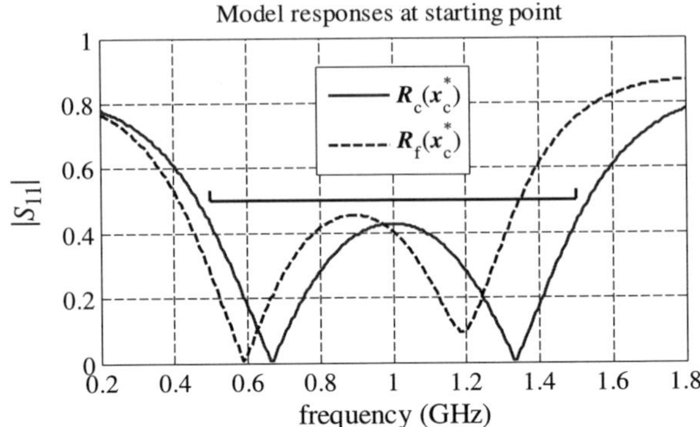

Fig. 11.28.   Coarse and fine model responses at the optimal coarse model design for the impedance transformer.

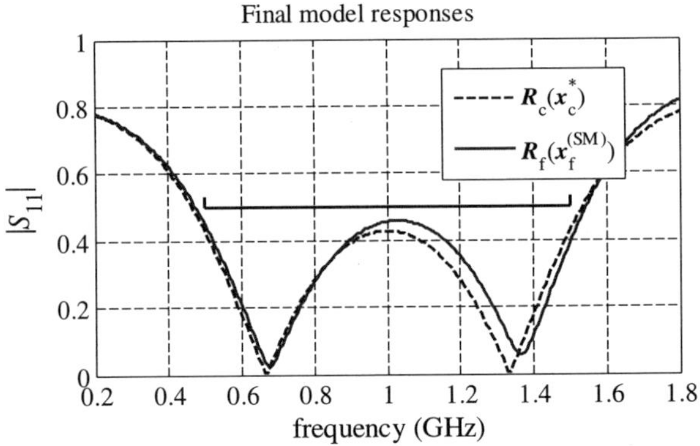

Fig. 11.29.   Results after applying the constrained Broyden-based input space mapping algorithm to the impedance transformer.

achieved with $h = 7$, since a larger number of hidden neurons in the output neural network deteriorates its generalization performance.

The linear-input neural-output mapped coarse model response at x_f^{SM} is compared in Fig. 11.31 with that of the fine model and the linear input mapped coarse model. It is seen that output mapping Q effectively eliminates the residual between $R_f(x_f^{SM})$ and $R_c(P(x_f^{SM}))$, as expected.

A comparison in the yield estimation around x_f^{SM} between the fine model, $R_f(x_f, \psi)$, the linear input mapped coarse model $R_c(P(x_f), \psi)$, and the linear-input neural-output mapped coarse model, $Q(R_c(P(x_f), \psi), x_f, \psi)$, is shown in Fig. 11.32 (Rayas–Sánchez *et al.*, 2006b), using maximum deviations of $\pm 5\%$ with respect to x_f^{SM} for 1,000 outcomes with

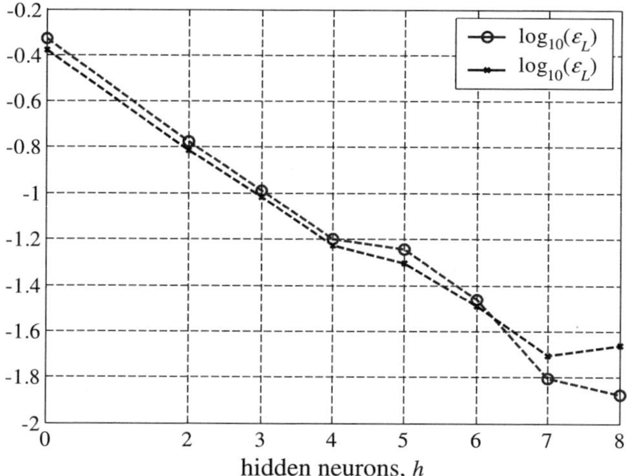

Fig. 11.30.   Evolution of learning and testing errors during training of the output neuromapping for the impedance transformer. Algorithm in Fig. 11.13 is used.

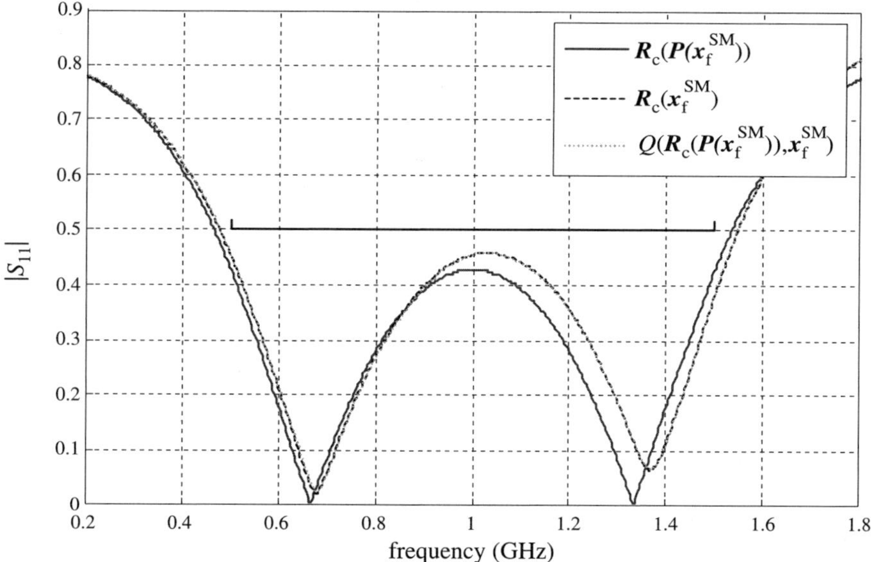

Fig. 11.31.   Results after training the output mapping $\boldsymbol{Q}$ for the impedance transformer.

 José E. Rayas-Sánchez

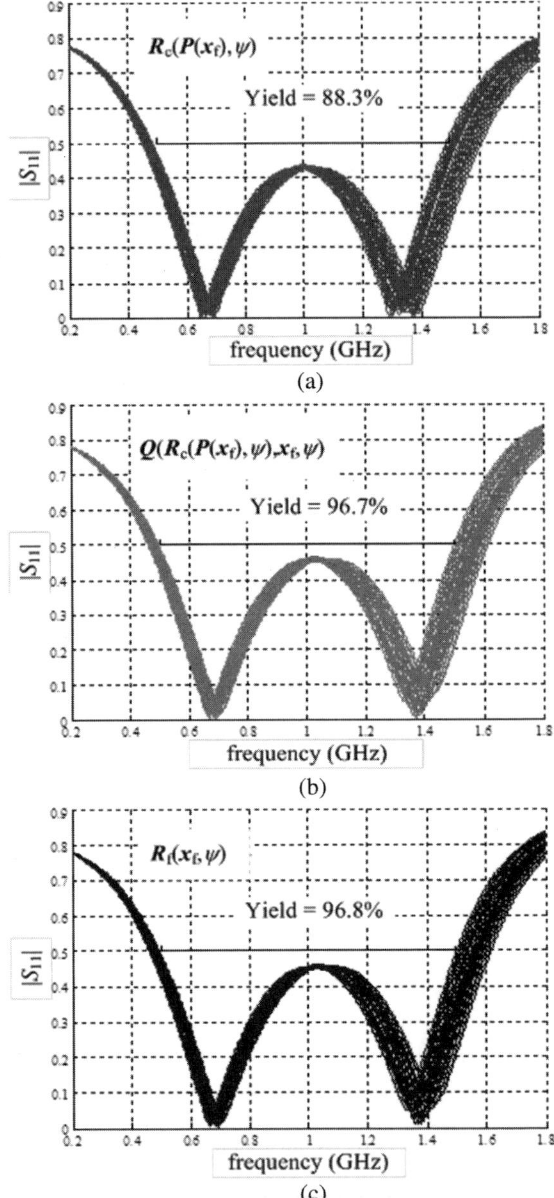

Fig. 11.32. Statistical analysis of the impedance transformer around x_f^SM (1,000 outcomes with uniform statistical distributions) using: (a) the linear input mapped coarse model, (b) the linear-input neural-output mapped coarse model, and (c) the fine model.

uniform statistical distributions. It is seen that the Monte Carlo responses around $x_{\mathrm{f}}^{\mathrm{SM}}$ of the linear input mapped coarse model approximately follow the corresponding fine model responses (compare Fig. 11.14(a) with Fig. 11.14(c)). This is an indication that the linear input mapped coarse model requires a small adjustment to match the fine model at $x_{\mathrm{f}}^{\mathrm{SM}}$ and around that point. This adjustment is provided by the output neural mapping Q. It is also seen in Fig. 11.32 that the yield predicted by the linear input mapped coarse model has a significant error with respect to the actual fine model yield. This error can be very large if the design specifications are hardly satisfied by the fine model. Finally, it is confirmed in Fig. 11.32 that the linear-input neural-output space mapping model predicts with a very high accuracy the fine model yield around $x_{\mathrm{f}}^{\mathrm{SM}}$, as expected, since the Monte Carlo analysis was realized within the training region of the output neural mapping Q.

11.5.3.4. *Example: microstrip notch filter*

Statistical analysis and yield estimation of the microstrip notch filter with mitered bends and open stubs illustrated in Fig. 11.6 is now realized. A detailed description of this filter, including its fine and coarse models, is provided in Section 11.3.2.3.

Taking the space mapped solution $x_{\mathrm{f}}^{\mathrm{SM}}$ found in Section 11.3.2.3, a tolerance region of $\pm 0.5\,\mathrm{mil}$ is used for all the physical design parameters around $x_{\mathrm{f}}^{\mathrm{SM}}$. An output neuromapping is now developed. Since $n = 3$ for this problem, only seven learning and six testing base points are used. The performance of the corresponding neural network, while being trained using the algorithm described in Section 11.4.1, is shown in Fig. 11.33. Again, learning and testing errors ε_{L} and ε_{T} are defined as in (11.19), being plotted in Fig. 11.33 in logarithmic scale. It is seen that the best generalization performance of the output neural network is obtained for this problem when four hidden neurons are employed ($h = 4$).

The linear-input neural-output mapped coarse model response at $x_{\mathrm{f}}^{\mathrm{SM}}$ is compared in Fig. 11.34 with that of the fine model and the input mapped coarse model. It is again confirmed that output neural mapping Q effectively eliminates the residual between $R_{\mathrm{f}}(x_{\mathrm{f}}^{\mathrm{SM}})$ and $R_{\mathrm{c}}(P(x_{\mathrm{f}}^{\mathrm{SM}}))$.

Finally, in Fig. 11.35 a comparison in the yield estimation around $x_{\mathrm{f}}^{\mathrm{SM}}$ between the linear input mapped coarse model $R_{\mathrm{c}}(P(x_{\mathrm{f}}), \psi)$ and the linear-input neural-output mapped coarse model, $Q(R_{\mathrm{c}}(P(x_{\mathrm{f}}), \psi), x_{\mathrm{f}}, \psi)$ is shown, using maximum deviations of $\pm 0.2\,\mathrm{mil}$ with respect to $x_{\mathrm{f}}^{\mathrm{SM}}$ for 50 outcomes with uniform statistical distributions. This example

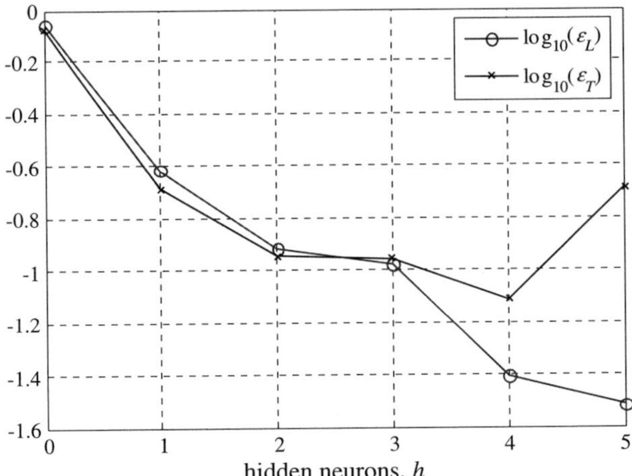

Fig. 11.33. Evolution of learning and testing errors during training of the output neuromapping for the microstrip notch filter. Algorithm in Fig. 11.13 is used.

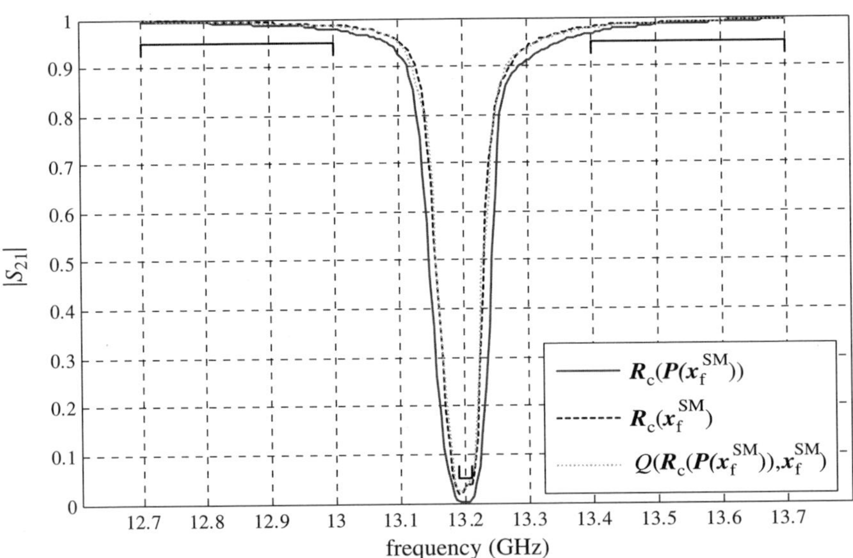

Fig. 11.34. Results after training the output mapping Q for the microstrip notch filter.

clearly illustrates how the linear input mapped coarse model can produce quite inaccurate yield predictions due to the existing residual between $R_c(P(x_f), \psi)$ and $R_f(x_f^{SM})$, and due to the fact that the design specifications are very tight for fine model response (see Fig. 11.34). The

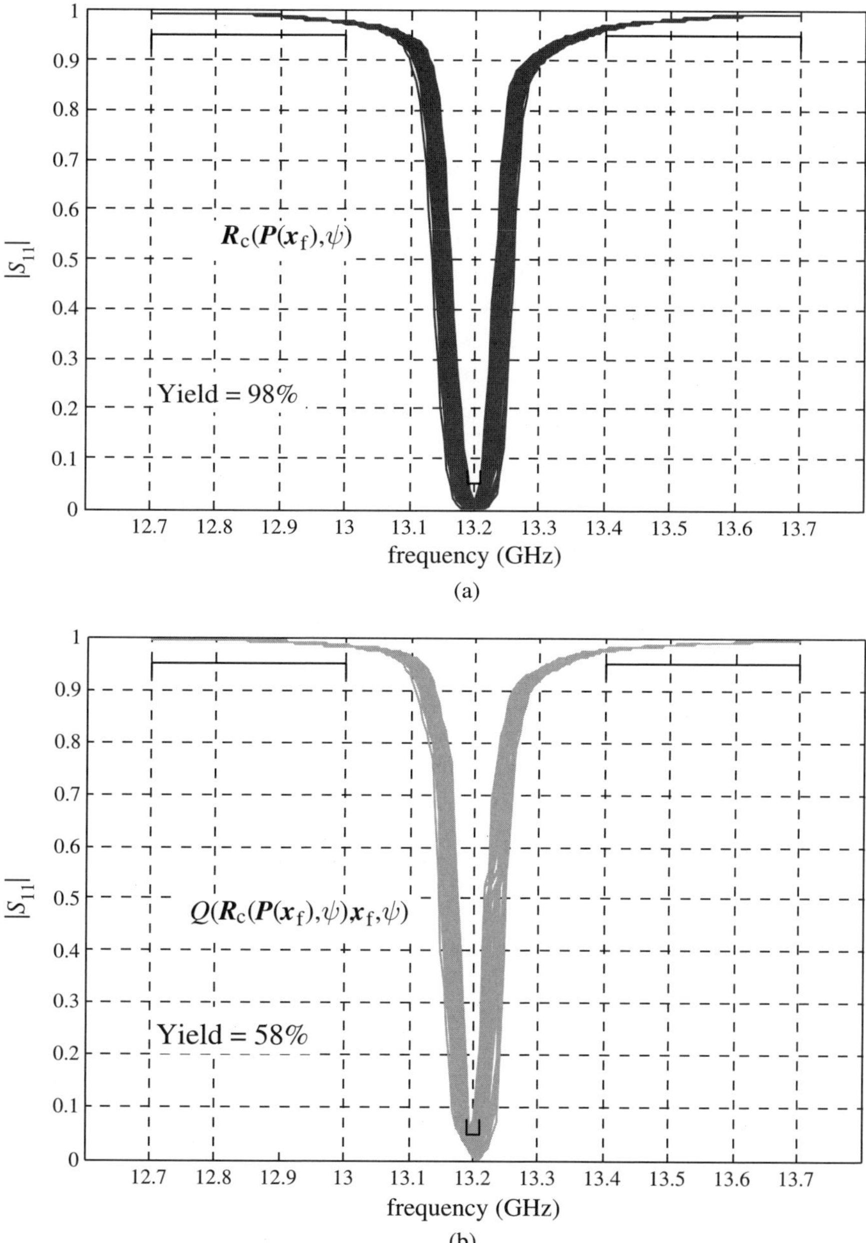

Fig. 11.35.   Statistical analysis of the microstrip notch filter around x_f^{SM} (50 outcomes with uniform statistical distributions) using: (a) the linear input mapped coarse model, and (b) the linear-input neural-output mapped coarse model.

problem is alleviated by the linear-input neural-output mapped coarse model.

11.6. Conclusion

Computationally inexpensive methods based on neural space mapping techniques for EM-based design optimization, statistical analysis, and yield predictions have been described and compared in this chapter. The process for developing an input space mapping approach to enhance available coarse models was described first. This can be done linearly, using a Broyden-based approach, and nonlinearly using neural networks. It was shown how these input space mapping techniques can be employed to perform efficient yield estimations. It was also illustrated how a linear input mapped coarse model can lead to insufficiently accurate yield estimations in some problems. Finally, the combination of linear input mappings with neural output mappings to develop more accurate and equally efficient surrogate models was described. This last strategy allows the transformation of conventional equivalent circuit models into accurate vehicles for EM-based statistical analysis and design. In all cases, the corresponding neural networks are trained using reduced sets of full-wave EM data. The design optimization and yield prediction of several problems, using commercially available CAD tools, illustrated the techniques.

References

APLAC version 8.10 (2005). APLAC Solutions Corporation, now AWR Corporation, 1960 E. Grand Avenue, Suite 430, El Segundo: CA 90245.

Bandler, J.W., Biernacki, R.M., Chen, S.H., Grobelny, P.A. and Ye, S. (1993). Yield-driven electromagnetic optimization via multilevel multidimensional models, *IEEE T. Microw. Theory*, **41**, 2269–2278.

Bandler, J.W., Biernacki, R.M., Chen, S.H. and Grobelny, P.A. (1994). A CAD environment for performance and yield driven circuit design employing electromagnetic field simulators, *IEEE Int. Symp. Circ. S.*, **1**, 145–148.

Bandler, J.W., Biernacki, R.M., Chen, S.H., Getsinger, W.J., Grobelny, P.A., Moskowitz, C. and Talisa, S.H. (1995a). Electromagnetic design of high-temperature superconducting microwave filters, *Int. J. Microwave Mill.*, **5**, 331–343.

Bandler, J.W., Biernacki, R.M., Chen, S.H., Hemmers, R.H. and Madsen, K. (1995b). Electromagnetic optimization exploiting aggressive space mapping, *IEEE T. Microw. Theory*, **41**, 2874–2882.

Bandler, J.W., Ismail, M.A., Rayas-Sánchez, J.E. and Zhang, Q.J. (1999). Neuromodeling of microwave circuits exploiting space mapping technology, *IEEE T. Microw. Theory*, **47**, 2417–2427.

Bandler, J.W., Rayas-Sánchez, J.E. and Zhang, Q.J. (2002). Yield-driven electromagnetic optimization via space mapping-based neuromodels, *Int. J. RF Microw, CE*, **12**, 79–89.

Bandler, J.W., Ismail, M.A., Rayas-Sánchez, J.E. and Zhang, Q.J. (2003). Neural inverse space mapping for EM-based microwave design, *Int. J. RF Microw. CE*, **13**, 136–147.

Bandler, J.W., Cheng, Q., Dakroury, S.A., Mohamed, A.S., Bakr, M.H., Madsen, K. and Søndergaard, J. (2004a). Space mapping: the state of the art. *IEEE T. Microw. Theory*, **52**, 337–361.

Bandler, J.W., Hailu, D.M., Madsen, K. and Pedersen, F.A. (2004b). Space-mapping interpolating surrogate algorithm for highly optimized EM-based design of microwave devices, *IEEE T. Microw. Theory*, **52**, 2593–2600.

Bakr, M.H., Bandler, J.W., Georgieva, N. and Madsen, K. (1999). A hybrid aggressive space-mapping algorithm for EM optimization, *IEEE T. Microw. Theory*, **47**, 2440–2449.

Bakr, M.H., Bandler, J.W., Ismail, M.A., Rayas-Sánchez, J.E. and Zhang, Q.J. (2000). Neural space mapping optimization for EM-based design. *IEEE T. Microw. Theory*, **48**, 2307–2315.

Biernacki, R.M., Bandler, J.W., Song, J. and Zhang, Q.J. (1989). Efficient quadratic approximation for statistical design, *IEEE T. Circuits Syst.*, **36**, 1449–1454.

Burrascano, P., Dionigi, M., Fancelli, C. and Mongiardo, M. (1998). A neural network model for CAD and optimization of microwave filters, *IEEE MTT-S*, Baltimore, MD, pp. 13–16.

Burrascano, P. and Mongiardo, M. (1999). A review of artificial neural networks applications in microwave CAD, *Int. J. RF Microw. CE*, **9**, 158–174.

Creech, G.L. and Zurada, J.M. (1999). Neural network modeling of GaAs IC material and MESFET device characteristics, *Int. J. RF Microw. CE*, **9**, 241–253.

em^{TM} Suite version 12.52 (2009), Sonnet Software, Inc., 100 Elwood Davis Road, North Syracuse: NY 13212.

Gutiérrez-Ayala, V. and Rayas-Sánchez, J.E. (2010). Neural input space mapping optimization based on nonlinear two-layer perceptrons with optimized nonlinearity, *Int. J. RF Microw. CE*, **20**, 512–526.

Haykin, S. (1999). *Neural Networks: A Comprehensive Foundation*, (2nd ed.), Prentice Hall, New Jersey: MA, pp. 213–218.

Koziel, S., Bandler, J.W., Mohamed, A.S. and Madsen, K. (2005). Enhanced surrogate models for statistical design exploiting space mapping technology, *IEEE MTT-S*, Long Beach: CA, pp. 1–4.

Koziel, S., Cheng, Q.S. and Bandler, J.W. (2008). Space mapping, *IEEE Microw. Mag.*, **9**, 105–122.

Matlab Optimization Toolbox, Version 7.4.1, The Mathworks, Inc., 3 Apple Hill Drive, Natick: MA 01760-2098.

OSA90/hope^{TM} and Empipe^{TM}, Version 4.0, Agilent Technologies, 1400 Fountaingrove Parkway, Santa Rosa: CA 95403-1799.

Rayas-Sánchez, J.E. (2004). EM-based optimization of microwave circuits using artificial neural networks: the state of the art, *IEEE T. Microw. Theory*, **52**, 420–435.

Rayas-Sánchez, J.E., Lara-Rojo, F. and Martínez-Guerrero, E. (2005). A linear inverse space mapping (LISM) algorithm to design linear and nonlinear RF and microwave circuits, *IEEE T. Microw. Theory*, **53**, 960–968.

Rayas-Sánchez, J.E. and Gutiérrez-Ayala, V. (2006a). EM-based statistical analysis and yield estimation using linear-input and neural-output space mapping, *IEEE MTT-S*, San Francisco: CA, pp. 1597–1600.

Rayas-Sánchez, J.E. and Gutiérrez-Ayala, V. (2006b). EM-based Monte Carlo analysis and yield prediction of microwave circuits using linear-input neural-output space mapping, *IEEE T. Microw. Theory*, **54**, 4528–4537.

Swidzinski, J.F. and Chang, K. (2000). Nonlinear statistical modeling and yield estimation technique for use in Monte Carlo simulations, *IEEE T. Microw. Theory*, **48**, 2316–2324.

Zaabab, A.H., Zhang, Q.J. and Nakhla, M.S. (1995). A neural network modeling approach to circuit optimization and statistical design, *IEEE T. Microw. Theory*, **43**, 1349–1358.

Zhang, Q.J. and Gupta, K.C. (2000). *Neural Networks for RF and Microwave Design*, (1st ed.), Artech House, Norwood: MA, p. 211.

Chapter 12

**Neural Network Inverse Modeling for Microwave
Filter Design**

Humayun Kabir, Ying Wang, Ming Yu, and Qi-Jun Zhang

In this chapter we present the systematic neural network (NN) inverse
modeling technique. Formulation of neural network inverse model and
model development is presented. We then describe advanced techniques to
improve the accuracy of inverse models. The technique will then be applied
to model and design waveguide filters (Kabir *et al.*, 2008).

12.1. Introduction

In recent years, neural network techniques have been recognized as a
powerful tool for microwave design and modeling problems. They have
been applied to various microwave design applications (Zaabab *et al.*, 1995;
Watson *et al.*, 1996, 1997; Markovic *et al.*, 2000; Devabhaktuni *et al.*,
2001; Bandler *et al.*, 2003; Davis *et al.*, 2003; Zhang *et al.*, 2003; Rizzoli
et al., 2004; Isaksson *et al.*, 2005; Garcia *et al.*, 2006; Kabir *et al.*, 2008).
The most common way of developing a device or circuit model is to
replicate the input and output relationship of the original electromagnetic
(EM) problem. A neural network model developed this way is known as
the forward model, where the model inputs are physical or geometrical
parameters and outputs are electrical parameters. For the design purpose,
the information is often processed in the reverse direction in order to find the
geometrical/physical parameters for given values of electrical parameters,
which is called the inverse problem. There are two methods to solve the
inverse problem, i.e., the optimization method and the inverse modeling
method. In the optimization method, the EM simulator, or the forward

311

model, is evaluated repetitively in order to find the optimal solutions of the geometrical parameters that can lead to a good match between modeled and specified electrical parameters. An example of such an approach is found in Vai *et al.* (1998). This method of inverse modeling is also known as the synthesis method.

The inverse problem can also be solved by using an inverse model whose inputs are electrical parameters and outputs are physical/geometrical dimensions. The formula for the inverse problem, i.e., to compute the geometrical parameters from given electrical parameters, is difficult to find analytically. Therefore, the neural network becomes a logical choice since it can be trained to learn from the data of the inverse problem. In order to formulate an inverse model, we define the input neurons of a neural network to be the electrical parameters of the modeling problem and the output neurons as the geometrical parameters. Training data for the neural network inverse model can be obtained simply by swapping the input and output data used to train the forward model. This method is called direct inverse modeling and an example of this approach is found in Selleri *et al.* (2002). Once training is completed, the inverse model provides inverse solutions immediately unlike the optimization method, where repetitive forward model evaluations are required. Therefore, a design method using an inverse model is faster than one using the optimization method.

12.2. Neural Network Inverse Model Formulation

In this section we describe the formulation of the neural network inverse model (Kabir *et al.*, 2008). Let n and m represent the number of inputs and outputs of the forward model. Let $\boldsymbol{x}$ be an n-vector containing the inputs and $\boldsymbol{y}$ be an m-vector containing the outputs of the forward model. Then the forward modeling problem is expressed as

$$\boldsymbol{y} = \boldsymbol{f}(\boldsymbol{x}), \tag{12.1}$$

where $\boldsymbol{x} = [x_1 \ x_2 \ x_3 \ldots x_n]^T$, $\boldsymbol{y} = [y_1 \ y_2 \ y_3 \ldots y_m]^T$, and f defines the input–output relationship. An example of a neural network diagram of a forward model and its corresponding inverse model is shown in Fig. 12.1. Note that two outputs and two inputs of the forward model are swapped to the input and output of the inverse model respectively. In general some or all of them can be swapped from input to output or vice versa.

Let us define a subset of $\boldsymbol{x}$ and a subset of $\boldsymbol{y}$. These subsets of input and output are swapped to the output and input respectively in order to

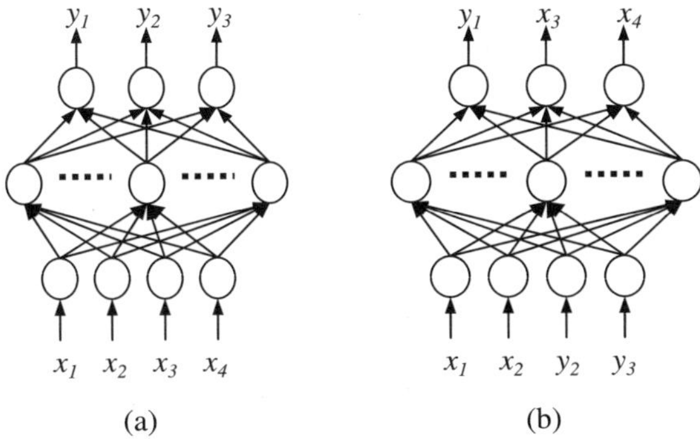

Fig. 12.1. Example illustrating neural network forward and inverse models: (a) forward model, (b) inverse model. The inputs x_3 and x_4 (output y_2 and y_3) of the forward model are swapped to the outputs (inputs) of the inverse model respectively (Kabir *et al.*, 2008).

form the inverse model. Let I_x be defined as an index set containing the indices of inputs of the forward model that are moved to the output of the inverse model,

$$I_x = \{i|\ \text{if } x_i \text{ becomes output of inverse model}\}. \tag{12.2}$$

Let I_y be the index set containing the indices of outputs of the forward model that are moved to the input of the inverse model,

$$I_y = \{i|\ \text{if } y_i \text{ becomes input of inverse model}\}. \tag{12.3}$$

Let $\bar{x}$ and $\bar{y}$ be vectors of inputs and outputs of the inverse model. The inverse model can be defined as

$$\bar{y} = \bar{f}(\bar{x}), \tag{12.4}$$

where $\bar{y}$ includes y_i if $i \notin I_y$ and x_i if $i \in I_x$; $\bar{x}$ includes x_i if $i \notin I_x$ and y_i if $i \in I_y$; and $\bar{f}$ defines the input–output relationship of the inverse model. For example the inputs x_3 and x_4 of Fig. 12.1(a) may represent the iris length and width of a filter, and outputs y_2 and y_3 may represent electrical parameters such as the coupling parameter and insertion phase. To formulate the inverse filter model we swap the iris length and width with the coupling parameter and insertion phase. For the example in Fig. 12.1 the inverse model is formulated as

$$I_x = \{3,4\} \tag{12.5}$$

$$I_y = \{2,3\} \tag{12.6}$$

$$\bar{x} = [\bar{x}_1 \ \bar{x}_2 \ \bar{x}_3 \ \bar{x}_4]^T = [x_1 \ x_2 \ y_2 \ y_3]^T \tag{12.7}$$

$$\bar{y} = [\bar{y}_1 \ \bar{y}_2 \ \bar{y}_3]^T = [y_1 \ x_3 \ x_4]^T. \tag{12.8}$$

12.2.1. *Example of Neural Network Inverse Model Formulation*

In this section we show an example of how to formulate an inverse model of an internal coupling iris of a dual-mode circular waveguide filter whose forward model inputs are circular cavity diameter D, center frequency f_o, vertical and horizontal coupling slot lengths L_v and L_h and outputs are coupling values M_{23} and M_{14}, and the loading effect of the coupling iris on the two orthogonal modes, respectively P_v and P_h. It is given that the input and output vectors of the forward model are defined as $x = [D \ L_v \ L_h \ f_o]^T$, $y = [M_{23} \ M_{14} \ P_v \ P_h]^T$, and $I_x = \{2, 3\}$, $I_y = \{1, 2\}$.

We formulate the inverse model by swapping the parameters according to the index of parameters defined by I_x and I_y. From the input vector x parameter number 2 and 3 are moved to the output of the inverse model, and from output vector y parameter number 1 and 2 are moved to the input of the inverse model. Thus the input vector of the inverse model is formulated as

$$\bar{x} = [D \ f_o \ M_{23} \ M_{14}]^T \tag{12.9}$$

and the output vector of the inverse model is formulated as

$$\bar{y} = [L_v \ L_h \ P_v \ P_h]^T. \tag{12.10}$$

The inverse model is formulated as

$$\bar{y} = \bar{f}(\bar{x})$$

$$[L_v \ L_h \ P_v \ P_h]^T = \bar{f}(D \ f_o \ M_{23} \ M_{14}). \tag{12.11}$$

12.3. Neural Network Inverse Model Development

In order to develop an inverse model, we train a neural network with inverse data. Usually data are generated by EM solvers in the forward way, e.g., physical parameters/geometrical dimensions are given and electrical parameters such as coupling parameters are computed. We then reorganize the input and output data by swapping the input and output parameters to obtain the training data of the inverse model. The neural network trained this way becomes the inverse model. Model training can be done with a training program such as NeuroModelerPlus (2005).

Developing an inverse model simply by swapping data and training a neural network with the entire data is simple, and is suitable when the problem is relatively easy, for example, when the original input–output relationship is smooth and monotonic, and/or if the numbers of inputs/outputs are small. On the other hand if the problem is complicated, then segmentation of the training data is utilized to improve the model accuracy. Segmentation of microwave structures has been reported in existing literature such as Cid and Zapata (2001), where a large device is segmented into smaller units. In this chapter, we apply the segmentation concepts over the range of model inputs to split the training data into multiple sections each covering a smaller range of input parameter space. Neural network models are trained for each section of data. A small amount of overlapping data is reserved between adjacent sections so that the connections between neighboring segmented models become smooth.

12.3.1. *Example of Neural Network Inverse Model Development*

We develop a neural network inverse model for the iris defined in Section 12.2.1. We then apply the segmentation technique to improve the accuracy and compare two results.

First we generate training data in the forward way using the EM model (Kabir *et al.*, 2008). The standard data generation procedure is followed where the input variables D, f_o, L_v, and L_h are varied to generate samples of P_v, P_h, M_{23}, and M_{14}. A total of 37,000 data samples are generated for the model. We then reorganize the data for training the inverse model. In this case the data samples of L_v and L_h are moved to the output side and M_{23} and M_{14} are moved to the input side. In the next step, we use the inverse data to train a neural network. NeuroModelerPlus (2005) is used to train the model. The average error of the inverse model before segmentation is 0.24% and the worst case error is 14.2%.

The segmentation technique is now applied to improve the accuracy of the model. The inverse data are divided into four sets according to the value of D. Then four separate neural network models are trained with the four sets of data. These four sub-models combined represent a more accurate inverse model than the inverse model without segmentation. In this way we reduce the average error of the inverse model from 0.24% to 0.17% and worst case error from 14.2% to 7.2%.

12.4. Non-Uniqueness of Input–Output Relationship in the Neural Network Inverse Model and its Detection

When the original forward input–output relationship is not monotonic, non-uniqueness becomes an inherent problem in the inverse model. In order to solve this problem, we start by addressing multivalued solutions in training data as follows (Kabir *et al.*, 2008):

If two different input values in the forward model lead to the same value of output then a contradiction arises in the training data of the inverse model, because the single input value in the inverse model has two different output values. Since we cannot train the neural network inverse model to match two different output values simultaneously, the training error cannot be reduced to a small value. As a result the trained inverse model will not be accurate. For this reason, it is important to detect the existence of multivalued solutions, which create contradictions in the training data.

Detection of multivalued solutions would be straightforward if the training data were generated by deliberately choosing different geometrical dimensions such that they lead to the same electrical value. However in practice, the training data are not sampled at exactly those locations. Therefore we develop numerical criteria to detect the existence of multi-valued solutions.

We assume I_x and I_y contain same number of indices, and that the indices in I_x (or I_y) are in ascending order. Let us define the distance between two samples of training data, sample number l and k as

$$d^{(k,l)} = \sqrt{\sum_{i=1}^{n}(\bar{x}_i^{(k)} - \bar{x}_i^{(l)})^2 / (\bar{x}_i^{\max} - \bar{x}_i^{\min})^2}, \qquad (12.12)$$

where, $\bar{x}_i^{\max}$ and $\bar{x}_i^{\min}$ are the maximum and minimum value of $\bar{x}_i$ respectively as determined from training data. We use a superscript to denote the sample index in training data. For example, $\bar{x}_i^{(k)}$ and $\bar{y}_i^{(k)}$ represent values of $\bar{x}_i$ and $\bar{y}_i$ in the kth training data respectively. Sample $\bar{x}^{(k)}$ is in the neighborhood of $\bar{x}^{(l)}$ if $d^{(k,l)} < \varepsilon$, where ε is a user-defined threshold whose value depends on the step size of data sampling. The maximum and minimum "slope" between samples within the neighborhood of $\bar{x}^{(l)}$ is defined as

$$S_{\max}^{(l)} = \sqrt{\max_{\substack{k \\ d^{(k,l)} < \varepsilon}} \frac{\sum_{i=1}^{m}\{(\bar{y}_i^{(k)} - \bar{y}_i^{(l)})/(\bar{y}_i^{\max} - \bar{y}_i^{\min})\}^2}{\sum_{i=1}^{n}\{(\bar{x}_i^{(k)} - \bar{x}_i^{(l)})/(\bar{x}_i^{\max} - \bar{x}_i^{\min})\}^2}} \qquad (12.13)$$

and

$$S_{\min}^{(l)} = \sqrt{\min_{k}{\atop d^{(k,l)} < \varepsilon} \frac{\sum_{i=1}^{m}\{(\bar{y}_i^{(k)} - \bar{y}_i^{(l)})/(\bar{y}_i^{\max} - \bar{y}_i^{\min})\}^2}{\sum_{i=1}^{n}\{(\bar{x}_i^{(k)} - \bar{x}_i^{(l)})/(\bar{x}_i^{\max} - \bar{x}_i^{\min})\}^2}}. \qquad (12.14)$$

Input sample $\bar{x}^{(l)}$ will have multivalued solutions if, within its neighborhood, the slope is larger than maximum allowed or the ratio of maximum and minimum slope is larger than the maximum allowed slope change. Mathematically, if

$$S_{\max}^{(l)} > S_M \qquad (12.15)$$

and

$$S_{\max}^{(l)} > S_{\min}^{(l)} > S_{R'}, \qquad (12.16)$$

then $\bar{x}^{(l)}$ has multivalued solutions in its neighborhood where S_M is the maximum allowed slope and $\boldsymbol{S}_R$ is the maximum allowed slope change.

We employ the simple criteria of (12.15) and (12.16) to detect possible multivalued solutions. A suggestion for ε can be at least twice the average step size of y in training data. A reference value for S_R can be approximately the inverse of a similarly defined "slope" between adjacent samples in the training data of the forward model. The value of S_M should be greater than 1. In the overall modeling method, conservative choices of ε, S_M, and S_R (larger ε, smaller S_M, and S_R) lead to more use of the derivative division procedure to be described in the next section, while aggressive choices of ε, S_M, and S_R lead to early termination of the overall algorithm (or more use of the segmentation procedure) when model accuracy is achieved (or not achieved). In this way, the choices of ε, S_M, and S_R mainly affect the training time of the inverse models, rather than model accuracy. The modeling accuracy is determined from segmentation or from the derivative division step to be described in the next section.

12.5. Method to Divide Training Data Containing Multivalued Solutions

In this section we describe a method to solve the multivalued problem in training data of the neural network inverse model (Kabir *et al.*, 2008). If the existence of multivalued solutions is detected in the training data, we perform data preprocessing to divide the data into different groups such that the data in each group do not have the problem of multivalued solutions. To do this, we need to develop a method to decide which data samples should be moved into which group. We divide the overall training data into groups based on derivatives of outputs vs. inputs of the forward model. Let us define the derivatives of inputs and outputs that have been exchanged to formulate the inverse model, evaluated at each sample, as

$$\left.\frac{\partial y_i}{\partial x_j}\right|_{\boldsymbol{x}=\boldsymbol{x}^{(k)}}, i \in I_y \quad \text{and} \quad j \in I_x, \tag{12.17}$$

where $k = 1, 2, 3, \ldots, N_s$ and N_s is the total number of training samples. The entire training data should be divided based on the derivative criteria such that training samples satisfying

$$\left.\frac{\partial y_i}{\partial x_j}\right|_{\boldsymbol{x}=\boldsymbol{x}^{(k)}} < \delta \tag{12.18}$$

belong to one group, and training samples satisfying

$$\left.\frac{\partial y_i}{\partial x_j}\right|_{\boldsymbol{x}=\boldsymbol{x}^{(k)}} > -\delta, \tag{12.19}$$

belong to a different group. The value for δ is zero by default. However to produce an overlapping connection at the break point between the two groups we can choose a small positive value for it. In that case a small amount of data samples whose absolute values of derivative are less than δ will belong to both groups. The value for δ other than the default suggestion of zero can be chosen as a value slightly larger than the smallest absolute value of derivatives of (12.17) for all training samples. Choice of δ only affects the accuracy of the sub-models at the connection region. The model accuracy for the rest of the region will remain unaffected.

This method exploits derivative information to divide the training data into groups. Therefore, accurate derivative is an important requirement for this method. Computation of derivatives of (12.17) is not a straightforward task since no analytical equation is available. We compute the derivatives by exploiting adjoint neural network technique (Xu *et al.*, 2003). We first train an accurate neural network forward model. After training is finished, its adjoint neural network can be used to produce the derivative information used in (12.18) and (12.19). The computed derivatives are employed to divide the training data into multiple smaller groups according to (12.18) and (12.19) using different combinations of i and j. Multiple neural networks are then trained with the divided data. Each neural network represents a sub-model of the overall inverse model.

12.6. Method to Combine Neural Network Inverse Sub-Models

When a multivalued solution exists in the training data, we divide the data and produce multiple inverse sub-models. The inverse sub-models provide more accurate solutions than a single inverse sub-model trained with the entire data. We need to combine the multiple inverse sub-models to reproduce the overall inverse model completely. For this purpose a mechanism is needed to select the right one among multiple inverse sub-models for a given input $\bar{x}$. Figure 12.2 shows the inverse sub-model combining method for a two sub-model system (Kabir *et al.*, 2008). For convenience of explanation, suppose $\bar{x}$ is a randomly selected sample of training data. Ideally if $\bar{x}$ belongs to a particular inverse sub-model then

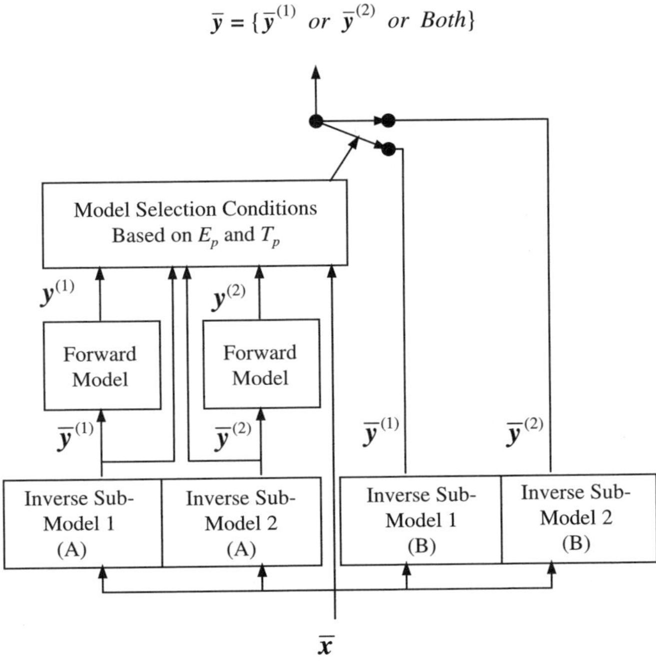

Fig. 12.2.   Diagram of inverse sub-model combining technique after derivative division for a two sub-model system (Kabir *et al.*, 2008).

the output from it should be the most accurate one among various inverse sub-models. Conversely the outputs from the other inverse sub-models should be less accurate if $\bar{x}$ does not belong to them. However, when using the inverse sub-models with general input $\bar{x}$ whose values are not necessarily equal to that of any training samples, the value from the sub-models is the unknown parameter to be solved. So we still do not know which inverse sub-model is the most accurate one.

To address this dilemma, we use the forward model to help decide which inverse sub-model should be selected. If we supply an output from the correct inverse sub-model to an accurate forward model we should be able to obtain the original data input to the inverse sub-model. For example, suppose $y = f(x)$ is an accurate forward model. Suppose the inputs and outputs of the inverse sub-model are defined such that $\bar{x} = y$ and $\bar{y} = x$. If the inverse sub-model is $\bar{y} = \bar{f}(\bar{x})$ true, then

$$f(\bar{f}(\bar{x})) = \bar{x} \tag{12.20}$$

is also true. Conversely, if $f(\bar{f}(\bar{x})) \neq \bar{x}$ then $\bar{f}(\bar{x})$ is a wrong inverse sub-model. In this way we can use a forward model to help determine which inverse sub-model should be selected for a particular value of input. In our method the input $\bar{x}$ is supplied to each inverse sub-model and the output from them is fed to the accurately trained forward model respectively, which generate different y. These outputs are then compared with the input data $\bar{x}$. The inverse sub-model that produces least error between y and $\bar{x}$ is selected and the output from the corresponding inverse sub-model is chosen as the final output of the overall inverse modeling problem. In cases where both the inverse sub-models produce error less than a specified value, both outputs are taken as a valid solution. This way we obtain all the valid solutions from the inverse sub-model. Once combined, the inverse sub-models together represent an accurate single inverse model.

The purpose of the model combining technique is to reproduce the original multivalued input–output relationship for the user. This method is an advanced one over the direct inverse modeling method since the latter produces only an inaccurate result when there are multivalued solutions (i.e., produces a single solution which may not match any of the original multivalues). This method can be used to provide a quick model to reproduce multivalued solutions in inverse EM problems. Using the solutions from this inverse model (including reproduced multivalued solutions), the user can proceed to a circuit design.

Table 12.1. Data for the spiral inductor model development.

D_i (um)	Q_{eff}
45	48
57	50
75	52.2
90	53.5
110	54.2
125	53.5
145	51
160	48.5
180	45.5
198	43
225	40

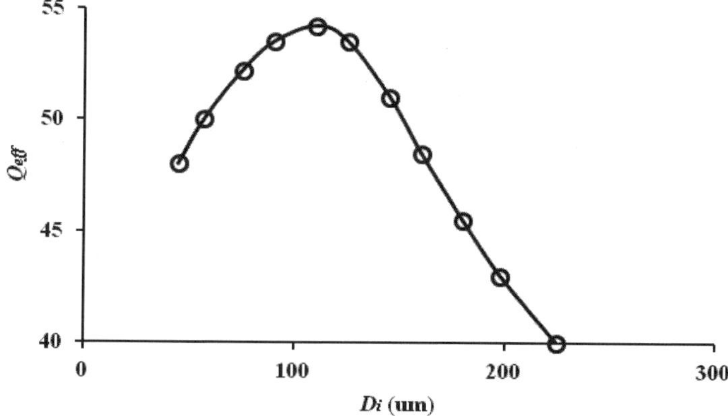

Fig. 12.3. Variations of the effective quality factor Q_{eff} of a spiral inductor model with respect to its inner diameter D_i (Bahl, 2003).

12.6.1. *Neural Network Inverse Model Formulation and Development of a Spiral Inductor*

Values of inner mean diameter (D_i) and corresponding values of effective quality factor (Q_{eff}) of a spiral inductor are listed in Table 12.1. The data is plotted in Fig. 12.3 showing the variation of the two (Bahl, 2003). Let us develop an inverse model of the spiral inductor which can be used to calculate the D_i for a given value of Q_{eff}. We will also employ advanced techniques as described in Section 12.4, 12.5, and 12.6 to improve the accuracy and will compare the results.

Table 12.2.   Data for the inverse
model of the spiral inductor.

Q_{eff}	D_i (um)
48	45
50	57
52.2	75
53.5	90
54.2	110
53.5	125
51	145
48.5	160
45.5	180
43	198
40	225

The input and output variables of the forward model are the inner mean diameter $x = [D_i]$ and the effective quality factor $y = [Q_{eff}]$ respectively. We formulate the input of the inverse model as $\bar{x} = [Q_{eff}]$ and the output as $\bar{y} = [D_i]$ as presented in Kabir *et al.* (2008).

The data for the inverse model is obtained by swapping the input and output data of the forward model. The inverse model data is shown in Table 12.2. We train a neural network model with the data. However the model error is high with a value over 14%. We use a utility of NeuroModelerPlus (2005) to detect the existence of multivalued solutions which uses (12.13)–(12.16) for that purpose. We supply the training data of Table 12.2 to NeuroModelerPlus and set values of parameters as $S_M = 80$, $S_R = 80$, and $\varepsilon = 0.01$. The program detects several contradictions in the data as expected. We plot the data of the inverse model as shown in Fig. 12.4(a). The plot clearly shows a non-unique input–output relationship since, in the range from $Q_{eff} = 47$ to $Q_{eff} = 55$, a single Q_{eff} value will produce two different D_i values. Therefore, we need advanced techniques to improve the model accuracy.

In the next step, we divide the training data according to the derivative (Kabir *et al.*, 2008). A neural network forward model is trained to learn the data in Table 12.1 and its adjoint neural network is used to compute the derivatives $\partial Q_{eff}/\partial D_i$. We compare all the values of derivatives and the lowest absolute value is found to be 0.018. The next large absolute value of derivative is 0.07. Therefore we choose the value of $\delta = 0.02$, which is in between 0.018 and 0.07. The training data are divided such that samples satisfying (12.18) are assigned into Group 1 and samples satisfying (12.19)

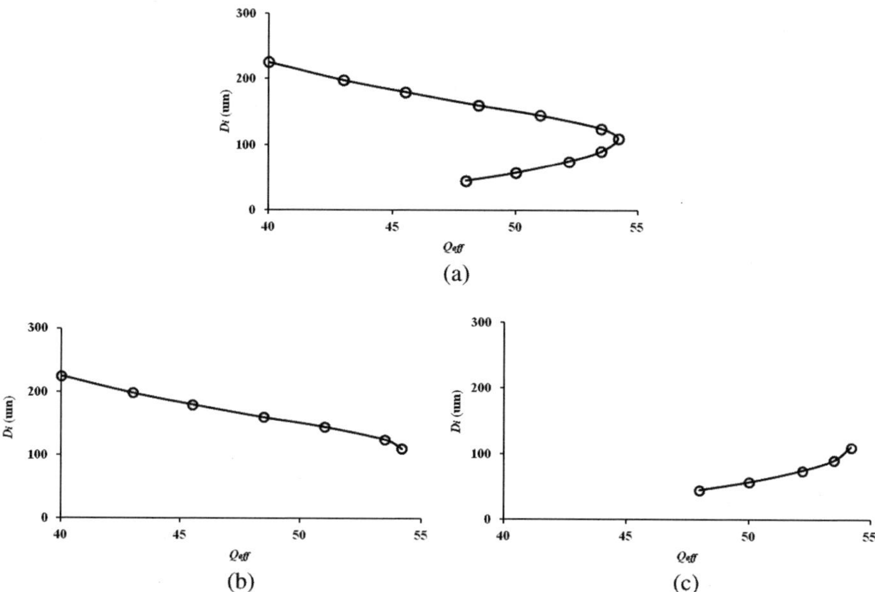

Fig. 12.4. (a) Non-unique relationship of an inverse spiral inductor model. Training data containing multivalued solutions are divided into groups according to the derivative: (b) Group 1 data with negative derivative; (c) Group 2 data with positive derivative. Within each group, the data are free of multivalued solutions, and consequently, the input–output relationship becomes unique (Kabir *et al.*, 2008).

Table 12.3. Two data groups for the inverse model of the spiral inductor.

Group 1		Group 2	
Q_{eff}	D_i (um)	Q_{eff}	D_i (um)
54.2	110	48	45
53.5	125	50	57
51	145	52.2	75
48.5	160	53.5	90
45.5	180	54.2	110
43	198		
40	225		

are assigned into Group 2. Table 12.3 presents the divided data samples. Figure 12.4(b) and (c) show the plots of the two groups, which confirm that each individual group becomes free of multivalued solutions after dividing the data according to the derivative information.

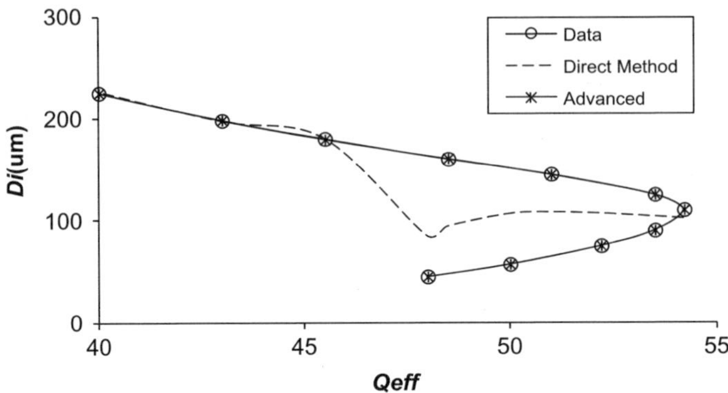

Fig. 12.5.   Comparison of inverse model outputs of the spiral inductor developed using the direct inverse modeling method and the advanced methodology (Kabir *et al.*, 2008).

Two inverse sub-models of the spiral inductor are trained using the two data groups shown in Table 12.3. The two individual sub-models become very accurate, and they are combined using the model combining technique. These two models combined become the accurate inverse model of the spiral inductor.

Figure 12.5 shows comparisons of outputs obtained from the inverse model directly trained with the data of Table 12.2 and the inverse model trained with the advanced techniques. It shows that in the first case, the model produces inaccurate results because of confusions over training data with multivalued solutions. In the latter case, the model delivers accurate solutions that match the data for the entire range reducing the average test error from 13.6% to 0.05%.

12.7. Overall Neural Network Inverse Modeling Methodology

In this section we describe the overall methodology of inverse modeling which combines all the aspects described in previous section (Kabir *et al.*, 2008). The inverse model of a microwave device may contain unique or non-unique behavior over various regions of interest. In the region with unique solutions direct segmentation can be applied and the training error is expected to be low. On the other hand, in the region with non-uniqueness, the model should be divided according to derivatives. If the overall problem is simple, the methodology will end with a simple inverse model directly trained with all data. In complicated cases, the methodology

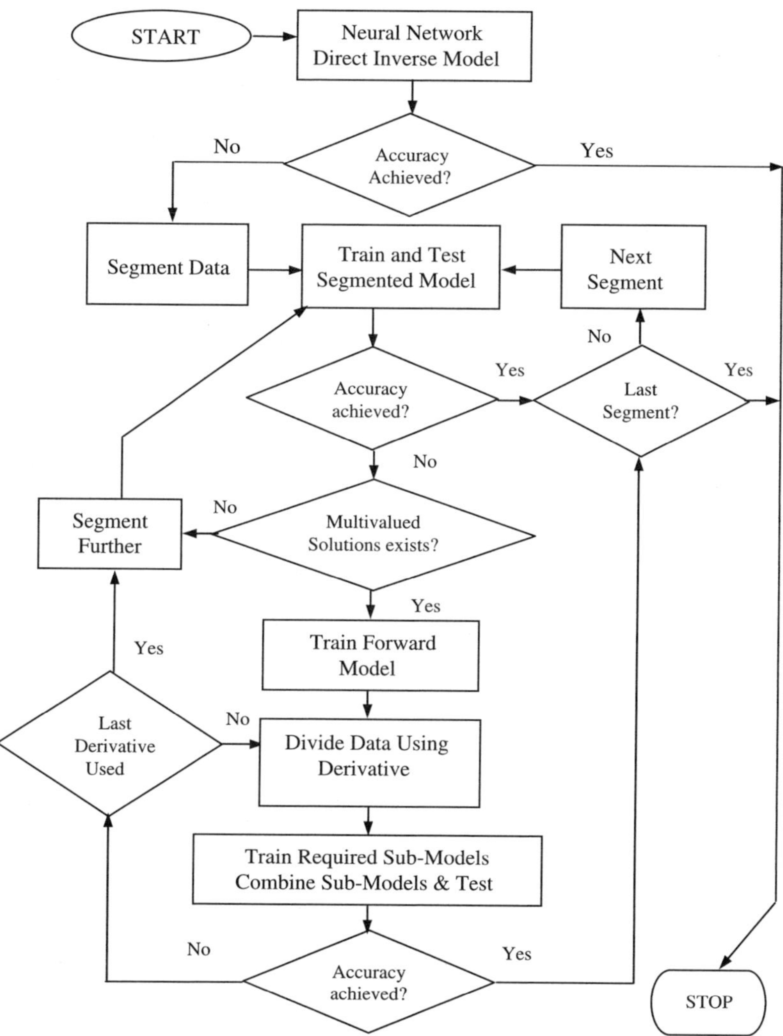

Fig. 12.6. Flow diagram of advanced inverse modeling methodology consisting of direct inverse modeling, segmentation, derivative dividing, and model combining techniques (Kabir *et al.*, 2008).

uses derivative division and sub-model combining method to increase model accuracy. This approach increases the overall efficiency of modeling. The flow diagram of the overall inverse modeling approach is presented in Fig. 12.6. The overall methodology is summarized in the following steps:

Step 1: Define the inputs and outputs of the model. Detailed formulation can be found in Section 12.2. Generate data using EM simulator or measurement. Swap the input and output data to obtain data for training the inverse model. Train and test the inverse model. If the model accuracy is satisfied then stop. Results obtained here is the direct inverse model.

Step 2: Segment the training data into smaller sections. If there have been several consecutive iterations between Steps 2 and 5, then go to Step 6.

Step 3: Train and test models individually with segmented data.

Step 4: If the accuracy of all the segmented models in Step 3 is satisfactory, stop. Else for the segments that have not reached accuracy requirements proceed to the next steps.

Step 5: Check for multivalued solutions in model's training data using (12.15) and (12.16). If none are found then perform further segmentation by going to Step 2.

Step 6: Train a neural network forward model.

Step 7: Using the adjoint neural network of the forward model divide the training data according to derivative criteria as described in Section 12.5.

Step 8: With the divided data, train necessary sub-models, for example two inverse sub-models.

Step 9: Combine all the sub-models that have been trained in Step 8 according to method in Section 12.6. Test the combined inverse sub-models. If the test accuracy is achieved then stop. Else go to Step 7 for further division of data according to derivative information in different dimensions, or if all the dimensions are exhausted, go to Step 2.

The algorithm increases efficiency by choosing the right techniques in the right order. For simple problems, the algorithm stops immediately after the direct inverse modeling technique. In this case no data segmentation or other techniques are used, and training time is short. The segmentation and subsequent techniques will be applied only when the directly trained model cannot meet accuracy criteria. In this way, more training time is needed only with more complexity in the model input–output relationship, such as the multivalued relationship.

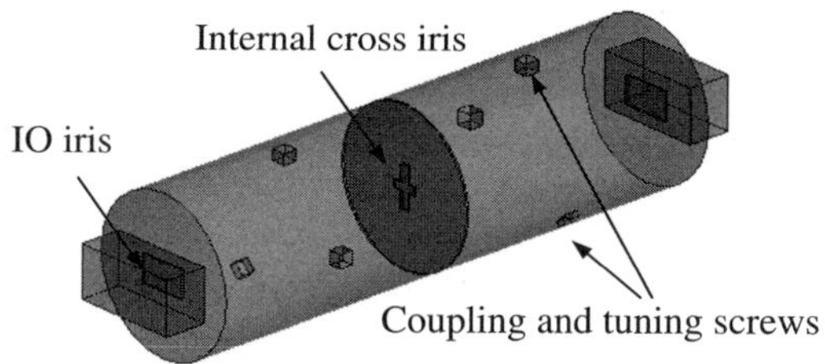

Fig. 12.7.   Diagram of a circular waveguide dual-mode filter (Yu and Wang, 2011).

12.7.1. *Inverse Iris Model Development of a Circular Waveguide Dual-Mode Filter*

In this section we develop a neural network inverse model for the calculation of input–output iris and coupling iris lengths for the circular waveguide dual-mode filter shown in Fig. 12.7. We employ advanced techniques to achieve less than 1% of the average model error.

The filter is decomposed into three different modules each representing a separate filter junction, namely the input–output (IO) iris, the internal coupling iris, and the tuning screws. Training data for neural networks are generated from physical parameters firstly through EM simulation (e.g., the mode-matching method) producing the generalized scattering matrix (GSM). Coupling values are then obtained from GSM through analytical equations (Wang *et al.*, 2006).

For the calculation of the iris lengths, two neural network inverse models need to be developed (Kabir *et al.*, 2008). The first neural network inverse model of the filter structure is developed for the internal coupling iris, which has been developed in Section 12.3.1.

The second inverse model of the filter is the IO iris model (Kabir *et al.*, 2008). The input parameters of the IO iris inverse model are circular cavity diameter D, center frequency f_o, and the coupling value R. The output parameters of the model are the iris length L, the loading effect of the coupling iris on the two orthogonal modes P_v and P_h, and the phase loading on the input rectangular waveguide P_{in}. The IO iris forward model is formulated as

$$\boldsymbol{x} = [D \ f_o \ L]^T \tag{12.21}$$

$$\boldsymbol{y} = [R \ P_v \ P_h \ P_{in}]^T. \tag{12.22}$$

The indices of input and output variables of the forward model that are swapped are defined as $I_x = \{3\}$ and $I_y = \{1\}$, respectively. The inverse model is defined as

$$\bar{y} = [x_3 \; y_2 \; y_3 \; y_4]^T = [L \; P_v \; P_h \; P_{in}]^T \tag{12.23}$$

$$\bar{x} = [x_1 \; x_2 \; y_1]^T = [D \; f_o \; R]^T. \tag{12.24}$$

We generate training data in the forward way using the EM model. Standard data generation procedure was followed where the input variables D, f_o, and L were varied to generate samples of R, P_v, P_h, and P_{in}. Four different sets of training data were generated according to the width of iris using the mode-matching method. In order to obtain the inverse training data, we swap the samples of L with the samples R in each of the four data sets. The model for each set is trained and tested separately using the direct inverse modeling method. The average test error of the model trained with entire data is more than 1%. Therefore, in the next step we segment the data into smaller sections. Models for these sections are trained separately, which reduces the worst-case error. The average error of the inverse model reduces from 1.2% to 0.4% and the worst case error reduces from 54% to 18.4%. We can improve the accuracy further by splitting the data set into more sections, and therefore achieve accurate results as required.

12.7.2. *Inverse Tuning Screw Model Development of a Circular Waveguide Dual-Mode Filter*

In this section we develop a neural network inverse model for the calculation of tuning screw lengths for the filter shown in Fig. 12.7. We will follow the advanced technique to detect contradictory data of the inverse model and to reduce model error. We will compare the test errors of inverse models developed with and without the advanced techniques.

The input parameters of the forward tuning screw model are circular cavity diameter D, center frequency f_o, the coupling between the two orthogonal modes in one cavity M_{12}, and the difference between the phase shift of the vertical mode and that of the horizontal mode across the tuning screws P. The outputs of the forward model are the phase shift of the horizontal mode across the tuning screw P_h, coupling screw length L_c, and the horizontal tuning screw length L_h. The forward tuning screw model is defined as

$$x = [D \; f_o \; L_h \; L_C]^T \tag{12.25}$$

$$y = [M_{12} \; P \; P_h]^T. \tag{12.26}$$

The indices of input and output variables of forward model that are swapped are defined as $I_x = \{2,3\}$ and $I_y = \{1,2\}$, respectively. Thus the inputs and outputs of the inverse model are formulated as (Kabir *et al.*, 2008)

$$\bar{y} = [y_3 \ x_3 \ x_4]^T = [P_h \ L_h \ L_C]^T \qquad (12.27)$$

$$\bar{x} = [x_1 \ x_2 \ y_1 \ y_2]^T = [D \ f_o \ M_{12} \ P]^T. \qquad (12.28)$$

We generate approximately 34,000 samples of training data by using the forward (i.e., EM) model. Input variables of the forward model are varied and corresponding output electrical parameters are generated. The inverse model data are generated by swapping data of L_h and L_c with M_{12} and P. We then train a neural network with the data to develop the inverse model of the tuning screws. The inverse model is trained directly using the entire training data. The training error is high even with many hidden neurons. Therefore, we proceed to segment the data into smaller sections. In this example we use the segmentation, which corresponds to two adjacent samples of frequency f_o and two adjacent samples of diameter D. Each segment of data is used to train a separate inverse model. Some of the segments produce accurate models with error less than 1% while others are still inaccurate. The segments that cannot reach the desired accuracy are checked for the existence of multivalued solutions individually. NeuroModelerPlus (2005) is used to detect the existence of multivalued solutions in the training data. For this example, neighborhood size $\varepsilon = 0.01$, maximum slope $S_M = 80$, and maximum slope change $S_R = 80$ are chosen (Kabir *et al.*, 2008). NeuroModelerPlus (2005) suggests that the data contain multivalued solutions. Therefore, we need to proceed to train a neural network forward model and apply the derivative division technique to divide the data.

To compute the derivative we train a neural network as a forward tuning screw model. Then the derivatives are computed using the adjoint neural network model through NeuroModelerPlus (2005). Considering $\delta = 0$ and applying the derivative $\partial P/\partial L_h$ to (12.18) and (12.19), we divide the data into Group 1 and Group 2 respectively. Two inverse sub-models are trained using Group 1 and Group 2 data. The sub-models altogether represent the accurate inverse model of the tuning screw. The results of the inverse models with and without use of the advanced methodology are compared in Table 12.4, showing the average, L2 and worst-case errors between model and test data. The results demonstrate that the advanced methodology improves model accuracy significantly.

Table 12.4.   Comparison of test errors between inverse
models developed with and without advanced methods
for the tuning screw model (Kabir *et al.*, 2008).

Modeling Method	Model Test Error (%)		
	Average	L2	Worst
Inverse	3.85	7.51	94.2
Advanced Inverse	0.40	0.59	5.10

12.8. Comparison of Filter Design Approach Using Forward and Inverse Model

We consider the application of neural network inverse modeling techniques for microwave filter design. Some results have been reported using neural network techniques to model microwave filters, including rectangular waveguide iris bandpass filter (Fedi *et al.*, 1999; Mediavilla, 2000; Cid and Zapata, 2001), low pass microstrip step filter (Nunez and Skrivervik, 2004), E-plane metal-insert filter (Burrascano *et al.*, 1998), coupled microstrip line bandpass filter (Ciminski, 2002), etc. In most cases, a forward neural network is developed to characterize a filter response.

In this section, we provide a brief description of how the neural network inverse models are applied to the microwave waveguide filter design as presented in Wang *et al.* (2006) and compare the method with the conventional design approach based on the forward model (EM model) (Kabir *et al.*, 2010). The filter design starts with synthesizing the ideal coupling matrix to satisfy filter specifications. The EM method for finding physical/geometrical parameters to realize the required coupling matrix is an iterative EM optimization procedure. This procedure performs EM analysis (mode-matching or finite element methods) on each waveguide junction of the filter to get the generalized scattering matrix (GSM). From GSM we extract coupling coefficients. We then modify the design parameters (i.e., the dimensions of filter) and re-perform EM analysis iteratively until the required coupling coefficients are realized. Figure 12.8(a) demonstrates the diagram of design approach using a forward type of model such as the EM model. The optimization process using a forward neural network model would speed up the process. However, the optimization process needs many iterations and it may suffer from convergence problems.

In Fig. 12.8(b) we show a diagram of the design process using a neural network inverse model. Note that the steps of generating an ideal coupling matrix from specifications are the same for approaches using both the

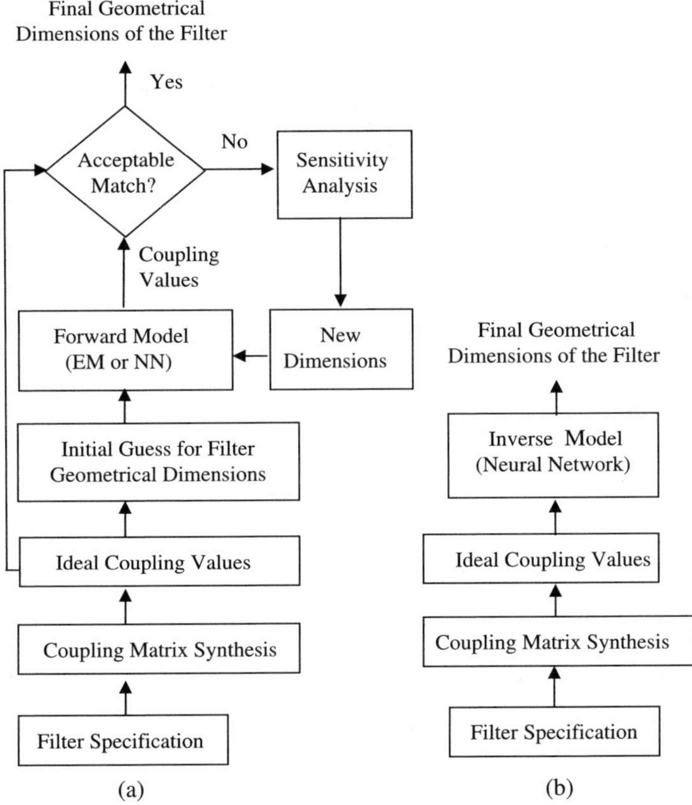

Fig. 12.8. Diagram showing filter design approach using (a) forward model such as EM solver or neural network forward model, and (b) neural network inverse model (Kabir *et al.*, 2010).

forward and inverse models. However, when using the inverse models, the filter dimensions are obtained in one step. Repetitive model evaluations can thus be avoided, in addition to the fact that there are no convergence issues.

12.9. Filter Design Using Developed Neural Network Inverse Models

We use the neural network inverse models that have been developed in Sections 12.3.1, 12.7.1, and 12.7.2 to design a four-pole filter with two transmission zeros (Kabir *et al.*, 2008). The layout of a four-pole filter is similar to that in Fig. 12.7. The filter center frequency is 11.06 GHz, bandwidth is 58 MHz, and cavity diameter is chosen to be 1.17 inch. The normalized ideal coupling values are

$$R_1 = R_2 = 1.07$$

$$M = \begin{bmatrix} 0 & 0.86 & 0 & -0.278 \\ 0.86 & 0 & 0.82 & 0 \\ 0 & 0.82 & 0 & 0.86 \\ -0.278 & 0 & 0.86 & 0 \end{bmatrix}. \tag{12.29}$$

The neural network inverse models developed in Sections 12.3.1, 12.7.1, and 12.7.2 are used to calculate irises and tuning screw dimensions. For example, in order to find the lengths of the vertical and horizontal slots for the internal coupling iris as shown in Fig. 12.7, the cavity diameter $D = 1.17$ inch and $f_o = 11.06$ GHz, together with the coupling values, are supplied to the inverse neural model in Section 12.3.1 as the inputs. The neural model outputs show that the two slot lengths are 0.299 inch and 0.212 inch respectively. The IO irises and tuning screws lengths are obtained similarly using the models in Sections 12.7.1 and 12.7.2. The dimensions are summarized in Table 12.5.

The filter has been manufactured and tuned by adjusting irises and tuning screws to match the ideal response. Figure 12.9 compares the measured and the ideal filter response. Dimensions of the tuned filter are also listed in Table 12.5 (Kabir *et al.*, 2008). Very good correlation can be seen between the initial dimensions provided by the neural network inverse models and the measured final dimensions of the fine-tuned filter.

Table 12.5. Comparison of the four-pole filter dimensions obtained from the neural network inverse models and measurement (Kabir *et al.*, 2008).

Filter Dimensions (Design Variables)	Value (inch)		
	NN Model	Measurement	Difference
IO irises	0.405	0.405	0000
M23 iris	0.299	0.297	−0.002
M14 iris	0.212	0.216	0.004
M11/M44 T-Screw	0.045	0.005	−0.040
M22/M33 T-Screw	0.133	0.135	0.002
M12/M34 C-Screw	0.111	0.115	0.004
Cavity length	1.865	1.864	−0.001

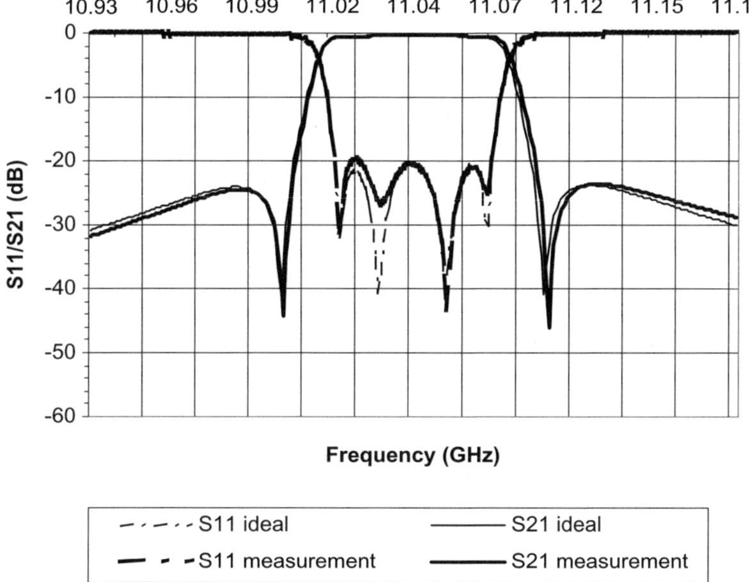

Fig. 12.9. Comparison of the ideal four-pole filter response with the measured filter response after tuning (Kabir *et al.*, 2008).

12.10. Discussion of the Neural Network Inverse Modeling Technique

The advantage of using the trained neural network inverse model is also demonstrated in terms of time compared to EM models. An EM simulator can be used for synthesis, which requires typically 10 to 15 iterations to generate inverse model dimensions. Comparisons of time to obtain dimensions using the EM and the trained neural network models are listed in Table 12.6. It shows that the time required by the neural network inverse model is negligible compared to EM models (Kabir *et al.*, 2008).

Table 12.6. Comparison of central processing unit (CPU) time spent on finding the design variables (filter dimensions) using the inverse model and the forward model (Kabir *et al.*, 2008).

Design Method	CPU Time (s)		
	IO Iris	Coupling Iris	Tuning Screw
Forward (EM)	1.5	120	240
Inverse (NN)	0.14E-3	0.1E-3	1.3E-3

The inverse technique is useful for highly repeated design tasks such as designing filters of different orders and different specifications. The technique is not suitable if the inverse model is used for the purpose of only one or a few particular designs, because the model training time will make the technique cost-ineffective. Therefore the technique should be applied to inverse tasks that will be re-used frequently. In such a case the benefit of using the model far outweighs the cost of training for four reasons: (i) training is a one-off investment, and the benefit of the model increases when the model is used over and over again, for example, filters with different orders but similar configurations can use the same set of models, (ii) conventional EM design is part of the design cycle, while neural network training is outside the design cycle, (iii) circuit design requires much human involvement, while neural network training is a machine-based computational task, (iv) neural network training can be done by a model developer and the trained model can be used by multiple designers.

The neural network approach cuts expensive design time by shifting much burden to off-line computer-based neural network training (Zhang and Ghupta, 2000). An even more significant benefit of the advanced technique is the new feasibility of interactive design and what-if analysis using the instant solutions of inverse neural networks, substantially enhancing design flexibility and efficiency.

References

Bahl, I. (2003). *Lumped Elements for RF and Microwave Circuits*, Artech House, Boston: MA.

Bandler, J.W., Ismail, M.A., Rayas-Sanchez, J.E. and Zhang, Q.J. (2003). Neural inverse space mapping (NISM) optimization for EM-base microwave design, *Int. J. RF Microw. CE*, **13**, 136–147.

Burrascano, P., Dionigi, M., Fancelli, C. and Mongiardo, M. (1998). A neural network model for CAD and optimization of microwave filters, *IEEE MTT-S*, Baltimore, MD, pp. 13–16.

Cid, J.M. and Zapata, J. (2001). CAD of rectangular-waveguide H-plane circuits by segmentation, finite elements and artificial neural networks, *IEE Electronic Letters*, **37**, 98–99.

Ciminski, A.S. (2002). Artificial neural networks modeling for computer-aided design of microwave filter, *Int. Conf. on Microwaves, Radar and Wireless Communications*, Poland, pp. 96–99.

Davis, B., White, C., Reece, M.A., Bayne, M.E. Jr, Thompson II, W.L., Richardson, N.L. and Walker, L. Jr (2003). Dynamically configurable pHEMT model using neural networks for CAD, *IEEE MTT-S*, Philadelphia, PA, pp. 177–180.

Devabhaktuni, V.K., Yagoub, M.C.E. and Zhang, Q.J. (2001). A robust algorithm for automatic development of neural-network models for microwave applications, *IEEE T. Microw. Theory*, **49**, 2282–2291.

Fedi, G., Gaggelli, A., Manetti, S. and Pelosi, G. (1999). Direct-coupled cavity filters design using a hybrid feedforward neural network – finite elements procedure, *Int. J. RF Microw. CE*, **9**, 287–296.

Garcia, J.P., Pereira, F.Q., Rebenaque, D.C., Tornero, J.L.G. and Melcon, A.A. (2006). A neural-network method for the analysis of multilayered shielded microwave circuits, *IEEE T. Microw. Theory*, **54**, 309–320.

Isaksson, M., Wisell, D. and Ronnow, D. (2005). Wide-band dynamic modeling of power amplifiers using radial-basis function neural networks, *IEEE T. Microw. Theory*, **53**, 3422–3428.

Kabir, H., Wang, Y., Yu, M. and Zhang, Q.J. (2008). Neural network inverse modeling and applications to microwave filter design, *IEEE T. Microw. Theory*, **56**, 867–879.

Kabir, H., Yu, M. and Zhang, Q.J. (2010). Recent advances of neural network-based EM-CAD, *Int. J. RF Microw. CE*, **20**, 502–511.

Markovic, V., Marinkovic, Z. and Males-Llic, N. (2000). Application of neural networks in microwave FET transistor noise modeling, *Proc. Neural Network Applications in Electrical Engineering*, Belgrade, Serbia, pp. 146–151.

Mediavilla, A., Tazon, A., Pereda, J. Lazaro, A., Santamaria, M.I. and Pantaleon, C. (2000). Neuronal architecture for waveguide inductive iris bandpass filter optimization, *Proc. IEEE-INNS-ENNS Int. Joint Conf. on Neural Networks*, Italy, pp. 395–399.

NeuroModelerPlus (2005), Department of Electronics, Carleton University, 1125 Colonel By Drive, Ottawa, Ontario, K1S 5B6, Canada.

Nunez, F. and Skrivervik, A.K. (2004). Filter approximation by RBF-NN and segmentation method, *IEEE MTT-S*, Fort Worth: TX, pp. 1561–1564.

Rizzoli, V., Costanzo, A., Masotti, D., Lipparini, A. and Mastri, F. (2004). Computer-aided optimization of nonlinear microwave circuits with the aid of electromagnetic simulation, *IEEE T. Microw. Theory*, **52**, 362–377.

Selleri, S., Manetti, S. and Pelosi, G. (2002). Neural network applications in microwave device design, *Int. J. RF Microw. CE*, **12**, 90–97.

Vai, M.M., Wu, S., Li, B. and Prasad, S. (1998). Reverse modeling of microwave circuits with bidirectional neural network models, *IEEE T. Microw. Theory*, **46**, 1492–1494.

Wang, Y., Yu, M., Kabir, H. and Zhang, Q.J. (2006). Effective design of cross-coupled filter using neural networks and coupling matrix, *IEEE MTT-S*, San Francisco: CA, pp. 1431–1434.

Watson, P.M. and Gupta, K.C. (1996). EM-ANN models for microstrip vias and interconnects in dataset circuits, *IEEE T. Microw. Theory*, **44**, 2495–2503.

Watson, P.M. and Gupta, K.C. (1997). Design and optimization of CPW circuits using EM-ANN models for CPW components, *IEEE T. Microw. Theory*, **45**, 2515–2523.

Xu, J., Yagoub, M.C.E., Ding, R. and Zhang, Q.J. (2003). Exact adjoint sensitivity analysis for neural-based microwave modeling and design, *IEEE T. Microw. Theory*, **51**, 226–237.

Yu, M. and Wang, Y. (2011). Synthesis and beyond, *IEEE Microw. Mag.*, **12**, 62–76.

Zaabab, A.H., Zhang, Q.J. and Nakhla, M.S. (1995). A neural network modeling approach to circuit optimization and statistical design, *IEEE T. Microw. Theory*, **43**, 1349–1358.

Zhang, Q.J. and Gupta, K.C. (2000). *Neural Networks for RF and Microwave Design*, Artech House, Boston: MA.

Zhang, Q.J., Gupta, K.C. and Devabhaktuni, V.K. (2003). Artificial neural networks for RF and microwave design-from theory to practice, *IEEE T. Microw. Theory*, **51**, 1339–1350.

Chapter 13

Simulation-Driven Design of Microwave Filters
for Space Applications

Elena Díaz Caballero, José Vicente Morro Ros,
Héctor Esteban González, Vicente Enrique Boria Esbert,
Carmen Bachiller Martín, and Ángel Belenguer Martinez

13.1. Introduction

Since the end of the 1960s, when the first Intelsat satellite was launched, the variety and quantity of satellite communications applications have grown exponentially. Being developed to render intercontinental voice services, the use of communications satellites has spread to many other applications over the last 40 years. Thanks to their global coverage, they are used, for instance, for broadcasting radio and television signals, for remote observation of the Earth, for radionavigation and for mobile communications.

As it is said in Kudsia *et al.* (1992), microwave devices, and particularly filters, are key devices within the satellite payload. A full revision of the different filter topologies can be found in Matthaei *et al.* (1980). However, the choice of a specific filter topology will depend on its application and the filtering needs.

Microwave filters can be used in a wide frequency range, from hundreds of megahertzs to 40 GHz, depending on the telecommunication service rendered by the payload. Navigation services and mobile communications via satellite usually operate in the L (1–2 GHz) and S (2–4 GHz) bands, whereas remote observation applications normally do it in the C band (4–8 GHz). In the commercial applications range, the increasing requirements related to quality of service have led to a continuous increase in the operating frequencies. So, while the first fixed service satellites (FSS) operated at low frequencies (6/4 GHz and 8/7 GHz), the most modern ones are already working in the Ku band (12–18 GHz), and a potential jump to the band of 30/50 GHz is expected in the next few years.

Besides their use in satellite communications, the microwave filters' relevance is due also to other telecommunications applications, such as wireless access systems (IEEE 802.11 (Wi-Fi)), Worldwide Interoperatibility for Microwave Access (WiMAX), Wireless Local Loop (WLL), Local Multipoint Distribution System (LMDS), Multipoint Video Distribution System (MVDS)) and mobile communications systems (Global System for Mobile Communications (GSM), Universal Mobile Telecommunication System (UMTS)).

If we focus on the satellite applications, the increasing number of services rendered by communications satellites has led to the congestion of the frequency spectrum. This fact results in an enormous demand for high-performance filters which must satisfy very stringent specifications. They must have high selectivity in order to avoid interferences, together with constant group delay and flat amplitude response in the passband in order to minimize the signal degradation. Moreover, filters have to handle high power preserving good thermal stability, and their mass and volume must be low. Therefore, one of the main problems when designing microwave filters lies in finding a compromise solution between the electric response of the filter (selectivity, insertion loss, group delay), the physical considerations of the device (mass, volume), the development time, and the fabrication cost. Some other design difficulties come with the high-power handling requirements: multipaction breakdown, ionization breakdown, passive inter-moldulation (PIM), and mechanical stability due to thermal variations.

Microwave filters for satellite applications have been historically implemented in metallic guide technology (Fig. 13.1(a)) due to its low losses and its high power-handling capabilities. However, the rising necessity for mass and volume reduction has led to the appearance of new filter topologies based on the use of dielectrics (Fig. 13.1(b)), which can reduce mass and volume by 50%. They also have better thermal stability for high-power applications.

Therefore, fast development of very accurate electromagnetic (EM) analysis tools for these technologies has taken place in the last few years. But still, an expert user is commonly needed for the analysis and design tasks, thus slowing down the design process of these filters and not always guaranteeing the optimum solution. This is the reason why the development of computer-aided design (CAD) tools based on accurate and efficient EM simulators is needed.

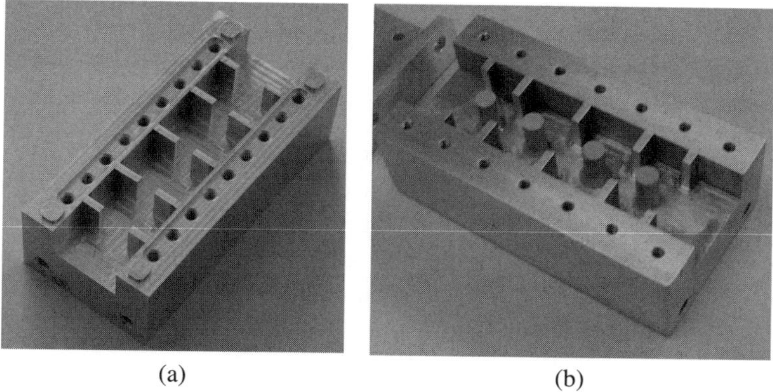

Fig. 13.1. H-plane coupled-cavities propagating filters: (a) without dielectric posts in the resonant cavities, (b) with dielectric posts in the resonant cavities.

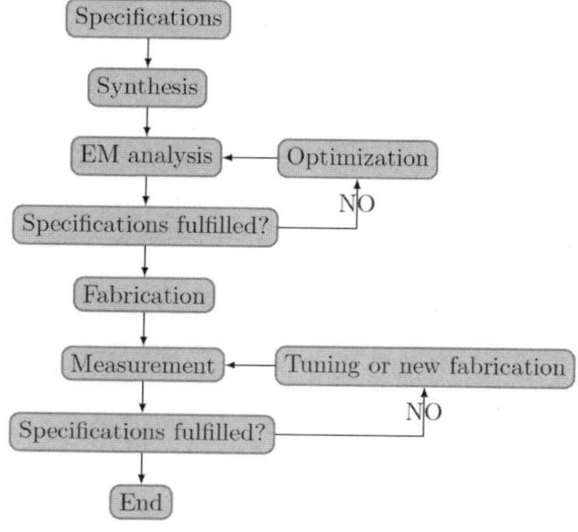

Fig. 13.2. Flowchart of a CAD process.

CAD strictly stands for any design process where a computer is used as a tool. However, the term CAD usually implies that the design process would have been impossible or much more difficult, expensive, and long had it not been for the use of the computer. Figure 13.2 shows the typical flowchart of a microwave device computer-aided design. Nowadays there are several techniques and methodologies for the CAD design of microwave filters. Many publications describe the state-of-the-art of the filter design process in Levy *et al.* (2002), letting the designers choose the most adequate method according to the filter topology and characteristics.

A microwave filter design process always begins with the given specifications set or goals to be achieved by the device, no matter the physical structure that we intend to design. Next the synthesis phase starts. It consists of two stages: first we obtain a circuit model whose response is the desired one by using the classical synthesis techniques (Zverev, 2005); next we find the equivalence between the circuit elements model and the real filter structure, obtaining the initial physical dimensions of the structure. If the structure is simple, the second step can be done from the circuit approximations for the real filter structure (Marcuvitz, 1986). However, this step can be combined with the optimization phase due to the complexity of some structures.

Once the initial structure is obtained, we must verify if it fulfills the design specifications. In order to do this, we will use an electromagnetic analysis tool. If verification fails, we should modify the initial structure by means of an optimization process till we get a new one. After this, we do the verification again and iterate till specifications are fulfilled. However, the use of a general purpose EM analysis tool in the optimization stage can become a big problem, because the analysis time can be prohibitive even if there are few optimization parameters.

Though optimization is a powerful tool, even the most elegant and robust strategy may be unable to find an acceptable solution if a good initial point is not provided.

When design satisfies specifications, the filter can be fabricated and tested. If measurements fulfill specifications, the design process is finished. Otherwise some modifications must be made and the design process should be completed again. A CAD tool should minimize the number of experimental iterations, drastically reducing the time and cost of the overall design process while increasing its quality.

This chapter aims to present a CAD procedure capable of meeting the main challenges facing every CAD tool: the obtention of a good initial point, the reduction of the computational cost of EM tools, and the development of a robust optimization strategy. In order to do so, this procedure must:

- Synthesize a good initial point. As mentioned above, this is a key element in any optimization process. However, the direct application of network synthesis theory to the design of microwave filters is possible just in a few cases, due to the limitations imposed by the structures to be fabricated. Therefore, several techniques will be used to make the synthesis stage more feasible.

- Reduce the optimization computational cost. The EM analysis of every possible solution must be done so that its electrical response can be compared with the specifications. As it is one of the main reasons for the computational cost of the process, the use of the space mapping philosophy is proposed. This technique reduces the computational effort of the design process by combining different simulation tools or models with different accuracy and efficiency.

- Improve the optimization process robustness. Traditionally, in the optimization phase all the parameters of the structure under design are modified at the same time. This approach does not guarantee the convergence of the solution. One remedy for this problem would be the division of the process in steps so that each step modifies just a few physical parameters.

- Improve the optimization process efficiency. In order to do so, the use of an adequate combination of different optimization algorithms is proposed instead of using a single one. At the beginning of the optimization process, robust but not very efficient algorithms (evolutionary computation, simplex) are used, and after some iterations, when the solution is close to the optimum, gradient methods (BFGS) are used to refine the solution.

With the purpose of verifying the developed design procedure, the following structures will be designed, fabricated, and measured:

- H-plane direct-coupled-cavity propagating filters (Fig. 13.1(a)).
- H-plane direct-coupled-cavity propagating filters with rounded corners (Fig. 13.3). Milling and molding are the cheapest fabrication techniques

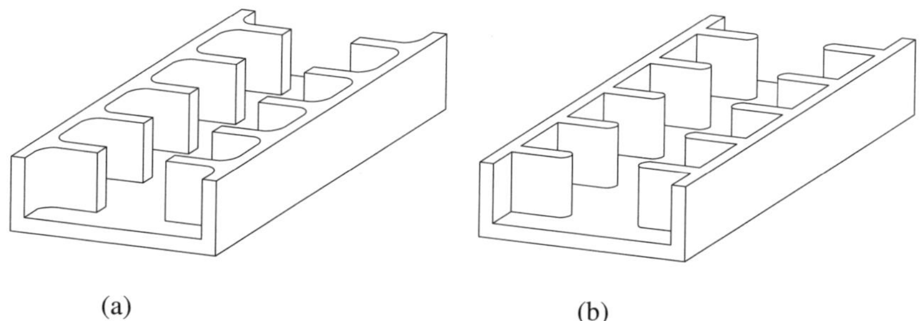

(a)   (b)

Fig. 13.3. H-plane direct-coupled-cavities propagating filters: (a) rounded corners in the resonant cavities, (b) rounded corners in the coupling windows.

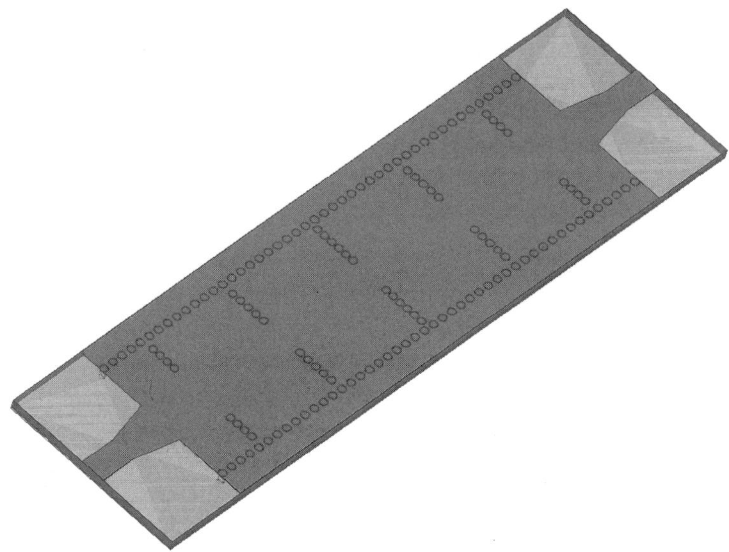

Fig. 13.4. SIW filter.

for the previous filters, but they respectively introduce rounded corners in the resonant cavities (Fig. 13.3(a)) and in the coupling windows (Fig. 13.3(b)), which modify the electric response. Considering them in the design process allows for the obtention of the desired response while minimizing the fabrication cost.

- H-plane direct-coupled-cavity propagating filters with dielectric resonators (Fig. 13.1(b)). This technology provides filters with remarkable reductions in terms of mass and volume when compared with conventional metallic filters, while preserving their frequency response.

- Substrate integrated waveguide (SIW) filters (Fig. 13.4). Based on a synthesized waveguide in a planar dielectric substrate with two rows of metallic vias, this low cost realization of the traditional waveguide circuit inherits the merits from both the microstrip for easy integration and the waveguide for low radiation losses.

So this chapter is organized as follows. In Section 13.2 the synthesis techniques used in the design process are presented. Next, Section 13.3 introduces the design techniques utilized in order to reduce the computational cost of the process while guaranteeing its robustness and accuracy. In Section 13.4, some examples of filters designed with the implemented design processes are discussed. Finally, some conclusions are briefly summarized in Section 13.5.

13.2. Initial Point Synthesis

Despite the significant progress in microwave filters design made in the last few decades, most of the synthesis strategies are based on Cohn's work (Cohn, 1957) originally proposed in the late 1950s. This procedure is based on the following steps:

- An equivalent circuit prototype, which fulfills certain ideal response (i.e., Chebyshev, Butterworth, Elliptical), is synthesized based on lumped and/or distributed elements. This prototype must be as similar as possible to the physical structure, while being simple enough to be analytically obtained.
- An equivalence is performed between the prototype elements and the different parts of the real structure, thereby achieving an initial estimation of the physical dimensions of the device.

The success of the synthesis procedure is based on the similarities between the prototype and the structure. In fact, when the equivalent prototype behaves like the real structure, the obtained dimensions are close to the final ones. This guarantees the convergence in the optimization phase. However, if the initial point is not good enough, the optimization phase can fail, obtaining a device which does not fulfill the specifications.

In order to model an H-plane filter in guided technology as those referred to in Section 13.1, we can use a circuit prototype based on inmitance inverters and resonators (see Fig. 13.5). In order to more accurately represent the structure, the characteristic impedances, Z_{0_i}, the propagation constants, β_i, and the waveguide wavelengths, λ_{g_i}, of the transmission lines coincide with those of the fundamental mode of its corresponding waveguide. A monomode prototype has been chosen due to the fact that the coupling among the higher-order modes is negligible.

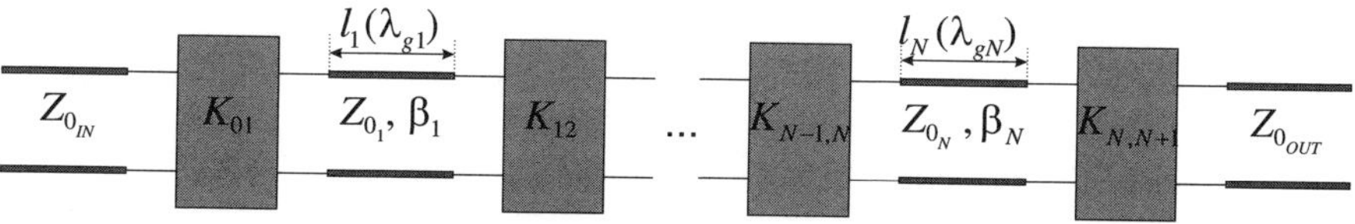

Fig. 13.5. Generalized prototype for bandpass microwave filters.

The inversion constants of the equivalent circuit prototype and the transmission lines lengths are (Matthaei *et al.*, 1980):

$$K_{0,1} = \sqrt{\frac{Z_{0_{IN}} \chi_1 \omega}{g_0 g_1}} \tag{13.1}$$

$$K_{i,i+1} = \omega \sqrt{\frac{\chi_i \chi_{i+1}}{g_i g_{i+1}}} \tag{13.2}$$

$$K_{N,N+1} = \sqrt{\frac{Z_{0_{OUT}} \chi_n \omega}{g_n g_{n+1}}} \tag{13.3}$$

$$l_i = \frac{\lambda_{g0_i}}{2}, \tag{13.4}$$

where χ_i is the slope parameter of the ith resonator. It is defined as:

$$\chi_i = \frac{\omega_0}{2} \left. \frac{dX_i}{d\omega} \right|_{\omega=\omega_0}. \tag{13.5}$$

Furthermore, g_i are the lowpass prototype elements, X_i is the ith resonator reactance, λ_{g0_i} is the waveguide wavelength λ_{g_i} of the ith resonator at the resonance frequency f_0, and Z_{0_i}, $Z_{0_{IN}}$, and $Z_{0_{OUT}}$ are, respectively, the characteristic impedances of the ith resonator, of the input waveguide and of the output waveguide. Lastly, ω represents the fractional filter bandwidth, defined as:

$$\omega = \frac{\lambda_{g1} - \lambda_{g2}}{\lambda_{g0}}, \tag{13.6}$$

where λ_{g1} and λ_{g2} are the lower and upper cutoff wavelengths, respectively, and λ_{g0} represents the resonance wavelength.

Finally, the extraction of the final device dimensions will depend on the specific type of H-plane filter considered, as will be seen in Section 13.4.

13.3. Design Techniques

This section will briefly describe the different design techniques implemented in the CAD tools used to design the advanced microwave filters presented in Section 13.4. First, there is a descriptive summary of basic optimization concepts in which the design of microwave devices is based. Later, some techniques, whose purpose is to improve the robustness, efficiency, and accuracy of the design process, are presented (space mapping, segmentation and hybrid optimization).

13.3.1. *Optimization Basic Concepts*

This section is a sweeping review of the basic concepts of the optimization of microwave devices.

13.3.1.1. *Error function*

The performance of a microwave filter is usually expressed as a frequency-dependent specification set, although time-dependent specifications could also be defined. In practice, a discrete set of m frequency points will be considered. These points must be representative enough of the whole interest frequency range. Although specifications can also define upper or lower limits, we will consider them to be just equality specifications. Thus, the error function is defined as the difference between the given specifications and the electromagnetic response calculated in each of those frequency points, as follows:

$$e_j(\mathbf{x}) = w_j \, |R_j(\mathbf{x}) - S_j| \quad j = 1, 2, \ldots, m, \tag{13.7}$$

where $\mathbf{x} \in \mathbb{R}^n$ is the vector of parameters to be designed, $\mathbf{R}(\mathbf{x}) \in \mathbb{R}^{m \times 1}$ is a vector with the electromagnetic response of the device in the m frequency points, $\mathbf{S} \in \mathbb{R}^{m \times 1}$ is a vector with the design specifications in the m frequency points, and $\mathbf{w} \in \mathbb{R}^{m \times 1}$ is a vector with positive weights, which allows us to emphasize the importance of some parts of the device response according to the design needs.

13.3.1.2. *Optimization of a microwave device*

From the definition of the error function, the device optimization problem can be mathematically expressed as:

$$\underset{\mathbf{x} \in \mathbb{R}^n}{\text{minimize}} \ U(\mathbf{x}), \tag{13.8}$$

$U(\mathbf{x})$ being the scalar objective function.

Function $U(\mathbf{x})$ will be generally subject to m restrictions, both equality (m_e restrictions) and inequality ones, which must be fulfilled during the optimization process and also by the optimum solution:

$$G_i(\mathbf{x}) = 0 \quad i = 1, \ldots, m_e (i \in E) \tag{13.9}$$

$$G_i(\mathbf{x}) \geq 0 \quad i = m_e + 1, \ldots, m (i \in I). \tag{13.10}$$

Thanks to the use of the objective function $U(\mathbf{x})$, we are able to convert a parameter extraction problem in a multidimensional minima search problem. In fact, its definition is essential, as it determines the shape of the multidimensional surface in which the optimization algorithm must find the minimum.

$U(\mathbf{x})$ is usually defined on the basis of an l_p-norm of $\mathbf{e}(\mathbf{x})$. The properties of the different norms used in the design of microwave devices will be described in the next paragraphs.

13.3.1.3. *The l_p-norms*

The l_p-norm (Temes and Zai, 1969) provide us with a scalar value which makes it possible to estimate the deviation between the device response and the specifications. The l_p-norm is defined as:

$$\|\mathbf{e}\|_p = \left[\sum_{j=1}^{m} |e_j|^p \right]^{1/p}.$$

$$(13.11)$$

l_1, l_2, and minimax, which are defined below, are the most common used l_p-norms:

$$\|\mathbf{e}\|_1 = \sum_{j=1}^{m} |e_j| \tag{13.12}$$

$$\|\mathbf{e}\|_2 = \left[\sum_{j=1}^{m} |e_j|^2 \right]^{1/2} \tag{13.13}$$

$$\|\mathbf{e}\|_\infty = \max_j |e_j|. \tag{13.14}$$

Although the previous norms are all derived from the same formula, they have different key characteristics. In fact, we can observe that the relative importance of the high error values increases with the norm order, p. The extreme case being the minimax norm ($\| \|_\infty$), where just the worst case is considered, ignoring the rest of the points.

13.3.1.4. *The Huber norm*

The Huber norm (Bandler *et al.*, 1993) combines the properties of l_1 and l_2 norms, low order, and continuity:

$$\|\mathbf{e}\|_H = H(\mathbf{e}) = \sum_{j=1}^{m} \rho_k(e_j) \tag{13.15}$$

$$\rho_k(e_j) = \begin{cases} \dfrac{e_j^2}{2} & \text{if} \quad |e_j| \le k \\[2ex] k|e_j| - \dfrac{k^2}{2} & \text{if} \quad |e_j| > k \end{cases} \qquad (13.16)$$

As we can see, the Huber norm coincides with the l_2-norm in $e_j \in [-k, k]$ and with l_1-norm out of that range. So the l_1-norm is used when we are far from the optimum point, while the l_2-norm is used when we are close to it.

13.3.2. *Space Mapping*

When simulating electromagnetic structures, we usually have to consider a compromise solution between low computational cost and accuracy. But the central processing unit effort of accurate simulations of complex structures can be prohibitive. The space mapping philosophy reduces the computational load of the design process by using different simulation tools (models) with different accuracy and efficiency. In their classic versions (space mapping (SM) and aggressive space mapping (ASM) (Bandler *et al.*, 1994, 1995)), two simulation tools are used: an efficient but not very accurate tool (coarse model), and a very accurate but not so efficient one (fine model). This way, the computational load is shifted to the coarse model, thus reducing the global computational cost, while accuracy is guaranteed by the fine model.

Despite the fact that classic SM implementations, especially ASM, have been widely used in the design of microwave devices, sometimes an adequate mapping between optimization and validation spaces cannot be found due to the non-uniquiness of parameter extraction (PE) process. This problem usually appears when the coarse and fine models are severely misaligned.

Therefore, at present much effort is dedicated to improving the robustness and performance of this philosophy. Some of these initiatives are the trust-region ASM (TRASM) (Bakr *et al.*, 1998), which integrates the trust-region concepts with the ASM, the hybrid ASM (HASM) (Bakr *et al.*, 1999), which uses TRASM and direct optimization, implicit SM (Bandler *et al.*, 2004), which obtains preassigned parameters to adjust the fine and coarse model, neural SM (Bakr *et al.*, 2000) and neural inverse space mapping (Bandler *et al.*, 2003), which use methodologies based on neural networks. In addition, multimodel ASM (MASM) (Morro *et al.*, 2005a) and multiple coarse model SM (Koziel and Bandler, 2008) progressively reduce the misalignment between the coarse and fine models by using auxiliary models in the validation and optimization space respectively.

The different SM approaches implemented in the design of the microwave filters presented in Section 13.1 are briefly described in the following sections.

13.3.2.1. *Aggressive space mapping (ASM)*

The original ASM describes the behavior of a system by two models in two spaces: the optimization space (OS), denoted by $\mathbf{X}_{os}$, and the validation space (VS), denoted by $\mathbf{X}_{em}$. We represent the design parameters in these spaces by vectors $\mathbf{x}_{os}$ and $\mathbf{x}_{em}$, respectively.

The objective of the ASM procedure is to find the optimum point $\mathbf{x}_{em}$ in VS that minimizes the following nonlinear function:

$$\mathbf{f}(\mathbf{x}_{em}) = \mathbf{P}(\mathbf{x}_{em}) - \mathbf{x}_{os}^*, \tag{13.17}$$

where $\mathbf{x}_{os}^*$ is the optimum point in OS, and $\mathbf{P}(\mathbf{x}_{em})$ is the point in OS which satisfies $\mathbf{R}_f(\mathbf{x}_{em}) = \mathbf{R}_c(\mathbf{P}(\mathbf{x}_{em}))$, $\mathbf{R}_f$ and $\mathbf{R}_c$ being the vectors with the responses of the fine and coarse models. The ASM procedure finishes when $\|\mathbf{f}(\mathbf{x}_{em})\|$ is below some threshold η near zero.

At each iteration j, the next VS parameter vector is obtained by a quasi-Newton iteration:

$$\mathbf{x}_{em}^{(j+1)} = \mathbf{x}_{em}^{(j)} + \mathbf{h}^{(j)}, \tag{13.18}$$

where $\mathbf{x}_{em}^{(0)} = \mathbf{x}_{os}^*$ and $\mathbf{h}^{(j)}$ solves the linear system:

$$\mathbf{B}^{(j)}\mathbf{h}^{(j)} = -\mathbf{f}^{(j)}. \tag{13.19}$$

$\mathbf{B}^{(j)}$ is an approximation to the Jacobian matrix and is obtained from $\mathbf{B}^{(j-1)}$ using the Broyden update:

$$\mathbf{B}^{(j)} = \mathbf{B}^{(j-1)} + \frac{\mathbf{f}^{(j)}\mathbf{h}^{(j-1)^T}}{\mathbf{h}^{(j-1)^T}\mathbf{h}^{(j-1)}}, \tag{13.20}$$

where $\mathbf{B}^{(0)}$ is the identity matrix, $\mathbf{f}^{(j)} = \mathbf{P}(\mathbf{x}_{em}^{(j)}) - \mathbf{x}_{os}^*$, and $\mathbf{P}(\mathbf{x}_{em}^{(j)})$ is obtained by solving the following parameter extraction problem:

$$\mathbf{P}(\mathbf{x}_{em}^{(j)}) = \mathbf{x}_{os}^{(j)} = \arg\min_{\mathbf{x}_{os}} \|\mathbf{R}_f(\mathbf{x}_{em}^{(j)}) - \mathbf{R}_c(\mathbf{x}_{os})\|. \tag{13.21}$$

13.3.2.2. *Multimodel ASM (MASM)*

In MASM, the behavior of the system to be designed is described by N models in N spaces: the optimization space, denoted by $\mathbf{X}_{os}$, and $N-1$ validation spaces, denoted by $\mathbf{X}_{em}|_i$, $i \in [1,\ldots,N-1]$. We represent the design parameters in these spaces by vectors $\mathbf{x}_{os}$ and $\mathbf{x}_{em}|_i$, respectively.

For $j > i$, the j-model in the validation space, $\mathbf{X}_{em}|_j$, is more accurate than the i-model, $\mathbf{X}_{em}|_i$, but requires more computational time.

The MASM proceeds in $N-1$ iterations. The ith iteration performs a traditional ASM between the coarse model and the i-model. This allows to gradually map the coarse model to the finest model through $N-1$ intermediate mappings, thus avoiding the possibility of being trapped in local minima when the coarse model and the finest model are severely misaligned. The coarse model, i.e., the fastest model, is always used in the optimization space in every iteration, and so efficiency is guaranteed. And the accuracy of the result is also guaranteed by the use of the finest VS model in the last iteration.

In order to take advantage of the improvement made in each iteration, the initial guess for the approximation of the Jacobian matrix is fixed to its value at the end of the previous iteration.

13.3.3. *Segmentation*

A segmentation strategy was proposed in Guglielmi (1994) and Alos and Guglielmi (1997) for the design of some filter structures, such as H-plane direct-coupled-cavity filters composed of N resonant cavities and $N+1$ coupling windows (Fig. 13.6). This strategy consists of designing at each step i only the parameters related to the ith cavity (dimensions of that cavity, as well as dimensions of the two adjacent coupling windows and of the previous cavity), and using the values obtained in the previous iterations for the first $i-1$ cavities' dimensions. At each step i, only the response of the first i cavities is simulated and compared with an objective response for that particular part of the structure.

This segmentation technique transforms a slow multidimensional design process into several efficient and robust design steps, where a small number of parameters are designed at the same time. Still, there is the risk that the coupling among all cavities (not just among adjacent ones) is not properly designed by means of this classical segmentation approach.

In order to solve this problem, new steps have been added to the original segmentation strategy in Morro *et al.* (2005b). The resulting segmentation strategy designs the filter through the following steps:

- Ordinary step. The parameters related to the ith cavity are designed by simulating the first i cavities and using the values obtained in previous iterations for the rest of the parameters of the first $i-1$ cavities. The error function is computed by comparing the response of the i first cavities with their ideal response.
- Coupling step. Every time that three consecutive cavities are designed, a new optimization process adjusts, at the same time, all the design parameters of the cavities previously designed. This step provides the small changes required in the parameter values due to couplings from near cavities.
- Central cavity step. For symmetric filters, when the central cavity is reached, and the first half of the filter has been designed step by step, a new optimization is performed considering the whole structure of the filter, but only the dimensions of the central cavity are finely adjusted, thus considering the coupling among all the cavities.
- Full structure step. A final step is made in order to refine the design and to take into account all possible interactions among cavities. The whole filter is simulated and all the dimensions are refined at the same time, the starting point being the result of the previous steps.

13.3.4. *Hybrid Optimization*

Both the efficiency and the robustness of the optimization process can be drastically improved by using an adequate combination of optimization algorithms instead of using a single one. However, the previous knowledge of the different optimization algorithms and their properties is necessary before suggesting a particular hybrid optimization strategy. For this reason, first we briefly discuss the different types of optimization algorithms, and then the hybrid strategy that will be applied in the design of microwave devices is described.

13.3.4.1. *Optimization algorithms*

Generally, the optimization algorithms can be classified into two main groups according to their search method: deterministic algorithms and probabilistic algorithms.

The deterministic algorithms search for the optimum point of the objective function heuristically from an initial set of values, i.e., the next candidate point to be verified is determined in a particular way from the information gathered by the algorithm. The main advantage of this type of algorithm is their efficiency. However, when the starting point is not good, they can be trapped in local minima.

The deterministic algorithms are usually grouped into two large families:

- Direct search strategies. In general terms, this algorithm family is based on the calculation of the objective function in multiple test points. In these algorithms, the search direction for new points is generally determined from the previously calculated values. Some examples of these direct search algorithms are random search, one-at-a-time search (Wilde, 1964), simplex methods (Nelder and Mead, 1965), and rotating coordinates (Rosenbrock, 1960).
- Gradient strategies. On the contrary, this algorithm family uses the information provided by the partial derivatives calculation to determine the search direction. These algorithms are appropriate when derivative information can be easily obtained, because calculating derivatives can be computationally costly. Some outstanding examples are steepest-descent, Broyden–Fletcher–Goldfarb–Shanno (BFGS) (Fletcher, 2000; Press *et al.*, 2007), and least squares (Aaron, 1956).

On the other hand, the probabilistic algorithms use random procedures to determine the next candidate point. This type of algorithm is usually very slow but commonly converges to the optimum solution. The probabilistic algorithms are typically grouped into three algorithm families:

- Evolutionary algorithms (EA) (Haupt and Haupt, 2004; Michalewicz, 2011). This is a family whose objective is to guide the stochastic search making use of mechanisms based on biology (mutation, crossing, and natural selection) to iteratively refine a set of possible solutions. Genetic algorithms, evolutionary programming and evolutionary strategies are some examples of these algorithms.
- Simulated annealing (SA) (Kirkpatrick *et al.*, 1983). Simulated annealing is an optimization method that can be applied to arbitrary search problems. This method simulates the annealing process in which a substance is heated above its melting temperature and then gradually cooled to produce a crystalline lattice, which minimizes its energy probability distribution. This crystalline lattice, composed of millions of atoms perfectly aligned, is an example of nature finding an optimal

structure. However, quickly cooling retards the crystal formation, and the substance becomes an amorphous mass with a higher than optimum energy state. The key to crystal formation is carefully controlling the rate of change of temperature.

- Swarm intelligence (SI) (Eberhart *et al.*, 2001; Dorigo *et al.*, 2006). This family of algorithms simulates the behavior of a biological social system, such as an ant colony or a flock of birds. When a swarm is searching for food, its individuals spread through the surrounding area, their movements being totally independent from each other. Each individual has a high degree of freedom or randomization in its movements, which makes it possible to find deposits of food. Therefore, sooner or later, one of them will find edible foodstuff and, being social, will warn the others. This way, the others could also come closer to the food source. Nowadays, much effort is dedicated to applying this type of algorithm to the design of microwave devices. Some of the most used algorithms are particle swarm optimization (PSO) and ant colony optimization (ACO).

13.3.4.2. *Hybrid optimization*

If only one gradient method is used, it may fail to reach the optimum if the starting point is far from it. On the other hand, the use of a robust method such as the simplex method or a genetic algorithm, largely used in circuit design, ensures convergence but at the cost of low efficiency.

The most suitable combination of optimization algorithms depends on the confidence that we have on the proximity of the starting point to the minimum. Therefore, a different hybridization strategy has been adopted for each step of the segmentation procedure (see types of segmentation steps in Section 13.3.3):

- Ordinary step. In this step the parameters related to the ith cavity are designed. Since this is the first time that these parameters are designed, there is a small probability that the starting point is very close to the minimum. Therefore, the optimization starts with a genetic algorithm or direct search with the coordinate rotation algorithm in order to approach the minimum. The search is then continued with the simplex method. At this point we are close to the minimum, and the BFGS method is used.
- Coupling step. Every time that three consecutive cavities are designed, the parameters of the first i cavities are all re-designed at the same time. Here we should be close to the minimum, so only simplex and BFGS are used.

- Central cavity step. When the central cavity is reached, the whole structure is simulated, but only the parameters related to the central cavity are designed. Since it is the first time that we simulate the whole structure, we might be far from the optimum, so three different optimization algorithms are used (direct search with coordinate rotation, simplex, and BFGS).
- Full structure step. In this final step, the whole structure is simulated and the parameters of all the cavities are re-designed together. In this case, only the simplex and BFGS algorithms are used to minimize the error.

The shift from one kind of algorithm to another is controlled by the parameter termination tolerance, $\mathbf{x}_{tol}$, the termination tolerance of the error function, f_{tol}, and the maximum number of function evaluations, nF_{max}, permitted for each method. f_{tol} is higher for the first algorithm, and its value is decreased in each subsequent algorithm.

13.4. Advanced Microwave Filters Design

13.4.1. *H-Plane Direct-Coupled-Cavity Propagating Filters*

The H-plane direct-coupled-cavity propagating filters (Fig. 13.6) are microwave devices widely used in satellite communications systems. Therefore, in the last few years much effort has been made in the development of tools that facilitate the automated design of these waveguide filters.

Next, one of these design tools (Morro *et al.*, 2005b) is described, as an example of a particular implementation of the previously presented design

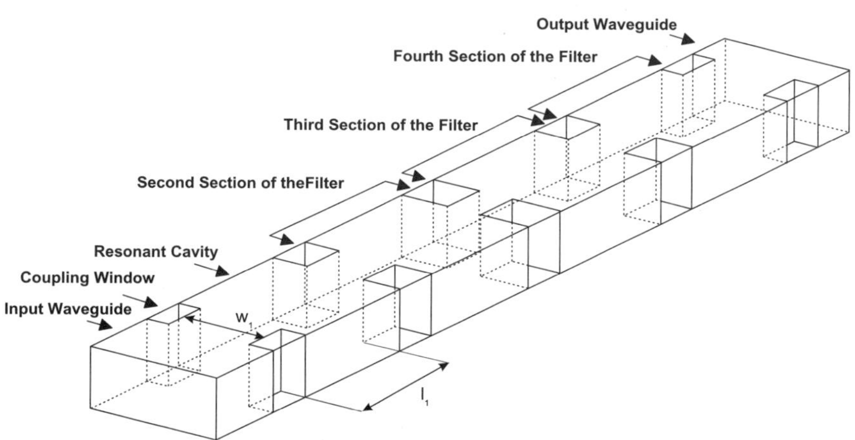

Fig. 13.6.   A four cavities H-plane rectangular waveguide filter.

procedure applied to this filter topology. This tool is robust, efficient, and accurate due to the use of a good starting point, ASM, segmentation, and hybrid optimization. At the end of its description, results obtained by using this tool for the design of a Ka-band filter are discussed.

13.4.1.1. *Initial point synthesis*

The proposed equivalent circuit prototype expressed in terms of inmitance inverters and resonators (Fig. 13.5) is only appropriate for bandpass filters with a narrow or moderate fractional bandwidth (lower than 3%). In such cases it takes into account the dispersive behavior of waveguide components. As a result, the ideal response can be easily recovered with a real structure by substituting specific waveguide elements for the two basic building blocks, i.e., resonators and inmitance inverters.

The ideal resonators can be easily replaced by half-wavelength sections of uniform waveguide. On the other hand, a coupling window terminated with two waveguide sections of appropriate length can be used to model the electrical behavior of an impedance inverter on a moderate frequency range. In fact, under fundamental-mode incidence, the normalized inverter parameter $\overline{K}$ of the proposed real structure can be determined from its reflection coefficient S_{11} as follows:

$$\overline{K}^2 = \frac{1 - |S_{11}|}{1 + |S_{11}|}, \tag{13.22}$$

and the lengths of the input and output terminating waveguides must be adjusted to obtain a π-radians phase for both the S_{11} and S_{22} scattering parameters, since the normalized inverter parameters $(\overline{K})$ required in the considered filters are always smaller than 1. These lengths must be added to the total length of the input and output waveguide resonators connected to the real impedance inverter. This procedure has been fully automated by using the ASM coarse simulator in combination with the Brent's root finding method (Press *et al.*, 2007).

13.4.1.2. *ASM implementation*

The ASM requires two different types of design optimization. First, an optimization must be performed in order to obtain $\mathbf{x}_{os}^*$. Next, another optimization finds the point $\mathbf{x}_{os} = \mathbf{P}(\mathbf{x}_{em})$ that minimizes (13.22) in each iteration.

The objective function is computed comparing the reflection or transmission coefficient (in logarithmic scale) of the corresponding segmented structure provided by the coarse and the fine tool respectively. The use of the logarithmic scale may lead to large errors at the frequencies where the reflection or transmission zeros occur, so the Huber function with $k = 5$ is used and, in order to increase the robustness against large errors, upper and lower thresholds are fixed so that errors exceeding these limits are ignored.

The same modal simulator has been used both as coarse and fine model in ASM. When used as fine model, the number of accessible modes, the number of basis functions in the method of moments (MoM), and the number of kernel terms in the integral equation technique (Gerini and Guglielmi, 1998) are high enough to obtain very accurate results. When used as a coarse model, a small number of modes are considered in order to obtain a faster simulator at the expense of poorer accuracy.

13.4.1.3. *Segmentation*

Table 13.1 summarizes the characteristics of each step in the implemented segmentation strategy (see Section 13.3.3). When we are far from the optimum point (ordinary, coupling, and central steps), the trasmission parameter, S_{21}, is used to caltulate the objective function. Whereas in the full structure step, the reflection parameter, S_{11}, will be used, as we are close to the minimum.

13.4.1.4. *Hybrid optimization*

The design process has been improved by using a suitable combination of optimization algorithms in each step of the segmentation strategy. Robust non-gradient methods (direct search and simplex) are used at the beginning and, after some iterations, an efficient gradient algorithm (Broyden–Fletcher–Goldfarb–Shanno (BFGS)) is used to refine the solution when we are close to the minimum.

Table 13.1. Characteristics of each step of the segmentation strategy.

Step	Structure Simulated	Design Parameters	Error Function	Performed
ordinary	first i cavities	dimensions of cavity i	S_{21}	for each cavity
coupling	first i cavities	dimensions of the first i cavities	S_{21}	each 3 cavities
central cavity	whole	dimensions of the central cavity	S_{21}	once
full structure	whole	dimensions of all the cavities	S_{11}	once

Table 13.2. Termination criteria for the hybrid optimization strategy.

		Direct Search	Simplex	BFGS
Ordinary/Central	nF_{max}	100	400	400
	x_{tol}	$tol/40$	$tol/40$	$tol/40$
	f_{tol}	10^{-4}	10^{-6}	10^{-9}
Coupling	nF_{max}	—	400	400
	x_{tol}	—	$tol/100$	$tol/100$
	f_{tol}	—	10^{-6}	10^{-9}
Full Structure	nF_{max}	—	400	400
	x_{tol}	—	$tol/100$	$tol/100$
	f_{tol}	—	10^{-1}	10^{-2}

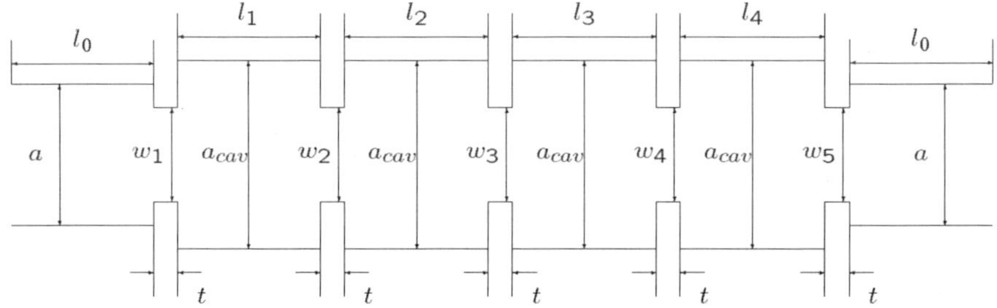

Fig. 13.7.   Four-cavities H-plane filter.

Table 13.2 shows the specific combination of algorithms for each segmentation step and the scheme followed to shift from one algorithm to another.

13.4.1.5. *Results*

This CAD tool has been used to design a four-pole bandpass filter in Ka-band in order to be used as front-end in an LMDS receptor. It must have a Chebyshev response centered at 28 GHz, with 800 MHz of bandwidth, and the following return losses specifications:

$$\begin{cases} L_R \leq 0.044dB, & \text{for } f \leq 27 \text{ GHz and } f \geq 28 \text{ GHz} \\ L_R \geq 20dB, & \text{for } 27.5 \leq f \leq 28.3 \text{ GHz} \end{cases} \quad (13.23)$$

The cavity lengths and the coupling window widths (see Fig. 13.7) have been chosen as design parameters. Input and output waveguides are considered standard WR-28 waveguides ($a = 7.112\,\text{mm}$, $b = 3.556\,\text{mm}$),

Table 13.3. Design parameters (mm) of the H-plane coupled-cavities filter.

a	b	w_1,w_5	w_2,w_4	w_3	l_0,l_5	l_1,l_4	l_2,l_3
7.112	3.556	4.768	3.573	3.361	10	4.868	5.686

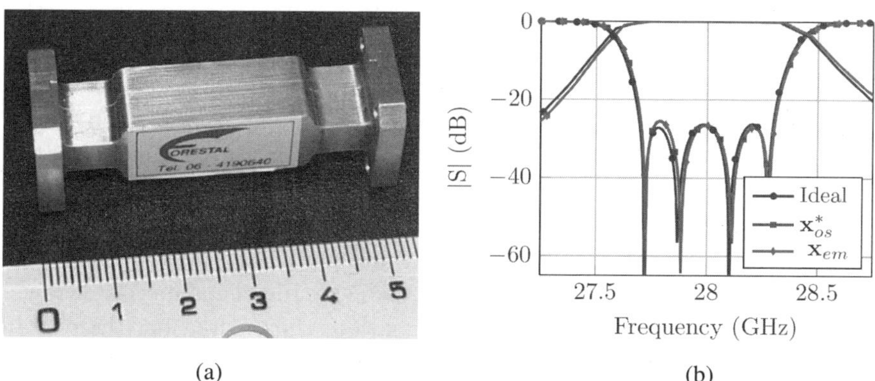

(a)   (b)

Fig. 13.8. (a) Manufactured prototype of the H-plane coupled-cavities filter and (b) the comparison among its ideal response, the coarse model response of $\mathbf{x}^*_{OS}$, and the fine model response of $\mathbf{x}_{em}$.

whereas the cavity widths are equal to $a_{cav} = 8.636\,\text{mm}$ (non-homogeneous structure). All the coupling windows' lengths have been fixed to 2.5 mm.

The design takes just 18 seconds,[1] while the final error in the validation space is $U(\mathbf{x_{em}}) = 0.096$. After two ASM iterations, the final design parameter values ($\mathbf{x_{em}}$) shown in Table 13.3 are obtained.

The designed filter fulfills the design specifications, as shown in Fig. 13.8(b). In fact, it can be observed that the response of the optimum structure in the optimization space, $\mathbf{x^*_{os}}$, is almost identical to the desired ideal response. Moreover, that optimum response found in the optimization space has been successfully recovered in the validation space by the final design solution, $\mathbf{x}_{em}$.

In order to validate the design, a prototype of the LMDS filter was manufactured using a low cost spark erosion technique (with mechanical tolerances of ± 20 microns) without silver plating and alignment pins. Figure 13.8(a) shows an external view of the prototype.

---

[1]Intel®Core ^{TM}2 Quad Q660 @ 2.4 GHz.

13.4.2. *H-Plane Direct-Coupled-Cavity Propagating Filters with Rounded Corners*

Passive waveguide filters are key elements in satellite communications systems. Coupled cavities H-plane filters manufactured with milling techniques are one of the most economic choices with a reasonable compromise of low losses, low volume, and high power handling capability. Since the diameter of the drill cannot be zero, it is impossible to mechanize perfect square corners with milling. When it is used to directly mechanize the cavities in a metallic body, rounded corners appear in the resonant cavities (Fig. 13.3(a)). However, if milling is used to mechanize a mold which is later used to cast the filter, the rounded corners appear in the coupling windows (Fig. 13.3(b)).

When the rounded corners are not considered in the design phase, a deviation appears between the measured and predicted electrical response. This deviation is traditionally corrected using tuning elements (i.e., screws) that must be adjusted by human intervention, thus increasing the overall production costs. So, in this section an automated CAD tool which takes into account the presence of rounded corners is proposed.

13.4.2.1. *Implemented CAD tool*

On the one hand, the design tool described in 13.4.1 is not able to analyze the effect of the presence of rounded corners. On the other hand, the design of filters with rounded corners requires the electromagnetic simulation of complex structures combining straight sections with curved sections, making the direct optimization impossible due to the its high computational cost.

All these things considered, a new robust and efficient CAD tool is presented here. It uses the MASM technique (Section 13.3.2.2), using the simulator utilized in the CAD tool just described in section 13.4.1 as the coarse model, and a tool able to analyze two-port H-plane structures with arbitrary geommetry (Belenguer *et al.*, 2009) as fine models. In this tool, the accuracy and efficiency can be adjusted by the density of cells in which the structure is divided: straight sections per wavelength unit (N_d) and curved stretches per radian (Ω).

In addition, as rounded corners are not considered in the optimization space, the new CAD tool uses the same initial point synthesis techniques, structure segmentation, and hybrid optimization strategy as the ones described in Section 13.4.1.

13.4.2.2. *Results*

This CAD tool has been used to design six-cavities bandpass filters to be manufactured with milling and die-casting techniques (see Fig. 13.12). They must have a Chebyshev response centered at 13.869 GHz, with 290 MHz of bandwidth, and the following return losses specifications:

$$\begin{cases} L_R \leq 3\,dB, & \text{para } f \leq 13.694\,\text{GHz y } f \geq 14.046\,\text{GHz} \\ L_R \geq 25\,dB, & \text{para } 13.725 \leq f \leq 14.015\,\text{GHz} \end{cases} \tag{13.24}$$

The cavity lengths and the coupling window widths (see Fig. 13.7) have been chosen as design parameters. Input and output waveguides, as well as the resonant cavities, are considered standard WR-75 waveguides ($a = 19.05\,\text{mm}$, $b = 9.525\,\text{mm}$). All the coupling windows' lengths have been fixed to 3.844 mm.

Two different fine models have been considered in the MASM algorithm with the following precisions:

- Fine model 1 (F1): $N_d = 50$, $\Omega = 12/\pi$.
- Fine model 2 (F2): $N_d = 100$, $\Omega = 24/\pi$.

Table 13.4 shows the results for the different designed filters, in particular the combination of the different fine models for different radii in the resonant cavities, r_c, or in the coupling windows, r_w. When just the coarse model C and the fine model F2 are used, the MASM turns into the classical ASM.

Figure 13.9 represents the comparison between the response of the optimum point $\mathbf{x}_{OS}^*$ in the coarse model and its response using the fine model F2 and considering different curvature radii. Figures 13.10 and 13.11

Table 13.4. Design results for filters with rounded corners.

Radius(mm)		Models	Fine Model Simulations	Final Error	Computational Cost
$r_w = 0$	$r_c = 3$	C+F2	2	1.692	3′33″
		C + F1 + F2	5 + 2	0.854	12′19″
	$r_c = 4$	C + F2	4	1.760	7′52″
		C + F1 + F2	5 + 3	0.256	16′25″
$r_c = 0$	$r_w = 1$	C + F2	4	0.477	4′24″
		C + F1 + F2	5 + 5	0.213	9′36″
	$r_w = 2$	C + F2	—	Doesn't converge	—
		C + F1 + F2	5 + 3	0.194	8′52″

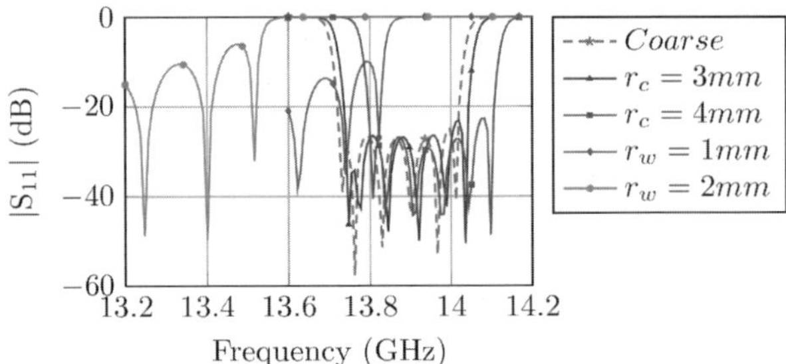

Fig. 13.9. Misalignment between the coarse and fine models for different curvature radii.

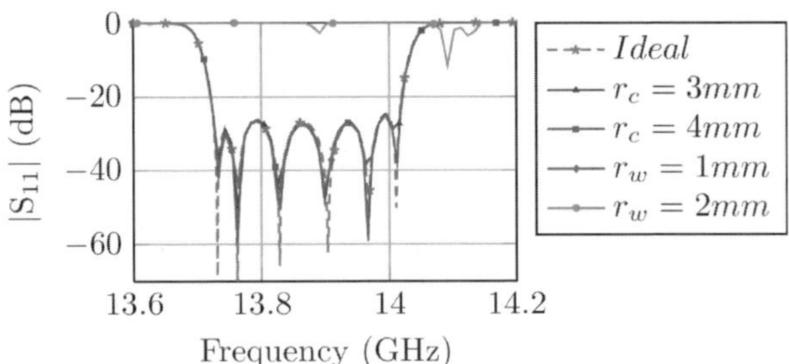

Fig. 13.10. Responses of the filters designed with ASM vs. ideal response.

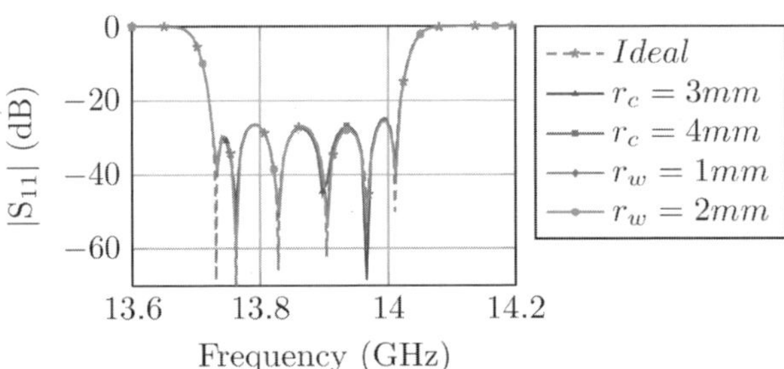

Fig. 13.11. Responses of the filters designed with MASM vs. ideal response.

show the ideal response vs. the responses of the filters designed with ASM and MASM respectively.

Considering Table 13.4 and Figs. 13.9–13.11, we can conclude that the classic ASM is more efficient than MASM when optimization and validation spaces are not very misaligned. However, if they are very misaligned, ASM does not converge to a valid solution, while MASM always converges and its solution is more accurate.

Finally, the manufactured prototypes are shown in Fig. 13.12.

13.4.3. *H-Plane Direct-Coupled-Cavity Propagating Filters with Dielectric Posts*

In this section, a CAD tool developed for the design of H-plane coupled-cavities filters with dielectric posts in their resonant cavities (see Fig. 13.1(b)) is described.

13.4.3.1. *Implemented CAD tool*

The main limitation when designing this new type of structures was the lack of a good synthesis technique able to provide a good starting point. So in Morro *et al.* (2007), the use of evolutionary computation together with hybrid optimization was proposed for this purpose. As evolutionary computation does not provide a deterministic solution, it was hybridized with different deterministic optimization algorithms (simplex, SQP).

After this first optimization stage, we have an initial estimation for the dimensions of the structure under design (see Fig. 13.13). Next, the optimization itself starts. Once again, segmentation and hybrid

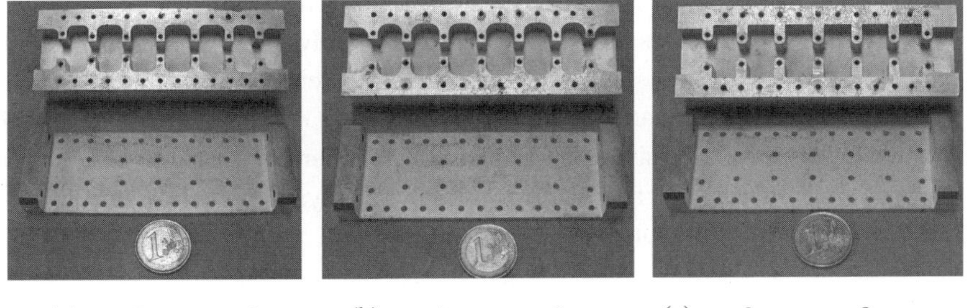

(a) r_c=3 mm,r_w=0 mm   (b) r_c=4 mm,rw=0 mm   (c) r_c=0 mm,r_w=2 mm

Fig. 13.12. Prototypes of H-plane-coupled cavities filters with rounded corners manufactured with milling techniques.

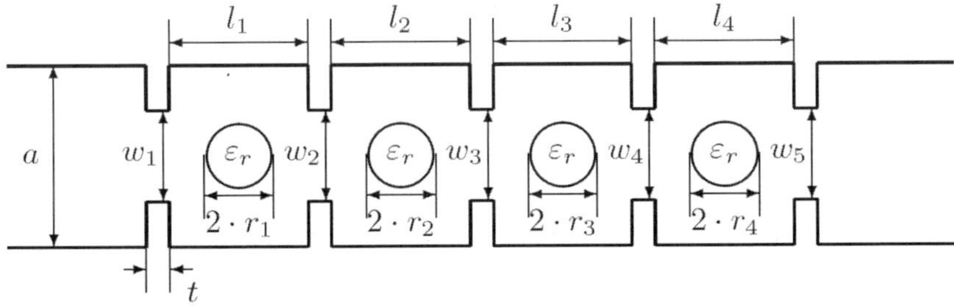

Fig. 13.13. Four-cavities H-plane filter with dielectric resonators.

optimization will be used. However, some special features are required due to the particular characteristics of these structures:

- Reduction of the number of design parameters. The diameter of all the dielectric posts has been fixed to the mean value of the radii obtained for the starting point. It simplifies this phase, makes it more robust and reduces the manufacturing costs.
- Introduction of restrictions in optimization algorithms. This need comes from the lack of a good starting point, which can lead to final values without physical sense (e.g., negative dimensions, cavity lengths smaller than the posts radius, etc). The constrained optimization algorithm SQP (sequential quadratic programming) has been included with this purpose, and also a penalization mechanism has been applied to the simplex algorithm.
- Multilevel optimization. One of the main limitations regarding the analysis tool that will be used to analyze filters with dielectric posts (Bachiller *et al.*, 2005) is its numerical stability when accuracy is needed. When extreme situations, such as coupling windows' widths close to the widths of the connecting waveguides, are generated by the optimization algorithms, the analysis technique requires the inversion of ill-conditioned matrices. In order to solve this limitation, each optimization stage is divided into two different hybrid optimizations: a first one, reducing the accuracy of the analysis module to avoid numerical instabilities; and a second one, increasing the accuracy in the analysis as we are now closer to the optimum solution.

13.4.3.2. *Results*

This CAD tool has been used to design a four-cavities bandpass filter for spatial applications in the X-band. It must have a Chebyshev response centered at 11 GHz, with 300 MHz of bandwidth, and the following return losses specifications:

$$\begin{cases} L_R \leq 3\,dB, & \text{para } f \leq 10.8\,\text{GHz y } f \geq 11.2\,\text{GHz} \\ L_R \geq 25\,dB, & \text{para } 10.85 \leq f \leq 11.15\,\text{GHz} \end{cases} \tag{13.25}$$

The cavity lengths, the dielectric post radii, and the coupling windows' widths (see Fig. 13.13) have been chosen as design parameters, while the cavity widths and the coupling windows' thicknesses are fixed. The relative electric permittivity of the dielectric posts is 19.5. Input and output

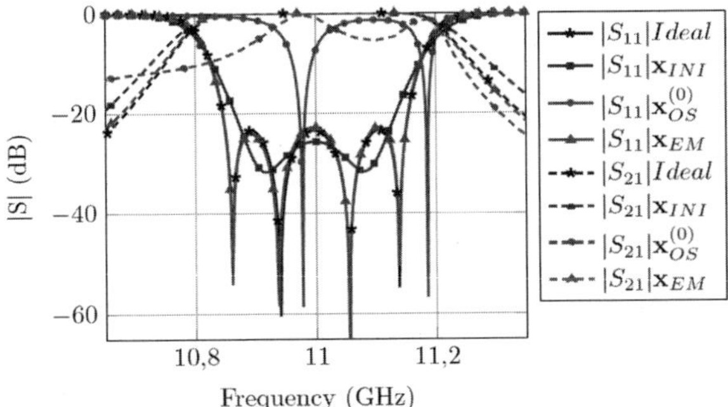

Fig. 13.14. Comparison among the ideal response, the coarse model response of $\mathbf{x}_{INI}$ and $\mathbf{x}_{OS}^{(0)}$, and the fine model response of $\mathbf{x}_{EM}$.

waveguides, as well as the resonant cavities, are considered standard WR-75 waveguides ($a = 19.05\,\text{mm}$, $b = 9.525\,\text{mm}$).

Figure 13.14 shows the comparison between the ideal response and the coarse model response in the starting point $\mathbf{x}_{INI}$, whose obtention process has taken 1 h 1 min. This point obtained in the synthesis phase will be used as starting point ($\mathbf{x}_{OS}^{(0)}$) in the multilevel optimization strategy except that the radius of the dielectric posts has been fixed to the average of the different radius values obtained in the synthesis. In Fig. 13.14 we can also observe that fixing the radius of the post to a single value causes a frequency shift in the response of $\mathbf{x}_{OS}^{(0)}$ and lowers the return losses.

At the end of the optimization strategy, which takes 5 h 31 min 3 s, the response of the designed filter $\mathbf{x}_{EM}$ perfectly matches the ideal response (see Fig. 13.14). Finally, the manufactured prototype is shown in Fig. 13.15.

13.4.4. *SIW Filters*

The development of substrate integrated waveguide technology (SIW) has opened new perspectives for circuits and systems in the microwave and millimeter-wave frequency ranges. Based on a synthesized waveguide in a planar dielectric substrate with two rows of metallic vias (Deslandes and Wu, 2006), this low cost realization of the traditional waveguide circuit inherits the merits from both the microstrip for easy integration and the waveguide for low radiation loss (Bozzi *et al.*, 2008).

Among the wide class of SIW components proposed in the literature, SIW filters have received particular attention, due to the possibility of achieving higher quality-factor and better selectivity, compared to classical

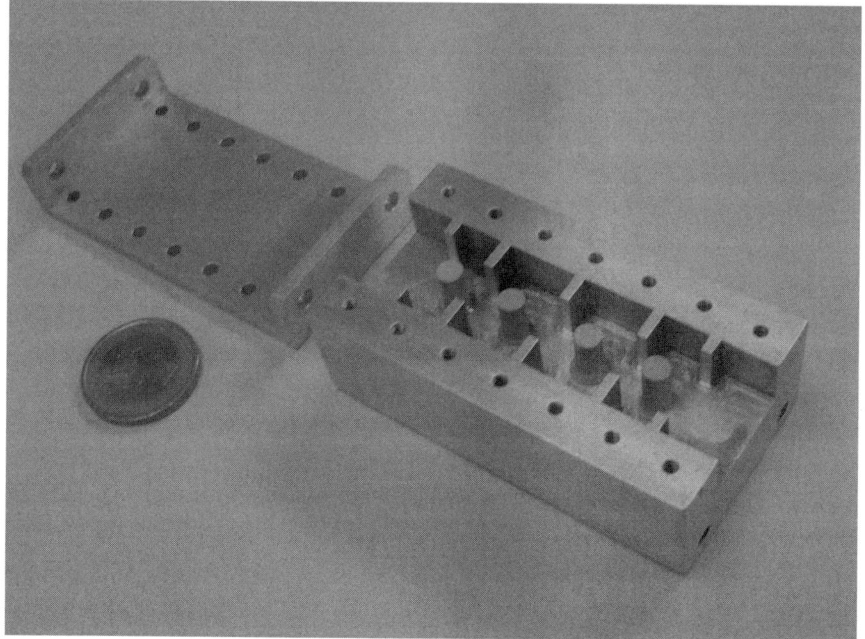

Fig. 13.15. Manufactured H-plane coupled-cavities filter with dielectric posts.

planar filters in microstrip and coplanar-waveguide technology. For these reasons, the efficient analysis and design of SIW devices has become a new challenge that has been the object of intense research in the last few years.

As long as radiation losses in the SIW structure are negligible, the two rows of via holes of the side walls of the SIW (separated a distance equal to w) can be substituted by authentic metallic walls separated a distance a_{eq} in the equivalent rectangular waveguide. Both parameters and others, such as the repetition period of the via holes s or their diameter d, can be determined following some design restrictions as in Diaz *et al.* (2011).

All these parameters must be calculated just once, at the beginning of the design process, being careful that the diameter of the metallic vias corresponds to one of the available commercial drills. Then, just the cavity lengths, L_i, and the post separation to the center, off_i (see Fig. 13.16), have been chosen as design parameters.

All these things considered, there are several approaches to be adopted for the SIW filter design. Two different approaches are mentioned below,

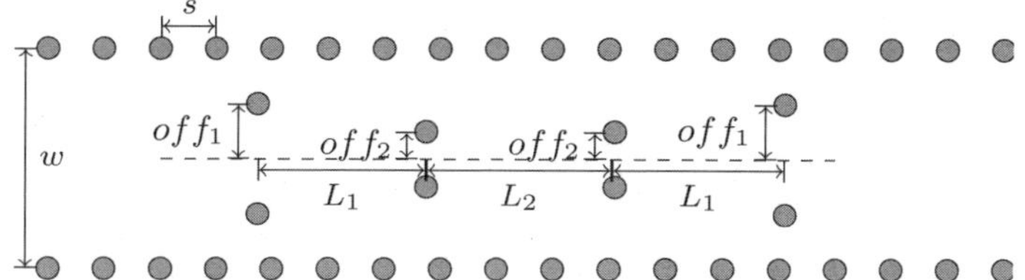

Fig. 13.16. SIW filter layout for the double post topology.

both of them taking advantage of the equivalent rectangular waveguide concept:

- In a first stage of the process, a good equivalent waveguide design can be obtained. Then the filter design should be mapped from waveguide to SIW technology as in Diaz *et al.* (2011). But there is a slight response mismatch after mapping to SIW, the SIW initial design should be optimized with an efficient SIW simulation tool as in Belenguer *et al.* (2011) until the design fulfills the specifications.
- An ASM approach can be implemented, so that the coarse model corresponds to an efficient rectangular waveguide simulator and the fine model corresponds to an accurate SIW simulator.

The following SIW filter design is based on the second approach, as it has proved to be more efficient than the first one.

13.4.4.1. *Implemented CAD tool*

Here an ASM design procedure for SIW filters with double post topology (see Fig. 13.16) is presented. However, this procedure is also valid for other topologies, such as off-centered posts, rectangular coupling windows, etc. Next, the coarse and fine models are briefly explained.

Coarse model in waveguide technology

In this ASM strategy, an accurate and efficient analysis technique (Bachiller *et al.*, 2010) for H-plane devices in rectangular waveguide with metallic or dielectric cylindric posts is going to be used as the coarse model in the optimization space. It solves the matching between cylindrical and guided waves, and uses the concept of transfer function or characterization matrix of a scattering object.

Fine model in SIW technology

A novel simulation tool for the efficient analysis of SIW devices with multiple accessing ports (Diaz Caballero *et al.*, 2012) is used as fine model in the validation space. In this method, fields emerging from each metallic via are expanded as cylindrical open space modes, and guided fields in the accessing ports are expressed as summations of progressive and regressive guided modes.

Initial point for x_{os}^* and parameter extraction

The first step in every ASM strategy is to obtain the optimum design which fulfills the specifications in the coarse model, x_{os}^*. A synthesis methodology based on a prototype with inverters and transmission lines has been applied obtaining an initial point for the coarse model, x_{os}^{ini}. An optimization process with the Nelder–Mead simplex algorithm has been performed in order to obtain the optimum x_{os}^*.

Later on, in order to obtain the $x_{os}^{(j)}$ in each ASM iteration, we have also used the Nelder–Mead simplex algorithm, first over the linear S_{11} response and then over the S_{11} in dB.

13.4.4.2. *Results*

As an example of practical application, we aim to achieve a bandpass filter with double post topology (see Fig. 13.16) for spatial applications in the C-band, whose ideal transfer function is the standard four-pole Chebyshev response, centered at 7 GHz, and with 3% of bandwidth. The return losses in the bandpass must be under 25 dB.

Figure 13.17(b) shows the comparison between the responses in the different steps of the ASM process in their respective fine or coarse spaces. The computational cost for obtaining the optimum design for the coarse model, x_{os}^*, from the x_{os}^{ini}, for 150 frequency points, is 28 min. Then two ASM iterations are necessary to arrive to the final x_{em} (i.e., the final four-pole SIW filter design. See Table 13.5). The total ASM process has taken 34 min 29 s. Finally, Fig. 13.17(a) shows the manufactured prototype.

Table 13.5. Final design parameters (x_{em}).

	L_1	L_2	off_1	off_2	off_3
x_{em}	13.6223	15.2128	6.0498	4.4929	4.2235

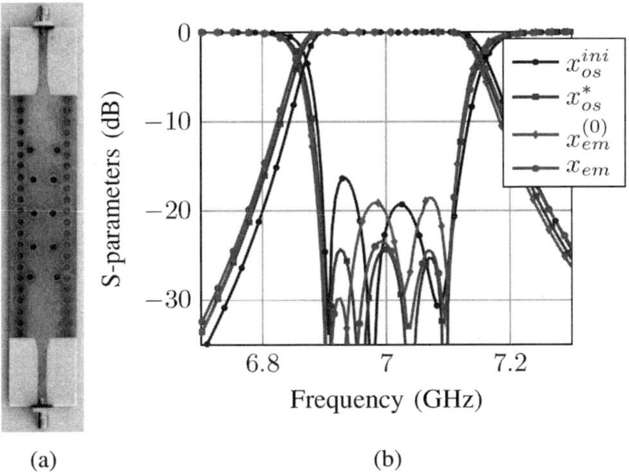

Fig. 13.17. (a) Fabricated SIW filter and (b) response in the different steps of the ASM process.

13.5. Conclusions

Microwave filters are key devices for satellite communications systems as well as for many other telecommunication applications. The increasing number of services rendered by these systems demands very stringent specifications. This need for increasingly more demanding performance of the filters has given rise to new filter topologies, as well as new CAD tools able to efficiently and accurately design all these types of microwave filters.

In this chapter, the complete flowchart of the full-wave electromagnetic driven CAD design process was discussed, and the different techniques involved in this process (initial point synthesis, optimization algorithms, segmentation, space mapping) was studied, and demonstrated through the design of several filters of different topologies (rectangular waveguide H-plane all-metallic filters, metallic filters with rounded corners, filters with dielectric resonators, and filters in substrate integrated waveguide technology).

After all this study, the need for a good starting point in order to ensure the convergence to a proper design that satisfies the specifications can be emphasized. The importance of segmentation in order to improve the convergence and efficiency of the design process by optimizing only a few design parameters at the same time has also been demonstrated. The space mapping philosophy has also proven to be of great help, since it allows the use of accurate simulators but leaves the computation burden to a fast but not so accurate simulator. Finally, the results also prove that the hybridization of several optimization algorithms improves the robustness and efficiency of the overall design process.

References

Aaron, M. (1956). The use of least squares in system design, *IRE T. Circuit Theory*, **3**, 224–231.

Alos, J. and Guglielmi, M. (1997). Simple and effective EM-based optimization procedure for microwave filters, *IEEE T. Microw. Theory*, **45**, 856–858.

Bachiller, C., Esteban, H., Boria, V. E., Morro, J. V., Rogla, L. J., Taroncher, M. and Belenguer, A. (2005). Efficient CAD tool of direct-coupled-cavities filters with dielectric resonators, in *Antennas and Propagation Society International Symposium, 2005 IEEE*, Vol. 1B, pp. 578–581.

Bachiller, C., Esteban, H., Mata, H., Valdes, M., Boria, V., Belenguer, A. and Morro, J. (2010). Hybrid mode matching method for the efficient analysis of metal and dielectric rods in h plane rectangular waveguide devices, *IEEE T. Microw. Theory*, **58**, 3634–3644.

Bakr, M., Bandler, J., Biernacki, R., Shao, H. C. and Madsen, K. (1998). A trust region aggressive space mapping algorithm for EM optimization, *IEEE T. Microw. Theory*, **46**, 2412–2425.

Bakr, M., Bandler, J., Georgieva, N. and Madsen, K. (1999). A hybrid aggressive space-mapping algorithm for EM optimization, *IEEE T. Microw. Theory*, **47**, 2440–2449.

Bakr, M., Bandler, J., Ismail, M., Rayas-Sanchez, J. and Zhang, Q. (2000). Neural space-mapping optimization for EM-based design, *IEEE T. Microw. Theory*, **48**, 2307–2315.

Bandler, J., Biernacki, R., Shao, H. C., Grobelny, P. and Hemmers, R. (1994). Space mapping technique for electromagnetic optimization, *IEEE T. Microw. Theory*, **42**, 2536–2544.

Bandler, J., Biernacki, R., Shao, H. C., Hemmers, R. and Madsen, K. (1995). Electromagnetic optimization exploiting aggressive space mapping, *IEEE T. Microw. Theory*, **43**, 2874–2882.

Bandler, J., Chen, S. H., Biernacki, R., Gao, L., Madsen, K. and Yu, H. (1993). Huber optimization of circuits: a robust approach, *IEEE T. Microw. Theory*, **41**, 2279–2287.

Bandler, J., Cheng, Q., Nikolova, N. and Ismail, M. (2004). Implicit space mapping optimization exploiting preassigned parameters, *IEEE T. Microw. Theory*, **52**, 378–385.

Bandler, J. W., Ismail, M. A., Rayas-Sánchez, J. E. and Zhang, Q. (2003). Neural inverse space mapping (NISM) optimization for EM-based microwave design, *Int. J. RF Microw. C.*, **13**, 136–147.

Belenguer, A., Esteban, H., Boria, V., Morro, J. and Bachiller, C. (2009). Efficient modal analysis of arbitrarily shaped h-plane two-port waveguide devices using the 2D parallel-plate Green's function, *IET Microw. Antenna. P.*, **3**, 62–70.

Belenguer, A., Esteban, H., Diaz, E., Bachiller, C., Cascon, J. and Boria, V. E. (2011). Hybrid technique plus fast frequency sweep for the efficient and accurate analysis of substrate integrated waveguide devices, *IEEE T. Microw. Theory*, **59**, 552–560.

Bozzi, M., Perregrini, L. and Wu, K. (2008). Modeling of conductor, dielectric, and radiation losses in substrate integrated waveguide by the boundary integral-resonant mode expansion method, *IEEE T. Microw. Theory*, **56**, 3153–3161.

Cohn, S. (1957). Direct-coupled-resonator filters, *Proceedings of the IRE*, **45**, 187–196.

Deslandes, D. and Wu, K. (2006). Accurate modeling, wave mechanisms, and design considerations of a substrate integrated waveguide, *IEEE T. Microw. Theory*, **54**, 2516–2526.

Diaz, E., Esteban, H., Belenguer, A., Boria, V., Morro, J. and Cascon, J. (2011). Efficient design of substrate integrated waveguide filters using a hybrid mom/mm analysis method and efficient rectangular waveguide design tools, *2011 Int. Conf. on Electromag. Advanced Applic.*, Torino, Italy (IEEE), pp. 456–459.

Diaz Caballero, E., Esteban, H., Belenguer, A. and Boria, V. (2012). Efficient analysis of substrate integrated waveguide devices using hybrid mode matching between cylindrical and guided modes, *IEEE T. Microw. Theory*, **60**, 232–243.

Dorigo, M., Birattari, M. and Stutzle, T. (2006). Ant colony optimization, *IEEE Comput. Intell. M.*, **1**, 28–39.

Eberhart, R.C., Shi, Y. and Kennedy, J. (2001). *Swarm Intelligence*, (1st ed.), Morgan Kaufmann, Waltham: MA.

Fletcher, R. (2000). *Practical Methods of Optimization*, Wiley, New York: NY.

Gerini, G. and Guglielmi, M. (1998). Efficient integral equation formulations for admittance or impedance representation of planar waveguide junctions, in *IEEE MTT-S International*, **3**, 1747–1750.

Guglielmi, M. (1994). Simple cad procedure for microwave filters and multiplexers, *IEEE T. Microw. Theory*, **42**, 1347–1352.

Haupt, R.L. and Haupt, S.E. (2004). *Practical Genetic Algorithms*, Wiley-Interscience, New York: NY.

Kirkpatrick, S., Gelatt, C.D. and Vecchi, M.P. (1983). Optimization by simulated annealing, *Science*, **220**, 671–680.

Koziel, S. and Bandler, J. (2008). Space mapping with multiple coarse models for optimization of microwave components, *IEEE Microw. Wirel. Co.*, **18**, 1–3.

Kudsia, C., Cameron, R. and Tang, W. (1992). Innovations in microwave filters and multiplexing networks for communications satellite systems, *IEEE T. Microw. Theory*, **40**, 1133–1149.

Levy, R., Snyder, R. and Matthaei, G. (2002). Design of microwave filters, *IEEE T. Microw. Theory*, **50**, 783–793.

Marcuvitz, N. (ed.) (1986). *Waveguide Handbook*, The Institution of Engineering and Technology, London.

Matthaei, G., Jones, E. and Young, L. (1980). *Microwave Filters, Impedance-Matching Networks, and Coupling Structures*, Artech House Publishers, Norwood: MA.

Michalewicz, Z. (2011). *Genetic Algorithms + Data Structures = Evolution Programs*, Springer-Verlag, Berlin.

Morro, J., Esteban, H., Boria, V., Bachiller, C. and Coves, A. (2005a). New multimodel aggressive space mapping technique for the efficient design of complex microwave circuits, *IEEE MTT-S*, **3**, 1613–1616.

Morro, J., Soto, P., Esteban, H., Boria, V., Bachiller, C., Taroncher, M., Cogollos, S. and Gimeno, B. (2005b). Fast automated design of waveguide filters using aggressive space mapping with a new segmentation strategy and a hybrid optimization algorithm, *IEEE T. Microw. Theory*, **53**, 1130–1142.

Morro, J. V., González, H. E., Bachiller, C. and Boria, V. E. (2007). Automated design of complex waveguide filters for space systems: A case study, *Int. J. RF Microw.*, **17**, 84–89.

Nelder, J. and Mead, R. (1965). A simplex method for function minimization, *Comput. J.*, **7**, 308–313.

Press, W.H., Teukolsky, S.A., Vetterling, W.T. and Flannery, B.P. (2007). *Numerical Recipes: The Art of Scientific Computing* (3rd ed.), Cambridge University Press, Cambridge.

Rosenbrock, H.H. (1960). An automatic method for finding the greatest or least value of a function, *Comput. J.*, **3**, 175–184.

Temes, G. and Zai, D. (1969). Least pth approximation, *IEEE T. Circuit Syst.*, **16**, 235–237.

Wilde, D.J. (1964). *Optimum Seeking Methods* (1st ed.), Prentice Hall, Englewood Cliffs: NJ.

Zverev, A.I. (2005). *Handbook of Filter Synthesis*, Wiley-Interscience, New York: NY.

Chapter 14

Time Domain Adjoint Sensitivities: The Transmission Line Modeling (TLM) Case

Mohamed H. Bakr and Osman S. Ahmed

In this chapter, we address the adjoint variable method (AVM) for accelerating gradient-based optimization. AVM efficiently estimates the sensitivities of a given objective function with respect to all parameters using at most one extra adjoint simulation regardless of their number. The AVM approach exploits only field values stored in both the original and adjoint simulations. This approach is more efficient than classical finite difference approaches whose cost scales linearly with the number of parameters. We illustrate the AVM approach using the transmission line modeling (TLM) method, which is widely used in time-domain modeling of high-frequency structures. The memory and computational cost of the TLM-based AVM approaches are addressed. Several suggested techniques for making this approach even more efficient are presented. Examples illustrate different aspects of this approach.

14.1. Adjoint Sensitivities: An Introduction

High-frequency structures are usually designed using computer tools. Optimization algorithms are usually included to determine the optimal dimensions and material properties to satisfy the given design specifications. One class of these algorithms is gradient-based optimization, where the available first-order sensitivity information (gradient) is utilized to guide the optimization iterations starting from an initial design. Traditionally, the required sensitivities are estimated using finite difference approaches which require repeated simulations of the high-frequency structure with perturbed parameter values. The computational cost of these estimates scales linearly

371

with the number of parameters. This can make the optimization approach prohibitive especially for structures with a large number of parameters, as in the case in microwave imaging or for structures with intensive simulation time.

The adjoint variable method (AVM) aims at efficiently estimating the sensitivities of a given response relative to all parameters. It requires at most one extra adjoint simulation. The computational cost of this adjoint simulation is identical to the original simulation. This approach was first introduced in structural engineering and circuit and control theories (Director and Rohrer, 1969; Bandler *et al.*, 1977; Haug *et al.*, 1986; Belegundu and Chandrupatla, 1999). A finite element method (FEM)-based approach was suggested in Garcia and Webb (1990), and Tortorelli and Michaleris (1994). The FEM is particularly suitable for AVM because the components of the resulting system of equations are analytical functions of the different material and geometrical variables of the problem. Recently, this approach has been introduced for other numerical electromagnetic (EM) techniques by Chung *et al.* The work (Chung *et al.*, 2001) triggered significant research in this area. A feasible adjoint sensitivity approach with the method of moments (MoM) approach was proposed in Georgieva *et al.* (2002). Other approaches for sensitivity analysis with the MoM, which is widely used for modeling planar structures, have been also reported (Ureel and De Zutter, 1996; Amarai, 2001; Soliman *et al.*, 2005). The first TLM-based AVM approach was developed in Bakr and Nikolova (2004a) and adapted for dispersive TLM boundaries in Bakr and Nikolova (2004b). These initial TLM-based papers addressed only conducting discontinuities. This approach was later developed to include dielectric discontinuities (Basl *et al.*, 2005). The TLM-based AVM approach was then extended to wideband adjoint sensitivity analysis of the *S*-parameters in Bakr and Nikolova (2005), Bakr *et al.* (2005) and Basl *et al.* (2008). The AVM technique was also applied to different numerical EM techniques (El Sabbagh *et al.*, 2006; Song *et al.*, 2008a; Swillam *et al.*, 2008a; Swillam *et al.*, 2008b; Nikolova *et al.*, 2009). The first-order adjoint sensitivities have been implemented in some commercial solvers (Ansoft HFSS ver. 13.1; CST Studio Suite ver. 2010.06 (2010)). They have been utilized in filter design (Bandler *et al.*, 2005; Khalatpour *et al.*, 2011), antenna design (Zhu *et al.*, 2007; Uchida *et al.*, 2009; Bakr *et al.*, 2011b), microwave imaging (Liu *et al.*, 2010), and design of photonic devices (Swillam *et al.*, 2008a).

All AVM approaches applied for computational EM techniques utilize the same concept. An extra adjoint simulation is performed. The field information in both the original and adjoint simulations are stored at specific subdomains related to each parameter. Using the stored field information, the sensitivities with respect to all parameters are estimated. This means that only one extra adjoint simulation is required regardless of the number of parameters.

For certain objective functions, there is no need for an adjoint simulation. All the required adjoint field information can be deduced from the original simulation. This "self-adjoint" approach applies to the case of network parameters such as the Z-parameters, the Y-parameters, and the S-parameters. The first TLM-based self-adjoint approach is presented in Bakr *et al.* (2005). This approach applied to TLM problems with only one dielectric. It was later extended in Basl *et al.* (2008) for problems with arbitrary dielectric distribution. The pioneering work in Bakr *et al.* (2005) and Basl *et al.* (2008) was also extended to other numerical techniques.

Other approaches can also help accelerate the AVM technique exploit space sampling of the stored field information (Song *et al.*, 2008b), spectral sampling (Song and Nikolova, 2008c), or impulse sampling in the case of TLM (Ahmed *et al.*, 2012). These techniques may apply for some objective functions but not for others. They can sometimes be combined to increase the saving in terms of memory or computational time.

In this chapter, we review the basic concepts of the AVM approach as applied to the transmission line modeling (TLM) approach. We start this section by briefly reviewing the TLM approach in Section 14.2. The TLM-based AVM approach is illustrated in Section 14.3. Different approaches for accelerating the TLM-based AVM approach are illustrated in Section 14.4. Finally, conclusions are given in Section 14.5.

14.2. Transmission Line Modeling (TLM)

The transmission line modeling (TLM) approach can be traced back to the early work of P.B. Johns (Johns, 1987). TLM is based on Huygen's principle which states that every point on a wave front acts as a secondary source (Hoefer, 1985). Using a mesh of connected transmission lines, the propagation of EM waves in arbitrary media can be modeled. The voltages and currents on this network represent the electric and magnetic fields. The parameters of the transmission lines are determined by the properties of the modeled medium. Time is sampled and the field information is available at multiples of the time step Δt.

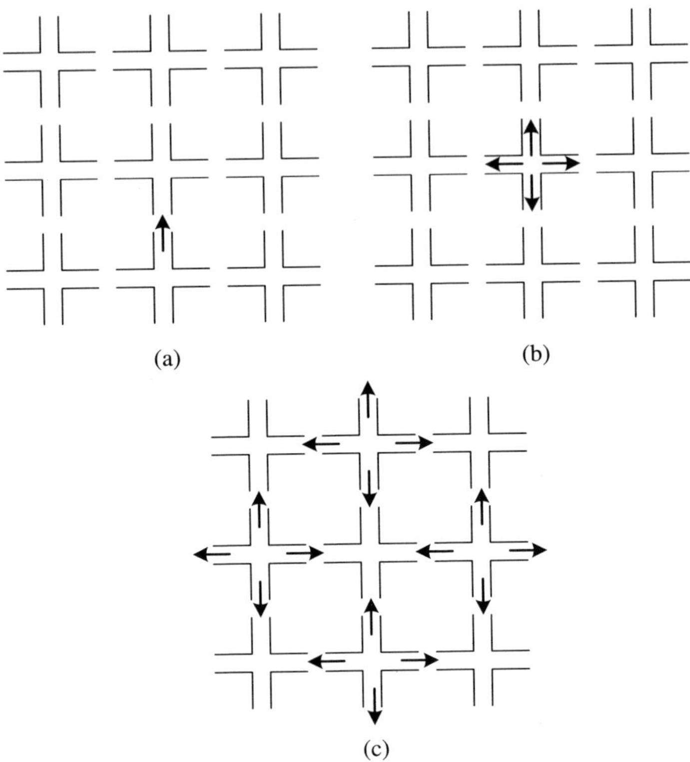

(a) (b)

(c)

Fig. 14.1. Illustration of the scattering and connection steps in a 2D TLM algorithm: (a) an impulse is incident on the jth node at a given time step, (b) the incident impulse is scattered into four reflected impulses, and (c) the reflected impulses propagate to neighboring nodes where they get scattered at the next time step.

Figure 14.1 illustrates the basic steps of the TLM approach. Assuming an incident impulse of 1.0 V on one of the four transmission lines making up the 2D node, this gives rise to scattering voltage impulses on all transmission lines. These scattered impulses then travel over the time interval Δt to neighboring cells where they get scattered themselves and so on. The process is repeated modeling the propagation of EM waves.

Figure 14.2 shows two TLM nodes; the 2D shunt node (Hoefer, 1985) and the symmetric condensed node (SCN) (Jin and Vahldieck, 1994) used for modeling 3D problems. The number of links per node L varies from one node to another. For the 2D shunt node, there are $L = 4$ incident voltage impulses on the node at the kth time step $\boldsymbol{V}_k = [V_1 \ V_2 \ V_3 \ V_4]^T$. The number of incident impulses for the SCN is $L = 12$. Parallel transmission

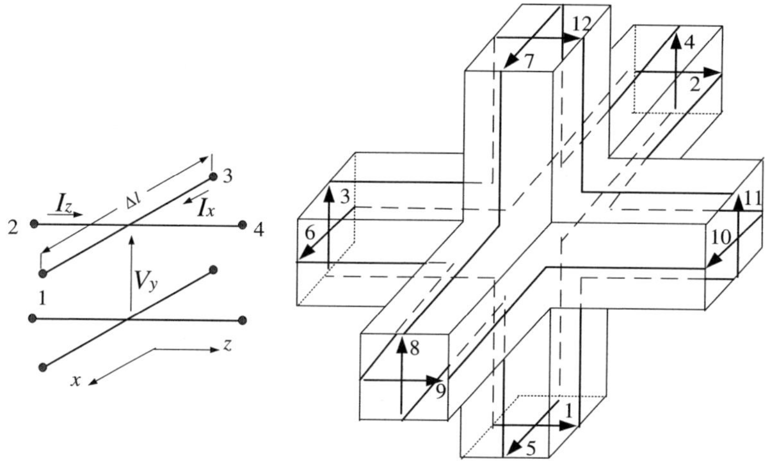

Fig. 14.2.   The 2D shunt node and the 3D symmetric condensed node (SCN).

lines (stubs) may be used in both cases to model extra permittivity or permeability.

We assume that the computational domain is discretized into a total of N nodes with a node size Δl. We denote by N_L the total number of TLM links associated with the N nodes where $N_L = L \times N$. For the jth node, the scattering relation is given by:

$$V_k^{R,j} = S^j V_k^j, \tag{14.1}$$

where V_k^j is the vector of incident impulses on the jth node at the kth time step, $V_k^{R,j}$ is the vector of reflected impulses of the jth node resulting from the scattering of the incident impulses V_k^j, and S^j is the scattering matrix of the jth node. For both 2D and 3D TLM problems with nondispersive boundaries, a single time step iteration for the whole domain is given by (Bakr and Nikolova, 2004a):

$$V_{k+1} = CSV_k + V_k^s, \tag{14.2}$$

where V_k is the vector of incident impulses at all nodes at the kth time step. S is the block diagonal scattering matrix of the whole domain whose jth diagonal matrix is S_j. C is the connection matrix that describes how transmission lines exchange voltage impulses. V_k^s is the vector of source excitation.

The components of the nodal scattering matrix S_j are functions of the characteristic impedances of the different transmission lines of the node.

These characteristic impedances are determined by the local permittivity
and permeability modeled by the node. The symmetric connection matrix
C is not a function of the material properties.

The computational domain is usually terminated by boundaries. All
voltage impulses reflected towards these boundaries give rise to other
incident impulses at the next time step. These boundaries include the
perfectly electrical conductor (PEC) wall, the perfect magnetic conductor
wall (PMC), and resistive walls.

In some cases, it may be of interest to model problems with dispersive
boundaries. For example, the wave impedance of the dominant mode
TE_{10} in a rectangular waveguide changes with frequency. In this case, the
boundaries cannot be modeled by a single reflection coefficient for all time
steps. The response of the boundaries at a certain time step depends on the
history of all impulses incident on that boundary. In this case, a general
formulation of the TLM step is given by (Bakr and Nikolova, 2004b):

$$V_{k+1} = CSV_k + V_k^s + \sum_{k'=0}^{k} G(k - k')V_{k'}^R. \qquad (14.3)$$

The last term in (14.3) represents the contribution of the dispersive
boundary. It represents a time convolution between the impulses incident
on the boundary and the temporal matrix $G(k) \in \Re^{N_L \times N_L}$. This matrix
is the kth time layer of the 3D Johns matrix (Eswarappa *et al.*, 1990).
The Johns matrix (named after B.P. Johns) describes the time-domain
impulse response of the dispersive boundaries. It is generated beforehand
through separate TLM simulations. The summation in (14.3) expresses
the time-domain convolution between the reflected impulses towards the
boundary and its discrete time-domain Green's function (Johns matrix).
An illustration of a TLM domain with both nondispersive and Johns matrix
boundaries is shown in Fig. 14.3.

The TLM technique can be also used to model dispersive materials (De
Mendez and Hoefer, 1994a; De Mendez and Hoefer, 1994b; Paul *et al.*, 1999).
In this case, the material properties are functions of frequency according
to a certain dispersive model. These models include the Drude, Debye,
Lorentz, and Cole–Cole models. The same formulation (14.2) applies with
a modified scattering matrix. The size of the incident voltages is expanded
to include the extra storage utilized to model the dispersive behavior. The
scattering matrix is also modified to accommodate the dispersion storage
update equation.

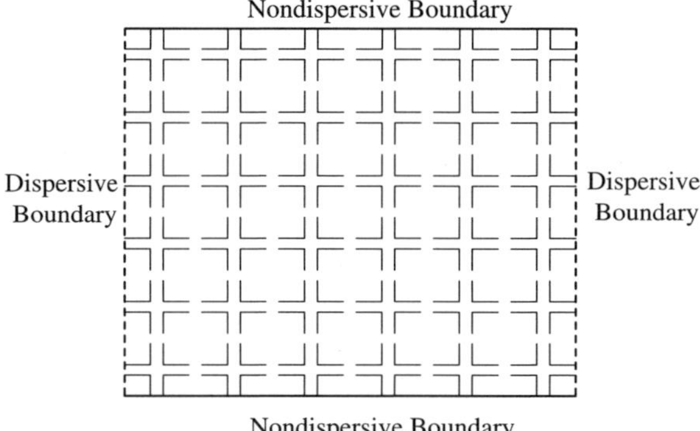

Fig. 14.3. Illustration of the classification of the links of the computational domain: links ending on a Johns matrix boundary are denoted as dispersive boundary links, while links connected to other internal links or to nondispersive boundaries are denoted as nondispersive boundary links.

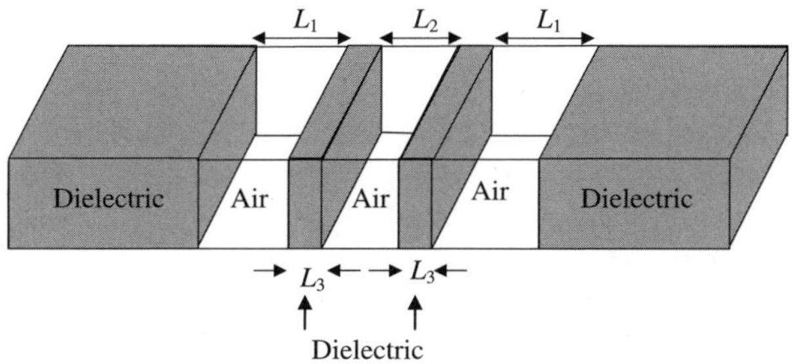

Fig. 14.4. The evanescent mode waveguide filter (Basl *et al.*, 2008).

The TLM modeling approach is illustrated through two examples. The first one consists of alternating dielectric and air sections (see Fig. 14.4, Basl *et al.* (2008)). The waveguide width is 60.0 mm. The dielectric used has $\varepsilon_r = 2.54$ and the cell size is $\Delta l = 1.0$ mm. The structure is symmetrical with respect to its two ports. The lengths of the air and dielectric sections are $[L_1\ L_2\ L_3]^T = [13.0\ 23.0\ 23.0]^T$ mm. Figure 14.5 shows the TLM response as compared to a mode matching (MM) frequency-domain solver. Both techniques agree well.

In the second example, a single subwavelength slit surrounded by chirped dielectric surface grating is modeled (See Fig. 14.6, Ahmed *et al.* (2010)). This structure focuses light at a certain focal point that is determined by the dielectric material and the chirped grating period. It can

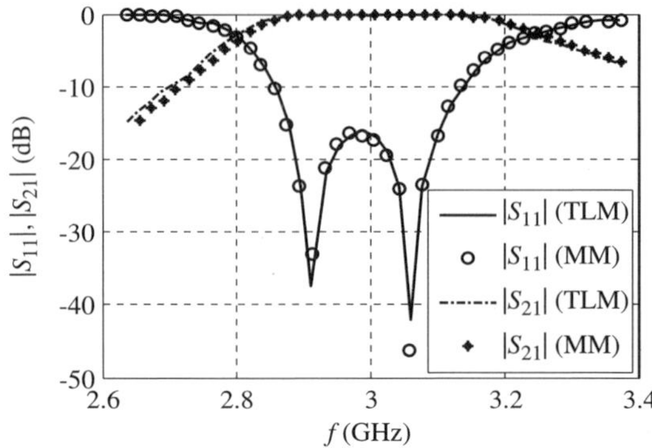

Fig. 14.5.   The response of the evanescent mode waveguide filter (Basl *et al.*, 2008).

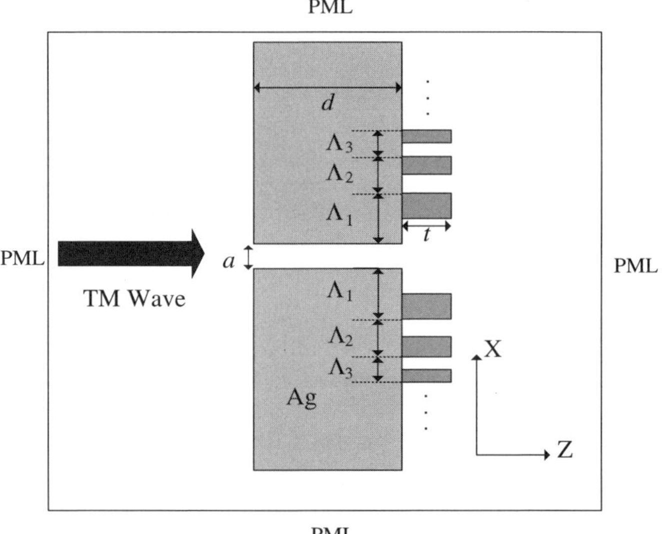

Fig. 14.6.   The schematic of surface plasmon polaritons (SPP) beam focusing using aperture dielectric grating (Ahmed *et al.*, 2010).

be used for focusing the light in nanoscale scanning optical microscopy. It consists of a silver slab with thickness $d = 1.0\,\mu$m and width $a = 100.0$ nm. Here, $t = 120.0$ nm. The structure also has 12 surface grating pitches at each side of the slit with a 0.5 fill factor. The periods of these gratings are given by $\Lambda = [327.0\ 294.0\ 272.0\ 258.0\ 249.0\ 242.0\ 238.0\ 234.0\ 231.0\ 229.0$

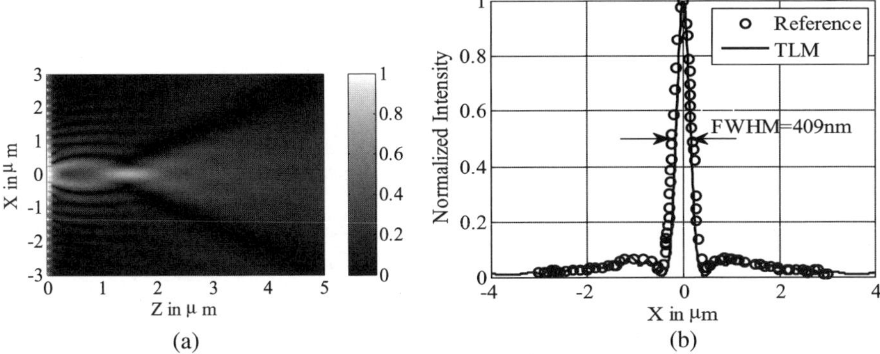

Fig. 14.7.   (a) The magnetic field distribution of the focused beam from subwavelength metallic slit in the presence of an aperture dielectric grating at an incidence wavelength of 532.0 nm. (b) A cross-section of the magnetic field intensity is shown and compared to previous results (Ahmed *et al.*, 2010).

228.0 226.0] nm. A silver dispersion model is utilized (Ahmed *et al.*, 2010). The grating is made of a dielectric material with a refractive index of 1.72.

The structure is simulated with a space step of $\Delta l = 5.0$ nm and a number of time steps $N_T = 8000$ utilizing different types of excitation. The field distribution ($|Hy|$) at an operating wavelength of 532.0 nm is shown in Fig. 14.7. The achieved focal point ($1.5\,\mu$m) and the full width at half maximum (FWHM = 409.0 nm) are identical to those reported in literature.

14.3. The TLM-Based AVM

One of the key problems to solve in computer-aided design (CAD) is to determine the optimal set of parameters of a given structure that satisfy the design specifications. This problem is given by:

$$\boldsymbol{x}^* = arg\ \min_{x}\ U(\boldsymbol{R}(\boldsymbol{V})). \tag{14.4}$$

Here, $\boldsymbol{x}$ is the vector of designable parameters which may include different dimensions or material properties. U is the objective function to be minimized, and $\boldsymbol{R}$ is the vector of responses (e.g. S_{11} at different frequencies). The components of $\boldsymbol{R}$ are obtained through an integration of the electric and magnetic fields of the form:

$$R(\boldsymbol{V}) = \int_0^{T_m} \psi(\boldsymbol{V}) dt. \tag{14.5}$$

T_m is the maximum simulation time where $T_m = N_T \Delta t$ and N_T is the number of time steps. The derivative of $R(\boldsymbol{V})$ with respect to the ith parameter x_i is given by:

$$\frac{\partial R}{\partial x_i} = \int_0^{T_m} \left(\frac{\partial \psi}{\partial \boldsymbol{V}} \right)^T \frac{\partial \boldsymbol{V}}{\partial x_i} dt. \tag{14.6}$$

The adjoint variable method (AVM) aims at efficiently estimating the gradient of a general objective function of the form (14.5). Traditionally, the sensitivities of (14.5) are estimated using finite difference approximations. For example, the accurate central finite difference (CFD) estimate is given by:

$$\frac{\partial R}{\partial x_j} \approx \frac{R(x_j + \Delta x_j) - R(x_j - \Delta x_j)}{2\Delta x_j}. \tag{14.7}$$

Here, two extra simulations are required per parameter to get an accurate sensitivity estimate. For a structure with n designable parameters ($\boldsymbol{x} \in \Re^n$), $2n$ extra simulations are required. The less accurate and less expensive forward finite differences (FFD) or backward finite differences (BFD) approaches require only n extra simulations.

The basic concept of AVM is illustrated in Fig. 14.8. In addition to the original simulation, another adjoint simulation is constructed. Using the responses of the original and adjoint simulations, the sensitivities of the objective function with respect to all parameters are estimated.

The AVM approach estimates the required sensitivities using at most one extra simulation regardless of n. Recently, it was introduced for computational EM. We explain in this section how the AVM is derived for a TLM simulation of the form (14.2) and an objective function of the form (14.5). For a sufficiently small time step Δt, (14.2) can be expressed as (Bakr and Nikolova, 2004):

$$V_{\mathrm{k}} + \left(\frac{\partial V}{\partial t} \right)_{\mathrm{k}} \Delta \mathrm{t} \approx CSV_k + V_{\mathrm{k}}^{\mathrm{s}}. \tag{14.8}$$

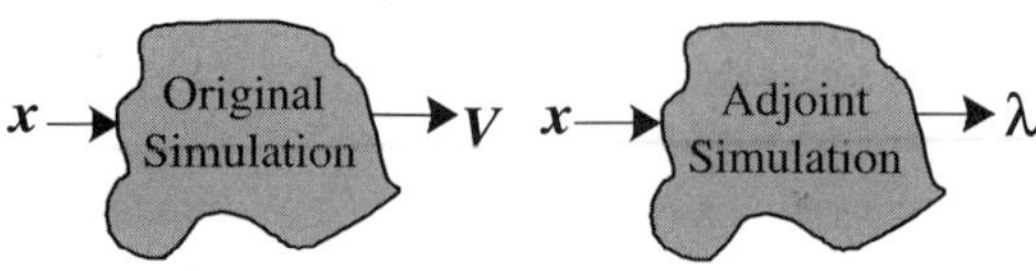

Fig. 14.8.   Illustration of the adjoint variable method (AVM).

Equation (14.8) can be then written in the form

$$\frac{\partial V}{\partial t} = A(x)V + \frac{V^s}{\Delta t},\tag{14.9}$$

where $A(x) = (C(x)S(x) - I)/\Delta t$ and I is the identity matrix. The matrix A is the system matrix and it is a function of the material properties. The subscript k is omitted to denote an arbitrary time t. Notice that the matrix A is never constructed in reality as all operations are carried out on a cell by cell basis.

Assuming a perturbation Δx_i of the ith parameter, this causes a perturbation ΔA_i of the matrix A. It also causes a perturbation of $\Delta V_i(t)$ in the vector V. It follows that for the perturbed system, (14.8) can be written as:

$$\frac{\partial (V + \Delta V_i)}{\partial t} = (A(x) + \Delta A_i)(V + \Delta V_i) + \frac{V^s}{\Delta t}.\tag{14.10}$$

Subtracting (14.9) from (14.10) we get:

$$\frac{\partial \Delta V_i}{\partial t} = \Delta A_i V + A(x)\Delta V_i + \Delta A_i \Delta V_i.\tag{14.11}$$

Dividing both sides of (14.11) by Δx_i we obtain the approximate differential expression:

$$\frac{\partial^2 V}{\partial t \partial x_i} \approx \frac{\Delta A_i}{\Delta x_i}V + A\frac{\partial V}{\partial x_i} + \Delta A_i \frac{\partial V}{\partial x_i}.\tag{14.12}$$

Notice that the excitation is assumed independent of the designable parameters.

The adjoint variable vector λ is defined through the equation:

$$\int_0^{T_m} \lambda^T \left(\frac{\partial^2 V}{\partial t \partial x_i} - \frac{\Delta A_i}{\Delta x_i}V - A\frac{\partial V}{\partial x_i} - \Delta A_i \frac{\partial V}{\partial x_i} \right) dt = 0.\tag{14.13}$$

Integrating (14.13) by parts we get:

$$\lambda^T \frac{\partial V}{\partial x_i} \Big|_0^{T_m} - \int_0^{T_m} \left(\frac{d\lambda^T}{dt} + \lambda^T (A + \Delta A_i) \right) \frac{\partial V}{\partial x_i} dt$$

$$= \int_0^{T_m} \lambda^T \frac{\Delta A_i}{\Delta x_i} V dt.\tag{14.14}$$

Up till this point the adjoint variable λ is arbitrary. Imposing some conditions on it makes it a unique vector. It is selected to have a terminal value of $\lambda(T_m) = \mathbf{0}$. Also, because the vector $\mathbf{V}$ has an initial zero value regardless of the values of the parameter $x_i, i = 1, 2, \ldots, n$, the first term in (14.14) vanishes. Equation (14.14) can thus be written as:

$$\int_0^{T_m} \left(\frac{d\lambda}{dt}^{\mathrm{T}} + \lambda^{\mathrm{T}}(A + \Delta A_{\mathrm{i}}) \right) \frac{\partial \mathbf{V}}{\partial \mathbf{x}_{\mathrm{i}}} dt = - \int_0^{T_m} \lambda^{\mathrm{T}} \frac{\Delta A_{\mathrm{i}}}{\Delta \mathbf{x}_{\mathrm{i}}} \mathbf{V} dt. \quad (14.15)$$

Comparing (14.6) with the left hand side of (14.14), we choose

$$\frac{d\lambda}{dt}^T + \boldsymbol{\lambda}^T (\boldsymbol{A} + \Delta \boldsymbol{A}_i) = \frac{\partial \psi}{\partial \boldsymbol{V}}. \quad (14.16)$$

Using the definition of the matrix $\boldsymbol{A}$, we write (14.16) in discrete time as

$$\lambda_{k-1} = \boldsymbol{S}^\lambda \boldsymbol{C}^\lambda \boldsymbol{\lambda}_k - \boldsymbol{V}_k^{s,\lambda}, \quad \lambda(T_m) = \mathbf{0}, \quad (14.17)$$

where $\boldsymbol{S}^\lambda = \boldsymbol{S}^T(\boldsymbol{x} + \Delta x_i \boldsymbol{e}_i)$ is the scattering matrix of the adjoint system, $\boldsymbol{C}^\lambda = \boldsymbol{C}^T(\boldsymbol{x} + \Delta x_i \boldsymbol{e}_i)$ is the connection matrix of the adjoint system, and $\boldsymbol{V}_k^{s,\lambda} = \Delta t (\partial \psi / \partial \boldsymbol{V})_{t=k\Delta t}$ is the adjoint excitation.

Equation (14.17) represents a TLM simulation that is running backward in time with known excitation. This simulation provides the values of the adjoint variable vector λ at all time steps. Using (14.6), (14.14), and (14.15), the sensitivity of the response R with respect to the ith parameter is given by:

$$\frac{\partial R}{\partial x_i} = - \int_0^{T_m} \boldsymbol{\lambda}^T \frac{\Delta \boldsymbol{A}_i}{\Delta x_i} \boldsymbol{V} dt \approx -\Delta t \sum_k \lambda_k^T \boldsymbol{\eta}_{i,k}. \quad (14.18)$$

The matrix $\Delta \boldsymbol{A}_i$ in (14.18) contains only few nonzero elements. We thus need only to store the impulses for the original and adjoint problems for a small number of mesh links at all time steps.

The main difficulty in applying (14.18) is that the adjoint problem in (14.17) is solved for the perturbed problem, which is parameter-dependent. To overcome this, we assume that the perturbation done in each parameter is small and does not affect in a significant way the distribution of the incident impulses. The adjoint impulses required in (14.18) are approximated by the values of the corresponding incident impulses for the unperturbed adjoint problem:

$$\lambda_{k-1} = \boldsymbol{S}^T(\boldsymbol{x}) \boldsymbol{C}^T(\boldsymbol{x}) \boldsymbol{\lambda}_k - \boldsymbol{V}_k^{s,\lambda}, \quad \lambda(T_m) = \mathbf{0}. \quad (14.19)$$

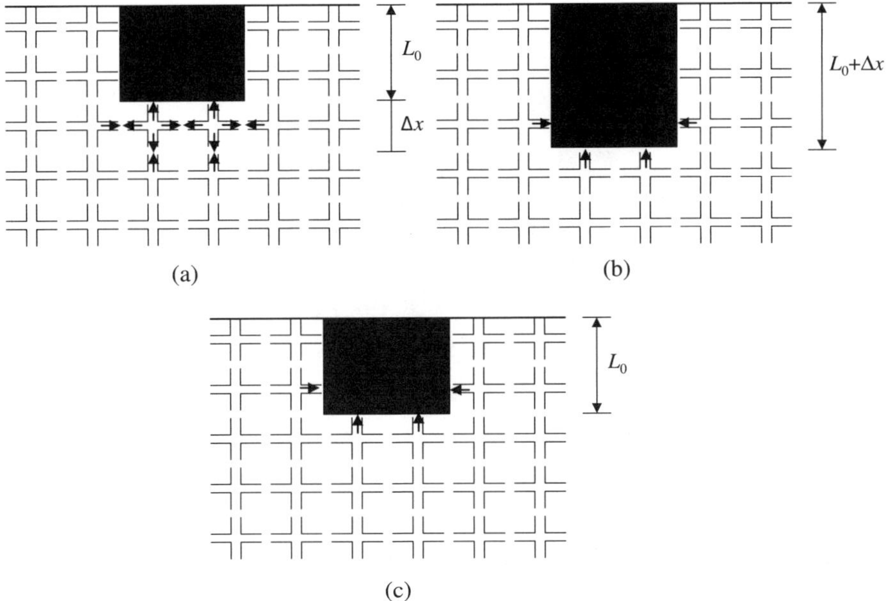

Fig. 14.9. Illustration of the links storage: (a) the arrowed links are the ones for which the matrix ΔA has nonzero components for a perturbation of $1\Delta x$ of the parameter L, (b) the arrowed links are the ones that should be stored during the adjoint analysis of the perturbed circuit, and (c) the adjoint impulses in (b) are approximated by their corresponding ones for the unperturbed circuit (Bakr and Nikolova, 2004a).

This approximation introduces very little error if Δx_i is sufficiently small. This approximation is illustrated for 2D TLM in Fig. 14.9.

It should be noted that for some parameters such as the material permittivity, permeability, and conductivity, the scattering matrix is an analytical function of these parameters. In this case, (14.18) can be cast in the form:

$$\frac{\partial R}{\partial x_i} = -\int_0^{T_m} \lambda^T \frac{\partial A}{\partial x_i} V \, dt. \tag{14.20}$$

Figure 14.10 shows the storage required for a problem with a large dielectric discontinuity. As evident from Figs. 14.9 and 14.10, the storage required for perfectly conducting discontinuities is much less than that required for dielectric discontinuities, as the field is stored only outside the discontinuity.

For problems with dispersive boundaries, it can be shown that the sensitivities of an objective function can be obtained using the following

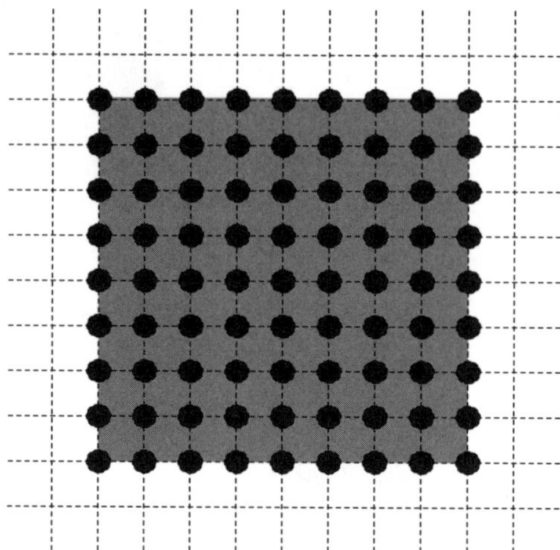

Fig. 14.10.   Stored field for a dielectric discontinuity.

adjoint system (Bakr and Nikolova, 2004b):

$$\lambda_{k-1} = \boldsymbol{S}^T(\boldsymbol{x}) \left[\boldsymbol{C}^T(\boldsymbol{x})\lambda_k + \sum_{k'=k}^{N_t} \boldsymbol{G}^T(k'-k)\lambda_{k'} \right] - \boldsymbol{V}_k^{s,\lambda}, \quad \lambda(\mathrm{T_m}) = \boldsymbol{0}.$$

$$(14.21)$$

The simulation (14.21) is also a backward running simulation with the convolution term reversed. The proof of (14.21) follows similar steps to those for (14.8)–(14.17) (Bakr and Nikolova, 2004b).

The AVM approach is illustrated through two examples. First, we apply the TLM-based AVM approach to estimate the sensitivities of the single-resonator filter shown in Fig. 14.11. The length and width of the waveguide are 7.8 cm and 6.0 cm, respectively. The utilized cell size has $\Delta l = 1.0$ mm. The designable parameters are $\boldsymbol{x} = [d \ W]^T$, where d is the distance between the waveguide discontinuities and W is their length. The AVM approach is tested using an energy function of the form (Bakr and Nikolova, 2004):

$$R = \Delta t \sum_{k=1}^{N_t} \sum_{i=1}^{Nx} \mathbf{V}_{3,k}^2(i, \mathbf{N_z}), \qquad (14.22)$$

where $V_{3,k}$ is the value of the incident field of the third node link at the kth time step. The waveguide is excited with a narrow Gaussian modulated

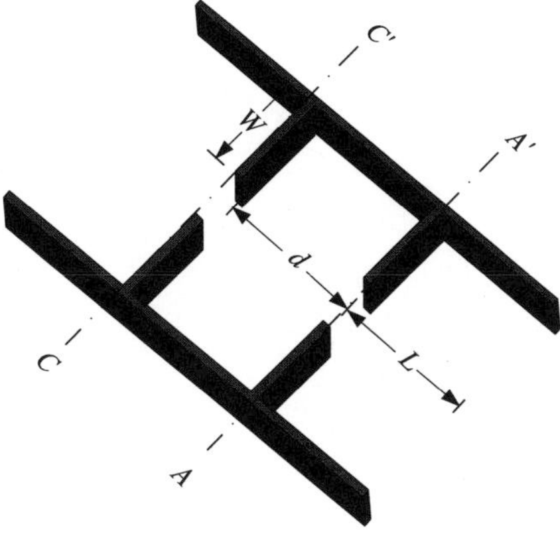

Fig. 14.11.   A single-resonator filter (Bakr and Nikolova, 2004a).

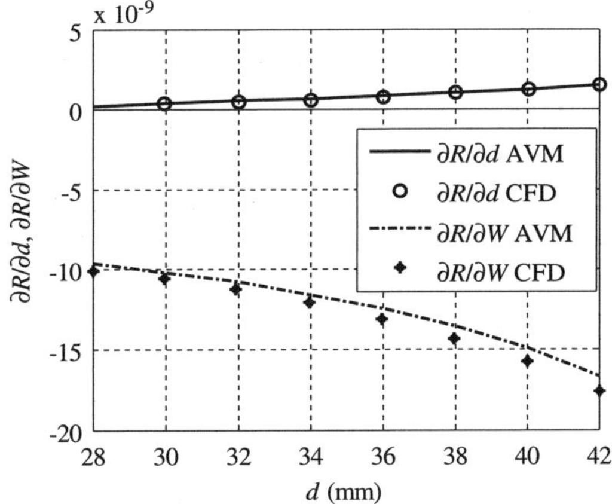

Fig. 14.12.   Objective sensitivities for the single-resonator filter example at $W = 14\Delta l$ with $\Delta l = 1.0\,\text{mm}$ for different values of d (Bakr and Nikolova, 2004a).

sinusoidal signal with a center frequency $\omega_o = 3.0\,\text{GHz}$. Figure 14.12 shows the adjoint sensitivities for a sweep of one of the parameters. Good match is achieved with central finite differences (CFD).

The TLM-based AVM approach is also applied to the six-section H-plane filter shown in Fig. 14.13. The waveguide length and width are $301\Delta l$ and $56\Delta l$, respectively, with $\Delta l = 0.6223\,\text{mm}$. The vector designable parameters is $\boldsymbol{x} = [L_1\ L_2\ L_3\ W_1\ W_2\ W_3\ W_4]^T$. Symmetry is employed and only half the structure is simulated. The objective function is the energy

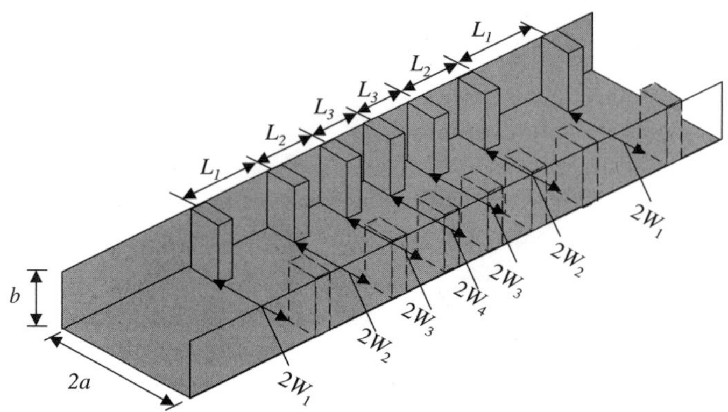

Fig. 14.13.   The six-section H-plane filter (Bakr and Nikolova, 2004b).

function given by (14.22). The waveguide is excited with the dominant mode field profile with a Gaussian-modulated sinusoidal centered at frequency $\omega_o = 5.5\,\text{GHz}$. The number of TLM time steps is $N_T = 3,000$. Here, a modal John's matrix was utilized (Bakr and Nikolova, 2004b).

The sensitivities of the objective function are estimated for a sweep of the parameter L_1. The other parameters are fixed at $[L_2\ L_3\ W_1\ W_2\ W_3\ W_4]^T = [26\Delta l\ 27\Delta l\ 21\Delta l\ 19\Delta l\ 18\Delta l\ 18\Delta l]^T$. The results are shown in Fig. 14.14 for the parameter L_2. Good match is obtained between the AVM approach and the more expensive central differences. Similar accurate results are obtained for all other parameters (Bakr and Nikolova, 2004b). The central difference approximation requires 14 extra TLM simulations, while the TLM-based AVM approach needs only one extra simulation.

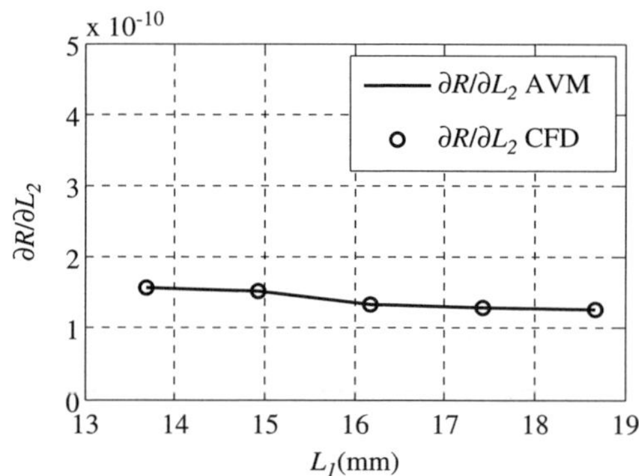

Fig. 14.14.   The sensitivities of the objective function relative to the spacing L_2 for a sweep of the parameter L_1 (Bakr and Nikolova, 2004b).

14.4. Memory Efficient Approaches

The basic steps of the TLM-based AVM discussed in Section 14.3 can be summarized by the following steps (Bakr and Nikolova, 2004a):

- Parameterization: determine the sets of link indices L_i whose connection and scattering matrices are affected by the perturbations $\Delta x_i, i = 1, 2, \ldots, n$.
- Original Analysis: carry out the original TLM analysis (14.2) or (14.3), store original impulses $\boldsymbol{\eta}_{i,k}$, and determine adjoint excitation $\boldsymbol{V}_k^{s,\lambda}$ at every time step.
- Adjoint Analysis: carry out the adjoint analysis (14.17) or (14.21) and store the adjoint impulses $\boldsymbol{\lambda}_k$.
- Sensitivities Estimation: evaluate (14.18) for $i = 1, 2, \ldots, n$.

A number of points to note regarding the original adjoint simulation approach. First, the storage for each parameter is determined by how many cells are affected by the change in that parameter. This storage represents the number of nonzero elements in the vectors $\boldsymbol{\eta}_{i,k}$ and $\boldsymbol{\lambda}_k$. If the parameter represents a material property for an electrically large dielectric, the storage per time step may be extensive. Second, the cost of carrying out the adjoint simulation is identical to that of the original simulation. Third, for each perturbed cell, we need to store all the impulses of that cell at all time steps. All these computational and memory aspects of the TLM-based AVM problem are illustrated in Fig. 14.15. Several approaches have been proposed to reduce the memory and computational cost of the TLM-based AVM approach. They are discussed in the next subsection.

14.4.1. *The S-Parameters Self-Adjoint AVM*

In the pioneering work of Garcia and Webb (1990), it was shown that for narrowband FEM simulations and for certain objective functions, there is no need to carry out an adjoint simulation. The adjoint field data needed for adjoint sensitivity calculations can be deduced from the original simulation. This work was extended in Bakr and Nikolova (2005), Bakr *et al.* (2005) and Basl *et al.* (2008) to the case of wideband TLM-based AVM. It was shown that wideband S-parameter sensitivities are efficiently calculated without carrying out any adjoint simulations. The N_p original simulations used to calculate the S-parameters of an N_p-port electromagnetic structure supply their wideband sensitivities as well. Through mathematical proofs that exploit the modal field profiles (Bakr *et al.*, 2005), it was shown that

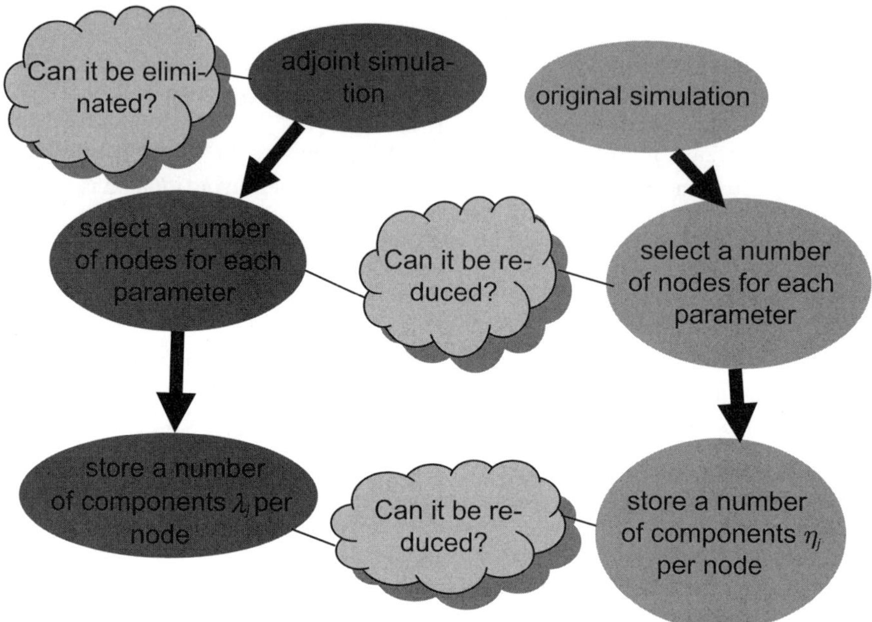

Fig. 14.15.   Computational and memory aspects of the TLM-based AVM.

the adjoint and original fields for a self-adjoint problem are related by (Bakr
et al., 2005; Basl *et al.*, 2008):

$$\lambda_{N_t-k} = Y \eta_{i,k}, \tag{14.23}$$

where Y is a constant matrix. This implies that the original and adjoint
fields required for sensitivity estimation are both available using only
the original simulation. There is no need for an adjoint simulation. This
approach requires that the excitation domain of the original problem and
its observation domain, where the objective is calculated, are the same. It
applies only to linear objective function where the adjoint source has the
same spatial profile as the original excitation. Both of these conditions
are satisfied for network parameters such as the Z-parameters, the Y-
parameters, and the S-parameters. This approach was later extended to
other numerical techniques (El Sabbagh *et al.*, 2006; Song *et al.*, 2008a;
Swillam *et al.*, 2008a; Swillam *et al.*, 2008b; Nikolova *et al.*, 2009).

The self-adjoint AVM (SAVM) approach is illustrated through the
patch antenna shown in Fig. 14.16. The substrate is made of foam with
relative permittivity $\varepsilon_r = 1.07$ (Basl *et al.*, 2008). The dielectric under

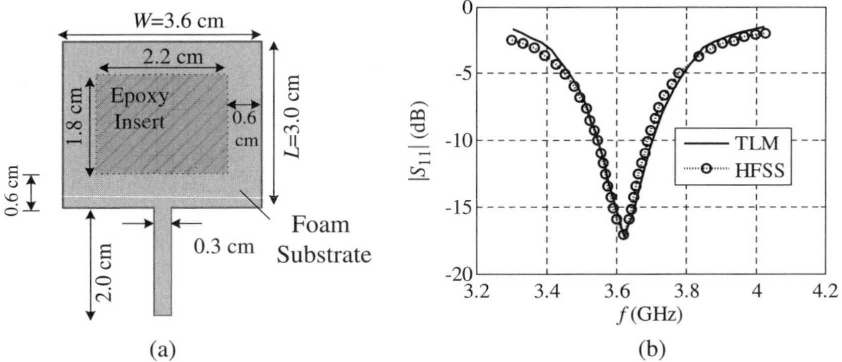

Fig. 14.16. (a) The patch antenna layout, and (b) the reflection co-efficient as compared with the commercial solver HFSS (Basl *et al.*, 2008).

the metal patch has a higher permittivity of $\varepsilon_r = 4.3$. The substrate's height is 1.5 mm. The rest of the dimensions are shown in Fig. 14.16. The computational domain has $\Delta l = 0.5$ mm. Zero reflection boundaries are used to simulate the open structure. The vector of designable parameters is $x = [W\ L]^T$. The in-house TLM simulator is first compared against the commercial FEM solver HFSS (Ansoft HFSS ver. 13.1) for accuracy of response as shown in Fig. 14.16. The sensitivities of the S-parameter ($|S_{11}|$) are calculated over the frequency band 3.3 to 4.0 GHz. The results are shown in Fig. 14.17. Good match is obtained between the self-adjoint AVM approach and the CFD sensitivities. For central difference sensitivities, four extra simulations are needed. For SAVM, no adjoint simulations are needed.

14.4.2. *Coarse Sampling*

As mentioned earlier, the storage required for the AVM approach depends on the number of cells affected by the change in each parameter. When the parameter affects the material properties for a large subset of the computational domain, a huge storage is required at each time step. This case was illustrated in Fig. 14.12, where the sensitivities of the response with respect to the dielectric constant of a region require storing the field information in both the original and adjoint simulations for the whole region. Such an extensive storage is not acceptable in many applications.

This storage can, however, be reduced using the following coarse sampling approach. The basic integral used for estimating the sensitivities

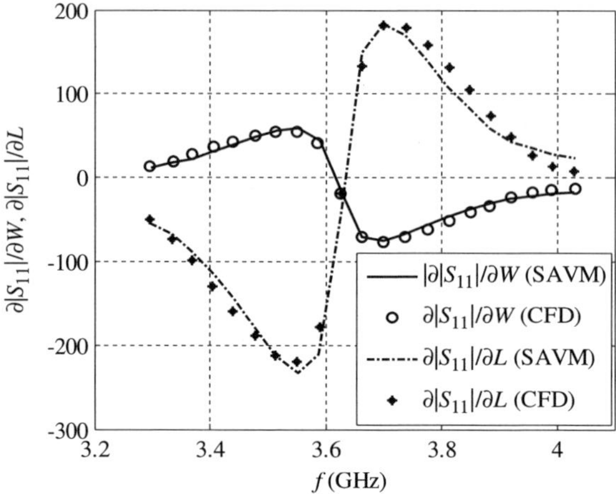

Fig. 14.17. Return loss sensitivities for the patch antenna (Basl *et al.*, 2008).

is given by:

$$\frac{\partial F}{\partial x_i} = - \int_\Omega \int_0^{T_m} \boldsymbol{\lambda}_i^T \boldsymbol{\eta}_i \, dt \, d\Omega. \tag{14.24}$$

This integral requires carrying out an inner product between the vector of original and adjoint fields over all affected cells over all time steps. This integral can be discretized in the following form:

$$\frac{\partial F}{\partial x_i} = \sum_k \sum_j \boldsymbol{\lambda}_{i,j}^T(k\Delta t)\boldsymbol{\eta}_{i,j}(k\Delta t)\Delta t \Delta\Omega. \tag{14.25}$$

The AVM approach utilizes $\Delta\Omega = \Delta l$ which implies sampling the field at each node. However, for sensitivity estimation, it was shown that such a fine discretization may not be required. The approach presented in Song *et al.* (2008) suggests making $\Delta\Omega$ a multiple of the simulation step size, i.e., $\Delta\Omega = m\Delta l, m = 1, 2, \ldots$. This implies sampling the field at a multiple of the cell size. The required storage is reduced by a factor of m in each dimension. This approach is illustrated in Fig. 14.18 for $m = 2, 4$, and 8.

The coarse spatial sampling approach is illustrated through the example shown in Fig. 14.19. The structure (shown in the inset) represents a dielectric slab with $\varepsilon_r = 20$ and $\sigma = 0.751\,\mathrm{S/m}$ placed inside a waveguide filled with air. The dominant mode is excited for this waveguide. The sensitivities of the S-parameters are estimated in Fig. 14.19 with respect to the discontinuity conductivity for different values of m. It is obvious

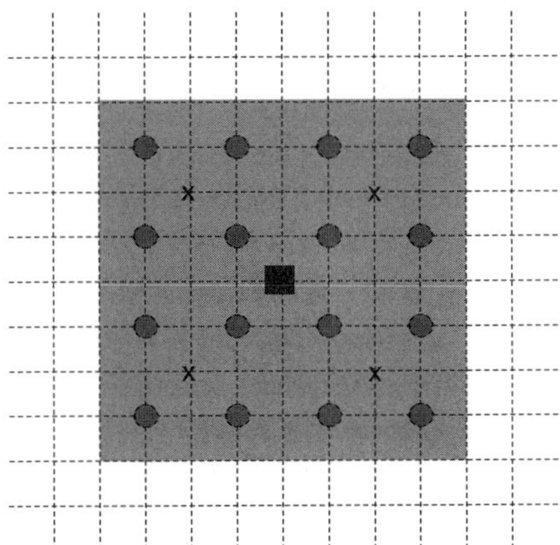

Fig. 14.18. Sampling the field at multiples of the step size; circles show sampling at every other cell, the crosses show sampling at every four cells, while the square shows sampling at every eight cells.

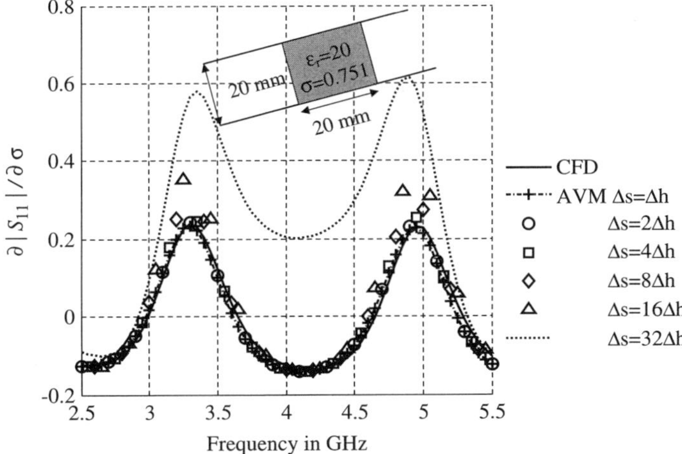

Fig. 14.19. Sensitivity analysis of the S-parameter ($|S_{11}|$) with respect to the conductivity of the discontinuity. The sensitivities are estimated for different sampling rates $m = 2, 4, 8, 16, 32$.

that the estimated sensitivities are relatively accurate for all $m < 32$. The accuracy sees a sudden drop for coarser sampling. It was concluded in Song *et al.* (2008a) that for a reasonable accuracy, $\Delta\Omega$ should not be larger than one quarter of a wavelength at the highest frequency.

14.4.3. *Spectral Sampling*

In some applications, the objective function is a linear transformation of the original fields. For example, the frequency-domain transform of a field quantity is given by:

$$\tilde{E}(f_0) = \int_0^{T_m} E(t)\exp(-2\pi f_0\sqrt{-1}t)dt. \tag{14.26}$$

In this case, the corresponding adjoint excitation is a complex sinusoidal that is given by (Bakr and Nikolova, 2005):

$$\boldsymbol{V}_s^\lambda = \frac{\partial E(t)}{\partial V}\exp(-2\pi\sqrt{-1}f_0t). \tag{14.27}$$

Using (14.24), the adjoint sensitivities of the spectral function (14.26) are estimated by carrying out the integral:

$$\frac{\partial\tilde{E}(f_0)}{\partial x_i} = -\int_\Omega\int_0^{T_m}\tilde{\boldsymbol{\lambda}}^T(f_0)\boldsymbol{\eta}dtd\boldsymbol{\Omega}, \tag{14.28}$$

where $\tilde{\lambda}$ is the adjoint response due to the sinusoidal excitation (14.27). But as this excitation is sinusoidal, $\tilde{\lambda}$ is also sinusoidal and (14.28) can be expressed as:

$$\frac{\partial\tilde{E}(f_0)}{\partial x_i} = -\int_\Omega\int_0^{T_m}\hat{\boldsymbol{\lambda}}^T(f_0)\exp(-2\pi f_0\sqrt{-1}t)\boldsymbol{\eta}dtd\boldsymbol{\Omega}, \tag{14.29}$$

where $\hat{\lambda}$ is the corresponding phasor. Rearranging the integrals in (14.29), we get:

$$\frac{\partial\tilde{E}(f_0)}{\partial x_i} = -\int_\Omega\hat{\boldsymbol{\lambda}}^T(f_0)\hat{\boldsymbol{\eta}}(f_0)d\boldsymbol{\Omega}. \tag{14.30}$$

Here, $\hat{\boldsymbol{\eta}}$ is the original field phasor. It follows from (14.30) that we do not need to store the time-domain field responses but rather store their phasors at each frequency. The sensitivities are obtained by carrying out an inner product between the phasors for each frequency. This significantly reduces the storage per cell. This approach is valid only for problems with linear materials and where a frequency-domain quantity is the objective function. The approach (14.26)–(14.30) summarizes the spectral domain approach addressed in Bakr and Nikolova (2005), and Song and Nikolova (2008c).

The spectral approach is illustrated through the example shown in Fig. 14.20. A small discontinuity of properties $\varepsilon_r = 30$ and $\sigma = 0.3\,\text{S/m}$ is

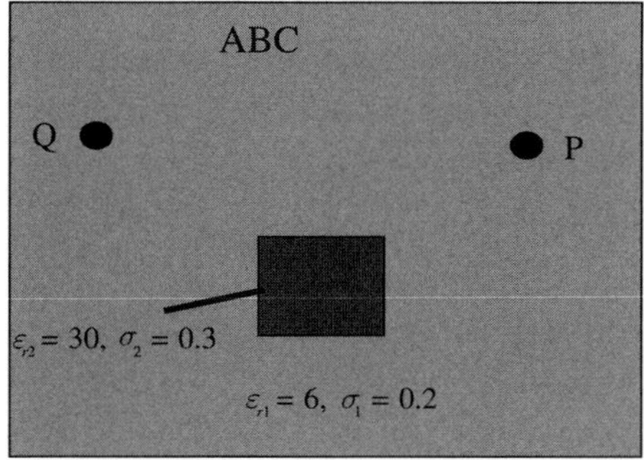

Fig. 14.20. An example for the spectral domain method; the field is excited at point P and the response is observed at points P and Q (Song and Nikolova, 2008c).

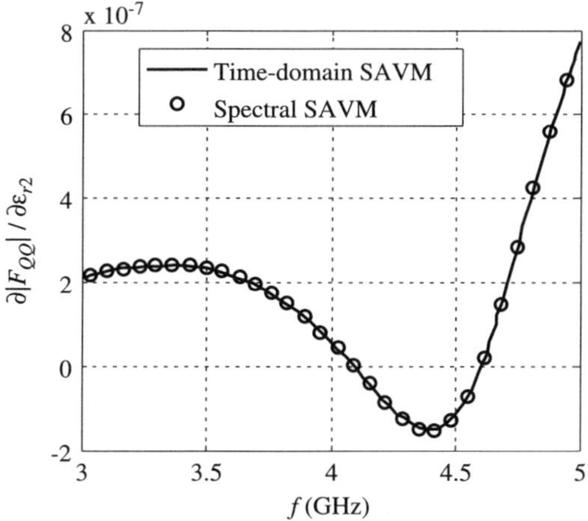

Fig. 14.21. Illustration of spectral self-adjoint sensitivity analysis (Song and Nikolova, 2008c).

inside a host medium with properties $\varepsilon_r = 6$ and $\sigma = 0.2\,\mathrm{S/m}$. A point source is excited at point P and the resultant field is observed at points P and Q (Song and Nikolova, 2008c). The sensitivities of the response with respect to the relative permittivity of the discontinuity are estimated using the time-domain self-adjoint approach (SAVM) and the spectral self-adjoint approach. The comparison between both approaches is shown in Fig. 14.21. A perfect match is achieved for both approaches even though the spectral approach requires only a fraction of the memory storage required using the time-domain approach.

14.4.4. *Impulse Sampling for TLM-Based AVM*

The TLM approach explained in Section 14.2 utilizes a number of incident impulses per cell in the scattering and connection processes. For S perturbed TLM cells, the number of extra storage is $L \times S$ per time step, where L is the number of links per cell. For lossy dielectric materials, the number of links L per cell is at least 15 for 3D TLM utilizing the symmetrical condensed node (SCN). This storage can be formidable especially for problems with large dielectric discontinuities.

A recent approach was proposed to reduce this storage (Ahmed *et al.*, 2012). The nodal scattering matrix $\boldsymbol{S}^j$ at each cell is expanded as the sum of two matrices; a parameter-dependent transmission matrix $(\boldsymbol{T}^j)$ and a constant matrix $(\boldsymbol{P}^j)$ that is independent of the optimization variables. Due to the symmetry of link contributions for each cell, the matrix $\boldsymbol{T}^j$ can be divided into three sub-matrices each with identical rows. Each sub-matrix is associated with a distinct field polarization;

$$\boldsymbol{T}^j = \begin{bmatrix} \boldsymbol{T}^j_x \\ \boldsymbol{T}^j_y \\ \boldsymbol{T}^j_z \end{bmatrix}. \tag{14.31}$$

By calculating the derivative of each sub-matrix with respect to the design parameters we can show that:

$$\frac{\partial \boldsymbol{T}^j}{\partial p_i} \boldsymbol{V}^j_k = \begin{bmatrix} \zeta^i_{j,k,x} \mathbf{1} \\ \zeta^i_{j,k,y} \mathbf{1} \\ \zeta^i_{j,k,z} \mathbf{1} \end{bmatrix}. \tag{14.32}$$

This implies that for 3D problems, only three unknowns should be stored per each cell. A storage reduction of 80% over the original AVM approach is achieved.

The impulse sampling approach is illustrated through the example shown in Fig. 14.22. A dielectric resonator antenna (DRA) with double segment is fed through a microstrip transmission line (Bakr and Ahmed, 2011a). The parameters for this problem are $\varepsilon_{sub} = 3.0$, $\varepsilon_r = 10.0$, $w = 7.875\,\text{mm}$, $d = 3.0\,\text{mm}$, $h = 3.175\,\text{mm}$, $\varepsilon_i = 20.0$, and $t = 0.6\,\text{mm}$. The utilized discretization is $\Delta l = 0.2\,\text{mm}$. The feeding microstrip line has a characteristic impedance of $50\,\Omega$. The sensitivity of the S-parameter (S_{11}) with respect to the parameters $\boldsymbol{x} = [\varepsilon_r \varepsilon_i d]^T$ are shown in Figs. 14.23–14.25. Here, we apply a self-adjoint approach combined with coarse space

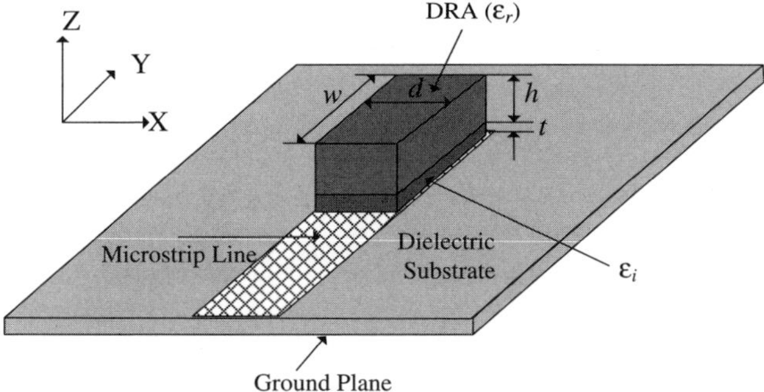

Fig. 14.22. The microstrip-fed dielectric resonator antenna (Bakr and Ahmed, 2011a).

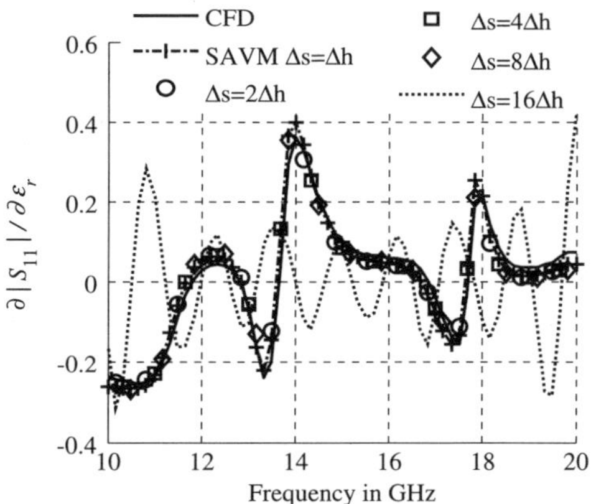

Fig. 14.23. The sensitivities of $|S_{11}|$ with respect to the parameter ε_r for the dielectric resonator example (Bakr and Ahmed, 2011a).

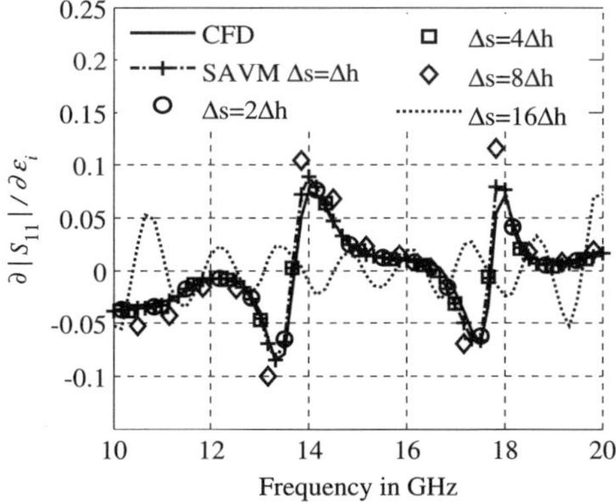

Fig. 14.24. The sensitivities of $|S_{11}|$ with respect to the parameter ε_i for the dielectric resonator example (Bakr and Ahmed, 2011a).

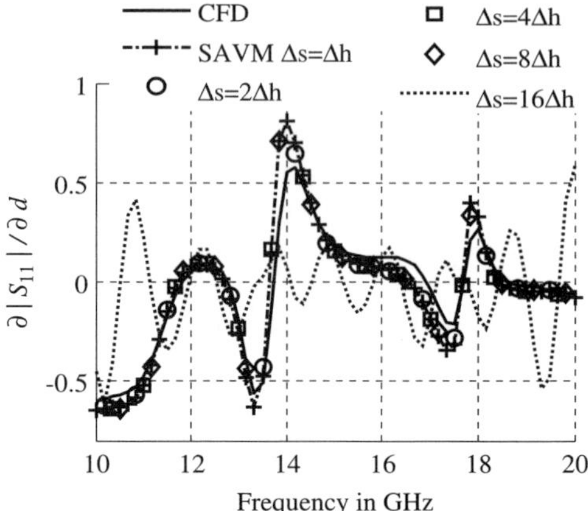

Fig. 14.25. The sensitivities of $|S_{11}|$ with respect to the parameter d for the dielectric resonator example (Bakr and Ahmed, 2011a).

sampling, spectral sampling, and impulse sampling. Good agreement is achieved with the more expensive central finite differences.

14.5. Conclusions

In this chapter we reviewed the theory of TLM-based adjoint variable method (AVM). We showed that the sensitivities of a given objective function with respect to all parameters are estimated using at most one extra adjoint simulation. This simulation has the same computational and memory cost as the original simulation. This approach was applied to different objective functions including the S-parameters. The self-adjoint approach was shown to eliminate the adjoint simulation altogether for certain objective functions including network parameters. The computational cost and memory storage requirements of the technique can be significantly reduced by utilizing efficient implementations including coarse space sampling, spectral sampling, and impulse sampling. The TLM-based AVM approach and its efficient implementations were illustrated through a number of examples.

References

Ahmed, O.S., Bakr, M.H. and Li X (2012). A memory-efficient implementation of TLM-based adjoint sensitivity analysis, *IEEE T. Antenn. Propag.*, **60**, 2122–2125.

Ahmed, O.S., Swillam, M.A., Bakr, M.H. and Li, X. (2010). Modeling and design of nano-plasmonic structures using transmission line modeling, *Opt. Express*, **18**, 21784–21797.

Amarai, S. (2001). Numerical cost of gradient computation with the method of moments and its reduction by means of novel boundary-layer concept, *IEEE IMS*, Phoenix: AZ, pp. 1945–1948.

Ansoft HFSS ver. 13.1, Ansoft Corporation, 225 West Station Square Drive, Suite 200, Pittsburgh: PA 15219, USA.

Bakr, M.H. and Nikolova, N.K. (2004a). An adjoint variable method for time domain TLM with fixed structured grids, *IEEE T. Microw. Theory*, 52, 554–559.

Bakr, M.H. and Nikolova, N.K. (2004b). An adjoint variable method for time domain TLM with wideband Johns matrix boundaries, *IEEE T. Microw. Theory*, 52, 678–685.

Bakr, M.H. and Nikolova, N.K. (2005). Efficient estimation of adjoint-variable S-parameter sensitivities with time domain TLM, *Int. J. Numer. Model. El.*, **18**, 171–187.

Bakr, M.H., Nikolova, N.K. and Basl, P.A.W. (2005). Self-adjoint S-parameter sensitivities for lossless homogeneous TLM problems, *Int. J. Numer. Model. El.*, **18**, 441–455.

Bakr, M.H. and Ahmed, O.S. (2011a). Fast and memory efficient time domain adjoint sensitivities and their applications, *IEEE-MTT-S IMS*, Baltimore: MD.

Bakr, M.H., Ghassemi, M. and Sangary, N. (2011b). Bandwidth enhancement of narrow band antennas exploiting adjoint-based geometry evolution, *Proceedings of the IEEE-APSURSI*, Washington: DC, pp. 2909–2911.

Bandler, J.W., Johns, P.B. and Rizk, M.R.M. (1977). Transmission-line modeling and sensitivity evaluation for lumped network simulation and design in the time domain, *J. Frankl. Inst.*, **304**, 15–32.

Bandler, J.W., Mohamed, A.S. and Bakr, M.H. (2005). TLM-based modeling and design exploiting space mapping, *IEEE T. Microw. Theory*, **53**, 2801–2811.

Basl, P.A.W., Bakr, M.H. and Nikolova, N.K. (2005). Efficient estimation of sensitivities in TLM with dielectric discontinuities, *IEEE Microw. Wirel. Co.*, **15**, 89–91.

Basl, P.A.W., Bakr, M.H. and Nikolova, N.K. (2008). Theory of self-adjoint S-parameter sensitivities for lossless nonhomogeneous transmission-line modeling problems, *IET Microw. Antenn. Propag.*, **2**, 211–220.

Belegundu, A.D. and Chandrupatla, T.R. (1999). *Optimization Concepts and Applications in Engineering*, Prentice-Hall, Upper Saddle River: NJ.

Chung, Y.S., Cheon, C., Park, I.H. and Hahn, S.Y. (2001). Optimal design method for microwave device using time domain method and design sensitivity analysis — part II: FDTD case, *IEEE T. Magn.*, **37**, 3255–3259.

CST Studio Suite ver. 2010.06, Computer Simulation Technology, Bad Nauheimer Str. 19, 64289 Darmstadt, Germany.

De Menezes, L. and Hoefer, W.J.R. (1994a). Modeling frequency dependent dielectrics in TLM, *IEEE AP-S*, Seattle: WA, pp. 1140–1143.

De Menezes, L. and Hoefer, W.J.R. (1994b). Modeling nonlinear dispersive media in 2-D-TLM, *Proc. 24th European Microwave Conf.*, Cannes, France, pp. 1739–1744.

Director, S.W. and Rohrer, R.A. (1969). The generalized adjoint network and network sensitivities, *IEEE T. Circuits Syst.*, **16**, 318–323.

El Sabbagh, M.A., Bakr, M.H. and Nikolova, N.K. (2006). Sensitivity analysis of the scattering parameters of microwave filters using the adjoint network method, *RF and Microwave Computer Aided Engineering*, **16**, 596–606.

Eswarappa, Costache, G.I. and Hoefer, W.J.R. (1990). Transmission line matrix modeling of dispersive wide-band absorbing boundaries with time-domain diakoptics for S-parameter extraction, *IEEE T. Microw. Theory*, **38**, 379–386.

Garcia, P. and Webb, J.P. (1990). Optimization of planar devices by the finite element method, *IEEE T. Microw. Theory*, **38**, 48–53.

Georgieva, N., Glavic, S., Bakr, M.H. and Bandler, J.W. (2002). Feasible adjoint sensitivity technique for em design optimization, *IEEE T. Microw. Theory*, **50**, 2751–2758.

Haug, E.J., Choi, K.K. and Komkov, V. (1986). *Design Sensitivity Analysis of Structural Systems*, Academic Press, Orlando: FL.

Hoefer, W.J.R. (1985). The transmission-line matrix method-theory and applications, *IEEE T. Microw. Theory*, **33**, 882–893.

Jin, H. and Vahldieck, R. (1994). Direct derivations of TLM symmetrical condensed node and hybrid symmetrical condensed node from Maxwell's equations using centered differencing and averaging, *IEEE T. Microw. Theory*, **42**, 2554–2561.

Johns, P.B. (1987). A symmetrical condensed node for the TLM method, *IEEE T. Microw. Theory*, **35**, 370–377.

Khalatpour, A., Amineh, R.K., Bakr, M.H., Nikolova, N.K. and Bandler, J.W. (2011). Accelerating input space mapping optimization with adjoint sensitivities, *IEEE Microw. Wirel. Co.*, **21**, 280–282.

Liu, L., Trehan, A. and Nikolova, N.K. (2010). Near-field detection at microwave frequencies based on self-adjoint response sensitivity analysis, *Inverse Probl.*, **26**, 105001–105029.

Nikolova, N.K., Zhu, X., Song, Y., Hasib, A. and Bakr, M.H. (2009). S-parameter sensitivities for electromagnetic optimization based on volume field solutions, *IEEE T. Microw. Theory*, **57**, 1526–1538.

Paul, J., Christopoulos, C. and Thomas, D.W.P. (1999). Generalized material models in TLM — part I: materials with frequency-dependent properties, *IEEE T. Antenn. Propag.*, **47**, 1528–1534.

Soliman, E.A., Bakr, M.H. and Nikolova, N.K. (2005). Accelerated gradient-based optimization of planar circuits, *IEEE T. Antenn. Propag.*, **53**, 880–883.

Song, Y., Li, Y., Nikolova, N.K. and Bakr, M.H. (2008a). Self-adjoint sensitivity analysis of lossy dielectric structures with electromagnetic time-domain simulators, *Int. J. Numer. Model. El.*, **21**, 117–132.

Song, Y., Nikolova, N.K. and Bakr, M.H. (2008b). Efficient time-domain sensitivity analysis using coarse grids, *Appl. Comput. Electrom.*, **23**, 5–15.

Song, Y. and Nikolova, N.K. (2008c). Memory efficient method for wideband self-adjoint sensitivity analysis, *IEEE T. Microw. Theory*, **56**, 1917–1927.

Swillam, M.A., Bakr, M.H. and Li, X. (2008a). Full wave sensitivity analysis of guided wave structures using FDTD, *J. Electromagnet. Wave*, **22**, 2135–2145.

Swillam, M.A., Bakr, M.H. and Li, X. (2008b). Full vectorial 3D sensitivity analysis and design optimization using BPM, *J. Lightwave Technol.*, **26**, 528–536.

Tortorelli, D.A. and Michaleris, P. (1994). Design sensitivity analysis: overview and review, *Inverse Probl. Eng.*, **1**, 71–105.

Uchida, N., Nishiwaki, S., Izui, K., Yoshimura, M., Nomura, T. and Sato, K. (2009). Simultaneous shape and topology optimization for the design of patch antennas, *3rd European Conference on Antennas and Propagation*, Berlin, pp. 103–107.

Ureel, J. and De Zutter, D. (1996). A new method for obtaining the shape sensitivities of planar microstrip structures by a full-wave analysis, *IEEE T. Microw. Theory*, **44**, 249–260.

Zhu, J., Bandler, J.W., Nikolova, N.K. and Koziel, S. (2007). Antenna optimization through space mapping, *IEEE T. Antenn. Propag.*, **55**, 651–658.

Chapter 15

Boundary Conditions for Two-Dimensional Finite-Element Modeling of Microwave Devices

Tian-Hong Loh and Christos Mias

An accurate solution to discontinuity problems in waveguiding structures plays an important role in the design of microwave devices. In particular, for the analysis of geometrically complex waveguiding structures, there exists high demand for the development of efficient electromagnetic (EM) field solvers that are able to accurately solve electrically large problems. For this reason research on the truncation of the computational domain in order to reduce computational overheads is of considerable importance.

This chapter presents a number of 2D Galerkin weighted residual finite-element time-domain (FETD), finite-element frequency-domain (FEFD), and envelope-finite-element (EVFE) formulations incorporating various port and open boundary conditions to model the electromagnetic behavior of various microwave devices. The port boundary conditions considered include an exact and mesh-efficient multimodal absorbing boundary termination condition (MABTC), and a first-order absorbing boundary condition (ABC). For an open waveguide, a perfectly matched layer (PML) is applied to terminate the open region. Their computational performances are compared. Several electromagnetic problems are characterized numerically.

The 2D model can be used in the analysis and design of microwave devices such as microwave filters (Loh and Mias, 2003; 2004a; Pozar, 2005), periodically loaded metallo-dielectric structures waveguide (Loh and Mias, 2003, 2004b), surface waveguiding structures (Loh and Mias, 2003), electromagnetic band-gap waveguides (Boscolo *et al.*, 2002; Moreno *et al.*, 2002; Chietera *et al.*, 2004), planar transmission lines (Zgainski and

401

Meunier, 1995), microstrip devices (Tsai *et al.*, 2002), monolithic microwave integrated circuits (MMICs) (Lee *et al.*, 1991; Polycarpou *et al.*, 1996), active nonlinear microwave circuits (Tsai *et al.*, 2002), microwave heating applicators (Hallac and Metaxas, 2003), optical propagation analysis (Bertolani *et al.*, 2003), antenna ports (Geuzaine *et al.*, 2000), Bragg grating (Cocinotta *et al.*, 2001), photonic crystal in optical range, etc. It can also serve as a benchmark for 3D analysis.

15.1. Overview and Basic Theory

Time-domain numerical techniques such as the finite-element time-domain (FETD) method (Dibben and Metaxas, 1994; Gedney and Navsariwala, 1995; Lee *et al.*, 1997; Rao, 1999; Jiao and Jin, 2001; Loh and Mias, 2003) or the finite-difference time-domain (FDTD) method (Yee, 1966; Taflove and Brodwin, 1975; Mur, 1981; Gedney, 1996; Taflove and Hugness, 2005) often require accurate absorbing boundary conditions at the limits of the computational domain for obtaining a unique solution to the computation domain. An example is the port boundary condition when dealing with waveguides containing discontinuities. For this example, the distance between the port and the discontinuities should be as short as possible in order to reduce the size of the computational domain and hence the numerical effort. This was achieved in the finite-element frequency-domain (FEFD) method using an exact multimodal condition (Jin, 2002).

In the time domain, a variety of local boundary conditions of varying accuracy have been proposed such as Mur's first-order absorbing boundary condition (ABC) (Mur, 1981), the Berenger's perfectly matched layer (PML) condition (Berenger, 1994), originally applied in the FDTD method and Bermani *et al.*'s numerically integrated convolved PML (Bermani *et al.*, 1999). For waveguide problems, an exact, in principle, multimodal absorbing port boundary condition has been proposed and successfully applied by Moglie *et al.* (1992) in the FDTD method, Pierantoni *et al.* (2002) in transmission line matrix (TLM) method, and Loh and Mias (2004a) and Lou and Jin (2005) in the FETD method. This time-domain boundary condition is obtained from the frequency-domain boundary condition using the inverse Laplace transform (ILT).

Both frequency- and time-domain finite-element (FE) approaches have their own merits (see Section 15.1.2.5). One may transform the results obtained from the time domain to the frequency domain and vice versa

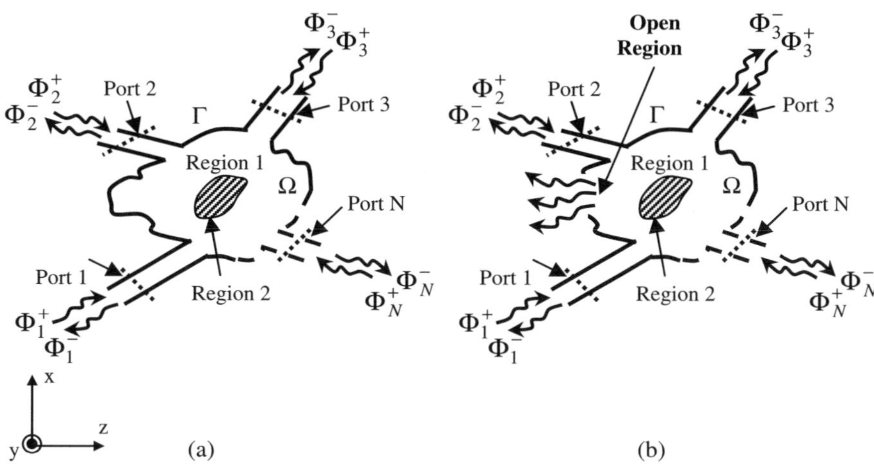

Fig. 15.1.   A general 2D *N*-port waveguiding device: (a) without open region (i.e., closed waveguide), (b) with open region (i.e., open waveguide).

via the discrete Fourier transform (DFT) pair (Eqs. 15.10 and 15.11). Time-domain techniques have been gaining importance as they can provide fast wideband solutions to electromagnetic problems and are applicable to nonlinear problems.

This chapter considers 2D electromagnetic analysis of microwave devices using various finite-element methods (FEMs) incorporating various port and open boundary conditions. As depicted in Fig. 15.1, the computation domain is assumed to be oriented in xz-plane, uniform in the y-axis, and having the medium within the finite computational domain Ω, bounded by the boundary Γ.

Only linear isotropic problems are considered in this chapter. Generalized formulations for both transverse electric (TE) and transverse magnetic (TM) polarizations are presented. This section outlines the basic governing equations used and the Galerkin weighted residual finite-element procedure.

15.1.1. *Basic Governing Equations*

In this chapter, the fields in the frequency domain have a time-harmonic variation $e^{j\omega t}$; the fields in the time domain have a modulated Gaussian transient variation (see (15.9)). The following formulations have as unknowns the time-varying electric or magnetic fields for the FETD or the

EVFE methods and the phasor electric or magnetic fields for the FEFD method.

15.1.1.1. *General time-domain time-varying field Maxwell's equations*

The general time-dependent Maxwell's equations that govern the electromagnetic field behavior of microwave devices are,

$$\nabla \times \mathbf{H} = \mathbf{J}_s + \mathbf{J}_c + \mathbf{J}_d = \mathbf{J}_s + \sigma \mathbf{E} + \varepsilon \frac{\partial \mathbf{E}}{\partial t} \tag{15.1}$$

$$\nabla \times \mathbf{E} = -\mu \frac{\partial \mathbf{H}}{\partial t} \tag{15.2}$$

$$\nabla \bullet \mathbf{D} = \rho_v \tag{15.3}$$

$$\nabla \bullet \mathbf{H} = 0, \tag{15.4}$$

where

$\mathbf{J}_s$ = impressed (source) electric current density (amperes/meter^2)
$\mathbf{J}_c$ = $\sigma \mathbf{E}$, conduction electric current density (amperes/meter^2)
$\mathbf{J}_d$ = $\varepsilon \frac{\partial \mathbf{E}}{\partial t}$, displacement electric current density (amperes/meter^2)
$\mathbf{E}$ = electric field strength (volts/meter)
$\mathbf{H}$ = magnetic field strength (amperes/meter)
$\mathbf{D}$ = $\varepsilon \mathbf{E}$, electric flux density (coulombs/meter^2)
ρ_v = electric charge density per unit volume (coulombs/meter^3)
ε = $\varepsilon_0 \varepsilon_r$, permittivity (farads/meter)
μ = $\mu_0 \mu_r$, permeability (henries/meter)
σ = conductivity (siemens/meter)

where the parameters ε, μ, and σ are the constitutive parameters of the medium. These parameters are tensors for anisotropic media (see Section 15.3.2), and scalars for isotropic media (Jin, 2002). For inhomogeneous media, they are functions of position, while for homogeneous media they are constant. For lossless[1] media, the conductivity is equal to zero. ε_0, μ_0, ε_r, and μ_r denote the free space permittivity (8.85×10^{-12} farads/m), free space permeability ($4\pi \times 10^{-7}$ henries/m), relative permittivity, and relative permeability, respectively. It is noted that throughout this chapter and Chapter 16, the relative permeability is assumed to be unity, $\mu_r = 1$.

---

[1]The medium is said to be lossless when the electromagnetic wave does not suffer any attenuation as it travels through the medium.

15.1.1.2. *General frequency-domain time-harmonic field Maxwell's equations*

The propagation properties of a time-harmonic electromagnetic field, such as its phase velocity, ν_p, and wavelength, λ, are governed by the angular frequency, ω, and the three constitutive parameters of the medium, ε, μ, and σ. The general frequency-dependent Maxwell's equations that govern the electromagnetic field behavior of microwave devices are,

$$\nabla \times \tilde{\mathbf{H}} = \tilde{\mathbf{J}}_s + \tilde{\mathbf{J}}_c + \tilde{\mathbf{J}}_d = \tilde{\mathbf{J}}_s + \sigma\tilde{\mathbf{E}} + j\omega\varepsilon\tilde{\mathbf{E}} \qquad (15.5)$$

$$\nabla \times \tilde{\mathbf{E}} = -j\omega\mu\tilde{\mathbf{H}}, \qquad (15.6)$$

where $e^{j\omega t}$ is assumed and suppressed (Jin, 2002). $\tilde{\mathbf{J}}_s$, $\tilde{\mathbf{J}}_c$, $\tilde{\mathbf{J}}_d$, $\tilde{\mathbf{E}}$, and $\tilde{\mathbf{H}}$ are the frequency (phasor) domain expressions of the parameters defined in Eqs. 15.1 to 15.4.

15.1.1.3. *Time-domain vector wave equation*

Beginning with the time-dependent Maxwell's equations (15.1 and 15.2), one obtains the following time-dependent field vector wave differential equation (Dibben and Metaxas, 1996; Wang and Itoh, 2001),

$$\nabla \times \left(\frac{1}{p}\nabla \times \mathbf{F}\right) + \alpha\frac{q}{c^2}\left(\frac{\partial \mathbf{F}}{\partial t}\right) + \frac{q^2}{c}\frac{\partial^2 \mathbf{F}}{\partial t^2} = \mathbf{U}, \qquad (15.7)$$

where $p = \mu_r$, $q = \varepsilon_r$, $\mathbf{U} = -\mu_0\partial\mathbf{J}_s/\partial t$ for $\mathbf{F} = \mathbf{E}$, and $p = \varepsilon_r$, $q = \mu_r$, $\mathbf{U} = \nabla \times (\mathbf{J}_s/\varepsilon_r)$ for $\mathbf{F} = \mathbf{H}$; c is the speed of light in free space. The relative permittivity, relative permeability, and conductivity are represented by ε_r, μ_r σ, respectively and are assumed to be constant within each finite element. The constant α is given by $\alpha = \sigma/\varepsilon_0\varepsilon_r$. The field is generated by an electric current density $\mathbf{J}_s$. It is noted that (15.7) can be further simplified to the time-domain scalar wave equation as in (15.17) for 2D problems.

15.1.1.4. *Frequency-domain vector wave equation*

Beginning with the time-harmonic form of Maxwell's equations (15.5 and 15.6), one obtains the following frequency-domain vector wave differential equation (Wang and Mittra, 1994; Polycarpou *et al.*, 1997; Mias *et al.*, 1999),

$$\nabla \times \left(\frac{1}{p} \nabla \times \mathbf{F} \right) + j\omega \frac{\alpha q}{c^2} \mathbf{F} - k_0^2 q \mathbf{F} = \mathbf{U}, \qquad (15.8)$$

where $p = \mu_r$, $q = \varepsilon_r$, $\mathbf{U} = -j\omega\mu_0 \tilde{\mathbf{J}}_s$ for $\mathbf{F} = \tilde{\mathbf{E}}$, and $p = \varepsilon_r$, $q = \mu_r$, $\mathbf{U} = \nabla \times (\tilde{\mathbf{J}}_s / \varepsilon_r)$ for $\mathbf{F} = \tilde{\mathbf{H}}$; $k_0 = \omega/c$. c, α, ε_r, and μ_r are the same as those defined in (15.7). It is noted that (15.8) can be further simplified to the frequency-domain scalar wave equation as in (15.26) for 2D problems.

15.1.1.5. *Boundary conditions*

Table 15.1 lists the boundary conditions associated with this chapter. A general 2D problem is assumed where the first-order ABC or an exact and mesh efficient modal absorbing boundary termination condition (MABTC) (Loh and Mias, 2004a) is applied at the port boundaries, Dirichlet and Neumann boundary conditions at the waveguide wall, and the PML is applied to terminate the open region (see Fig. 15.5 in Section 15.3). Furthermore, it is assumed that the waveguide is filled with isotropic, linear, and homogenous medium, the waveguide medium over which the first-order ABC and MABTC are applied is lossless and isotropic, and the PML region is lossy and anisotropic. The developed boundary condition formulation

Table 15.1.   List of boundary conditions associated with this chapter.

Working Field Variable	Boundary	Boundary Conditions for 2D Scalar Field
Magnetic Field, H_y	PMC	$H_y = 0^*$
	PEC	$\partial H_y / \partial n = 0^\dagger$
	Open region	PML (see Section 15.3.2)
	Waveguide Ports	First-order ABC (see Section 15.3.3)
		MABTC (see Section 15.3.1)
Electric Field, E_y	PMC	$\partial E_y / \partial n = 0^\dagger$
	PEC	$E_y = 0^*$
	Open region	PML (see Section 15.3.2)
	Waveguide Ports	First-order ABC (see Section 15.3.3)
		MABTC (see Section 15.3.1)

ABC: Absorbing boundary condition.
MABTC: Modal absorbing boundary termination condition.
PEC: Perfect electric conductor (also called electric wall).
PMC: Perfect magnetic conductor (also called magnetic wall).
PML: Perfectly matched layer (Berenger, 1994; Taflove and Hugness, 2005).
*: Homogeneous Dirichlet boundary condition for the problems considered.
†: Homogeneous Neumann boundary condition for the problems considered.

and the employed methodology can be applied to both closed[2] and open[3] homogeneous waveguide problems. The finite element implementation of the boundary conditions is shown in Section 15.3.

15.1.1.6. *Sampling theorem and Gaussian pulse*

The sampling theorem (Meade and Dillon, 1991) states that:

> A continuous signal with frequency components in the range $f = f_c - f_b$ to $f = f_c + f_b$, can be reconstructed from a sequence of equally spaced samples, provided that the sampling frequency exceed $2(f_c + f_b)$ samples/second.

Therefore, for the time-domain method, to obey sampling theorem the time step is chosen to be at least $2(f_c + f_b)$ the sampling rate. A Gaussian pulse is often being employed because of its bandlimited nature in both the time and frequency domain. Equation 15.9 shows the general form of a modulated Gaussian pulse:

$$f(t) = \exp\left[-\frac{(t - t_0)^2}{b^2}\right]\sin(\omega_c(t - t_0)), \qquad (15.9)$$

where $\omega_c = 2\pi f_c$, f_c, denotes the carrier frequency, the constants t_0 and b control the time-shift[4] and the pulse width, respectively. The formulae for the discrete Fourier transform (DFT) pair are given as (Meade and Dillon, 1991):

$$F(k) = \sum_{n=0}^{N-1} f[n]e^{-j(2\pi/N)nk}, \quad k = 0, 1, 2, \ldots, \quad N - 1 \qquad (15.10)$$

$$f[n] = \frac{1}{N}\sum_{k=0}^{N-1} F[k]e^{j(2\pi/N)kn}, \quad n = 0, 1, 2, \ldots, \quad N - 1, \qquad (15.11)$$

where N denotes the total number of sampling points in the time domain and the frequency domain. $f[n]$ is the discrete-time representation of the $f(t)$ and $F[k]$ is the discrete-frequency representation of the $F(s)$, where $F(s)$ denotes the frequency-domain model of $f(t)$.

---

[2]There is no electromagnetic radiation from closed homogeneous waveguides.
[3]There is power loss due to electromagnetic radiation in open homogeneous waveguides.
[4]For $t_0 = 0$, the Gaussian pulse has its peak at $t = 0$.

Figure 15.2 shows the effects of the constant t_0. It is noted that increasing the constant t_0 will further delay the pulse in the time domain and increase the peak-to-minimum amplitude difference in the frequency domain. In this chapter, $t_0 = 4\sigma_t$ and $b = \sqrt{2}\sigma_t$ are chosen, where σ_t denotes the time-delay constant. Figure 15.3 shows the effects of the constant σ_t. It is noted that the increase in the constant σ_t will decrease the pulse width in time domain and hence increase its frequency bandwidth and vice versa. Furthermore, for obtaining the same frequency resolution Δf as in the frequency-domain method, the time-domain method must run for a total time of $t_{total} = 1/\Delta f$ (Meade and Dillon, 1991).

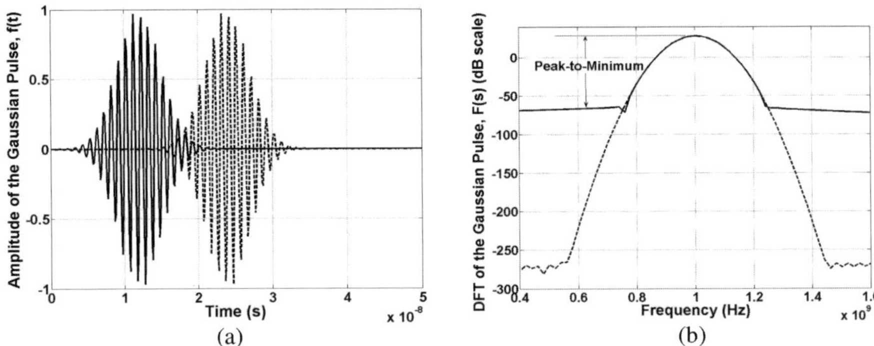

Fig. 15.2. Behavior of Gaussian pulse for various t_0: (a) time-domain plot, (b) frequency domain plot. Note that $f_c = 1\,\text{GHz}$, $b = 4.24\,\text{ns}$. Solid line: $t_0 = 12\,\text{ns}$; dash line: $t_0 = 24\,\text{ns}$.

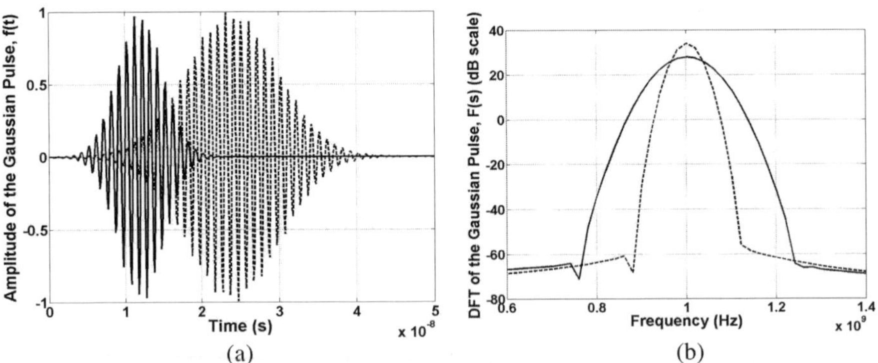

Fig. 15.3. Behavior of Gaussian pulse for various σ_t: (a) time-domain plot, (b) frequency-domain plot. Note that $f_c = 1\,\text{GHz}$, $t_0 = 4\sigma_t$, and $b = \sqrt{2}\sigma_t$. Solid line: $\sigma_t = 3\,\text{ns}$; dash line: $\sigma_t = 6\,\text{ns}$.

15.1.2. Galerkin Weighted Residual Finite-Element Procedure

Figure 15.4 illustrates the basic steps of the Galerkin weighted residual finite-element method to formulate the solution of the electromagnetic boundary value problems. Each step now detailed.

15.1.2.1. Galerkin weighted residual formulation

The wave equations have a general form (Jin, 2002),

$$\ell u = v \tag{15.12}$$

where ℓ is a differential operator, u is the unknown quantity, and v is the excitation or forcing function. The residual, r of (15.12) is defined as follows (Jin, 2002):

$$r = \ell u - v. \tag{15.13}$$

Within the computation domain Ω, the weighted residual method enforces the following condition (Mias *et al.*, 1999; Jin, 2002.),

$$R = \int_{\Omega} W r d\Omega = \sum_{e} R^e = 0, \tag{15.14}$$

where R denotes the weighted residual integral (following the assembly of the global matrix using the local and global relations on an element-by-element basis), W is some arbitrary weighting function. In the Galerkin weighted residual method, the weighting function is selected to be the same as those used for the expansion of the approximate solution, which usually leads to the most accurate solution (Jin, 2002). It is noted that W and r can either be scalar or vector functions. The same procedure is followed for both the time-domain and the frequency-domain finite-element methods.

15.1.2.2. Incorporation of the boundary conditions

The solution of the boundary-value problems can only proceed once the appropriate boundary conditions have been correctly applied. To define the boundary condition, one can use Green's theorem for 2D problems and use the vector form of Green's theorem for 3D problems. This leads to the general form of boundary conditions on the boundary domain, as follows:

$$\Re u = \vartheta, \tag{15.15}$$

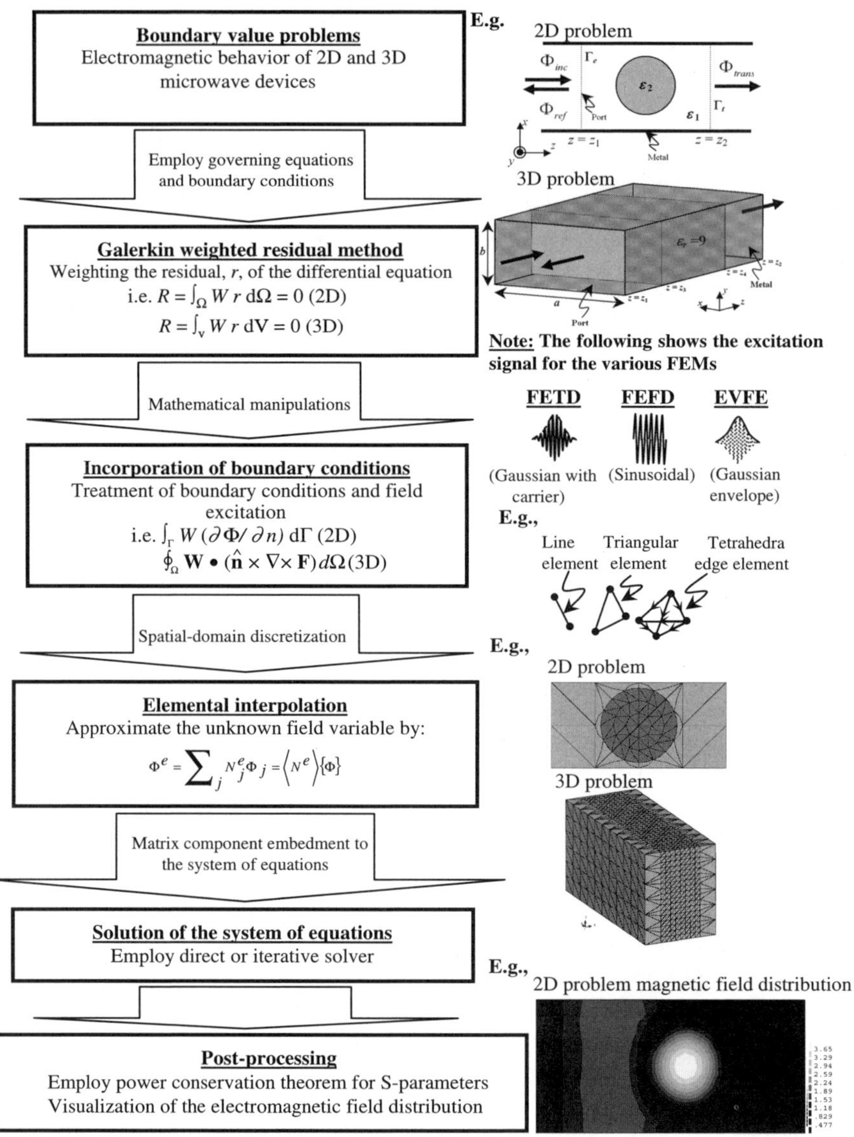

Fig. 15.4. The Galerkin weighted residual finite-element procedure. Note: the details of parameters are defined in detail later in the chapter.

where $\Re$ is a differential operator, u is the unknown quantity, and ϑ is the excitation or forcing function along that boundary. The detailed formulation of implementing the boundary conditions listed in Table 15.1 will be presented in the rest of the sections.

15.1.2.3. *Spatial-domain discretisation and mesh generation*

To employ the FEM method, the computational spatial-domain must be a finite domain. This step divides the computational spatial-domain into a finite number of subdomains or elements. Each element is associated with certain number of degrees of freedom (DOF) or unknowns to be solved. The finite-element solution approaches the exact solution as the number of degree of freedom increases. Time and memory constraints dictate the number of finite elements employed and consequently the accuracy of results.

The choice of element type is an important consideration for the finite-element discretization. The commonly used surface and volume element types are triangular and tetrahedral elements due to their discretization flexibility over the computational domain. One can choose to either employ scalar nodal-based elements or vector edge-based elements for these element types. A rule of thumb for the discretization (Mias, 1995) should be obeyed for obtaining accurate solutions. For a 2D mesh of second-order scalar nodal-based finite elements at least four second-order finite elements are required per wavelength whereas for a mesh of first-order finite elements at least ten first-order finite elements are required per wavelength. To create the finite-element mesh, one can develop or use mesh generation software which provides the following:

1. A list of the entire nodal point coordinates — node number and their Cartesian coordinates.
2. A list of all the element topology definition — element number, element type, material property number, and list of node numbers associated with a particular element.
3. A list of all the material properties — user-defined material parameters.

15.1.2.4. *Elemental field interpolation*

The unknown field variable at any point within each element is then approximated by a set of interpolation functions (also called expansion or basis functions), which are polynomial with a selected order. It is noted that the higher-order polynomial is more accurate but usually results in a more complicated formulation (Jin, 2002). Hence, the simpler first- (linear) or second- (quadratic) order polynomial is widely used. For the eth element, the polynomial has a general form of

$$\Phi^e = \sum_j N_j^e \Phi_j = \langle N^e \rangle \{\Phi\}, \tag{15.16}$$

where $\langle N^e \rangle$ is a row vector representing the elemental field interpolation functions, $\{\Phi\}$ is a column vector representing the spatial unknown values of the field variable Φ, j signifies nodes relating to the eth element but counted on a global basis, and N_j^e represents appropriate elemental field interpolation functions of the jth node, which has unity value at node j and zero value at the rest of the nodes. Note that the highest order of N_j^e is referred to as the order of the element. An important feature of the functions N_j^e is that it is nonzero only within the eth element but vanishes outside this element (Jin, 2002). It is noted that, in time domain, the variable Φ_j, in (15.16), is time-dependent, and it is discretized in the time domain further by employing the Newmark-beta method (Newmark, 1959; Zienkiewicz, 1977) (see Section 15.2), whereas in the frequency domain, the variable Φ_j, in (15.16), is a constant and needs no further discretization.

15.1.2.5. *Solution of the system of equations*

The methods for solving matrix equations can be categorized into two groups: direct methods (such as the Gaussian elimination method) and iterative methods (such as the bi-conjugate gradient (BCG) method) (Jin, 2002).

Direct solvers usually suffer from memory requirement to the extent that large problems cannot be solved at a reasonable cost even on state-of-the-art parallel machines (Chen *et al.*, 2002). Hence the iterative algorithms, whose memory requirements are a small fraction of that required by a direct solver, are preferred. However, this method can suffer from the convergence rate of the BCG for large and ill-conditioned problems unless proper matrix preconditioning is performed. The choice of solver depends on the computer memory capacity available.

Note that in the FEFD technique the resulting matrix equation has a left-hand-side (LHS) matrix consisting of frequency-dependent terms and hence a matrix assembly, inversion, and solution, is required for each frequency step. On the other hand the LHS matrix of the FETD method needs to be assembled and inverted only once for the EVFE first-order ABC technique (at $t = 0$) (see Section 15.3.3.3) and twice for the FETD-MABTC technique (at $t = 0$ and $t = t_c$) (see Section 15.3.1.2).

15.1.2.6. *Post-processing*

Upon solving the system of equations, this step makes use of the final solutions obtained to extract further useful information such as the normalized reflected/transmitted power and the field distribution. The Poynting theorem is used to derive the normalized reflected/transmitted power formulations. One can employ mesh-view software to view the field distribution everywhere within the bounded computational region.

15.2. Finite-Element Formulation

This section derives the 2D FETD, FEFD, and EVFE formulation following the standard Galerkin weighted residual finite-element procedures outlined in Section 15.1.2.

15.2.1. *Finite-Element Time-Domain Formulation*

Referring to the 2D problems as depicted in Fig. 15.1, the second-order scalar wave equation (also called scalar Helmholtz equation) regarding the transverse component of the field is

$$\nabla \cdot \frac{1}{p} \nabla \Phi - \alpha \frac{q}{c^2} \frac{\partial \Phi}{\partial t} - \frac{q}{c^2} \frac{\partial^2 \Phi}{\partial t^2} = V, \tag{15.17}$$

where $p = \mu_r$, $q = \varepsilon_r$, $V = \mu_0 \partial \mathrm{J}_{sy}/\partial t$ for $\Phi = E_y$ (transverse electric (TE) polarization); $p = \varepsilon_r$, $q = \mu_r$, $V = -1/\varepsilon_r(\partial \mathrm{J}_{sx}/\partial z - \partial \mathrm{J}_{sz}/\partial x)$ for $\Phi = H_y$ (transverse magnetic (TM) polarization). c, α, ε_r, μ_r and $\mathbf{J}_s$ are the same as those defined in (15.7).

The continuity conditions at the interface of two homogeneous regions (say Region 1 and Region 2 refer to Fig. 15.1) are (Mias, 1995; Jin, 2002.):

$$\Phi_1 = \Phi_2 \tag{15.18}$$

$$\frac{1}{p_1} \nabla \Phi_1 \bullet \hat{\mathbf{n}} = \frac{1}{p_2} \nabla \Phi_2 \bullet \hat{\mathbf{n}}. \tag{15.19}$$

At the waveguide wall, assumed to be a perfect electric conductor of infinite conductivity, the Dirichlet and Neumann boundary conditions apply:

$$\text{Dirichlet}: \quad \Phi = 0 \quad \text{(for TE polarization)} \tag{15.20}$$

$$\text{Neumann}: \quad \nabla \Phi \bullet \hat{\mathbf{n}} = 0 \quad \text{(for TM polarization)}, \tag{15.21}$$

where $\hat{\mathbf{n}}$ is the outward unit vector normal to the boundary, Γ. One notes that $\nabla \Phi \bullet \hat{\mathbf{n}} = \partial \Phi / \partial n$ denotes the outward normal derivative.

The Galerkin weighted residual method (Silvester and Ferrari, 1996; Jin, 2002.) (see Section 15.1.2), is then employed to solve (15.17). This is achieved by weighting (15.17) with nodal-based weighting functions N_i, which are chosen from the set of field interpolation functions employed over each element following a finite-element descretization of the spatial domain. At a particular time t, the unknown field at any point within an element can be expressed as

$$\Phi^e(x, z, t) = \sum_j N_j^e(x, z)\Phi_j(t) = \langle N^e(x, z)\rangle\{\Phi(t)\}, \tag{15.22}$$

where $\langle N^e(x, z)\rangle$ is a row vector representing the nodal-based elemental field interpolation functions, $\{\Phi(t)\}$ is a column vector representing, at the particular time t, the nodal-based spatial unknown values of the field variable Φ. By integrating the weighted (15.17) over the domain of interest and applying the differentiation product rule followed by the divergence theorem, a system of ordinary differential equations is obtained as follows (Dibben and Metaxas, 1996; Lee *et al.*, 1997; Rao, 1999; Wang and Itoh, 2001),

$$[T]\frac{\partial^2\{\Phi(t)\}}{\partial t^2} + [B]\frac{\partial\{\Phi(t)\}}{\partial t} + [S]\{\Phi(t)\} = \{F\}, \tag{15.23}$$

where the elemental matrices are defined as:

$$T_{ij}^e = \frac{q^e}{c^2}\iint_{\Omega^e} N_i^e N_j^e d\Omega$$

$$B_{ij}^e = \frac{\alpha^e q^e}{c^2}\iint_{\Omega^e} N_i^e N_j^e d\Omega$$

$$S_{ij}^e = \frac{1}{p^e}\iint_{\Omega^e} \nabla N_i^e \cdot \nabla N_j^e d\Omega$$

$$F_i^e = -\iint_{\Omega^e} N_i^e V d\Omega + \frac{1}{p^e}\oint_{\Gamma^e} N_i^e \frac{\partial \Phi}{\partial n} d\Gamma.$$

Employing the following Newmark-beta formulation (Newmark, 1959; Zienkiewicz, 1977) to discretize (15.23) in the time domain:

$$
\begin{cases}
\dfrac{\partial^2\{\Phi(t)\}}{\partial t^2} = \dfrac{1}{\Delta t^2}[\{\Phi\}^{n+1} - 2\{\Phi\}^n + \{\Phi\}^{n-1}] \\[2mm]
\dfrac{\partial\{\Phi(t)\}}{\partial t} = \dfrac{1}{2\Delta t}[\{\Phi\}^{n+1} - \{\Phi\}^{n-1}] \\[2mm]
\{\Phi(t)\} = \beta\{\Phi\}^{n+1} + (1-2\beta)\{\Phi\}^n + \beta\{\Phi\}^{n-1}
\end{cases}
\qquad (15.24)
$$

where, at time step $t = n\Delta t$, the discrete-time representation of $\{\Phi(t)\}$ is represented as $\{\Phi(t)\} = \{\Phi(n\Delta t)\} = \{\Phi\}^n$ and β is a parameter that has to be set. It was shown in Gedney and Navsariwala (1995) that unconditional stability is achievable by choosing the parameter $\beta \geq 1/4$, and that by choosing $\beta = 1/4$, the solution error can be minimized. Hence, $\beta = 1/4$ is chosen here, which leads to the following unconditionally stable two-step update scheme:

$$
\left(\frac{[T]}{\Delta t^2} + \frac{[B]}{2\Delta t} + \frac{[S]}{4}\right)\{\Phi\}^{n+1} = \left(\frac{2[T]}{\Delta t^2} - \frac{[S]}{2}\right)\{\Phi\}^n
$$

$$
+ \left(-\frac{[T]}{\Delta t^2} + \frac{[B]}{2\Delta t} - \frac{[S]}{4}\right)\{\Phi\}^{n-1}
$$

$$
+ \left(\frac{\{F\}^{n+1}}{4} + \frac{\{F\}^n}{2} + \frac{\{F\}^{n-1}}{4}\right).
$$

$$(15.25)$$

15.2.2. *Finite-Element Frequency-Domain Formulation*

Referring to the 2D problem as depicted in Fig. 15.1, and beginning with the time-harmonic form of Maxwell's equations (15.5 and 15.6), one obtains the following frequency-domain second-order scalar Helmholtz wave equation regarding the transverse component of the field

$$
\nabla \cdot \frac{1}{p}\nabla\Phi + k_0^2 q\Phi - \frac{j\omega}{c^2}\alpha q\Phi = V, \qquad (15.26)
$$

where $V = j\omega\mu_0\tilde{J}_{sy}$ for $\Phi = \tilde{E}_y$ (TE polarization); $V = -1/\varepsilon_r(\partial\tilde{J}_{sx}/\partial z - \partial\tilde{J}_{sz}/\partial x)$ for $\Phi = \tilde{H}_y$ (TM polarization); $k_0 = \omega/c$. c, p, q, α, ε_r, μ_r, and $\tilde{\mathbf{J}}_s$ are the same as those defined in (15.7). The boundary conditions (15.18)–(15.21) still apply.

The Galerkin weighted residual method (Silvester and Ferrari, 1996; Jin, 2002) (see Section 15.1.2), is then employed to solve (15.26). This is achieved by weighting (15.26) with nodal-based weighting functions N_i. By dividing the spatial domain into nodal-based elements, at a particular frequency f, the unknown field at any point within an element can be expressed as

$$\Phi^e(x, z, s) = \sum_j N_j^e(x, z)\,\Phi_j(s) = \langle N^e(x, z)\rangle\{\Phi(s)\}, \qquad (15.27)$$

where $\langle N^e(x, z)\rangle$ is a row vector representing the nodal-based elemental field interpolation functions, $s = j\omega$, $\{\Phi(s)\}$ is a column vector representing, at the particular frequency f, the nodal-based spatial unknown values of the field variable Φ. By integrating the weighted (15.26) over the domain of interest and applying the differentiation product rule followed by the divergence theorem, a system of equations is obtained as follows:

$$([S] + j\omega[B] - \mathrm{k}_0^2[T])\{\Phi(s)\} = \{F\}, \qquad (15.28)$$

where the elemental matrices are defined as

$$S_{ij}^e = \frac{1}{p^e}\iint_{\Omega^e}\nabla N_i^e \cdot \nabla N_j^e\, d\Omega$$

$$B_{ij}^e = \iint_{\Omega^e}\frac{\alpha^e q^e}{c^2}N_i^e N_j^e\, d\Omega$$

$$T_{ij}^e = \iint_{\Omega^e} q^e N_i^e N_j^e\, d\Omega$$

$$F_i^e = -\iint_{\Omega^e} N_i^e V\, d\Omega + \oint_{\Gamma^e}\frac{1}{p^e}N_i^e\frac{\partial\Phi}{\partial n}\, d\Gamma.$$

15.2.3. *Envelope-Finite-Element Formulation*

Let us refer to the 2D problem as depicted in Fig. 15.1, and begin with the time-dependent field second-order scalar Helmholtz wave equation (15.17). To de-embed the signal carrier from the time-domain wave equation, one can assume the field component Φ and the current density $\mathbf{J_s}$ having a solution of the form (Wang and Itoh, 2001):

$$\Phi(x, z, t) = \phi(x, z, t)e^{j\omega_c t} \qquad (15.29)$$

$$\mathbf{J_s}(x, z, t) = \mathbf{U}(x, z, t)e^{j\omega_c t}, \qquad (15.30)$$

where ω_c denotes the carrier frequency, $\phi(x, z, t)$ and $\mathbf{U}(x, z, t)$ are the time-varying complex envelopes of the field and the excitation current at the carrier frequency, respectively. Referring to (15.29) and (15.30), it is noted that the EVFE method becomes the FEFD method (see Section 15.2.2) when the envelope is independent of time (i.e., $\phi(x, z, t) \Rightarrow \phi(x, z)$ and $\mathbf{U}(x, z, t) \Rightarrow \mathbf{U}(x, z)$). When $\omega_c = 0$, it is of course the standard FETD when Φ and $\mathbf{J_s}$ are the unknowns.

Substituting (15.29) and (15.30) into (15.17) and dividing both sides by $e^{j\omega_c t}$, one obtains:

$$\nabla \cdot \frac{1}{p}\nabla\phi - \frac{q}{c^2}\left[(j\alpha\omega_c - \omega_c^2) + (\alpha + 2j\omega_c)\frac{\partial}{\partial t} + \frac{\partial^2}{\partial t^2}\right]\phi = \Psi, \qquad (15.31)$$

where $p = \mu_r$, $q = \varepsilon_r$, and $\Psi = \mu_0(j\omega_c U_y + \partial U_y/\partial t)$ for $\Phi = E_y$ (TE polarization); $p = \varepsilon_r$, $q = \mu_r$, $\Psi = -1/\varepsilon_r(\partial U_x/\partial z - \partial U_z/\partial x)$ for $\Phi = H_y$ (TM polarization). c, α, ε_r, and μ_r are the same as those defined in (15.7). The boundary conditions (15.18)–(15.21) are considered.

The Galerkin weighted residual method (Jin, 2002; Silvester and Ferrari, 1996) (see Section 15.1.2), is then employed to solve (15.31). This is achieved by weighting (15.31) with nodal-based weighting functions N_i. Following a finite-element discretization, the spatial domain is divided into a finite number of nodal-based elements. At a particular time t, the unknown field envelope at any point within an element can be expressed as:

$$\phi^e(x, z, t) = \sum_j N_j^e(x, z)\phi_j(t) = \langle N^e(x, z)\rangle\{\phi(t)\}, \qquad (15.32)$$

where the row vector $\langle N^e(x, z)\rangle$, the subscript j, and the term N_j^e are the same as those defined in (15.16). $\{\phi(t)\}$ is a column vector representing, at the particular time t, the nodal-based spatial unknown values of the field variable, ϕ. By integrating the weighted (15.31) over the domain of interest and applying the differentiation product rule followed by the divergence theorem, a system of ordinary differential equations is obtained as follows (Wang and Itoh, 2001):

$$[T]\frac{\partial^2\{\phi(t)\}}{\partial t^2} + [B]\frac{\partial\{\phi(t)\}}{\partial t} + [S]\{\phi(t)\} = \{F\}, \qquad (15.33)$$

where the elemental matrices are defined as

$$T_{ij}^e = \iint_{\Omega^e} \frac{q^e}{c^2} N_i^e N_j^e d\Omega$$

$$B_{ij}^e = \iint_{\Omega^e} (\alpha^e + 2j\omega_c)\frac{q^e}{c^2} N_i^e N_j^e d\Omega$$

$$S_{ij}^e = \iint_{\Omega^e} \left[\frac{1}{p^e} \nabla N_i^e \cdot \nabla N_j^e - \frac{q^e}{c^2}(\omega_c^2 - j\alpha^e \omega_c) N_i^e N_j^e \right] d\Omega$$

$$F_i^e = -\iint_{\Omega^e} N_i^e \Psi d\Omega + \frac{1}{p^e} \oint_{\Gamma^e} N_i^e \frac{\partial \phi}{\partial n} d\Gamma.$$

Employing the Newmark-beta formulation (15.24) to discretize (15.33) leads to the following two-step recurrence relation approximation:

$$\left(\frac{[T]}{\Delta t^2} + \frac{[B]}{2\Delta t} + \frac{[S]}{4} \right) \{\phi\}^{n+1} = \left(\frac{2[T]}{\Delta t^2} - \frac{[S]}{2} \right) \{\phi\}^n$$

$$+ \left(-\frac{[T]}{\Delta t^2} + \frac{[B]}{2\Delta t} - \frac{[S]}{4} \right) \{\phi\}^{n-1}$$

$$+ \left(\frac{\{F\}^{n+1}}{4} + \frac{\{F\}^n}{2} + \frac{\{F\}^{n-1}}{4} \right).$$

$$(15.34)$$

One should note that although (15.34) has a similar form to (15.25) the unknown function to be solved in EVFE is complex, whereas in FETD the unknown function is real.

15.3. Boundary Conditions

This section presents formulations for the following boundary conditions employed in this chapter (see Section 15.1.1.5 for the assumption on the medium of the computational domain):

(1) The modal absorbing boundary termination condition (MABTC) (Loh and Mias, 2004a).
(2) The perfectly matched layer (PML) condition.
(3) The first-order absorbing boundary condition (first-order ABC).

For a 2D problem, a general open parallel-plate waveguide is considered as shown in Fig. 15.5. It is assumed that the waveguide is uniform in the y-direction ($\partial/\partial y = 0$) and that the z-axis is the axis of wave propagation where an incident guided wave is propagating from left to right.

As depicted in Fig. 15.5, the contour part Γ of the finite-element boundary integral on the right hand side of (15.28) (for FEFD) and (15.23) (for FETD) consists of a perfect electric conductor (PEC) boundary Γ_c, an excitation boundary Γ_e positioned at $z = z_1$, and a termination boundary Γ_t positioned at $z = z_2$. There is no contribution from the Γ_c boundary

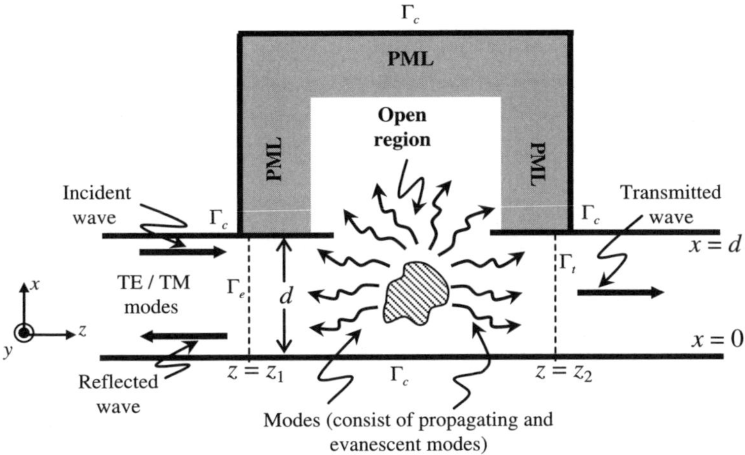

Fig. 15.5.   Sketch of an open parallel-plate waveguide incorporating discontinuities.

integral as along this boundary there exists a homogeneous Neumann boundary condition[5] for TM mode; a homogeneous Dirichlet boundary condition[6] for TE mode. At the truncation boundaries, Γ_e and Γ_t, either a first-order ABC or a MABTC boundary condition can be employed. The open region is truncated by PEC-backed PML boundary condition (i.e., a radiation boundary condition).

The discontinuity can be either geometrical, material, or both. If a discontinuity exists only a portion of the power carried by the incident wave can pass through the discontinuity. As higher-order modes are excited by the discontinuity the field near the discontinuity has no specific form and it is very complicated (Jin, 2002).

The PML boundary condition is an artificial absorber, which is analytically reflectionless at its surface for all incident angles and often PEC-backed. However, it introduces an unnecessary number of unknowns to be solved, while the MABTC gives exact results without a need for the careful selection of the PML complex tensors parameters.

For the waveguide problems operating within the monomode frequency region, in which only the dominant mode propagates, the first-order ABC

---

[5]This requires the normal derivative of the unknown field to vanish at the boundary. It is also call natural boundary condition because it is usually satisfied implicitly and automatically in the solution process.

[6]This prescribes and imposes the field at the boundary explicitly. It is also called essential boundary condition because it must be imposed explicitly.

can be employed at the port boundaries, which must be positioned at a sufficiently large distance from the discontinuity (Jin, 2002) so that all the higher-order modes excited by the waveguide discontinuity can die out before reaching this port boundary in order to obtain accurate results. A major drawback of this approach is that placing the port boundaries far away from the discontinuity results in a large computational region and hence a large number of unknowns to be solved for. On the other hand, the MABTC boundary condition (Loh and Mias, 2004a) is exact, permits multiple-mode propagation, and allows the port boundaries to be placed as close to the discontinuity as possible provided a sufficient number of higher-order modes is taken into account.

15.3.1. *Multimodal Absorbing Boundary Termination Condition*

15.3.1.1. *MABTC in the frequency domain*

In the frequency domain, the following relation holds for the modal characteristic wave impedance with m variations in the x-axis for a TM wave (designated TM_m) and a TE wave (designated TE_m), respectively (Collin, 1991; Moglie *et al.*, 1992; Ramo *et al.*, 1994):

$$Z_m^{\mathrm{TM}}(s) = \pm \frac{\tilde{E}_{\mathrm{xTM}_m}(\boldsymbol{r}, s)}{\tilde{H}_{\mathrm{yTM}_m}(\boldsymbol{r}, s)} = \frac{\eta}{s}\sqrt{s^2 + \omega_{c_m}^2} \quad \text{(for TM mode)} \quad (15.35)$$

$$Z_m^{\mathrm{TE}}(s) = \mp \frac{\tilde{E}_{\mathrm{yTE}_m}(\boldsymbol{r}, s)}{\tilde{H}_{\mathrm{xTE}_m}(\boldsymbol{r}, s)} = \frac{\eta s}{\sqrt{s^2 + \omega_{c_m}^2}} \quad \text{(for TE mode),} \quad (15.36)$$

where $\eta = \sqrt{\mu_0\mu_r/(\varepsilon_0\varepsilon_r)}$, $\boldsymbol{r} = (x, z)$, $s = j\omega$, $\omega_{c_m} = m\pi c/(d\sqrt{\mu_r\varepsilon_r})$ denotes the cut-off frequency of the mth mode, c denotes the speed of light, and d denotes the separation distance between the two plates. μ_r and ε_r are the same as those defined in Section 15.1.1.1. The signs $\pm$ (or corresponding $\mp$) are for waves propagating in the same or opposite direction to the normal outward unit direction respectively. Referring to (15.35) and (15.36), an imaginary part arises in the case of a waveguide medium with finite losses over which the port boundaries lie and for waves that are evanescent in the direction of propagation (Collin, 1991). Using the time-harmonic form of Maxwell's equations (15.5 and 15.6) and the modal characteristic wave impedance of mode m (15.35 and 15.36), one obtains the mth mode outward normal derivative of the field

$$\frac{\partial \Phi_m(\boldsymbol{r}, s)}{\partial n} = \pm\gamma_m \Phi_m(\boldsymbol{r}, s), \quad (15.37)$$

where $\Phi_m = \tilde{H}_{y_m}$ for TM mode m, $\Phi_m = \tilde{E}_{y_m}$ for TE mode m. The variables $\gamma_m = s\sqrt{\mu_r \varepsilon_r} Z_m(s)/c$ and $Z_m(s) = \sqrt{s^2 + \omega_{c_m}^2}/s$ denote the propagation constant and the modal characteristic wave impedance, respectively. The $\pm$ signs correspond to wave propagation in the opposite $(+)$ or the same $(-)$ direction to the normal outward unit vector.

Assuming single-mode incidence, the frequency-domain field equation at the excitation boundary Γ_e, where $z = z_1$, has the following form (Jin, 2002):

$$\Phi(x, z_1, s) = \Phi^{\text{inc}}(x, z_1, s) + \sum_{m=0}^{\infty} \Phi_m^{\text{ref}}(x, z_1, s)$$

$$= h_{\text{inc}}(x)\Phi^{\text{inc}}(z_1, s) + \sum_{m=0}^{\infty} a_m h_m(x)\Phi_m^{\text{ref}}(z_1, s), \qquad (15.38)$$

where a_m are the modal coefficients, and the functions $h_m(x)$ are given by

$$h_m(x) = \sqrt{\frac{\nu_m}{d}} \cos\frac{m\pi x}{d}, \quad \nu_m = \begin{cases} 1, & m = 0 \\ 2, & m \neq 0 \end{cases} \qquad \text{(for TM mode)}$$

$$h_m(x) = \sqrt{\frac{\nu_m}{d}} \sin\frac{m\pi x}{d}, \quad \nu_m = 2, m > 0 \qquad \text{(for TE mode).}$$

Note that for TE modes $m > 0$ and although the summations in (15.38) start at $m = 0$, this is only true for TM modes. For TE modes, one should ignore the $m = 0$ index.

To obtain the modal coefficients a_m one can make use of the orthogonality property (Kreyszig, 2010). Hence, multiplying both sides of (15.38) with $h_k(x)$ and integrating from 0 to d,

$$a_k = \frac{1}{\Phi_k^{\text{ref}}(z_1, t)} \int_0^d [\Phi(x, z_1, t) - \Phi^{\text{inc}}(x, z_1, t)]h_k(x)dx. \qquad (15.39)$$

Taking the partial derivative of (15.38) with respect to n at $z = z_1$, noting $\hat{\mathbf{n}} = -\hat{z}$, and making use of (15.37) and (15.39), one obtains

$$\left.\frac{\partial \Phi(\mathbf{r}, s)}{\partial n}\right|_{z=z_1} = -\sum_{m=0}^{\infty} \gamma_m h_m(x)\left[\int_0^d \Phi(x', z, s)h_m(x')dx'\right]\Bigg|_{z=z_1}$$

$$+ 2\gamma_{\text{inc}}[\Phi^{\text{inc}}(\mathbf{r}, s)]|_{z=z_1}. \qquad (15.40)$$

Similarly, the frequency-domain field equation at the termination truncation boundary Γ_t, where $z = z_2$, has the following form:

$$\Phi(x, z_2, s) = \sum_{m=0}^{\infty} \Phi_m^{\text{tran}}(x, z_2, s) = \sum_{m=0}^{\infty} b_m h_m(x) \Phi_m^{\text{tran}}(z_2, s).$$

$$(15.41)$$

Following the aforementioned procedure and noting $\hat{\mathbf{n}} = \hat{z}$, one obtains the modal coefficients, b_m, and the partial derivative of (15.41) with respect to n at $z = z_2$ as follows:

$$b_m = \frac{1}{\Phi_m^{\text{tran}}(z_2, s)} \int_0^d \Phi(x, z_2, s) h_m(x) dx \qquad (15.42)$$

$$\left.\frac{\partial \Phi(\mathbf{r}, s)}{\partial n}\right|_{z=z_2} = -\sum_{m=0}^{\infty} \gamma_m h_m(x) \left[\int_0^d \Phi(x', z, s) h_m(x') dx'\right]\Bigg|_{z=z_2}.$$

$$(15.43)$$

Applying (15.40) and (15.43) and rearranging both sides of (15.28), one obtains the following Galerkin weighted residual FEFD-MABTC formulation

$$[M]\{\Phi\} = \{K\}, \qquad (15.44)$$

where

$$[M] = [S] - k_0^2 [T] + [Q]$$

$$Q_{ij} = \frac{1}{p} \sum_{m=0}^{\infty} \gamma_m \int_\Gamma N_i h_m(x) d\Gamma \int_0^d N_j h_m(x') d\Gamma'$$

$$K_i = \frac{2\gamma_{\text{inc}}}{p} \Phi^{\text{inc}}(z_1, s) \int_\Gamma N_i h_{\text{inc}}(x) d\Gamma,$$

where $p = \mu_r$ for $\Phi = \tilde{E}_y$ (TE polarization) and $p = \varepsilon_r$ for $\Phi = \tilde{H}_y$ (TM polarization).

15.3.1.2. *MABTC in the time domain*

The derivation of the MABTC in the time domain starts from the time-harmonic form of Maxwell's equations (15.5 and 15.6) and MABTC in the frequency domain (Moglie *et al.*, 1992; Jin, 2002). Applying on both sides of (15.37) the inverse Laplace transform (ILT), one obtains the following time-domain modal equation:

$$\frac{\partial \Phi_m(\boldsymbol{r}, t)}{\partial z} = \mp \frac{\sqrt{\mu_r \varepsilon_r}}{c} \frac{\partial}{\partial t} [Z_m(t) * \Phi_m(\boldsymbol{r}, t)], \qquad (15.45)$$

where the mth mode time-domain modal characteristic impedance is

$$Z_m(t) = \delta(t) + \omega_{c_m} y_m(t) \qquad (15.46)$$

and the function $y_m(t)$ is defined as

$$y_m(t) = \int_0^{\omega_{c_m} t} \mathrm{J}_0(u) du - \mathrm{J}_1(\omega_{c_m} t), \qquad (15.47)$$

where J_0 and J_1 are the zero-order and first-order, respectively, Bessel functions of the first kind. The detail derivation of (15.46) is shown in the appendix of Loh and Mias (2004a); Loh (2004). Following a similar procedure to the MABTC derivation in the frequency domain, the time-domain field equation at the excitation boundary Γ_e, where $z = z_1$, has the following form:

$$\begin{aligned}
\Phi(x, z_1, t) &= \Phi^{\mathrm{inc}}(x, z_1, t) + \sum_{m=0}^{\infty} \Phi_m^{\mathrm{ref}}(x, z_1, t) \\
&= h_{\mathrm{inc}}(x) \Phi^{\mathrm{inc}}(z_1, t) + \sum_{m=0}^{\infty} a_m h_m(x) \Phi_m^{\mathrm{ref}}(z_1, t),
\end{aligned} \qquad (15.48)$$

where $h_m(x)$ is the same as those defined in (15.38) and the modal coefficients a_m are given by

$$a_m = \frac{1}{\Phi_m^{\mathrm{ref}}(z_1, t)} \int_0^d [\Phi(x, z_1, t) - \Phi^{\mathrm{inc}}(x, z_1, t)] h_m(x) dx. \qquad (15.49)$$

Taking the partial derivative of (15.48) with respect to n at $z = z_1$, noting $\hat{\mathbf{n}} = -\hat{\mathbf{z}}$, and making use of (15.45) and (15.49), one obtains:

$$\begin{aligned}
\left. \frac{\partial \Phi(\boldsymbol{r}, t)}{\partial n} \right|_{z=z_1} = {}&- \sum_{m=0}^{\infty} \frac{h_m(x) \sqrt{\mu_r \varepsilon_r}}{c} \\
&\times \frac{\partial}{\partial t} \left[Z_m(t) * \left[\int_0^d \Phi(x', z, t) h_m(x') dx' \right] \right] \Bigg|_{z=z_1} \\
&+ \frac{2\sqrt{\varepsilon_r}}{c} \frac{\partial}{\partial t} [Z_{\mathrm{inc}}(t) * \Phi^{\mathrm{inc}}(x, z, t)]|_{z=z_1}.
\end{aligned} \qquad (15.50)$$

Similarly, the time-domain field equation at the termination truncation boundary Γ_t, where $z = z_2$, has the following form:

$$\Phi(x, z_2, t) = \sum_{m=0}^{\infty} \Phi_m^{\text{tran}}(x, z_2, t) = \sum_{m=0}^{\infty} b_m h_m(x) \Phi_m^{\text{tran}}(z_2, t). \qquad (15.51)$$

Following the aforementioned procedure and noting $\hat{\mathbf{n}} = \hat{\mathbf{z}}$, one obtains the modal coefficients b_m, and the partial derivative of (15.51) with respect to n at $z = z_2$ as follows:

$$b_m = \frac{1}{\Phi_m^{\text{tran}}(z_2, t)} \int_0^d \Phi(x, z_2, t) h_m(x) dx \qquad (15.52)$$

$$\left. \frac{\partial \Phi(\mathbf{r}, t)}{\partial n} \right|_{z=z_2} = -\sum_{m=0}^{\infty} \frac{h_m(x) \sqrt{\mu_r \varepsilon_r}}{c}$$

$$\times \frac{\partial}{\partial t} \left[Z_m(t) * \left[\int_0^d \Phi(x', z, t) h_m(x') dx' \right] \right] \Bigg|_{z=z_2}. \qquad (15.53)$$

Upon substituting (15.50) and (15.53) into the boundary integral of (15.23) and rearranging both sides of (15.23), an expression is obtained that requires the first-order derivative of the following convolution term to be evaluated:

$$C(t) = Z_m(t) * \Phi(t). \qquad (15.54)$$

Recently, an efficient FETD code, employing a Floquet modal absorbing boundary condition to model scattering from periodic structures, has been proposed using recursive convolution (Cai and Mias, 2007), which is based on the ability to accurately approximate functions, over the entire computation time, using a summation of exponential functions. The recursive convolution scheme can readily be applied to the MABTC following the approach in Cai and Mias (2007). Here, however, an alternative approach to implement the convolution is followed. Substituting (15.46) into (15.54), one obtains:

$$C(t) = \Phi(t) + \omega_{c_m} [y_m(t) * \Phi(t)]$$

$$= \Phi(t) + \omega_{c_m} \int_0^t y_m(\tau) \Phi(t - \tau) d\tau, \qquad (15.55)$$

where $y_m(t) = \int_0^{\omega_{c_m} t} \mathrm{J}_0(u) du - \mathrm{J}_1(\omega_{c_m} t)$.

Using the trapezoidal integration rule, one obtains the following discrete-time representation form:

$$C^n = \Phi^n + \omega_{c_m} \sum_{k=0}^{n-1} \left[\Phi^{n-k} y_m^k + \Phi^{n-k-1} y_m^{k+1} \right] \frac{\Delta t}{2}, \tag{15.56}$$

where $y_m^k = y_m(k\Delta t)$ and $C^n = C(n\Delta t)$ are the discrete-time representation of $y_m(t)$ and $C(t)$ respectively. Note that $y_0(t) = 0$.

Using the three-point recurrence scheme unconditionally stable Newmark-beta formulation in (15.24) to discretize the time-domain functions as in Lee (1995) and Maradei (2001), the first-order time derivative of the convolution term can be expressed as:

$$\left. \frac{\mathrm{d}[C(t)]}{\mathrm{d}t} \right|_{t=n\Delta t} = \frac{C^{n+1} - C^{n-1}}{2\Delta t}. \tag{15.57}$$

Using (15.56) and (15.57), one obtains the discrete-time representation form of the first-order time derivative of the convolution term in terms of the unknown field as follows:

$$\left. \frac{\mathrm{d}[C(t)]}{\mathrm{d}t} \right|_{t=n\Delta t} = \frac{\Phi^{n+1} - \Phi^{n-1}}{2\Delta t} + \Psi^n, \tag{15.58}$$

where

$$\Psi^n = \frac{\omega_{c_m}}{4} \left\{ 2\Phi^n y_m^1 + \Phi^{n-1} y_m^2 \right.$$

$$\left. + \sum_{k=0}^{n-2} [\Phi^{n-k-1}(y_m^{k+2} - y_m^k) + \Phi^{n-k-2}(y_m^{k+3} - y_m^{k+1})] \right\}. \tag{15.59}$$

However, it is noted that $y_m^0 = 0$, hence the term, $\Phi^{n+1} y_m^0$ is discarded from (15.59). From (15.46) and (15.47), the time-domain modal characteristic impedance $Z_m(t)$ and the function $y_m(t)$ are functions of ω_{c_m}, t, and the Bessel function. As $t \to \infty$, one has $\lim_{t \to \infty} \mathrm{J}_1(t) = 0$ and $\int_0^\infty \mathrm{J}_0(t)dt = 1$ (Jeffrey, 2002) hence the value of $y_m(t)$ tends towards unity. Figure 15.6 shows how $y_m(t)$ varies with time for different modes ($m = 1,2,3$). From Fig. 15.6, in order to speed up the computations for $m > 0$, it is reasonable to assume the following:

$$Z_m(t) = \begin{cases} \delta(t) + \omega_{c_m} y_m(t), & t < t_c \\ \delta(t) + \omega_{c_m} u(t), & t \geq t_c \end{cases}, \tag{15.60}$$

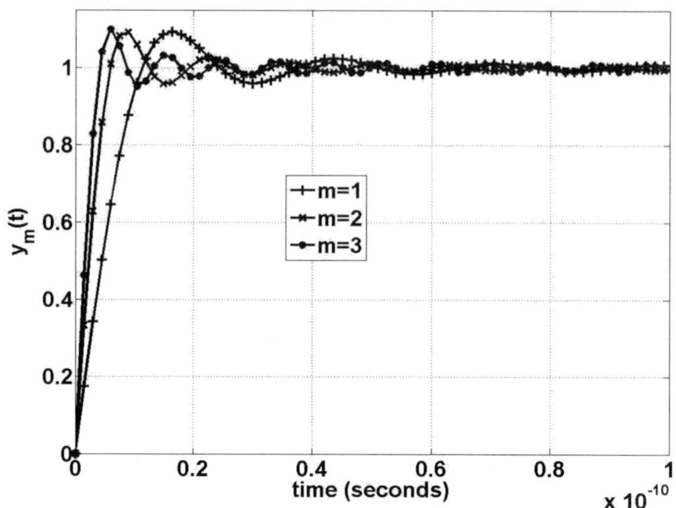

Fig. 15.6. Plot of $y_m(t)$ versus time for different modes ($m = 1,2,3$) with time step, $\Delta t = 1.5\,\mathrm{ps}$ and separation distance between the two plates, $d = 4\,\mathrm{cm}$.

where t_c is the "*approximated convolution cut-off time*" and $u(t)$ is the unit step function. The choice of the t_c value involves a trade-off between accuracy and computational effort. Therefore, for $t \geq t_c$, following the aforementioned procedures, one obtains:

$$\frac{d[C(t)]}{dt}\bigg|_{t=n\Delta t} = \frac{\Phi^{n+1} - \Phi^{n-1}}{2\Delta t} + \frac{\omega_{c_m}}{4}(\Phi^{n+1} + 2\Phi^n + \Phi^{n-1}). \quad (15.61)$$

One observes from Fig. 15.6 that the larger the ω_{c_m} the faster the convergence of $y_m(t)$, and vice versa. This implies that the value of t_c can be chosen by the convergence of $y_1(t)$, while $y_0(t) = 0$ for all time t.

Applying (15.60) in (15.25) and finally, by rearranging both sides of (15.25), one obtains the following Galerkin weighted residual FETD-MABTC formulation:

For $t < t_c$,

$$[M]\{\Phi\}^{n+1} = [P]\{\Phi\}^n + [Q]\{\Phi\}^{n-1} - \sum_{m=0}^{\infty} [R_0]\{\Psi\}^n$$

$$+\frac{\{K\}^{n+1}}{4} + \frac{\{K\}^n}{2} + \frac{\{K\}^{n-1}}{4}, \quad (15.62)$$

where

$$[M] = \frac{[T]}{\Delta t^2} + \frac{1}{2\Delta t}([B] + [R_0]) + \frac{[S]}{4}$$

$$[P] = \frac{2[T]}{\Delta t^2} - \frac{[S]}{2}$$

$$[Q] = -\frac{[T]}{\Delta t^2} + \frac{1}{2\Delta t}([B] + [R_0]) - \frac{[S]}{4}$$

$$K_i^n = \frac{2\sqrt{\mu_r \varepsilon_r}\, g_i^{\mathrm{inc}^n}}{pc} \int_\Gamma N_i h_{inc}(x)\, d\Gamma$$

$$R_{oij} = \frac{\sqrt{\mu_r \varepsilon_r}}{pc} \int_\Gamma N_i h_m(x)\, d\Gamma \int_0^d N_j h_m(x')\, d\Gamma'$$

$$g_i^{\mathrm{inc}^n} = (\Phi_i^{\mathrm{inc}^{n+1}} - \Phi_i^{\mathrm{inc}^{n-1}})/(2\Delta t) + \frac{\omega_{c_{\mathrm{inc}}}}{4} \left\{ 2\Phi_i^{\mathrm{inc}^n} y_{\mathrm{inc}}^1 \right.$$

$$+ \Phi_i^{\mathrm{inc}^{n-1}} y_{\mathrm{inc}}^2 + \sum_{k=0}^{n-2} \left[\Phi_i^{\mathrm{inc}^{n-k-1}} (y_{\mathrm{inc}}^{k+2} - y_{\mathrm{inc}}^k) \right.$$

$$\left. \left. + \Phi_i^{\mathrm{inc}^{n-k-2}} (y_{\mathrm{inc}}^{k+3} - y_{\mathrm{inc}}^{k+1}) \right] \right\},$$

where $p = \mu_r$ for $\Phi = E_y$ (TE polarization), $p = \varepsilon_r$ for $\Phi = H_y$ (TM polarization), and c is the speed of light in free space.

For $t \geq t_c$,

$$[M]\{\Phi\}^{n+1} = [P]\{\Phi\}^n + [Q]\{\Phi\}^{n-1} + \frac{\{K\}^{n+1}}{4} + \frac{\{K\}^n}{2} + \frac{\{K\}^{n-1}}{4}, \tag{15.63}$$

where

$$[M] = \frac{[T]}{\Delta t^2} + \frac{1}{2\Delta t}([B] + [R_0]) + \frac{1}{4}([S] + [R_1])$$

$$[P] = \frac{2[T]}{\Delta t^2} - \frac{1}{2}([S] + [R_1])$$

$$[Q] = -\frac{[T]}{\Delta t^2} + \frac{1}{2\Delta t}([B] + [R_0]) - \frac{1}{4}([S] + [R_1])$$

$$R_{1_{ij}} = \frac{\sqrt{\mu_r \varepsilon_r}}{pc} \sum_{m=0}^{\infty} \omega_{c_m} \int_{\Gamma} N_i h_m(x) d\Gamma \int_0^d N_j h_m(x') d\Gamma'$$

$$g_i^{\mathrm{inc}^n} = (\Phi_i^{\mathrm{inc}^{n+1}} - \Phi_i^{\mathrm{inc}^{n-1}})/(2\Delta t)$$

$$+\frac{\omega_{c_{\mathrm{inc}}}}{4}(\Phi_i^{\mathrm{inc}^{n+1}} + 2\Phi_i^{\mathrm{inc}^n} + \Phi_i^{\mathrm{inc}^{n-1}}).$$

The expressions for $\{K\}$ and $[R_0]$ are the same as for $t < t_c$. It is assumed that both input and output ports have the same plate separation distance.

15.3.2. *Perfectly Matched Layer*

15.3.2.1. *PML in the frequency domain*

Let $[\varepsilon]$, $[\sigma_{\mathrm{E}}]$, $[\mu]$, and $[\sigma_{\mathrm{M}}]$, be diagonal tensors of permittivity, electric conductivity, permeability, and magnetic conductivity, respectively. In the frequency domain, within the lossy and anisotropic PML region, the following general time-harmonic form of Maxwell's equations are employed (Sacks *et al.*, 1995):

$$\nabla \times \tilde{\mathbf{H}} = j\omega[\varepsilon]\tilde{\mathbf{E}} + [\sigma_{\mathrm{E}}]\tilde{\mathbf{E}} = j\omega\bar{\bar{\varepsilon}}\tilde{\mathbf{E}} \tag{15.64}$$

$$\nabla \times \tilde{\mathbf{E}} = -j\omega[\mu]\tilde{\mathbf{H}} - [\sigma_{\mathrm{M}}]\tilde{\mathbf{H}} = -j\omega\bar{\bar{\mu}}\tilde{\mathbf{H}}, \tag{15.65}$$

where complex diagonal tensors of permittivity, $\bar{\bar{\varepsilon}}(\omega)$, and permeability, $\bar{\bar{\mu}}(\omega)$, are defined as follows (Sacks *et al.*, 1995; Gedney, 1996; Taflove and Hugness, 2005):

$$\bar{\bar{\mu}}(\omega) = \mu_0[\Lambda(\omega)], \quad \bar{\bar{\varepsilon}}(\omega) = \varepsilon_0[\Lambda(\omega)], \tag{15.66}$$

where

$$[\Lambda(\omega)] = \begin{bmatrix} \Lambda_x & 0 & 0 \\ 0 & \Lambda_y & 0 \\ 0 & 0 & \Lambda_z \end{bmatrix} = \begin{bmatrix} \dfrac{s_y s_z}{s_x} & 0 & 0 \\ 0 & \dfrac{s_x s_z}{s_y} & 0 \\ 0 & 0 & \dfrac{s_x s_y}{s_z} \end{bmatrix},$$

$$s_{x,y,z} = 1 + \frac{\sigma_{x,y,z}}{j\omega\varepsilon_0},$$

where $\sigma_{x,y,z}$ denotes the conductivity along x-axis, y-axis, and z-axis, respectively. Within the 2D PML region (see Fig. 15.7) of the FEFD method

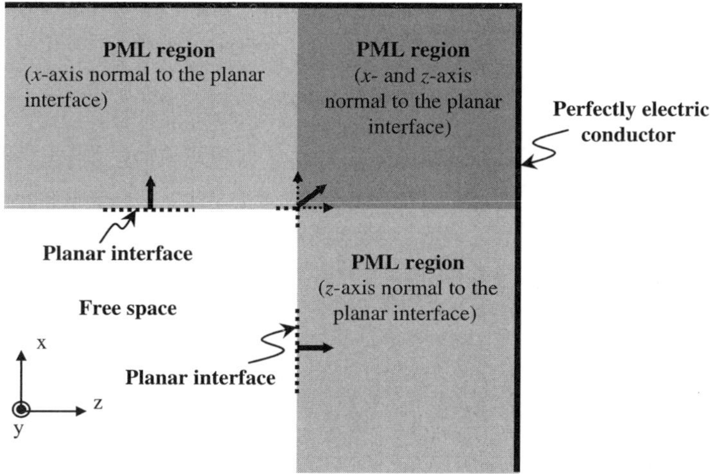

Fig. 15.7. Two-dimensional perfectly matched layer region.

the choice of the parameters in (15.66) is as follows:

1. For the part of the PML region in which the z-axis is normal to the planar interface between the free space and the PML absorber, $\sigma_x = \sigma_y = 0$, i.e., $s_x = s_y = 1$.
2. For the part of the PML region in which the x-axis is normal to the planar interface between the free space and the PML absorber, $\sigma_y = \sigma_z = 0$, i.e., $s_y = s_z = 1$.
3. For PML corner region, $\sigma_y = 0$, i.e., $s_y = 1$.

From the above, one notes that for a 2D problem having field variation in x- and z-axis, the implementation of PML to FEFD requires that $s_y = 1$ ($\sigma_y = 0$). The effectiveness of the PML layer can be optimized by carefully selecting its conductivity (Gedney, 1996).

Beginning with the general time-harmonic form of Maxwell's equations (15.64 and 15.65), one obtains, within the PML region, the following frequency-domain second-order scalar Helmholtz wave equation (Sacks *et al.*, 1995):

$$\frac{\partial}{\partial x}\left(\frac{1}{p_z}\frac{\partial \Phi}{\partial x}\right) + \frac{\partial}{\partial z}\left(\frac{1}{p_x}\frac{\partial \Phi}{\partial z}\right) + \omega^2 q_y \Phi = 0, \qquad (15.67)$$

where p_i, and q_i ($i = x$, y, z), are the elements of the tensors in (15.66) defined as $\bar{\bar{p}} = \bar{\bar{\varepsilon}}$, $\bar{\bar{q}} = \bar{\bar{\mu}}$ for $\Phi = \tilde{H}_y$ (TM polarization); $\bar{\bar{p}} = \bar{\bar{\mu}}$, $\bar{\bar{q}} = \bar{\bar{\varepsilon}}$ for $\Phi = \tilde{E}_y$ (TE polarization).

Substituting (15.66) into (15.67) and following the same procedure as in Section 15.2.2, one obtains, in matrix form, the following system of equations for the PML region:

$$[M]\{\Phi\} = \{K\}, \tag{15.68}$$

where

$$[M] = \frac{s_z}{s_x}[S_x] + \frac{s_x}{s_z}[S_z] - k_0^2 s_x s_z [T]$$

$$K_i = \oint_\Gamma N_i \left(\frac{s_z}{s_x} \frac{\partial \Phi}{\partial x} \hat{x} + \frac{s_x}{s_z} \frac{\partial \Phi}{\partial z} \hat{z} \right) \bullet \hat{n} d\Gamma$$

$$S_{x_{ij}} = \iint_\Omega \frac{\partial N_i}{\partial x} \frac{\partial N_j}{\partial x} d\Omega$$

$$S_{z_{ij}} = \iint_\Omega \frac{\partial N_i}{\partial z} \frac{\partial N_j}{\partial z} d\Omega.$$

The elements of $[T]$ and the term k_0 are the same as those defined in (15.28) and (15.26), respectively. Equation (15.68) will be incorporated into the global matrix. It should be noted that there will be no boundary integral contributions from (15.68) to the global matrix because of the interface conditions, as the boundary integral of the PML region cancels the boundary integral of air region at their interface, due to the fact that normal derivative operator $\partial/\partial n$ acts in opposite directions on the interface. In addition, the PEC backing of the PML region means that $\partial\Phi/\partial n = 0$ (for TM polarization) or $\Phi = 0$ (for TE polarization), hence the boundary integral value along the PEC boundary is zero.

15.3.2.2. *PML in the time domain*

To obtain the PML parameters in the time domain one can inverse Laplace transform the elements of the tensors in (15.66). To implement the PML in the time domain, the approach in Bermani *et al.* (1999) is followed. Instead of using the tensor components proposed by Kuzuoglu and Mittra (1996), those of (15.66) are employed as proposed by Gedney (1996)) resulting in an uniaxial perfectly matched anisotropic absorber. Hence a convolution integral involving these tensor components and the unknown field is obtained. The convolution integral is evaluated recursively without a need to store the fields of all the past time steps. Within the corner region, if one lets $s_x = s_z = 1 + \sigma/j\omega\varepsilon_0$, where σ denotes the conductivity, the difficulty in the computational implementation of the convolution integral can be significantly reduced (see (15.75)). Beginning with the general time-dependent form of Maxwell's equations as in Bermani *et al.* (1999):

$$\nabla \times \mathbf{E}(\boldsymbol{r}, t) = -\frac{\partial}{\partial t}(\overline{\overline{\mu}}(t) * \mathbf{H}(\boldsymbol{r}, t)) \tag{15.69}$$

$$\nabla \times \mathbf{H}(\boldsymbol{r}, t) = \frac{\partial}{\partial t}(\overline{\overline{\varepsilon}}(t) * \mathbf{E}(\boldsymbol{r}, t)), \tag{15.70}$$

where $\boldsymbol{r} = (x, z)$, and

$$\overline{\overline{\mu}}(t) = L^{-1}[\overline{\overline{\mu}}(\omega)] = \mu_0[\Lambda(t)], \quad \overline{\overline{\varepsilon}}(t) = L^{-1}[\overline{\overline{\varepsilon}}(\omega)] = \varepsilon_0[\Lambda(t)] \tag{15.71}$$

are the time-domain correspondence of the generalized complex diagonal tensor given in (15.66), one obtains, within the PML region, the following time-dependent second-order scalar Helmholtz wave equation:

$$\frac{\partial}{\partial x}\left(p_z^{-1}(t) * \frac{\partial \Phi(\boldsymbol{r}, t)}{\partial x}\right) + \frac{\partial}{\partial z}\left(p_x^{-1}(t) * \frac{\partial \Phi(\boldsymbol{r}, t)}{\partial z}\right)$$

$$-\frac{1}{c^2}q_y(t) * \frac{\partial^2 \Phi(\boldsymbol{r}, t)}{\partial t^2} = 0, \tag{15.72}$$

where c is the speed of light, Φ is the transverse field component, and p_i and q_i $(i = x, y, z)$ are the elements of the tensors in (15.71). They are defined as $\Phi = H_y$, $\overline{\overline{p}} = \overline{\overline{\varepsilon}}$, $\overline{\overline{q}} = \overline{\overline{\mu}}$ (TM polarization), and $\Phi = E_y$, $\overline{\overline{p}} = \overline{\overline{\mu}}$, $\overline{\overline{q}} = \overline{\overline{\varepsilon}}$ (TE polarization). Note that in (15.71), $\Lambda_x = L^{-1}(s_z/s_x)$, $\Lambda_y = L^{-1}(s_x s_z)$, $\Lambda_z = L^{-1}(s_x/s_z)$ where $s_{x,z}$ are the same as those defined in (15.66) and L^{-1} denotes the inverse Laplace transform operator.

Substituting (15.71) into (15.72) and using the same procedures as in Section 15.2.1, the following equations, in matrix form, for the PML region are obtained:

For PML region having the z-axis normal to the air-PML interface:

$$[M]\{\Phi\}^{n+1} = [P]\{\Phi\}^n + [Q]\{\Phi\}^{n-1} - [S_x]\{f\}^n + [S_z]\{g\}^n$$

$$+\frac{\{K\}^{n+1}}{4} + \frac{\{K\}^n}{2} + \frac{\{K\}^{n-1}}{4}. \tag{15.73}$$

For PML region having the x-axis normal to the air-PML interface:

$$[M]\{\Phi\}^{n+1} = [P]\{\Phi\}^n + [Q]\{\Phi\}^{n-1} - [S_x]\{f\}^n + [S_z]\{g\}^n$$

$$+ \frac{\{K\}^{n+1}}{4} + \frac{\{K\}^n}{2} + \frac{\{K\}^{n-1}}{4}. \tag{15.74}$$

For PML corner region (one lets $s_x = s_z = 1 + \sigma/j\omega\varepsilon_0$):

$$[M_c]\{\Phi\}^{n+1} = [P_c]\{\Phi\}^n + [Q_c]\{\Phi\}^{n-1}$$

$$+ \frac{\{K\}^{n+1}}{4} + \frac{\{K\}^n}{2} + \frac{\{K\}^{n-1}}{4}, \tag{15.75}$$

where

$$[M] = \frac{[T]}{c^2 \Delta t^2} + \frac{[B]}{2\Delta t} + \frac{[S]}{4},$$

$$[M_c] = \frac{[T]}{c^2 \Delta t^2} + \frac{[B]}{2\Delta t} + \frac{[S]}{4} + \left(\frac{\sigma}{\varepsilon_0}\right)^2 \frac{[T]}{4}$$

$$[P] = \frac{2[T]}{c^2 \Delta t^2} - \frac{[S]}{2}, \quad [P_c] = \frac{2[T]}{c^2 \Delta t^2} - \frac{[S]}{2} + \left(\frac{\sigma}{\varepsilon_0}\right)^2 \frac{[T]}{4}$$

$$[Q] = -\frac{[T]}{c^2 \Delta t^2} + \frac{[B]}{2\Delta t} - \frac{[S]}{4},$$

$$[Q_c] = -\frac{[T]}{c^2 \Delta t^2} + \frac{[B]}{2\Delta t} - \frac{[S]}{4} - \left(\frac{\sigma}{\varepsilon_0}\right)^2 \frac{[T]}{4}$$

$$S_{x_{ij}} = \frac{\sigma_a}{\varepsilon_0} \iint_\Omega \frac{1}{p_z} \frac{\partial N_i}{\partial x} \frac{\partial N_j}{\partial x} d\Omega$$

$$S_{z_{ij}} = \frac{\sigma_a}{\varepsilon_0} \iint_\Omega \frac{1}{p_x} \frac{\partial N_i}{\partial z} \frac{\partial N_j}{\partial z} d\Omega$$

$$K_i^n = p_0 \oint_\Gamma N_i \left(P_z^{-1}(t) * \frac{\partial \Phi^n}{\partial x} \hat{x} + P_x^{-1}(t) * \frac{\partial \Phi^n}{\partial z} \hat{z} \right) \bullet \hat{n} d\Gamma$$

for (15.73) $P_x^{-1}(t) = \dfrac{1}{P_0}\left(\delta(t) - \dfrac{\sigma_z}{\varepsilon_0} e^{\frac{\sigma_z}{\varepsilon_0}t} u(t)\right)$ and

$$P_z^{-1}(t) = \dfrac{1}{P_0}\left(\delta(t) + \dfrac{\sigma_z}{\varepsilon_0} u(t)\right)$$

for (15.74) $P_x^{-1}(t) = \dfrac{1}{P_0}\left(\delta(t) + \dfrac{\sigma_z}{\varepsilon_0} u(t)\right)$ and

$$P_z^{-1}(t) = \dfrac{1}{P_0}\left(\delta(t) - \dfrac{\sigma_z}{\varepsilon_0} e^{\frac{\sigma_z}{\varepsilon_0}t} u(t)\right)$$

for (15.75) $P_x^{-1}(t) = P_z^{-1}(t) = \dfrac{1}{P_0}\delta(t)$

$$f^n = f^{n-1} + (\Phi_j^n + \Phi_j^{n-1})\Delta t/2$$

$$g^n = e^{-\frac{\sigma_a}{\varepsilon_0}\Delta t} g^{n-1} + \left(\Phi_j^n + e^{-\frac{\sigma_a}{\varepsilon_0}\Delta t}\Phi_j^{n-1}\right)\Delta t/2,$$

where $[T]$, $[B]$, and $[S]$ are the same as those in (15.28). $\sigma_a = \sigma_z$ for (15.73), $\sigma_a = \sigma_x$ for (15.74). The constant p_0 is defined as $p_0 = \varepsilon_0$ for TM polarization ($\Phi = H_y$) and $p_0 = \mu_0$ for TE polarization ($\Phi = E_y$). It is noted that K_i^n will have no contribution to the boundary integral term in (15.73)–(15.75), and are hence ignored, as at the interface between the free space and the PML absorber the continuity condition is satisfied (15.19) and the PML is backed by a PEC on which the homogeneous conditions $\partial\Phi/\partial n = 0$ (for $\Phi = H_y$) and $\Phi = 0$ (for $\Phi = E_y$) apply. It was also shown that by properly choosing the PML parameters the lossy uniaxial medium is "perfectly matched" to an isotropic medium (Gedney, 1996).

15.3.3. *First-Order Absorbing Boundary Condition*

15.3.3.1. *First-Order ABC in the frequency domain*

The following outward normal derivative equation of the field in the frequency domain can be employed as the first-order ABC (Jin, 2002):

$$\frac{\partial\Phi(\boldsymbol{r}, s)}{\partial n} = \pm jk_0\sqrt{\mu_r\varepsilon_r}\,\Phi(\boldsymbol{r}, s), \tag{15.76}$$

where $\boldsymbol{r} = (x, z)$, $s = j\omega$. k_0, ε_r, are defined in (15.8). Note that this is only valid for the TM_0 mode i.e. $\Phi = \tilde{H}_y$. Although its implementation in FEFD can start from (15.76), here an FEFD implementation is presented that starts from (15.40) and (15.43) by considering only the TM_0 mode. Hence one obtains from (15.40) at $z = z_1$,

$$\left.\frac{\partial \tilde{H}_y(\boldsymbol{r}, s)}{\partial n}\right|_{z=z_1} = -h_{\text{inc}}(x) j k_0 \sqrt{\mu_r \varepsilon_r} \int_0^d \tilde{H}_y(x', z, s) h_{\text{inc}}(x') dx' \Big|_{z=z_1}$$

$$+ 2 j k_0 \sqrt{\mu_r \varepsilon_r} \tilde{H}_y^{\text{inc}}(\text{r}, s)|_{z=z_1}. \tag{15.77}$$

From (15.43) one obtains at $z = z_2$,

$$\left.\frac{\partial \tilde{H}_y(\boldsymbol{r}, s)}{\partial n}\right|_{z=z_2} = -h_{\text{inc}}(x) j k_0 \sqrt{\mu_r \varepsilon_r} \int_0^d \tilde{H}_y(x', z, s) h_{\text{inc}}(x') dx'|_{z=z_2}. \tag{15.78}$$

Applying (15.77) and (15.78) and rearranging both sides of (15.28), one obtains the following Galerkin weighted residual FEFD first-order ABC formulation

$$[M]\{H_y\} = \{K\}, \tag{15.79}$$

where

$$[M] = [S] - k_0^2[T] + [Q]$$

$$Q_{ij} = j k_0 \sqrt{\frac{\mu_r}{\varepsilon_r}} \int_\Gamma N_i h_{\text{inc}}(x) d\Gamma \int_0^d N_j h_{\text{inc}}(x') d\Gamma'$$

$$K_i = 2 j k_0 \sqrt{\frac{\mu_r}{\varepsilon_r}} H_y^{\text{inc}}(z_1, s) \int_\Gamma N_i h_{\text{inc}}(x) d\Gamma.$$

15.3.3.2. *First-order ABC in the time domain*

The Mur's first-order absorbing boundary condition (ABC) can be applied in the time domain (Mur, 1981), for the port truncation boundaries, based on the following outward normal derivative of the time-domain travelling-wave assumption (Wang and Itoh, 2001):

$$\frac{\partial \Phi(\boldsymbol{r}, t)}{\partial n} = \pm \frac{\sqrt{\mu_r \varepsilon_r}}{c} \frac{\partial \Phi(\boldsymbol{r}, t)}{\partial t}, \tag{15.80}$$

where c is the speed of light; the definition of $\boldsymbol{r}$ and the $\pm$ signs are given in (15.6). Again note that this only valid for the TM_0 mode, $\Phi = H_y$, and that an FETD implementation is presented that starts from (15.50) and (15.53). Noting that $Z_{\text{inc}}(t) = Z_0(t) = \delta(t)$, $\delta(t) * f(t) = f(t)$, and $h_{\text{inc}} = h_0$, it follows from (15.50) at the excitation boundary Γ_e, where $z = z_1$ that:

$$\frac{\partial H_y(\boldsymbol{r},t)}{\partial n}\bigg|_{z=z_1} = -\frac{h_{\text{inc}}(x)\sqrt{\mu_r \varepsilon_r}}{c}\left[\frac{\partial}{\partial t}\int_0^d H_y(x',z,t)h_{\text{inc}}(x')dx'\right]\bigg|_{z=z_1}$$

$$+ \frac{2\sqrt{\mu_r \varepsilon_r}}{c}\left[\frac{\partial}{\partial t}H_y^{\text{inc}}(\boldsymbol{r},t)\right]\bigg|_{z=z_1}. \qquad (15.81)$$

In addition, starting from (15.53) one obtains at $z = z_2$:

$$\frac{\partial H_y(\boldsymbol{r},t)}{\partial n}\bigg|_{z=z_2} = -\frac{h_{\text{inc}}(x)\sqrt{\mu_r \varepsilon_r}}{c}\left[\frac{\partial}{\partial t}\int_0^d H_y(x',z,t)h_{\text{inc}}(x')dx'\right]\bigg|_{z=z_2}. \qquad (15.82)$$

Applying (15.81) and (15.82) and rearranging both sides of (15.25), one obtains the following Galerkin weighted residual FETD first-order ABC formulation:

$$[M]\{H_y\}^{n+1} = [P]\{H_y\}^n + [Q]\{H_y\}^{n-1}$$

$$+ \frac{\{K\}^{n+1}}{4} + \frac{\{K\}^n}{2} + \frac{\{K\}^{n-1}}{4}, \qquad (15.83)$$

where

$$[M] = \frac{[T]}{\Delta t^2} + \frac{[B_1]}{\Delta t} + \frac{[S]}{4}$$

$$[P] = \frac{2[T]}{\Delta t^2} - \frac{[S]}{2}$$

$$[Q] = -\frac{[T]}{\Delta t^2} + \frac{[B_1]}{2\Delta t} - \frac{[S]}{4}$$

$$B_{1_{ij}} = \frac{1}{c}\sqrt{\frac{\mu_r}{\varepsilon_r}}\int_\Gamma N_i h_{\text{inc}}(x)d\Gamma \int_0^d N_j h_{\text{inc}}(x')d\Gamma'$$

$$K_i^n = \frac{2}{c}\sqrt{\frac{\mu_r}{\varepsilon_r}}[(H_{y_i}^{\text{inc}^{n+1}} - H_{y_i}^{\text{inc}^{n-1}})/2\Delta t]\int_\Gamma N_i h_{\text{inc}}(x)d\Gamma.$$

15.3.3.3. *First-order ABC in envelope-finite-element domain*

This section presents the formulations for implementing the first-order absorbing boundary condition (ABC) to the EVFE technique noting that from (15.29) $H_y(x, z, t) = \phi(x, z, t)e^{j\omega_c t}$. Starting from the derivation of equations in Section 15.3.3.2 one can show that at $z = z_1$:

$$
\left. \frac{\partial \phi(\boldsymbol{r}, s)}{\partial n} \right|_{z=z_1} = -\frac{h_{\mathrm{inc}}(x)}{c} \left[\left(\frac{\partial}{\partial t} + j\omega_c \right) \int_0^d \phi(x', z, t) h_{\mathrm{inc}}(x') dx' \right] \Bigg|_{z=z_1}
$$

$$
+ \frac{2}{c} \left[\left(\frac{\partial}{\partial t} + j\omega_c \right) \phi^{\mathrm{inc}}(\boldsymbol{r}, s) \right] \Bigg|_{z=z_1} , \tag{15.84}
$$

and at $z = z_2$:

$$
\left. \frac{\partial \phi(\boldsymbol{r}, s)}{\partial n} \right|_{z=z_2} = -\frac{h_{\mathrm{inc}}(x)}{c} \left[\left(\frac{\partial}{\partial t} + j\omega_c \right) \int_0^d \phi(x', z, t) h_{\mathrm{inc}}(x') dx' \right] \Bigg|_{z=z_2} . \tag{15.85}
$$

Upon employing the Newmark-beta formulation (see (15.24)) to discretize in time domain the time-varying complex envelope of the field, ϕ, in (15.84) and (15.85) which, in turn, are substituted in (15.34). Finally, by rearranging both sides of (15.34), one obtains the following Galerkin weighted residual EVFE first-order ABC formulation:

$$
[M]\{\phi\}^{n+1} = [P]\{\phi\}^n + [Q]\{\phi\}^{n-1} + \frac{\{K\}^{n+1}}{4} + \frac{\{K\}^n}{2} + \frac{\{K\}^{n-1}}{4}, \tag{15.86}
$$

where

$$
[M] = \frac{[T]}{\Delta t^2} + \frac{[B_1]}{\Delta t} + \frac{[S_1]}{4}
$$

$$
[P] = \frac{2[T]}{\Delta t^2} - \frac{[S_1]}{2}
$$

$$
[Q] = -\frac{[T]}{\Delta t^2} + \frac{[B_1]}{2\Delta t} - \frac{[S_1]}{4}
$$

$$
B_{1_{ij}} = \iint_\Omega \frac{2j\omega_c \mu_r}{c^2} N_i N_j d\Omega + \frac{1}{c\varepsilon_r} \int_\Gamma N_i h_{\mathrm{inc}}(x) d\Gamma \int_0^d N_j h_{\mathrm{inc}}(x') d\Gamma'
$$

$$S_{1_{ij}} = \iint_{\Omega} \left[\frac{1}{\varepsilon_r} \nabla N_i \cdot \nabla N_j - \frac{\omega_c^2 \mu_r}{c^2} N_i N_j \right] d\Omega$$

$$+ \frac{j\omega_c}{c\varepsilon_r} \int_{\Gamma} N_i h_{\mathrm{inc}}(x) d\Gamma \int_0^d N_j h_{\mathrm{inc}}(x') d\Gamma'$$

$$K_i^n = \frac{2 f_i^{\mathrm{inc}^n}}{c\varepsilon_r} \int_{\Gamma} N_i h_{\mathrm{inc}}(x) d\Gamma + \frac{2 j\omega_c g_i^{\mathrm{inc}^n}}{c\varepsilon_r} \int_{\Gamma} N_i h_{\mathrm{inc}}(x) d\Gamma$$

$$f_i^{\mathrm{inc}^n} = (\phi_i^{\mathrm{inc}^{n+1}} - \phi_i^{\mathrm{inc}^{n-1}})/2\Delta t$$

$$g_i^{\mathrm{inc}^n} = (\phi_i^{\mathrm{inc}^{n+1}} + 2\phi_i^{\mathrm{inc}^n} + \phi_i^{\mathrm{inc}^{n-1}})/4.$$

15.4. Numerical Results

The 2D finite-element methods discussed in this chapter have been coded in FORTRAN 77 using quadratic nodal-based triangular elements for area discretization and quadratic nodal-based line elements at the domain boundaries. The code embeds the components of local elements to the corresponding global location in the final matrix equation. The matrix equation is solved either iteratively using the bi-conjugate gradient method (Volakis *et al.*, 1998) or directly using the multifrontal sparse Gaussian elimination solvers, MA47 (for FETD method) and ME47 (for FEFD and EVFE methods) provided from the Harwell Subroutine Library (AEA Technology, Oxfordshire, UK).

Several 2D electromagnetic problems of research interest are characterized. For validating the accuracy, correct functioning and versatility of the developed formulations the calculated results between various finite-element methods are compared, where available, with analytical results.

In all the following examples the magnetic field (for TM polarization) is chosen as the working variable. In the FEFD code a TM_0 (i.e., transverse electromagnetic (TEM)) mode is sent. In the FETD code a modulated Gaussian pulse is sent (as (15.9)). In the EVFE code a modulated Gaussian pulse is sent and it is of the form:

$$H_y^{inc}(t) = \exp \left[-\frac{(t - 4\sigma_t)^2}{2\sigma_t^2} \right] e^{j\omega_c t}, \tag{15.87}$$

where $\omega_c = 2\pi f_c$, and f_c and σ_t denote the carrier frequency and the time-delay constant, respectively. In all examples, M+1 $(0 \leq m \leq M)$ modes, have been employed for MABTC terminations, where M $= 3$. The boundary integrals were evaluated numerically using a five-point Gaussian–Legendre quadrature integration (Silvester and Ferrari, 1996). Unless otherwise specified, the same frequency resolution, Δf, is applied. To achieve this, in the FETD and EVFE methods, a total time of $t_{total} = 1/\Delta f$ is required. The same mesh structure is employed for all the aforementioned FE methods, in order to compare their computational performance. The density of elements obeys the rule of thumb (Mias, 1995) for a mesh of second-order finite elements i.e., at least four second-order elements are required per wavelength. The Poynting theorem is employed for the evaluation of the normalized reflected power ($\mathrm{P}^{\mathrm{ref}}_{\mathrm{norm}}$) and normalized transmitted power ($\mathrm{P}^{\mathrm{tran}}_{\mathrm{norm}}$) in decibel scale obtained as follows:

$$\mathrm{P}^{\mathrm{ref}}_{\mathrm{norm}} = \sum_{m=0}^{\infty} \mathrm{P}^{\mathrm{ref}}_{\mathrm{norm}m} = \frac{\sum_{m=0}^{\infty} \eta_m |a_m \Phi^{\mathrm{ref}}{}_m(z_1, s)|^2}{\eta_{\mathrm{inc}} |\Phi^{\mathrm{inc}}(z_1, s)|^2} \tag{15.88}$$

$$\mathrm{P}^{\mathrm{tran}}_{\mathrm{norm}} = \sum_{m=0}^{\infty} \mathrm{P}^{\mathrm{tran}}_{\mathrm{norm}m} = \frac{\sum_{m=0}^{\infty} \eta_m |b_m \Phi^{\mathrm{tran}}{}_m(z_2, s)|^2}{\eta_{\mathrm{inc}} |\Phi^{\mathrm{inc}}(z_2, s)|^2} \tag{15.89}$$

where $\eta_k = \eta_{\mathrm{TM}_k}$ for TM polarisation and $\eta_k = -1/\eta_{\mathrm{TE}_k}$ for TE polarisation. The terms η_{TM_k} and η_{TE_k} are defined as $\eta_{\mathrm{TM}_k} = \gamma_k/j\omega\varepsilon$ and $\eta_{\mathrm{TE}_k} = j\omega\mu/\gamma_k$, where γ_k is defined in (15.37).

15.4.1. *Periodic Dielectric-Layer Waveguide*

It is known that periodic dielectric layers can be used to design microwave filters (Balanis, 1989; Collin, 1991; Yeh, 2005), which have a variety of microwave applications (Pozar, 2005). A consistent thickness of $\lambda_g/2$ or $\lambda_g/4$ for both the air-gap and the dielectric layer corresponds respectively to the periodic dielectric-layer band-pass or band-stop filter design (Yeh, 2005), where $\lambda_g = \lambda_0/\sqrt{\varepsilon_r}$ is the corresponding dominant-mode waveguide wavelength, $\lambda_0 = c/f$, c is the speed of light in free space, and f is the stop-band centre frequency. The number of periods of the dielectric-layer influences the attenuation characteristics of the stop-band region and the number of the side bands.

Figure 15.8 shows a five-period dielectric-layer band-stop filter. The length of the air gap is d_1, the thickness of the dielectric slab is d_2, the relative permittivity of the air and the dielectric slabs are ε_1 and ε_2, respectively. These parameters are chosen such that at 10 GHz there is band-stop behavior. As depicted in Fig. 15.8, the waveguide is 4 mm wide and having a length $\overline{z_1 z_2} = 45$ mm (for a five-period dielectric-layer

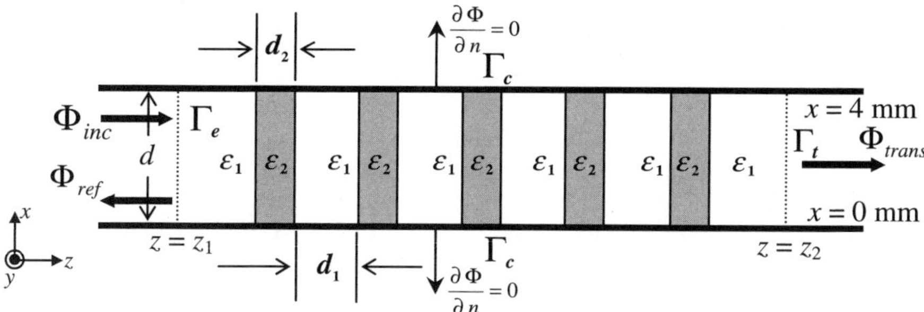

Fig. 15.8. Sketch of a five-period dielectric-layer band-stop filter. Height of the waveguide, $d = 4\,\text{mm}$, air-gap length, $d_1 = 7.5\,\text{mm}$; slab thickness, $d_2 = 2.371708\,\text{mm}$; $\varepsilon_1 = 1$; $\varepsilon_2 = 10$ (dielectric layer).

Table 15.2. Computation times of the waveguides in Fig. 15.8(a) on an AMD 1.6 GHz PC machine.

Problem	Technique	Frequency Resolution	Run Time (HH:MM:SS)	Memory
Fig. 15.8(a) (WITHOUT	FEFD-MABTC	0.1 GHz	00:01:57 (NF $= 171$)	10 MB
DEFECT) 519 DOF	FETD-MABTC (With (15.60) applied where $t_c = 1\,\text{ns}$)	0.1 GHz	00:00:50 ($\Delta t = 1.5\,\text{ps}$, NT $= 6667$)	6.2 MB
	EVFE first-order ABC	0.1 GHz	00:00:29 ($\Delta t = 3\,\text{ps}$, NT $= 3334$)	11.6 MB

NT $=$ Total number of time steps.
NF $=$ Total number of frequency steps.
DOF $=$ Degrees of freedom (unknowns).

structure). The distance between the port boundaries and their adjacent discontinuity is 1.57 mm. Furthermore, a defect introduced by removing the middle dielectric layer of the five-period dielectric-layer design is also modeled. The numerical modeling results obtained from the various finite-element methods are compared with those obtained from the analytical solution (namely the ABCD method) from Yeh (2005). A comparison of their computational performances is given in Table 15.2. The meshes used in the computations are shown in Fig. 15.9.

The waveguide is modeled in the frequency range from 5 GHz to 22 GHz. The frequency step between successive frequency computations is 0.1 GHz. To obey the sampling theorem, the time step is chosen to be

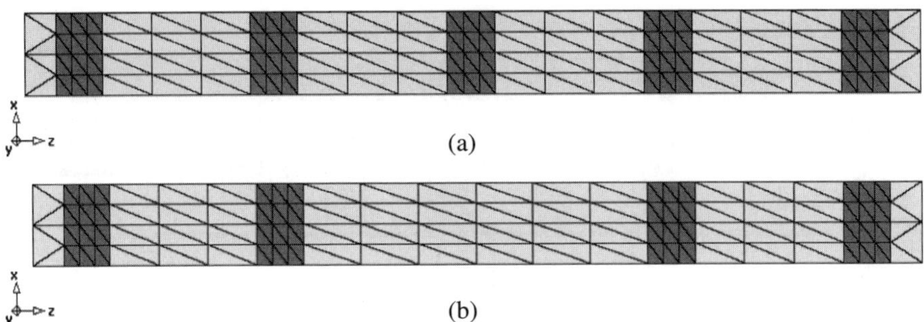

Fig. 15.9.　The FE meshes of the periodic dielectric-layer band-stop filters: (a) without defect, (b) with defect.

1.5 ps. The carrier frequency is $f_c = 13.5\,\text{GHz}$, the time-delay constant is $\sigma_t = 75\,\text{ps}$, and the total time is $t_{total} = 10\,\text{ns}$ (for $\Delta f = 0.1\,\text{GHz}$). Applying the discrete Fourier transform (DFT) to the time-domain waveform, an excellent agreement is found between the analytical solution, FETD, EVFE, and FEFD results, as shown in Fig. 15.10. The transmission maximum in the stop-band region appears due to the defect.

It is noted that the waveguide is operating in dominant mode hence EVFE has good agreement with the other techniques, despite the fact that the waveguide port is terminated with first-order ABC. The time step for the EVFE technique was chosen so as to obtain reasonably good results within the frequency range of interest. Among the finite-element techniques, EVFE has the lowest computation time; however this technique can only reduce computation time without compromising the result accuracy when the signal envelope-to-carrier ratio is very small (Wang and Itoh, 2001).

15.4.2. *Metamaterial Loaded Waveguide*

In recent years, there has been a growing interest in fabricated structures and composite materials that either mimic known material responses or qualitatively have new, physically realizable response functions that do not occur or may not be readily available in nature (Engheta and Ziolkowski, 2006). These devices are generically called metamaterials. Numerous metamaterial papers have been published in various journals and conferences. The underlying research interest in metamaterials is the potential to tailor and manipulate their electromagnetic properties for a variety of applications.

A category of metamaterials are the electromagnetic band gap (EBG) structures that appeared in the late 1990s and found applications at radio

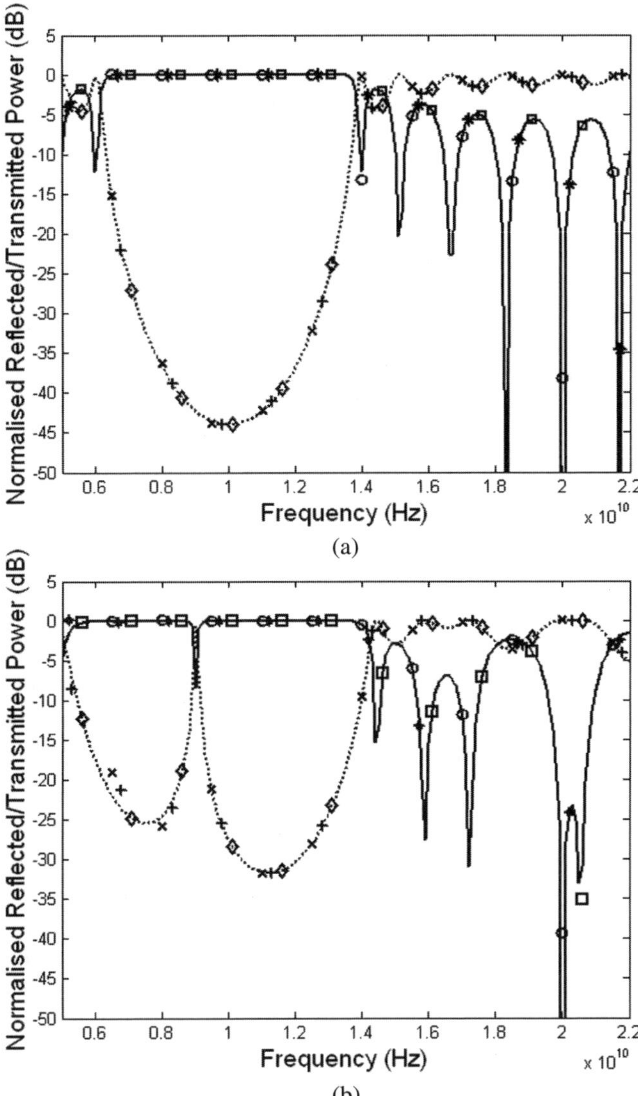

Fig. 15.10. Simulation results of the five-period dielectric-layer band-stop filter. Comparison of the normalized reflected/transmitted power obtained by FETD-MABTC with those obtained by FEFD-MABTC, EVFE first-order ABC, and analytical solution: (a) without defect, (b) with defect. Solid line: P_n^{ref} in analytical solution; *: P_n^{ref} in FETD method; □: P_n^{ref} in FEFD method; o: P_n^{ref} in EVFE method; dotted line: P_n^{tran} in analytical solution; +: P_n^{tran} in FETD method; ◊: P_n^{tran} in FEFD method; x: P_n^{tran} in EVFE method, where P_n^{ref} is the normalized reflected power and P_n^{tran} is the normalized transmitted power.

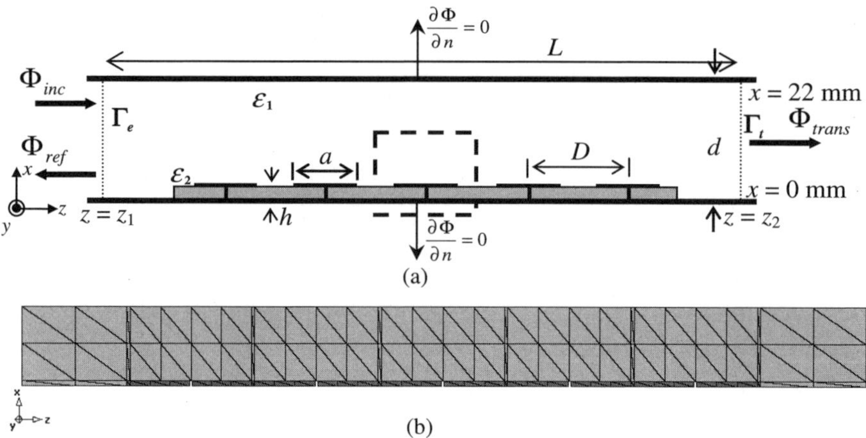

Fig. 15.11.   Sketch of a five-period mushroom-like metallo-dielectric EBG waveguide: height of the waveguide, $d = 22\,\text{mm}$; waveguide length, $L = 198.654\,\text{mm}$; height of the dielectric layer, $h = 1.6\,\text{mm}$, the patch length, $a = 35\,\text{mm}$, the periodicity, $D = 35.913458\,\text{mm}$, $\varepsilon_1 = 1$; $\varepsilon_2 = 4.2$ (dielectric layer).

and microwave frequencies (Scherer *et al.*, 1999; Engheta and Ziolkowski, 2006). Numerous designs have been investigated. To model surface wave guidance by an EBG structure, Fig. 15.11 presents a finite array of five mushroom-like patches (like those proposed by Sievenpiper *et al.* (1999)), situated on a dielectric layer ($\varepsilon_\text{r} = 4.2$) in an air-filled parallel-plate waveguide and the corresponding finite-element mesh. Uniform geometry is assumed in the transverse direction.

The waveguide is modeled in the frequency range from 50 MHz to 5 GHz. In the FEFD technique, the frequency step between successive frequency computations is $\Delta f = 1\,\text{MHz}$. In the FETD method, the sampling rate, carrier frequency, time delay constant, the approximated convolution cut-off time, and total time, are 8 ps, 2.5 GHz, 0.25 ns, 100 ns, and $1\,\mu\text{s}$ (for $\Delta f = 1\,\text{MHz}$), respectively. Upon applying the DFT to the time-domain waveform, very good agreement between the FETD and FEFD results is demonstrated in Fig. 15.12.

15.4.3. *Open Corrugated Surface Waveguiding Structures*

Surface wave guidance by a metallo-dielectric periodic structure has a variety of applications in microwave and optical devices (Collin, 1991). Although, the theory of the propagation of surface waves (Elliott, 1954; Harvey, 1959; Lee and Jones, 1971; Collin, 1991) along corrugated surfaces,

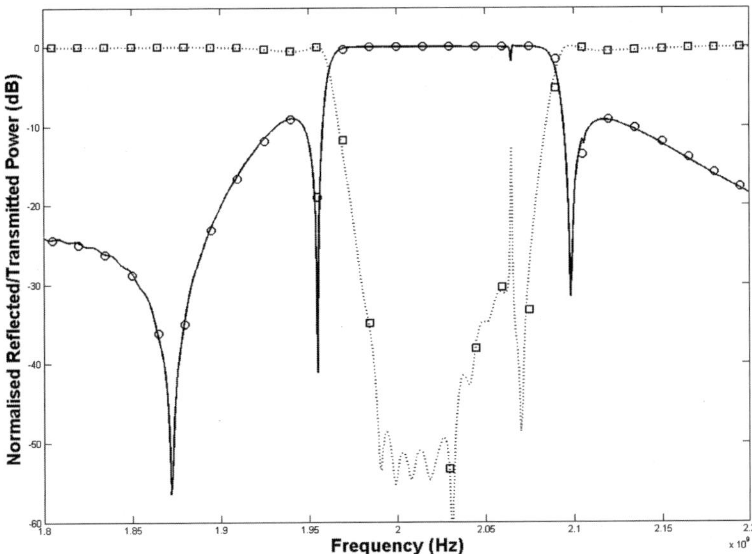

Fig. 15.12.   Simulation results of the mushroom-like metallo-dielectric EBG waveguide (TM polarization). Comparison of the normalized reflected/transmitted power obtained by FETD-MABTC with the one obtained by the FEFD-MABTC method. Solid line: P_n^{ref} in FETD method; dotted line: P_n^{tran} in FETD method; o: P_n^{ref} in FEFD method; □: P_n^{tran} in FEFD method, where P_n^{ref} is the normalized reflected power; P_n^{tran} is the normalized transmitted power.

or guided waves through periodic structures, has been studied for many years, recently there has been a renewed interest and research in the application of periodic structures in microwave devices and in radio wave propagation in buildings. The corrugations work by supporting a surface wave which can direct, for example, radio wave energy around the building, making it appear more transparent to incoming electromagnetic radiation.

As depicted in Fig. 15.13, a metallic infinitely thin teeth-corrugated surface waveguide structure with ($\theta = 20°$) and without ($\theta = 90°$) linear tapering profile, situated between two parallel-plate waveguides with an open region, is considered. The linear tapering profile is created by a planar cut through an array of uniform corrugations at an angle θ with respect to the horizontal axis. The tapering profile is used in order to improve the coupling of the waveguide mode onto the corrugated surface and hence to achieve maximum transmitted power within the pass-band region. In addition, a lossy material is introduced to the 20° tapering structures which

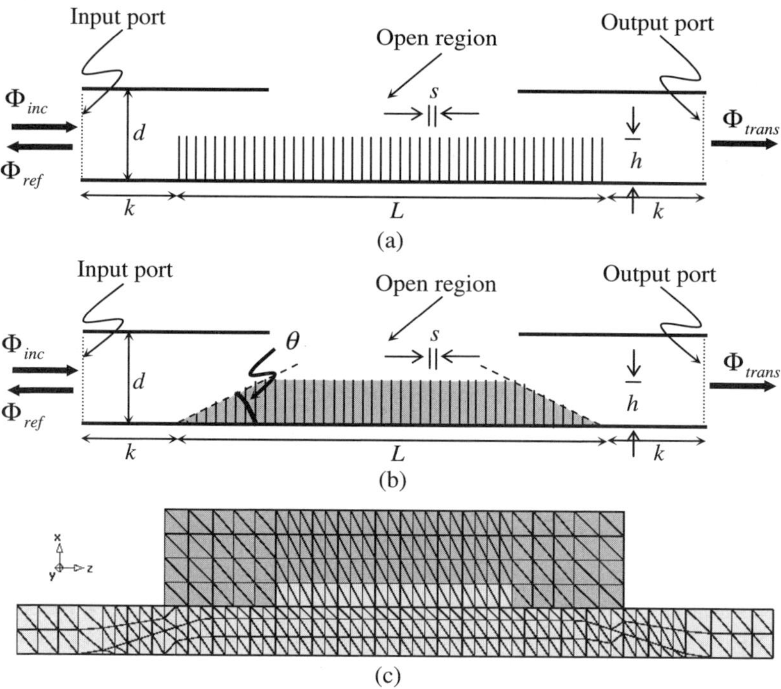

Fig. 15.13.   Sketch of a metallic infinitely thin teeth-corrugated surface waveguide: (a) without 20° tapering profile, (b) with 20° tapering profile, (c) the FE mesh of (b). The port boundaries are truncated by MABTC whereas the open region is truncated by PML. The length of the corrugated surfaces is $L = 470$ mm, the distance between the port boundaries and the corrugation discontinuity is $k = 50$ mm, the height of the waveguide is $d = 40$ mm, the corrugation height is $h = 30$ mm, the corrugation period is $s = 10$ mm; $\varepsilon_1 = 1$. A lossy material is introduced to (b) within the shaded region where $\varepsilon_r = 1.0$, $\sigma = 1$ mS/m.

is filled within the shaded region, as depicted in Fig. 15.13(b), where $\varepsilon_\mathrm{r} = 1$, $\sigma = 1$ mS/m (i.e., conductivity of fresh water[7] (Ulaby, 2000).

In the FEFD code, the frequency range 1.5 GHz to 3.5 GHz is chosen with a frequency step Δf of 0.01 GHz. In the FETD code, the carrier frequency and the time-delay constant are, respectively, $f_c = 2.5$ GHz and $\sigma_t = 0.5$ ns. The time step, Δt, and the cut-off time, t_c, are chosen to be 2.5 ps and 10 ns. The stop time, $t_{\mathrm{stop}} = 0.1\,\mu$s (for $\Delta f = 0.01$ GHz) is

---

[7]In real life, such a material might not exist. The objective here is to consider the losses due to the conductivity of fresh water such as rain on the corrugation structures when they are exposed to an outdoor environment.

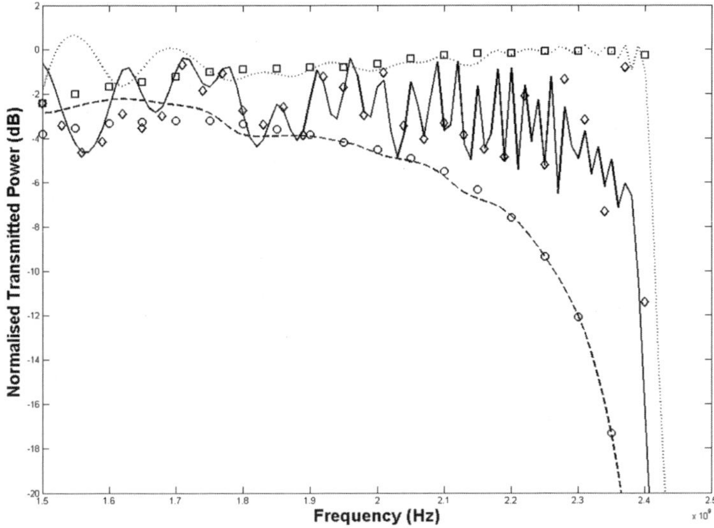

Fig. 15.14. Simulation results of the metallic infinitely thin teeth-corrugated surface waveguide with/without tapering (TM polarization). Comparison of the normalized transmitted power obtained by FETD-MABTC with those obtained by FEFD-MABTC. Solid line: P_n^{tran} in FETD method (without tapering without lossy material); dotted line: P_n^{tran} in FETD method (with 20° tapering without lossy material); dashed line: P_n^{tran} in FETD method (with 20° tapering with lossy material); $\Diamond$: P_n^{tran} in FEFD method (without tapering without lossy material); $\Box$: P_n^{tran} in FEFD method (with 20° tapering without lossy material); o: P_n^{tran} in FEFD method (with 20° tapering with lossy material), where P_n^{tran} is the normalized transmitted power.

chosen. Figure 15.14 shows a comparison of the normalized transmitted power obtained from each of the methods versus frequency within the range 1.5 GHz to 2.5 GHz. The results from the two methods are in good agreement. Note that the ripple for the FETD-MABTC results observed at the low frequency end is a result of truncation error due to the early truncation time t_{stop} of the corresponding transient response in the time domain at which time the transient response did not sufficiently decay to zero. The results can be improved by increasing t_{stop}. The results presented intend to demonstrate that FETD has a disadvantage in simulating structures with highly resonant behavior. In addition, the results also indicate that:

(1) Introducing the 20° tapering linear profile (see Fig. 15.13 (b)) to the uniform corrugated surface waveguide (see Fig. 15.13 (a)) improves the launching of the surfaces waves.

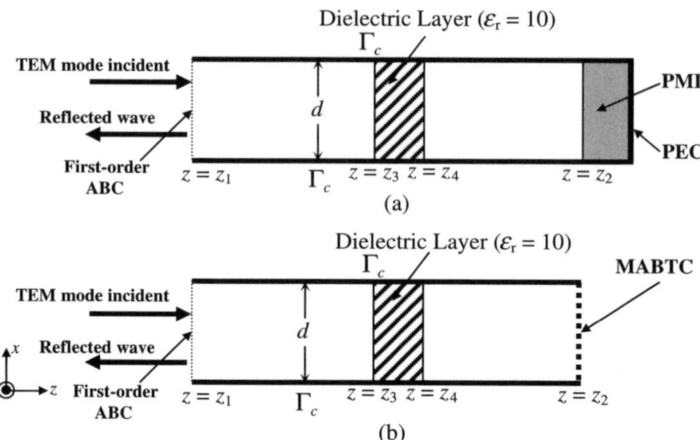

Fig. 15.15. Sketch of a single slab dielectric band-stop filter situated in parallel-plate waveguide terminated with various boundary conditions: (a) output port terminated with a PML truncation, (b) output port terminated with a MABTC truncation. Note that the input port for both (a) and (b) employs Mur's first-order ABC truncation.

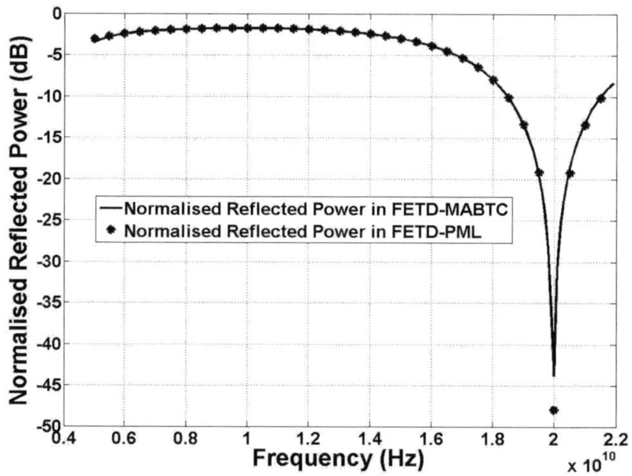

Fig. 15.16. Simulation results of the single slab dielectric band-stop filter situated in parallel-plate waveguide (TM polarization). Comparison of the normalized reflected power obtained from FETD-MABTC with FETD-PML.

(2) Introduction of a lossy material (such as rain) within the corrugated surfaces structure (see Fig. 15.13 (b)) provides an insight of the losses due to such material if the corrugated surface structure is exposed to the elements of nature.

Table 15.3. Computational requirements of the FETD-PML and the FETD-MABTC for the single slab dielectric 10 GHz band-stop filter situated in parallel-plate waveguide (see Fig. 15.15).

Numerical Setup	Sampling	Memory	Run Time (HH:MM:SS)
FETD first-order ABC (at input port) — MABTC (at output port)* (87 D.O.F.[†])	1.5 ps	1.7 MB	00:00:23
FETD first-order ABC (at input port) — PML (at output port) PML thickness = 5 mm PML layer consists of 5 cells $\sigma_z = 1.4$ (137 D.O.F.)	0.375 ps[‡]	1.8 MB	00:03:17

*:With (15.60) applied where $t_c = 10\%$ of t_{total}.
[†]:degree of freedom (unknowns).
[‡]: Sampling interval is 0.375 ps in the PML case in order to achieve stability (Nehrbass *et al.*, 1996).

15.4.4. *Single Slab Dielectric Band-Stop Filter*

In this subsection, a single slab dielectric band-stop filter situated in a parallel-plate waveguide is considered for the comparison of the computational requirements between the FETD-PML and FETD-MABTC codes, where PML and MABTC are employed to truncate the output port and first-order ABC is employed to truncate the input port, as depicted in Fig. 15.15.

The waveguide is 4 mm wide, $\overline{z_1 z_2} = 6$ mm and the length of dielectric layer ($\varepsilon_r = 10$) is $\overline{z_3 z_4} = 2.37$ mm. To obey the sampling theorem, the time step is chosen to be 1.5 ps. The carrier frequency is $f_c = 13.5$ GHz, the time-delay constant is $\sigma_t = 75$ ps, and the total time is $t_{total} = 10$ ns (for $\Delta f = 0.1$ GHz). Applying the discrete Fourier transform (DFT) to the time-domain waveform, a very good agreement is observed between the FETD and FEFD results (see Fig. 15.16).

The bi-conjugate gradient method (Volakis *et al.*, 1998) is employed to solve the final matrix equation iteratively. Table 15.3 shows the computational requirements of the FETD-PML and the approximated FETD-MABTC code when applied to this problem. Note that the convolution is not performed recursively.

References

Balanis, C.A. (1989). *Advanced Engineering Electromagnetics*, Wiley, New York: NY.

Berenger, J.P. (1994). A perfectly matched layer for the absorption of electromagnetic waves, *J. Comput. Phys.*, **114**, 185–200.

Bermani, E., Caorsi, S. and Raffetto, M. (1999). Causal perfectly matched anisotropic absorbers for the numerical solution of unbounded problems in the time domain, *Microw. Opt. Techn. Let.*, **21**, 295–299.

Bertolani, A., Cucinotta, A., Selleri, S., Vincetti, L. and Zoboli, M. (2003). Overview on finite-element time-domain approaches for optical propagation analysis, *Opt. Quant. Electron.*, **35**, 1005–1023.

Boscolo, S., Midrio, M. and Someda, C.G. (2002). Coupling and decoupling of electromagnetic waves in parallel 2-D photonic crystal waveguides, *IEEE J. Quantum Elect.*, **38**, 47–53.

Cai, Y. and Mias, C. (2007). Fast finite element time domain — floquet modal absorbing boundary condition modelling of periodic structures using recursive convolution, *IEEE T. Antenn. Propag.*, **55**, 2550–2558.

Chen, R.S., Yung, K.N., Chan, C.H., Wang, D.X. and Jin, J.M. (2002). An algebraic domain decomposition algorithm for the vector finite-element analysis of 3D electromagnetic field problems, *Microw. Opt. Techn. Let.*, **34**, 414–417.

Chietera, G., Bouk, A.H., Poletti, F., Poli, F., Selleri, S. and Cucinotta, A. (2004). Numerical design for efficiently coupling conventional and pohotonic-crystal waveguides, *Microw. Opt. Techn. Let.*, **42**, 196–199.

Collin, R.E. (1991). *Field Theory of Guided Waves*, (2nd ed.), IEEE Press, New York: NY.

Cucinotta, A., Selleri, S., Vincetti, L. and Zoboli, M. (2001). Impact of the cell geometry on the spectral properties of photonic crystal structures, *Appl. Phys. B-Lasers O.*, **73**, 595–600.

Dibben, D.C. and Metaxas, A.C. (1994). Finite element time domain analysis of multimode applicators using edge elements, *J. Microwave Power EE*, **29**, 242–251.

Dibben, D.C. and Metaxas, R. (1996). Time domain finte element analysis of multimode microwave applicators, *IEEE Trans. Magn.*, **32**, 942–945.

Elliott, R.S. (1954). On the theory of corrugated plane surfaces, *IRE T. Antennas Propag.*, **2**, 71–81.

Engheta, N. and Ziolkowski, R.W. (2006). *Metamaterials — Physics and Engineering Explorations*, Wiley, New York: NY.

Gedney, S.D. (1996). An anisotropic perfectly matched layer-absorbing medium for the truncation of FDTD lattices, *IEEE T. Antenn. Propag.*, **44**, 1630–1639.

Gedney, S.D. and Navsariwala, U. (1995). An unconditionally stable finite element time-domain solution of the vector wave equation, *IEEE Microw. Guided W.*, **5**, 332–334.

Geuzaine, C., Meys, B., Beauvois, V. and Legros, W. (2000). A FETD approach for the modeling of antennas, *IEEE T. Mag.*, **36**, 892–896.

Hallac, A. and Metaxas, R. (2003). Finite element time domain analysis of microwave heating applicators, *IEICE T. Electron*, **E86C**, 2357–2364.

Harvey, A. (1959). Periodic and guiding structures at microwave frequencies, *IRE T. Microw. Theory*, **8**, 30–61.

Jeffrey, A. (2002). *Advanced Engineering Mathematics*, Harcount Academic Press, New York: NY.

Jiao, D. and Jin, J.M. (2001). Time-domain finite-element modeling of dispersive media, *IEEE Microw. Wirel. Co.*, **11**, 220–222.

Jin, J.M. (2002). *The Finite Element Method in Electromagnetics*, (2nd ed.), Wiley, New York: NY.

Kreyszig, E. (2010). *Advanced Engineering Mathematics*, (10th ed.), Wiley, New York: NY.

Kuzuoglu, M. and Mittra, R. (1996). Frequency dependence of the constitutive parameters of causal perfectly matched anisotropic absorbers, *IEEE Microw. Guided W.*, **6**, 447–449.

Lee, J.F., Sun, D. and Cendes, Z.J. (1991). Full-wave analysis of dielectric waveguides using tangential vector finite elements, *IEEE T. Microw. Theory*, **39**, 1262–1271.

Lee, J.F. (1995). Whitney elements time domain (WETD) method, *IEEE T. Magn.*, **31**, 1325–1329.

Lee, J.F., Lee, R. and Cangellaris, A. (1997). Time-domain finite-element methods, *IEEE T. Antenn. Propag.*, **45**, 430–442.

Lee, S. and Jones, W. (1971). Surface waves on two-dimensional corrugated surfaces, *Radio Sci.*, **6**, 811–818.

Loh, T.H. and Mias, C. (2003). Time and frequency domain finite element modelling of periodic structures, *12th International Conference on Antennas and Propagation*, Exeter, UK, **1**, pp. 312–315.

Loh, T.H. and Mias, C. (2004a). Implementation of an exact modal absorbing boundary termination condition for the application of the finite-element time-domain technique to discontinuity problems in closed homogeneous waveguides, *IEEE T. Microw. Theory*, **52**, 882–888.

Loh, T.H. and Mias, C. (2004b). Photonic band gap waveguide simulation using FETD- and FEFD- methods with multi-modal boundary terminations, *IEE 5th International Conference on Computation in Electromagnetics*, Stratford-upon-Avon, UK, pp. 149–150.

Loh, T.H. (2004). An exact port boundary condition for the finite-element time-domain modelling of microwave devices. Ph.D. Thesis, University of Warwick, Coventry.

Lou, Z. and Jin, J.M. (2005). An accurate waveguide port boundary condition for the time-domain finite-element method, *IEEE T. Microw. Theory*, **53**, 3014–3023.

Maradei, F. (2001). A frequency-dependent WETD formulation for dispersive materials, *IEEE T. Magn.*, **37**, 3303–3306.

Meade, M.L. and Dillon, C.R. (1991). *Signals and Systems — Model and Behaviour*, Kluwer Academic Publisher, Norwell: MA, pp. 148–150.

Mias, C. (1995). Finite element modelling of the electromagnetic behaviour of spatially periodic structures. Ph.D. Thesis, Cambridge University, Cambridge.

Mias, C., Webb, J.P. and Ferrari, R.L. (1999). Finite element modelling of electromagnetic waves in doubly and triply periodic structures, *IEE Proc. Optoelectron.*, **146**, 111–118.

Moglie, F., Rozzi, T., Marcozzi, P. and Schiavoni, A. (1992). A new termination condition for the Application of FDTD techniques to discontinuity problems in close homogeneous waveguide, *IEEE Microw. Guided W.*, **2**, 475–477.

Moreno, E., Erni, D. and Hafner, C. (2002). Modeling of discontinuities in photonic crystal waveguides with the multiple multipole method, *Phys. Rev. E*, **66**, 0366181–03661812.

Mur, G. (1981). Absorbing boundary conditions for finite-difference approximation of the time-domain electromagnetic-field equations, *IEEE T. Electromagn. C.*, **EMC-23**, 1073–1077.

Nehrbass, J.W., Lee, J.F. and Lee, R. (1996). Stability analysis for perfectly matched layered absorbers, *Electromagnetics*, **16**, 385–397.

Newmark, N.M. (1959). A method of computation for structural dynamics, *J. Eng. Mech. Div-ASCE*, **85**, 67–94.

Pierantoni, L., Tomassoni, C. and Rozzi, T. (2002). A new termination condition for the application of the TLM method to discontinuity problems in closed homogeneous waveguide, *IEEE T. Microw. Theory*, **50**, 2513–2518.

Polycarpou, A.C., Lyons, M.R. and Balanis, C.A. (1996). Finite element analysis of MMIC waveguide structures with anisotropic substrates, *IEEE T. Microw. Theory*, **44**, 1650–1663.

Polycarpou, A.C., Trikas, P.A. and Balanis, C.A. (1997). The finite-element method for modeling circuits and interconnects for electronic packaging, *IEEE T. Microw. Theory*, **45**, 1868–1874.

Pozar, D.M. (2005). *Microwave Engineering*, (3rd ed.), Wiley, New York: NY.

Ramo, S., Whinnery, J.R. and Duzer, T.V. (1994). *Fields and Waves in Communication Electronics*, (3rd ed.), Wiley, New York: NY.

Rao, S.M. (1999). *Time Domain Electromagnetics*, Academic Press, Orlando: FL.

Sacks, Z.S., Kingsland, D.M., Lee, R. and Lee, J.F. (1995). A perfectly matched anisotropic absorber for use as an absorbing boundary condition, *IEEE T. Antenn. Propag.*, **43**, 1460–1463.

Scherer, A., Doll, T., Yablonovitch, E., Everitt, H. and Higgins, A. (1999). Guest editorial, *IEEE T. Microw. Theory*, **47**, 2057–2058.

Sievenpiper, D., Zhang, L., Broas, R.F.J., Alexopolous, N.G. and Yablonovitch, E. (1999). High-impedance electromagnetic surface with a forbidden frequency band, *IEEE T. Microw. Theory*, **47**, 2059–2074.

Silvester, P.P. and Ferrari, R.L. (1996). *Finite Elements for Electrical Engineers*, Cambridge, Cambridge University Press.

Taflove, A. and Brodwin, M.E. (1975). Numerical solution of steady-state electromagnetic scattering problems using the time-dependent Maxwell's equation, *IEEE T. Microw. Theory*, **MTT-23**, 623–638.

Taflove, A. and Hugness, S.C. (2005). *Computational Electrodynamics: The Finite-Difference Time-Domain Method*, Artech House, Norwood: MA.

Tsai, H.P., Wang, Y.X. and Itoh, T. (2002). An unconditionally stable extended (USE) finite-element time-domain solution of active nonlinear microwave circuits using perfectly matched layers, *IEEE Trans. Microw. Theory*, **50**, 2226–2232.

Ulaby, F.T. (2000). *Fundamentals of Applied Electromagnetics*, Prentice Hall, New Jersey: MA.

Volakis, J.L., Chatterjee, A. and Kempel, L.C. (1998). *Finite Element Method for Electromagnetics: Antennas, Microwave Circuits, and Scattering Applications*, IEEE Press, New York: NY.

Wang, J.S. and Mittra, R. (1994). Finite element analysis of MMIC structures and electronic packages using absorbing boundary conditions, *IEEE T. Microw. Theory*, **42**, 441–449.

Wang, Y.X. and Itoh, T. (2001). Envelope-finite-element (EVFE) technique — a more efficient time domain scheme, *IEEE T. Microw. Theory*, **49**, 2241–2247.

Yee, K.S. (1966). Numerical solutions of initial boundary value problems involving Maxwell's equation in isotropic media, *IEEE T. Antenn. Propag.*, **AP-14**, 302–307.

Yeh, P. (2005). *Optical Waves in Layered Media*, Wiley, New York: NY.

Zgainski, F. and Meunier, G. (1995). Analysis of microwave planar circuits MIC2D: a user-friendly software based on a time domain finite element method, *IEEE T. Magn.*, **31**, 1650–1653.

Zienkiewicz, O.C. (1977). A new look at the Newmark, Houboult and other time stepping formulas, A weighted residual approach, *Earthq. Eng. Struct. D*, **5**, 413–418.

Chapter 16

Boundary Conditions for Three-Dimensional Finite-Element Modeling of Microwave Devices

Tian-Hong Loh and Christos Mias

This chapter presents 3D Galerkin weighted residual finite-element time-domain (FETD) and finite-element frequency-domain (FEFD) formulations incorporating various port boundary conditions for the exact electromagnetic analysis of microwave devices. Generalized formulations for both transverse electric (TE) and transverse magnetic (TM) polarizations are considered. Several 3D electromagnetic problems are characterized numerically. The numerical results obtained are validated through FETD/FEFD comparisons, with analytical results, where available, and with results obtained from the commercial software Computer Simulation Technology Microwave Studio (CST, 2012). An overview of the employed boundary conditions, finite-element techniques, their applications for modeling of microwave devices, basic governing equations, and Galerkin weighted residual finite-element procedure has been discussed in Chapter 15.

The port boundary conditions considered focus on an exact and mesh-efficient modal absorbing boundary termination condition (MABTC) (Loh and Mias, 2004). The 3D MABTC formulation is an extension of the 2D MABTC shown in Chapter 15, for which the MABTC, at the waveguide ports, is obtained by the convolution of the modal characteristic impedance with the modal profile of the waveguide modes following an inverse Laplace transform (ILT). A time-truncated approximate version of the MABTC is also presented, which results in a more memory- and time-efficient FETD implementation compared with the FETD code without the approximation. The developed boundary condition formulation and the methodology can

453

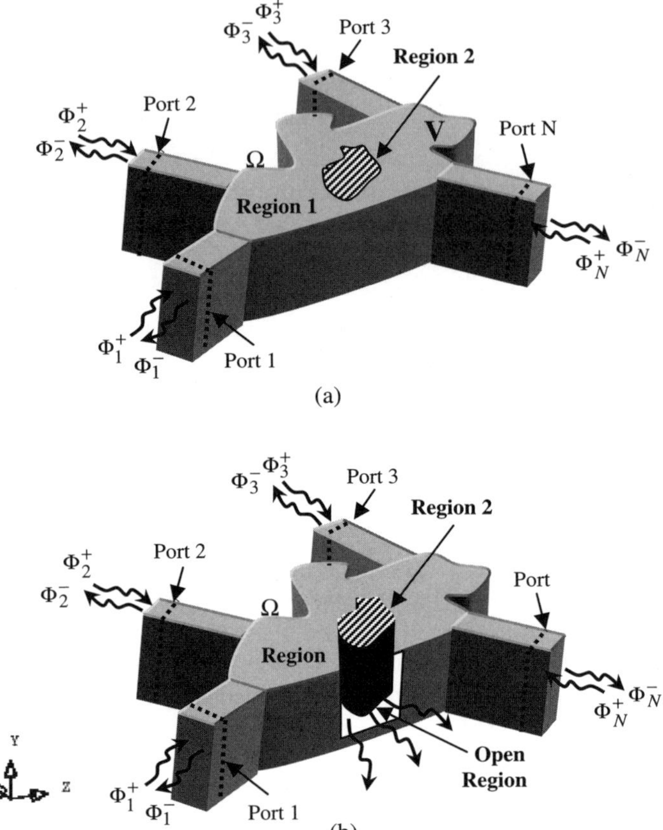

Fig. 16.1.   A general 3D N-port waveguide device: (a) without open region (i.e., closed waveguide), (b) with open region (i.e., open waveguide).

be applied to closed[1] and open[2] homogenous waveguide problems (see Fig. 16.1). Assuming the MABTC is exact, the solution accuracy for open structures depends on the absorption performance of other methodologies to truncate the computational domain, such as perfectly matched layers (PML) (Berenger, 1994; Taflove and Hugness, 2005).

---

[1]In the closed homogeneous waveguide, there is no electromagnetic radiation.

[2]In the open homogeneous waveguide, there is power loss due to electromagnetic radiation.

In the formulations that follow it is assumed that materials of the finite-element computation domain of volume V, shown in Fig. 16.1, which is enclosed by a surface Ω, are linear and isotropic.

16.1. Finite-Element Formulation

This section derives the 3D FETD, and FEFD formulation, following the standard Galerkin weighted residual finite-element procedures outlined in Chapter 15, Section 15.1.2.

16.1.1. *Finite-Element Time-Domain Formulation*

Referring to the 3D problems as depicted in Fig. 16.1, the time-dependent field vector wave differential equation is given in (15.7). The continuity conditions at the boundary of two homogeneous regions (say Region 1 and Region 2 refer to Fig. 16.1) are (Jin, 2002):

$$\hat{\mathbf{n}} \times \mathbf{F}_1 = \hat{\mathbf{n}} \times \mathbf{F}_2, \tag{16.1}$$

where $\mathbf{F} = \mathbf{E}$ or $\mathbf{H}$. From (16.1) and Maxwell's equations one can obtain

$$\frac{1}{p_1}\hat{\mathbf{n}} \times (\nabla \times \mathbf{F}_1) = \frac{1}{p_2}\hat{\mathbf{n}} \times (\nabla \times \mathbf{F}_2), \tag{16.2}$$

where $\mathbf{F} = \mathbf{H}$ or $\mathbf{E}$.

The waveguide walls are assumed to be perfect electric conductors, hence the following Dirichlet (16.3) and Neumann (16.4) boundary conditions apply:

$$\text{Dirichlet:} \quad \hat{\mathbf{n}} \times \mathbf{F} = 0 \qquad (\text{for } \mathbf{F} = \mathbf{E}) \tag{16.3}$$

$$\text{Neumann:} \quad \hat{\mathbf{n}} \times (\nabla \times \mathbf{F}) = 0 \quad (\text{for } \mathbf{F} = \mathbf{H}), \tag{16.4}$$

where $\hat{\mathbf{n}}$ is the outward surface unit vector normal to the surface boundary.

The Galerkin weighted residual method (Silvester and Ferrari, 1996; Jin, 2002) (see Chapter 15, Section 15.1.2), is then employed to solve (15.7). This is achieved by taking the scalar product of (15.7) with edge-based weighting functions, $\mathbf{N}_i$, which are chosen from the set of field interpolation functions employed over each element following a finite-element discretization of the spatial domain. At a particular time t, the unknown vector field at any point within an element can be expressed as:

$$\mathbf{F}^e(x,y,z,t) = \sum_j \mathbf{N}_j^e(x,y,z)F_j(t) = \langle \mathbf{N}^e(x,y,z)\rangle\{F(t)\}, \qquad (16.5)$$

where $\langle \mathbf{N}^e(x,y,z)\rangle$ is a row vector representing the edge-based elemental field interpolation functions. $\{F(t)\}$ is a column vector representing, at the particular time t, the edge-based spatial unknown values of the field variable, $\mathbf{F}$. The subscript j signifies edges relating to the eth element but counted on a global basis. $\mathbf{N}_j^e$ represents an appropriate edge-based elemental field interpolation function at jth edge. It is important to note that $\mathbf{N}_j^e$ is nonzero only within the eth element and vanishes outside the element (Jin, 2002). By integrating the weighted (15.7) over the domain of interest and applying the vector identity of (16.6) followed by the divergence theorem of (16.7) (i.e., vector form of Green's theorem) and the triple scalar product of (16.8),

$$\nabla \bullet (\mathbf{A} \times \mathbf{B}) = \mathbf{B} \bullet \nabla \times \mathbf{A} - \mathbf{A} \bullet \nabla \times \mathbf{B} \qquad (16.6)$$

$$\iiint_V (\nabla \bullet \mathbf{A})dV = \oint_\Omega \mathbf{A} \bullet d\Omega = \oint_\Omega \mathbf{A} \bullet \hat{n}d\Omega \qquad (16.7)$$

$$\mathbf{A} \bullet (\mathbf{B} \times \mathbf{C}) = -\mathbf{B} \bullet (\mathbf{A} \times \mathbf{C}), \qquad (16.8)$$

where $\hat{n}$ is the outward unit vector normal to the surface boundary Ω, one obtains a system of ordinary differential equation as follows (Dibben and Metaxas, 1996; Lee *et al.*, 1997; Rao, 1999; Wang and Itoh, 2001):

$$[T]\frac{\partial^2\{F(t)\}}{\partial t^2} + [B]\frac{\partial\{F(t)\}}{\partial t} + [S]\{F(t)\} = \{g\}, \qquad (16.9)$$

where the matrices are assembled from elemental entries of the form:

$$T_{ij}^e = \frac{q^e}{c^2}\iiint_{V^e} \mathbf{N}_i^e \bullet \mathbf{N}_j^e dV$$

$$B_{ij}^e = \frac{\alpha^e q^e}{c^2}\iiint_{V^e} \mathbf{N}_i^e \bullet \mathbf{N}_j^e dV$$

$$S_{ij}^e = \frac{1}{p^e}\iiint_{V^e} (\nabla \times \mathbf{N}_i^e)\cdot(\nabla \times \mathbf{N}_j^e)dV$$

$$g_i^e = \iiint_{V^e} \mathbf{N}_i^e \bullet \mathbf{U}dV - \frac{1}{p^e}\int_{\Omega^e} \mathbf{N}_i^e \bullet (\hat{n} \times \nabla \times \mathbf{F})d\Omega.$$

Referring to Chapter 15, Section 15.2.1, to solve (16.9), one needs to discretize the temporal derivatives in the time domain based on the

Newmark-beta formulation (Newmark, 1959; Zienkiewicz, 1977):

$$\begin{cases} \dfrac{\partial^2 \{F(t)\}}{\partial t^2} = \dfrac{1}{\Delta t^2}[\{F\}^{n+1} - 2\{F\}^n + \{F\}^{n-1}] \\[2ex] \dfrac{\partial \{F(t)\}}{\partial t} = \dfrac{1}{2\Delta t}[\{F\}^{n+1} - \{F\}^{n-1}] \\[2ex] \{F(t)\} = \beta\{F\}^{n+1} + (1 - 2\beta)\{F\}^n + \beta\{F\}^{n-1} \end{cases} \quad , \qquad (16.10)$$

where, at time step $t = n\Delta t$, the discrete-time representation of $\{F(t)\}$ is represented as $\{F(t)\} = \{F(n\Delta t)\} = \{F\}^n$ and β is the Newmark parameters. In this chapter, $\beta = 0.25$. This choice (see Chapter 15, Section 15.2.1), leads to the following unconditionally stable two-step update scheme:

$$\left(\frac{[T]}{\Delta t^2} + \frac{[B]}{2\Delta t} + \frac{[S]}{4}\right)\{F\}^{n+1}$$

$$= \left(\frac{2[T]}{\Delta t^2} - \frac{[S]}{2}\right)\{F\}^n + \left(-\frac{[T]}{\Delta t^2} + \frac{[B]}{2\Delta t} - \frac{[S]}{4}\right)\{F\}^{n-1}$$

$$+ \left(\frac{\{g\}^{n+1}}{4} + \frac{\{g\}^n}{2} + \frac{\{g\}^{n-1}}{4}\right). \qquad (16.11)$$

16.1.2. *Finite-Element Frequency-Domain Formulation*

Referring to the 3D problem as depicted in Fig. 16.1, the frequency-domain vector wave differential equation is given in (15.8). The Galerkin weighted residual method (Silvester and Ferrari, 1996; Jin, 2002) (see Chapter 15, Section 15.1.2), is employed to solve (15.8). This is achieved by taking the scalar product of (15.8) with edge-based weighting functions, $\mathbf{N}_i$, which are chosen from the set of field interpolation functions employed over each element following a finite-element discretization of the spatial domain. At a particular frequency f, the unknown vector field at any point within an element can be expressed as

$$\mathbf{F}^e(x, y, z, s) = \sum_j \mathbf{N}_j^e(x, y, z)F_j(s) = \langle \mathbf{N}^e(x, y, z)\rangle\{F(s)\}, \qquad (16.12)$$

where $\langle \mathbf{N}^e(x, y, z)\rangle$ is a row vector representing the edge-based vector elemental field interpolation functions, and $s = j\omega$, $\{F(s)\}$ is a column vector representing, at the particular frequency f, the edge-based spatial unknown values of the field variable, $\mathbf{F}$. The subscript j signifies edges relating to the eth element but counted on a global basis and $\mathbf{N}_j^e$ represents an appropriate edge-based elemental field interpolation function at jth edge. It is important to note that $\mathbf{N}_j^e$ is nonzero only within the eth element but vanish outside the element (Jin, 2002). By integrating the weighted (15.8) over the domain of interest and applying the vector identity (16.6), followed by the divergence theorem (16.7) and the triple scalar product (16.8), one obtains the following system of equations, in matrix form:

$$([S] + j\omega[B] - k_0^2[T])\{F(s)\} = \{g\}, \tag{16.13}$$

where the matrices are assembled from elemental entries of the form:

$$T_{ij}^e = q^e \iiint_{V^e} \mathbf{N}_i^e \bullet \mathbf{N}_j^e dV$$

$$B_{ij}^e = \frac{\alpha^e q^e}{c^2} \iiint_{V^e} \mathbf{N}_i^e \bullet \mathbf{N}_j^e dV$$

$$S_{ij}^e = \frac{1}{p^e} \iiint_{V^e} (\nabla \times \mathbf{N}_i^e) \cdot (\nabla \times \mathbf{N}_j^e) dV$$

$$g_i^e = \iiint_{V^e} \mathbf{N}_i^e \bullet \mathbf{U} dV - \frac{1}{p^e} \int_{\Omega^e} \mathbf{N}_i^e \bullet (\hat{\mathbf{n}} \times \nabla \times \mathbf{F}) d\Omega.$$

16.2. Boundary Conditions

For a 3D problem, a general rectangular waveguide with discontinuity is considered as shown in Fig. 16.2. Let a be the width and b be the height of the waveguide along the x- and y-axes respectively. The z-axis is the axis of wave propagation. As depicted in Fig. 16.2 on the boundaries, the

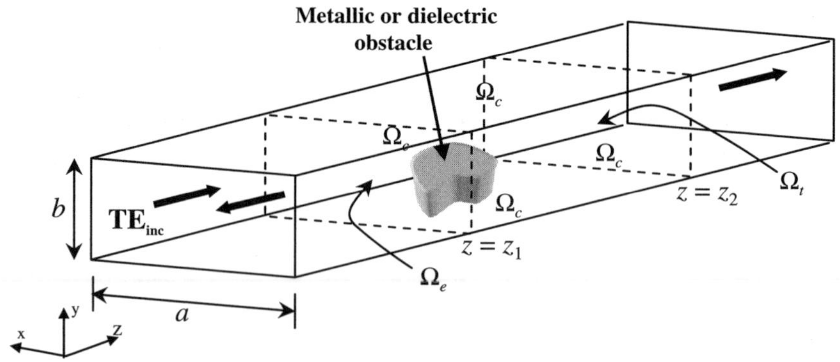

Fig. 16.2.   Sketch of a rectangular waveguide incorporating discontinuity.

Table 16.1. List of boundary conditions associated with this chapter.

Working Field Variable	Boundary	Boundary Conditions for 3D Vector Field
Magnetic Field, **H**	PMC	$\hat{\mathbf{n}} \times \mathbf{H} = 0^*$
	PEC	$\hat{\mathbf{n}} \times \nabla \times \mathbf{H} = 0^\dagger$
	Waveguide Ports	MABTC (see Section 16.2)
Electric Field, **E**	PMC	$\hat{\mathbf{n}} \times \nabla \times \mathbf{E} = 0^\dagger$
	PEC	$\hat{\mathbf{n}} \times \mathbf{E} = 0^*$
	Waveguide Ports	MABTC (see Section 16.2)

MABTC: Modal absorbing boundary termination condition.
PEC: Perfect electric conductor (also called electric wall).
PMC: Perfect magnetic conductor (also called magnetic wall).
*: Homogeneous Dirichlet boundary condition for the problems considered.
†: Homogeneous Neumann boundary condition for the problems considered.

tangential field component ($\mathbf{H}_t$ or $\mathbf{E}_t$) is chosen as the working field variable. The surface area Ω, of the finite element boundary integral on the right-hand side of (16.9) and (16.13), completely encloses the region of interest as shown in Fig. 16.2. The incident guided wave is propagating along the positive z-axis direction.

Table 16.1 lists the boundary conditions associated with this chapter. Waveguide discontinuities are either geometrical, material, or both. From Fig. 16.2, the surface boundary Ω consists of a perfect electric conductor (PEC) surface boundary Ω_c at the metal walls, an excitation surface boundary Ω_e positioned at the plane where $z = z_1$, and a termination surface boundary Ω_t positioned at the plane where $z = z_2$. A solution is obtained by solving the governing differential equation together with the appropriate boundary conditions within the computational domain (continuity conditions — (16.1) and (16.2)), at the waveguide walls (either homogeneous Dirichlet (16.3) or Neumann (16.4) boundary condition), and at the ports boundaries (MABTC (see Section 16.2)). There is no contribution from the Ω_c boundary integral to the system of equations as for $\mathbf{F} = \mathbf{E}$ one has $\hat{\mathbf{n}} \times \mathbf{F} = 0$ and for $\mathbf{F} = \mathbf{H}$ one has $\hat{\mathbf{n}} \times \nabla \times \mathbf{F} = 0$.

In addition to the drawback in the use of the Mur's first-order ABC (Mur, 1981) discussed in Chapter 15, Section 15.3, the wave velocity ν_p used in the time-domain formulation $\hat{\mathbf{n}} \times \nabla \times \mathbf{F} = \pm\hat{\mathbf{n}} \times \hat{\mathbf{n}} \times \frac{1}{\nu_p} \frac{\partial \mathbf{F}}{\partial t}$ (Tsai *et al.*, 2002) is constant where

$$\nu_p = \frac{c}{\left(\sqrt{\mu_r \varepsilon_r} \sqrt{1 - \left(\frac{\omega_{c_{mn}}}{\omega_c} \right)^2} \right)}, \quad \omega_c = 2\pi f_c, \text{ and } \omega_{c_{mn}}$$

denotes the cut-off frequency of the mnth mode (16.14 and 16.15). The parameter f_c denotes the carrier frequency of the Gaussian pulse (16.58). Hence, the solution is exact only at a single particular frequency f_c (Olivier and McNamara, 1994). It is assumed that the waveguide is filled with isotropic, linear, and homogenous medium, the waveguide medium over which the MABTC applied is lossless and isotropic. Their detailed formulations for incorporating with finite-element methods are given below.

16.2.1. *MABTC in the Frequency Domain*

In the frequency domain, the following relation holds for the modal characteristic wave impedance with the indices m and n along the x-axis and y-axis directions respectively for a transverse-magnetic (TM) wave (designated TM_{mn}) and a transverse-electric (TE) wave (designated TE_{mn}) (Collin, 1991; Moglie *et al.*, 1992; Ramo *et al.*, 1994):

$$Z_{mn}^{\text{TM}}(s) = \pm\frac{\text{E}_{\text{xTM}_{mn}}(\boldsymbol{r},s)}{\text{H}_{\text{yTM}_{mn}}(\boldsymbol{r},s)} = \mp\frac{\text{E}_{\text{yTM}_{mn}}(\boldsymbol{r},s)}{\text{H}_{\text{xTM}_{mn}}(\boldsymbol{r},s)} = \eta\Xi_{1_{mn}}(s) \quad (16.14)$$

$$Z_{mn}^{\text{TE}}(s) = \pm\frac{\text{E}_{\text{xTE}_{mn}}(\boldsymbol{r},s)}{\text{H}_{\text{yTE}_{mn}}(\boldsymbol{r},s)} = \mp\frac{\text{E}_{\text{yTE}_{mn}}(\boldsymbol{r},s)}{\text{H}_{\text{xTE}_{mn}}(\boldsymbol{r},s)} = \eta\Xi_{2_{mn}}(s), \quad (16.15)$$

where

$$\eta = \sqrt{\mu_0\mu_r/(\varepsilon_0\varepsilon_r)}, \quad \boldsymbol{r} = (x,y,z), \quad s = j\omega, \quad \Xi_{1_{mn}}(s) = \sqrt{s^2 + \omega_{c_{mn}}^2}/s,$$

$$\Xi_{2_{mn}}(s) = s/\sqrt{s^2 + \omega_{c_{mn}}^2}.$$

The variable $\omega_{c_{mn}} = c\sqrt{(m\pi/a)^2 + (n\pi/b)^2}/\sqrt{\mu_r\varepsilon_r}$ denotes the cut-off frequency of the mnth mode, the parameters a and b denote the width and height of the rectangular waveguide, and c denotes the speed of light. The signs $\pm$ (or corresponding $\mp$) are for waves propagating in the same or opposite direction to the normal outward unit vector $\hat{\mathbf{n}}$ direction respectively. Let us assume for simplicity the waveguide medium over which the port boundaries lie is free space then it is obvious that the $\Xi_1(s)$ and $\Xi_2(s)$ functions become purely imaginary in the case of evanescent modes (Collin, 1991).

Using the time-harmonic form of Maxwell's equations (15.5 and 15.6), the modal characteristic wave impedance of mode mn (16.14 and 16.15), and applying the following vector formula:

$$\hat{\mathbf{n}} \times \hat{\mathbf{n}} \times \mathbf{A} = -\mathbf{A}_t, \quad (16.16)$$

one obtains the mnth mode outward normal derivative of the field

$$\hat{\mathbf{n}} \times (\nabla \times \mathbf{F}_{m,n}(\mathbf{r},s)) = \pm\hat{\mathbf{n}} \times \hat{\mathbf{n}} \times [\chi_{mn}\mathbf{F}_{m,n}(\mathbf{r},s)], \tag{16.17}$$

where the parameters are presented below in a tabular form:

$\mathbf{F}_{mn}$	$\chi_{mn}(s)$
$\mathbf{E}_{\mathrm{TM}_{mn}}$	$\chi_{mn}^{\mathrm{TM}}(s) = -k_0^2\mu_r\varepsilon_r/\varsigma_{mn}$
$\mathbf{E}_{\mathrm{TE}_{mn}}$	$\chi_{mn}^{\mathrm{TE}}(s) = \varsigma_{mn}$
$\mathbf{H}_{\mathrm{TM}_{mn}}$	$\chi_{mn}^{\mathrm{TM}}(s) = \varsigma_{mn}$
$\mathbf{H}_{\mathrm{TE}_{mn}}$	$\chi_{mn}^{\mathrm{TE}}(s) = -k_0^2\mu_r\varepsilon_r/\varsigma_{mn}$

$\varsigma_{mn} = s\sqrt{\mu_r\varepsilon_r}\kappa_{mn}(s)/c$, $\kappa_{mn}(s) = \sqrt{s^2 + \omega_{c_{mn}}^2}/s$, where the right-hand side (RHS) is with $+$ sign or with $-$ sign if the travelling wave propagates in the opposite or the same direction, respectively, with respect to $\hat{\mathbf{n}}$. The parameters $\mathbf{r}$ and s are the same as those in (16.14) and (16.15).

At the waveguide ports of the 3D problems considered, the total field within the computational domain can be represented by TE and TM modes, which form a complete set of basis functions that can represent any Maxwellian field. Jin (2002) derived the formulation for the electric field. In this subsection, the generalized MABTC formulation is presented following similar procedures as in Jin (2002). The frequency-domain field equation at the excitation boundary Ω_e, where $z = z_1$, has the following form:

$$\begin{aligned}
\mathbf{F}(x,y,z_1,s) &= \mathbf{F}_{pq}^{\mathrm{inc}}(x,y,z_1,s) + \sum_{m=0}^{\infty}\sum_{n=0}^{\infty}\mathbf{F}_{mn}^{\mathrm{ref}}(x,y,z_1,s) \\
&= \mathbf{e}_{pq}^{\mathrm{TE}}(x,y)F_{\mathrm{TE}_{pq}}^{\mathrm{inc}}(z_1,s) + \mathbf{e}_{pq}^{\mathrm{TM}}(x,y)F_{\mathrm{TM}_{pq}}^{\mathrm{inc}}(z_1,s) \\
&\quad + \sum_{m=0}^{\infty}\sum_{n=0}^{\infty}a_{mn}\left[\mathbf{e}_{mn}^{\mathrm{TE}}(x,y) + \hat{z}e_{z_{mn}}^{\mathrm{TE}}(x,y)\right]F_{\mathrm{TE}_{mn}}^{\mathrm{ref}}(z_1,s) \\
&\quad + \sum_{m=1}^{\infty}\sum_{n=1}^{\infty}b_{mn}\left[\mathbf{e}_{mn}^{\mathrm{TM}}(x,y) + \hat{z}e_{z_{mn}}^{\mathrm{TM}}(x,y)\right]F_{\mathrm{TM}_{mn}}^{\mathrm{ref}}(z_1,s),
\end{aligned}$$

$$\tag{16.18}$$

where the case $m = n = 0$ is ignored. F_{TM} and F_{TE} denote the field variable F for TM and TE mode respectively. The parameters a_{mn} and b_{mn} are TE- and TM-modal coefficients, respectively. For the incident mode $m = p$ and

$n = q$. The TE mode vector function $\mathbf{e}_{pq}^{\mathrm{TE}}$ and the TM mode vector function $\mathbf{e}_{pq}^{\mathrm{TM}}$ are defined as (Harrington, 1961; Jin, 2002):

$$\left.\begin{aligned}
\mathbf{e}_{mn}^{\mathrm{TE}}(x,y) &= N_{mn}\left(\hat{x}\frac{n}{b}\cos\frac{m\pi x}{a}\sin\frac{n\pi y}{b} - \hat{y}\frac{m}{a}\sin\frac{m\pi x}{a}\cos\frac{n\pi y}{b}\right) \\
\mathbf{e}_{mn}^{\mathrm{TM}}(x,y) &= N_{mn}\left(\hat{x}\frac{m}{a}\cos\frac{m\pi x}{a}\sin\frac{n\pi y}{b} + \hat{y}\frac{n}{b}\sin\frac{m\pi x}{a}\cos\frac{n\pi y}{b}\right) \\
e_{z_{mn}}^{\mathrm{TM}}(x,y) &= \frac{N_{mn}\pi}{\gamma_{mn}}\left[\left(\frac{m}{a}\right)^2 + \left(\frac{n}{b}\right)^2\right]\sin\frac{m\pi x}{a}\sin\frac{n\pi y}{b}
\end{aligned}\right\} \quad \text{for } \mathbf{F} = \mathbf{E}$$

$$(16.19)$$

$$\left.\begin{aligned}
\mathbf{e}_{mn}^{\mathrm{TE}}(x,y) &= N_{mn}\left(\hat{x}\frac{n}{b}\sin\frac{m\pi x}{a}\cos\frac{n\pi y}{b} + \hat{y}\frac{m}{a}\cos\frac{m\pi x}{a}\sin\frac{n\pi y}{b}\right) \\
\mathbf{e}_{mn}^{\mathrm{TM}}(x,y) &= N_{mn}\left(\hat{x}\frac{m}{a}\sin\frac{m\pi x}{a}\cos\frac{n\pi y}{b} - \hat{y}\frac{n}{b}\cos\frac{m\pi x}{a}\sin\frac{n\pi y}{b}\right) \\
e_{z_{mn}}^{\mathrm{TE}}(x,y) &= \frac{N_{mn}\pi}{\gamma_{mn}}\left[\left(\frac{m}{a}\right)^2 + \left(\frac{n}{b}\right)^2\right]\cos\frac{m\pi x}{a}\cos\frac{n\pi y}{b}
\end{aligned}\right\} \quad \text{for } \mathbf{F} = \mathbf{H}.$$

$$(16.20)$$

N_{mn} and γ_{mn} are the normalization factor and propagation constant for the mnth mode respectively and are defined as:

$$N_{mn} = \sqrt{v_m v_n}\Big/\sqrt{n^2\frac{a}{b} + m^2\frac{b}{a}}, \quad v_m = \begin{cases}1 & m = 0 \\ 2 & m \neq 0\end{cases}, \quad v_n = \begin{cases}1 & n = 0 \\ 2 & n \neq 0\end{cases}$$

$$(16.21)$$

$$\gamma_{mn} = \begin{cases} j\sqrt{k_0^2 - \left(\frac{m\pi}{a}\right)^2 - \left(\frac{n\pi}{b}\right)^2} & \text{if } \left(\frac{m\pi}{a}\right)^2 + \left(\frac{n\pi}{b}\right)^2 \leq k_0^2 \\ \sqrt{\left(\frac{m\pi}{a}\right)^2 + \left(\frac{n\pi}{b}\right)^2 - k_0^2} & \text{if } \left(\frac{m\pi}{a}\right)^2 + \left(\frac{n\pi}{b}\right)^2 > k_0^2 \end{cases}.$$

$$(16.22)$$

Note that $e_{z_{mn}}^{\mathrm{TE}}(x,y) = 0$ and $e_{z_{mn}}^{\mathrm{TM}}(x,y) = 0$ for $\mathbf{F} = \mathbf{E}$ and $\mathbf{F} = \mathbf{H}$, respectively. In addition, for the TM_{mn} mode $m \neq 0$ and $n \neq 0$ (Pozar, 2005). The following orthogonality relations hold for $\mathbf{e}_{mn}^{\mathrm{TE}}$ and $\mathbf{e}_{mn}^{\mathrm{TM}}$ (Kreyszig, 2010):

$$\int_0^a \int_0^b \mathbf{e}_{mn}^{\mathrm{TE}} \bullet \mathbf{e}_{mn}^{\mathrm{TM}}\, dx\, dy = 0 \qquad (16.23)$$

$$\int_0^a \int_0^b \mathbf{e}_{mn}^{\mathrm{TE}} \bullet \mathbf{e}_{m'n'}^{\mathrm{TE}}\, dx\, dy = \delta_{mm'}\delta_{nn'} \qquad (16.24)$$

$$\int_0^a \int_0^b \mathbf{e}_{mn}^{\mathrm{TM}} \bullet \mathbf{e}_{m'n'}^{\mathrm{TM}}\, dx\, dy = \delta_{mm'}\delta_{nn'}, \qquad (16.25)$$

where the Kronocker delta function $\delta_{kk'}$ is defined as $\delta_{kk'} = 1$ for $k = k'$ and $\delta_{kk'} = 0$ for $k \neq k'$. These orthogonality properties are used to obtain the TE-modal coefficients, a_{mn}, and TM-modal coefficients, b_{mn}. Hence, taking the scalar product of the reflected field in (16.18) with $\mathbf{e}_{pq}^{\text{TE}}$ and $\mathbf{e}_{pq}^{\text{TM}}$ respectively and integrating over the cross-sectional area of the waveguide leads to the following relations:

$$a_{pq} = \frac{1}{F_{\text{TE}_{pq}}^{\text{ref}}(z_1, s)} \int_0^a \int_0^b \mathbf{e}_{pq}^{\text{TE}} \bullet [\mathbf{F}(x, y, z, s) - \mathbf{F}^{\text{inc}}(x, y, z, s)]|_{z=z_1} dy\,dx$$

$$(16.26)$$

$$b_{pq} = \frac{1}{F_{\text{TM}_{pq}}^{\text{ref}}(z_1, s)} \int_0^a \int_0^b \mathbf{e}_{pq}^{\text{TM}} \bullet [\mathbf{F}(x, y, z, s) - \mathbf{F}^{\text{inc}}(x, y, z, s)]|_{z=z_1} dy\,dx.$$

$$(16.27)$$

Taking the curl of (16.18) and crossing it with $\hat{\mathbf{n}} = -\hat{z}$, and making use of (16.16), (16.17), (16.26), and (16.27), one obtains the frequency-domain boundary termination condition for the excitation truncation surface boundary Ω_e at $z = z_1$ as:

$$\begin{aligned}
\hat{\mathbf{n}} \times \nabla \times \mathbf{F} = &+ \sum_{m=0}^{\infty} \sum_{n=0}^{\infty} \chi_{mn}^{\text{TE}} \mathbf{e}_{mn}^{\text{TE}}(x, y) \int_0^a \int_0^b \mathbf{e}_{mn}^{\text{TE}} \bullet \mathbf{F}_t|_{z=z_1} dy'\,dx' \\
&+ \sum_{m=1}^{\infty} \sum_{n=1}^{\infty} \chi_{mn}^{\text{TM}} \mathbf{e}_{mn}^{\text{TM}}(x, y) \int_0^a \int_0^b \mathbf{e}_{mn}^{\text{TM}} \bullet \mathbf{F}_t|_{z=z_1} dy'\,dx' \\
&- 2\chi_{pq} \mathbf{F}_t^{\text{inc}}(x, y, z_1, s),
\end{aligned}$$

$$(16.28)$$

where $\mathbf{F}_t$ is the tangential field component. Similarly, the frequency-domain field equation at the termination truncation boundary Ω_t, where $z = z_2$, has the following form:

$$\begin{aligned}
\mathbf{F}(x, y, z_2, s) = &\sum_{m=0}^{\infty} \sum_{n=0}^{\infty} \mathbf{F}_{mn}^{\text{tran}}(x, y, z_2, s) \\
= &+ \sum_{m=0}^{\infty} \sum_{n=0}^{\infty} c_{mn}[\mathbf{e}_{mn}^{\text{TE}}(x, y) + \hat{z} e_{z_{mn}}^{\text{TE}}(x, y)] F_{\text{TE}_{mn}}^{\text{tran}}(z_2, s) \\
&+ \sum_{m=1}^{\infty} \sum_{n=1}^{\infty} d_{mn}[\mathbf{e}_{mn}^{\text{TM}}(x, y) + \hat{z} e_{z_{mn}}^{\text{TM}}(x, y)] F_{\text{TM}_{mn}}^{\text{tran}}(z_2, s).
\end{aligned}$$

$$(16.29)$$

Following the aforementioned procedures and noting that $\hat{\mathbf{n}} = \hat{z}$, one obtains the TE- and TM-modal coefficients, c_{pq} and d_{pq}, and the frequency-domain boundary termination condition at $z = z_2$ as follows,

$$c_{mn} = \frac{1}{F_{\mathrm{TE}_{mn}}^{\mathrm{tran}}(z_2, s)} \int_0^a \int_0^b \mathbf{e}_{mn}^{\mathrm{TE}} \bullet \left[\mathbf{F}(x, y, z, s)\right]\big|_{z=z_2} dy\, dx \qquad (16.30)$$

$$d_{mn} = \frac{1}{F_{\mathrm{TM}_{mn}}^{\mathrm{tran}}(z_2, s)} \int_0^a \int_0^b \mathbf{e}_{mn}^{\mathrm{TM}} \bullet \left[\mathbf{F}(x, y, z, s)\right]\big|_{z=z_2} dy\, dx \qquad (16.31)$$

$$\hat{\mathbf{n}} \times \nabla \times \mathbf{F} = + \sum_{m=0}^{\infty} \sum_{n=0}^{\infty} \chi_{mn}^{\mathrm{TE}} \mathbf{e}_{mn}^{\mathrm{TE}}(x, y) \int_0^a \int_0^b \mathbf{e}_{mn}^{\mathrm{TE}} \bullet \mathbf{F}_t\big|_{z=z_2} dy'\, dx'$$
$$+ \sum_{m=1}^{\infty} \sum_{n=1}^{\infty} \chi_{mn}^{\mathrm{TM}} \mathbf{e}_{mn}^{\mathrm{TM}}(x, y) \int_0^a \int_0^b \mathbf{e}_{mn}^{\mathrm{TM}} \bullet \mathbf{F}_t\big|_{z=z_2} dy'\, dx'.$$

$$(16.32)$$

Applying (16.28) and (16.32) and rearranging both sides of (16.13), one obtains the following Galerkin weighted residual FEFD-MABTC global matrix equation:

$$[M]\{\Phi\} = \{K\}, \qquad (16.33)$$

where $[M] = [S] - k_0^2[T] + [R_{\mathrm{TE}}] + [R_{\mathrm{TM}}]$ and the matrices are assembled from the following elemental contributions:

$$R_{\mathrm{TE}_{ij}} = \frac{1}{p} \sum_{m=0}^{\infty} \sum_{n=0}^{\infty} \chi_{mn}^{\mathrm{TE}} \int_\Omega \mathbf{N}_i \bullet \mathbf{e}_{mn}^{\mathrm{TE}}(x, y) d\Omega \int_0^a \int_0^b \mathbf{N}_j \bullet \mathbf{e}_{mn}^{\mathrm{TE}}(x', y') d\Omega'$$

$$R_{\mathrm{TM}_{ij}} = \frac{1}{p} \sum_{m=1}^{\infty} \sum_{n=1}^{\infty} \chi_{mn}^{\mathrm{TM}} \int_\Omega \mathbf{N}_i \bullet \mathbf{e}_{mn}^{\mathrm{TM}}(x, y) d\Omega \int_0^a \int_0^b \mathbf{N}_j \bullet \mathbf{e}_{mn}^{\mathrm{TM}}(x', y') d\Omega'$$

$$K_i = \frac{2\chi_{pq}}{p} F_{t_{pq}}^{\mathrm{inc}}(z_1, s) \int_\Omega \mathbf{N}_i \bullet \mathbf{e}_{pq}^{\mathrm{TE/TM}}(x, y) d\Omega$$

$p = \mu_r$ for $\mathbf{F} = \mathbf{E}$ and $p = \varepsilon_r$ for $\mathbf{F} = \mathbf{H}$.

16.2.2. *MABTC in the Time Domain*

One may start the derivation of the MABTC in the time domain from the time-harmonic form of Maxwell's equations (15.5)–(15.6), and the modal absorbing boundary condition (Moglie *et al.*, 1992; Jin, 2002) in the frequency domain. Applying on both sides of (16.17) the inverse Laplace transform (ILT), the following time-domain modal equation is obtained:

$$\hat{\mathbf{n}} \times (\nabla \times \mathbf{F}_{mn}(\mathbf{r},t)) = \pm\hat{\mathbf{n}} \times \hat{\mathbf{n}} \times \left[\frac{\sqrt{\mu_r \varepsilon_r}}{c} \frac{\partial}{\partial t}[\xi_{mn}(t) * \mathbf{F}_{mn}(\mathbf{r},t)] \right],$$

$$(16.34)$$

where the right-hand side (RHS) is with $+$ sign or with $-$ sign if the travelling wave propagates in the opposite or the same direction respectively with respect to $\hat{\mathbf{n}}$. The parameters are presented in a tabular form as:

$\mathbf{F}_{mn}$	$\xi_{mn}(t)$
$\mathbf{E}_{\mathrm{TM}_{mn}}$	$\xi_{mn}^{\mathrm{TM}}(t) = \Xi_{2_{mn}}(t)$
$\mathbf{E}_{\mathrm{TE}_{mn}}$	$\xi_{mn}^{\mathrm{TE}}(t) = \Xi_{1_{mn}}(t)$
$\mathbf{H}_{\mathrm{TM}_{mn}}$	$\xi_{mn}^{\mathrm{TM}}(t) = \Xi_{1_{mn}}(t)$
$\mathbf{H}_{\mathrm{TE}_{mn}}$	$\xi_{mn}^{\mathrm{TE}}(t) = \Xi_{2_{mn}}(t)$

The corresponding time-domain modal characteristic impedance is given as:

$$\Xi_{1_{mn}}(t) = \delta(t) + \omega_{c_{mn}} y_{1_{mn}}(t) \qquad (16.35)$$

$$\Xi_{2_{mn}}(t) = \delta(t) + \omega_{c_{mn}} y_{2_{mn}}(t), \qquad (16.36)$$

and the function, $y_{1_{mn}}(t)$ and $y_{2_{mn}}(t)$ are defined as:

$$y_{1_{mn}}(t) = \int_0^{\omega_{c_{mn}} t} J_0(u)du - J_1(\omega_{c_{mn}} t) \qquad (16.37)$$

$$y_{2_{mn}}(t) = -J_1(\omega_{c_{mn}} t), \qquad (16.38)$$

where $\delta(t)$ is the Dirac delta function and the functions J_0 and J_1 are the zero-order and first-order, respectively, Bessel functions of the first kind. The functions $\Xi_{1_{mn}}(t)$ and $\Xi_{2_{mn}}(t)$ are ILT of $\Xi_{1_{mn}}(s)$ and $\Xi_{2_{mn}}(s)$ in (16.14) and (16.15). Their detailed derivations are given in Loh (2004).

Similar to the derivation of the MABTC in the frequency domain in Section 16.2.1, the time-domain total field vector equation (Moglie *et al.*, 1992) at the excitation truncation surface boundary Ω_e, at $z = z_1$, can be expressed as follows:

$$
\begin{aligned}
\mathbf{F}(x,y,z_1,t) &= \mathbf{F}^{\mathrm{inc}}(x,y,z_1,t) + \sum_{m=0}^{\infty}\sum_{n=0}^{\infty}\mathbf{F}^{\mathrm{ref}}_{mn}(x,y,z_1,t) \\
&= \mathbf{e}^{\mathrm{TE}}_{\mathrm{inc}}(x,y)F_{\mathrm{TE}}{}^{\mathrm{inc}}(z_1,t) + \mathbf{e}^{\mathrm{TM}}_{\mathrm{inc}}(x,y)F_{\mathrm{TM}}{}^{\mathrm{inc}}(z_1,t) \\
&\quad + \sum_{m=0}^{\infty}\sum_{n=0}^{\infty} a_{mn}[\mathbf{e}^{\mathrm{TE}}_{mn}(x,y) + \hat{z}\,e_{z\,mn}^{\mathrm{TE}}(x,y)]F_{\mathrm{TE}_{mn}}{}^{\mathrm{ref}}(z_1,t) \\
&\quad + \sum_{m=1}^{\infty}\sum_{n=1}^{\infty} b_{mn}[\mathbf{e}^{\mathrm{TM}}_{mn}(x,y) + \hat{z}\,e_{z\,mn}^{\mathrm{TM}}(x,y)]F_{\mathrm{TM}_{mn}}{}^{\mathrm{ref}}(z_1,t),
\end{aligned}
$$

$$(16.39)$$

where $\mathbf{e}^{\mathrm{TE}}_{mn}, e_{z\,mn}^{\mathrm{TE}}, \mathbf{e}^{\mathrm{TM}}_{mn}, e_{z\,mn}^{\mathrm{TM}}$, are defined in (16.19) for $\mathbf{F}=\mathbf{E}$ and in (16.20) for $\mathbf{F}=\mathbf{H}$. For the incident mode $m=p$ and $n=q$. Using the orthogonality relations, (16.23)–(16.25), and following a similar procedure as in Section 16.2.1, the TE-modal coefficient, a_{mn}, and TM-modal coefficients, b_{mn}, are given by:

$$
a_{mn} = \frac{1}{F_{\mathrm{TE}_{mn}}{}^{\mathrm{ref}}(z_1,t)}\int_0^a\int_0^b \mathbf{e}^{\mathrm{TE}}_{mn}\bullet[\mathbf{F}(x,y,z,t)-\mathbf{F}^{\mathrm{inc}}(x,y,z,t)]|_{z=z_1}\,dy\,dx
$$

$$(16.40)$$

$$
b_{mn} = \frac{1}{F_{\mathrm{TM}_{mn}}{}^{\mathrm{ref}}(z_1,t)}\int_0^a\int_0^b \mathbf{e}^{\mathrm{TM}}_{mn}\bullet[\mathbf{F}(x,y,z,t)-\mathbf{F}^{\mathrm{inc}}(x,y,z,t)]|_{z=z_1}\,dy\,dx.
$$

$$(16.41)$$

Taking the vector product of $\hat{\mathbf{n}} = -\hat{z}$ with the curl of (16.39) and making use of (16.16), (16.34), (16.40), and (16.41), one obtains the time-domain boundary termination condition for the excitation truncation boundary Ω_e at $z = z_1$ as:

$$
\begin{aligned}
\hat{\mathbf{n}}\times\nabla\times\mathbf{F} \\
= +\frac{\sqrt{\mu_r\varepsilon_r}}{c}\sum_{m=0}^{\infty}\sum_{n=0}^{\infty}\mathbf{e}^{\mathrm{TE}}_{mn}(x,y)\frac{\partial}{\partial t}&\left[\xi^{\mathrm{TE}}_{mn}(t)*\int_0^a\int_0^b \mathbf{e}^{\mathrm{TE}}_{mn}\bullet\mathbf{F}_t|_{z=z_1}\,dy'dx'\right] \\
+\frac{\sqrt{\mu_r\varepsilon_r}}{c}\sum_{m=1}^{\infty}\sum_{n=1}^{\infty}\mathbf{e}^{\mathrm{TM}}_{mn}(x,y)\frac{\partial}{\partial t}&\left[\xi^{\mathrm{TM}}_{mn}(t)*\int_0^a\int_0^b \mathbf{e}^{\mathrm{TM}}_{mn}\bullet\mathbf{F}_t|_{z=z_1}\,dy'dx'\right] \\
-\frac{2\sqrt{\mu_r\varepsilon_r}}{c}\frac{\partial}{\partial t}&[\xi_{\mathrm{inc}}(t)*\mathbf{F}^{\mathrm{inc}}_t(x,y,z_1,t)],
\end{aligned}
$$

$$(16.42)$$

where $\mathbf{F}_t$ is the tangential field component. Similarly, the time-domain total field vector equation at the termination truncation boundary Ω_t where $z = z_2$, has the following form:

$$\mathbf{F}(x, y, z_2, t) = \sum_{m=0}^{\infty} \sum_{n=0}^{\infty} \mathbf{F}_{mn}^{\text{tran}}(x, y, z_2, t)$$

$$= \sum_{m=0}^{\infty} \sum_{n=0}^{\infty} c_{mn} [\mathbf{e}_{mn}^{\text{TE}}(x, y) + \hat{z} e_{z\,mn}^{\text{TE}}(x, y)] F_{\text{TE}\,mn}^{\text{tran}}(z_2, t)$$

$$+ \sum_{m=1}^{\infty} \sum_{n=1}^{\infty} d_{mn} [\mathbf{e}_{mn}^{\text{TM}}(x, y) + \hat{z}\, e_{z\,mn}^{\text{TM}}(x, y)] F_{\text{TM}\,mn}^{\text{tran}}(z_2, t).$$

$$(16.43)$$

Following the aforementioned procedure and noting that $\hat{\mathbf{n}} = \hat{z}$, one obtains the following expressions for the TE- and TM-modal coefficients, c_{mn} and d_{mn}, and the time-domain boundary termination condition for the termination truncation boundary: Ω_t at $z = z_2$:

$$c_{pq} = \frac{1}{F_{\text{TE}_{pq}}^{\text{tran}}(z_2, t)} \int_0^a \int_0^b \mathbf{e}_{pq}^{\text{TE}} \bullet [\mathbf{F}(x, y, z, t)]|_{z=z_2} dy dx \quad (16.44)$$

$$d_{pq} = \frac{1}{F_{\text{TM}_{pq}}^{\text{tran}}(z_2, t)} \int_0^a \int_0^b \mathbf{e}_{pq}^{\text{TM}} \bullet [\mathbf{F}(x, y, z, t)]|_{z=z_2} dy dx \quad (16.45)$$

$$\hat{\mathbf{n}} \times \nabla \times \mathbf{F}$$

$$= +\frac{\sqrt{\mu_r \varepsilon_r}}{c} \sum_{m=0}^{\infty} \sum_{n=0}^{\infty} \mathbf{e}_{mn}^{\text{TE}}(x, y) \frac{\partial}{\partial t} \left[\xi_{mn}^{\text{TE}}(t) * \int_0^a \int_0^b \mathbf{e}_{mn}^{\text{TE}} \bullet \mathbf{F}_t|_{z=z_2} dy' dx' \right]$$

$$+ \frac{\sqrt{\mu_r \varepsilon_r}}{c} \sum_{m=1}^{\infty} \sum_{n=1}^{\infty} \mathbf{e}_{mn}^{\text{TM}}(x, y) \frac{\partial}{\partial t} \left[\xi_{mn}^{\text{TM}}(t) * \int_0^a \int_0^b \mathbf{e}_{mn}^{\text{TM}} \bullet \mathbf{F}_t|_{z=z_2} dy' dx' \right].$$

$$(16.46)$$

Upon substituting (16.42) and (16.46) into the boundary integral of (16.9) and rearranging both sides of (16.9), an expression is obtained that requires the first-order derivative of the following convolution term to be evaluated:

$$C(t) = \xi_{mn}(t) * F_t(t), \quad (16.47)$$

where $\xi_{mn}(t) = \xi_{mn}^{\mathrm{TE}}(t)$ for TE_{mn} mode and $\xi_{mn}(t) = \xi_{mn}^{\mathrm{TM}}(t)$ for TM_{mn} mode. This is achieved recursively as follows.

Substituting $\xi_{mn}(t)$ defined in (16.34)–(16.38) into (16.47) results in:

$$
\begin{aligned}
C(t) &= F_t(t) + \omega_{c_{mn}} [y_{mn}(t) * F_t(t)] \\
&= F_t(t) + \omega_{c_{mn}} \int_0^t y_{mn}(\tau) F_t(t-\tau) d\tau,
\end{aligned}
\tag{16.48}
$$

where, referring to (16.34)–(16.38), the parameters are presented in a tabular form as:

$\mathbf{F}_{mn}$	$y_{mn}(t)$
$\mathbf{E}_{\mathrm{TM}_{mn}}$	$y_{mn}^{\mathrm{TM}}(t) = y_{2_{mn}}(t)$
$\mathbf{E}_{\mathrm{TE}_{mn}}$	$y_{mn}^{\mathrm{TE}}(t) = y_{1_{mn}}(t)$
$\mathbf{H}_{\mathrm{TM}_{mn}}$	$y_{mn}^{\mathrm{TM}}(t) = y_{1_{mn}}(t)$
$\mathbf{H}_{\mathrm{TE}_{mn}}$	$y_{mn}^{\mathrm{TE}}(t) = y_{2_{mn}}(t)$

Using the trapezoidal integration rule, leads to the following discrete-time representation form:

$$
C^n = F_t^n + \omega_{c_{mn}} \sum_{k=0}^{n-1} \left[F_t^{n-k} y_{mn}{}^k + F_t^{n-k-1} y_{mn}{}^{k+1} \right] \frac{\Delta t}{2},
\tag{16.49}
$$

where $y_{mn}^k = y_{mn}(k\Delta t)$ and $C^n = C(n\Delta t)$ are the discrete-time representations of $y_{mn}(t)$ and $C(t)$, respectively. Note that $y_{1_{00}}(t) = y_{2_{00}}(t) = 0$. Using the unconditionally stable three-point recurrence scheme Newmark-beta formulation given in (16.10) to discretize the time-domain functions (Lee, 1995; Maradei, 2001), the first-order time derivative of the convolution term can be expressed as:

$$
\left. \frac{\mathrm{d}[C(t)]}{\mathrm{d}t} \right|_{t=n\Delta t} = \frac{C^{n+1} - C^{n-1}}{2\Delta t}.
\tag{16.50}
$$

Using (16.49) and (16.50), one obtains the discrete-time representation form of the first-order time derivative of the convolution term in (16.42) and (16.46) in terms of the unknown field as follows:

$$
\left. \frac{\mathrm{d}[C(t)]}{\mathrm{d}t} \right|_{t=n\Delta t} = \frac{F_t^{n+1} - F_t^{n-1}}{2\Delta t} + \Psi^n,
\tag{16.51}
$$

where $\Psi^n = \Psi^n_{\text{TE}}$ for TE mode and $\Psi^n = \Psi^n_{\text{TM}}$ for TM mode,

$$\Psi^n = \frac{\omega_{c_{mn}}}{4} \left\{ 2F_t^n y_{mn}{}^1 + F_t^{n-1} y_{mn}{}^2 \right.$$

$$\left. + \sum_{k=0}^{n-2} [F_t^{n-k-1}(y_{mn}{}^{k+2} - y_{mn}{}^k) + F_t^{n-k-2}(y_{mn}{}^{k+3} - y_{mn}{}^{k+1})] \right\}.$$

$$(16.52)$$

However, it is noted that $y_{mn}{}^0 = 0$, hence the term $F_t^{n+1} y_{mn}{}^0 = 0$ is discarded from (16.52). From (16.34)–(16.38), and 16.48, the time-domain TE/TM modal characteristic impedance $Z_{mn}(t)$ and the corresponding function $y_{mn}(t)$ are functions of $\omega_{c_{mn}}$, t, and the Bessel function. As $t \to \infty$, one has $\lim_{t\to\infty} J_1(t) = 0$ and $\int_0^\infty J_0(t)dt = 1$ (Jeffrey, 2002), hence the value of $y_{mn}(t)$ tends towards unity for TE modes and towards zero for the TM modes when $\mathbf{F} = \mathbf{E}$, and vice versa when $\mathbf{F} = \mathbf{H}$. Figure 16.3 shows, for $\mathbf{F} = \mathbf{E}$, how $y_{mn}(t)$ varies with time for different chosen modes (TE_{10}, TE_{01}, TE_{12}, TM_{11}, and TM_{23} modes).

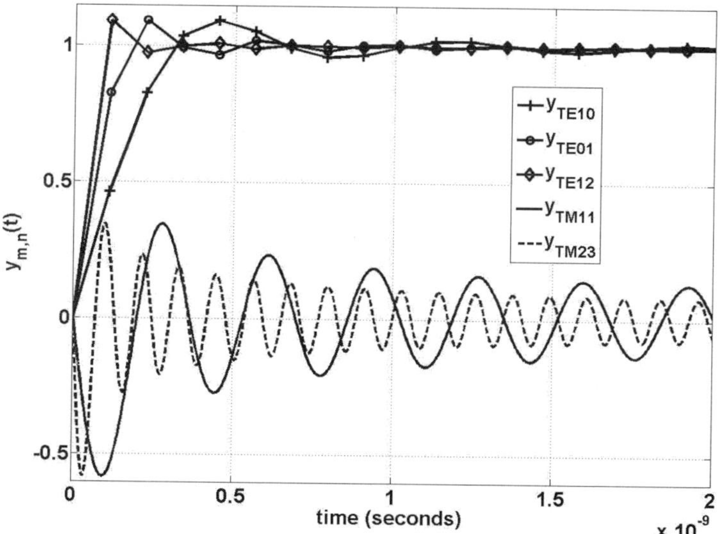

Fig. 16.3. Plot of $y_{mn}(t)$ versus time for different chosen TE/TM modes (TE_{10}, TE_{01}, TE_{12}, TM_{11}, and TM_{23} modes), for $\mathbf{F} = \mathbf{E}$, with time step, $\Delta t = 2.5\,\text{ps}$, the height and the width of a rectangular waveguide, $a = 110\,\text{mm}$ and $b = 55\,\text{mm}$, respectively.

From Fig. 16.3, in order to speed up the computations for $m > 0$ and $n > 0$, it is reasonable to assume, for $\mathbf{F} = \mathbf{E}$ (16.53) and for $\mathbf{F} = \mathbf{H}$ (16.54), respectively, the following:

$$Z_{mn}(t) = \begin{cases} \delta(t) + \omega_{c_{mn}} y_{mn}(t), & t < t_c \\ \delta(t) + \omega_{c_{mn}} u(t), & t \geq t_c \quad \text{(for TE modes)} \\ \delta(t), & t \geq t_c \quad \text{(for TM modes)}, \end{cases} \quad (16.53)$$

$$Z_{mn}(t) = \begin{cases} \delta(t) + \omega_{c_{mn}} y_{mn}(t), & t < t_c \\ \delta(t), & t \geq t_c \quad \text{(for TE modes)} \\ \delta(t) + \omega_{c_{mn}} u(t), & t \geq t_c \quad \text{(for TM modes)}, \end{cases} \quad (16.54)$$

where t_c is the *"approximated convolution cut-off time"* and $u(t)$ is the unit step function. The choice of the t_c value involves a trade-off between accuracy and computational effort. Therefore, for $t \geq t_c$, following the aforementioned procedures, one obtains:

$$\frac{d[C(t)]}{dt}\bigg|_{t=n\Delta t} = \frac{F_t^{n+1} - F_t^{n-1}}{2\Delta t} + \frac{\omega_{c_{mn}}}{4}(F_t^{n+1} + 2F_t^n + F_t^{n-1}). \quad (16.55)$$

One observes from Fig. 16.3 that the larger the $\omega_{c_{mn}}$ the faster the convergence of $y_{mn}(t)$, and vice versa. This implies that, for $\mathbf{F} = \mathbf{E}$, the value of t_c can be chosen by the convergence of $y_{10}^{\text{TE}}(t)$ and $y_{11}^{\text{TM}}(t)$ while $y_{1_{00}}(t) = y_{2_{00}}(t) = 0$ for all time t.

Applying (16.53) and (16.54) in (16.11) and rearranging both sides of (16.11) results in the following 3D Galerkin weighted residual 3D FETD-MABTC final formulations.

For $t < t_c$,

$$[M]\{F\}^{n+1} = [P]\{F\}^n + [Q]\{F\}^{n-1} - \sum_{m=0}^{\infty} \sum_{n=0}^{\infty} [R_{\text{TE}}]\{\Psi_{\text{TE}}\}^n +$$

$$- \sum_{m=1}^{\infty} \sum_{n=1}^{\infty} [R_{\text{TM}}]\{\Psi_{\text{TM}}\}^n + \frac{\{K\}^{n+1}}{4} + \frac{\{K\}^n}{2} + \frac{\{K\}^{n-1}}{4},$$

$$(16.56)$$

where

$$[M] = \frac{[T]}{\Delta t^2} + \frac{1}{2\Delta t}\left([B] + \sum_{m=0}^{\infty} \sum_{n=0}^{\infty} [R_{\text{TE}}] + \sum_{m=1}^{\infty} \sum_{n=1}^{\infty} [R_{\text{TM}}]\right) + \frac{[S]}{4}$$

$$[P] = \frac{2[T]}{\Delta t^2} - \frac{[S]}{2}$$

$$[Q] = -\frac{[T]}{\Delta t^2} + \frac{1}{2\Delta t}\left([B] + \sum_{m=0}^{\infty}\sum_{n=0}^{\infty}[R_{\mathrm{TE}}] + \sum_{m=1}^{\infty}\sum_{n=1}^{\infty}[R_{\mathrm{TM}}]\right) - \frac{[S]}{4}$$

$$K_i^n = \frac{2\sqrt{\mu_r\varepsilon_r}\,f_i^{\mathrm{inc}^n}}{pc}\int_{\Omega}\mathbf{N}_i\bullet\mathbf{e}_{\mathrm{inc}}^{\mathrm{TE/TM}}(x,y)d\Omega$$

$$R_{\mathrm{TE}_{ij}} = \frac{\sqrt{\mu_r\varepsilon_r}}{pc}\int_{\Omega}\mathbf{N}_i\bullet\mathbf{e}_{mn}^{\mathrm{TE}}(x,y)d\Omega\int_0^a\int_0^b\mathbf{N}_j\bullet\mathbf{e}_{mn}^{\mathrm{TE}}(x',y')d\Omega'$$

$$R_{\mathrm{TM}_{ij}} = \frac{\sqrt{\mu_r\varepsilon_r}}{pc}\int_{\Omega}\mathbf{N}_i\bullet\mathbf{e}_{mn}^{\mathrm{TM}}(x,y)\,d\Omega\int_0^a\int_0^b\mathbf{N}_j\bullet\mathbf{e}_{mn}^{\mathrm{TM}}(x',y')d\Omega'$$

$$f_i^{\mathrm{inc}^n} = (F_{t_i}^{\mathrm{inc}^{n+1}} - F_{t_i}^{\mathrm{inc}^{n-1}})/(2\Delta t) + \frac{\omega_{c_{\mathrm{inc}}}}{4}\left\{2F_{t_i}^{\mathrm{inc}^n}y_{\mathrm{inc}}{}^1 + F_{t_i}^{\mathrm{inc}^{n-1}}y_{\mathrm{inc}}{}^2\right.$$

$$+ \sum_{k=0}^{n-2}\left[F_{t_i}^{\mathrm{inc}^{n-k-1}}(y_{\mathrm{inc}}{}^{k+2} - y_{\mathrm{inc}}{}^k)\right.$$

$$\left.\left. + F_{t_i}^{\mathrm{inc}^{n-k-2}}(y_{\mathrm{inc}}{}^{k+3} - y_{\mathrm{inc}}{}^{k+1})\right]\right\}.$$

$p = \mu_r$ for $\mathbf{F} = \mathbf{E}$, $p = \varepsilon_r$ for $\mathbf{F} = \mathbf{H}$; c is the speed of light in free space, μ_r and ε_r are, respectively, the relative permeability and relative permittivity.

For $t \geq t_c$,

$$[M]\{F_t\}^{n+1} = [P]\{F_t\}^n + [Q]\{F_t\}^{n-1} + \frac{\{K\}^{n+1}}{4} + \frac{\{K\}^n}{2} + \frac{\{K\}^{n-1}}{4},$$

$$(16.57)$$

where

$$[M] = \frac{[T]}{\Delta t^2} + \frac{1}{2\Delta t}\left([B] + \sum_{m=0}^{\infty}\sum_{n=0}^{\infty}[R_{\mathrm{TE}}] + \sum_{m=1}^{\infty}\sum_{n=1}^{\infty}[R_{\mathrm{TM}}]\right) +$$

$$+ \frac{1}{4}\left([S] + \sum_{m=0}^{\infty}\sum_{n=0}^{\infty}[W_{\mathrm{TE}}] + \sum_{m=1}^{\infty}\sum_{n=1}^{\infty}[W_{\mathrm{TM}}]\right)$$

$$[P] = \frac{2[T]}{\Delta t^2} - \frac{1}{2}\left([S] + \sum_{m=0}^{\infty}\sum_{n=0}^{\infty}[W_{\mathrm{TE}}] + \sum_{m=1}^{\infty}\sum_{n=1}^{\infty}[W_{\mathrm{TM}}]\right)$$

$$[Q] = -\frac{[T]}{\Delta t^2} + \frac{1}{2\Delta t}\left([B] + \sum_{m=0}^{\infty}\sum_{n=0}^{\infty}[R_{\mathrm{TE}}] + \sum_{m=1}^{\infty}\sum_{n=1}^{\infty}[R_{\mathrm{TM}}]\right) +$$

$$-\frac{1}{4}\left([S] + \sum_{m=0}^{\infty}\sum_{n=0}^{\infty}[W_{\mathrm{TE}}] + \sum_{m=1}^{\infty}\sum_{n=1}^{\infty}[W_{\mathrm{TM}}]\right)$$

$$W_{\mathrm{TE}_{ij}} = \frac{\sqrt{\mu_r\varepsilon_r}}{pc}\omega_{c_{mn}}\int_{\Omega}\mathbf{N}_i\bullet\mathbf{e}_{mn}^{\mathrm{TE}}(x,y)d\Omega\int_0^a\int_0^b\mathbf{N}_j\bullet\mathbf{e}_{mn}^{\mathrm{TE}}(x',y')d\Omega'$$

$$W_{\mathrm{TM}_{ij}} = \frac{\sqrt{\mu_r\varepsilon_r}}{pc}\omega_{c_{mn}}\int_{\Omega}\mathbf{N}_i\bullet\mathbf{e}_{mn}^{\mathrm{TM}}(x,y)d\Omega\int_0^a\int_0^b\mathbf{N}_j\bullet\mathbf{e}_{mn}^{\mathrm{TM}}(x',y')d\Omega'$$

$$f_i^{\mathrm{inc}^n} = (F_{t_i}^{\mathrm{inc}^{n+1}} - F_{t_i}^{\mathrm{inc}^{n-1}})/(2\Delta t) + \frac{\omega_{c_{\mathrm{inc}}}}{4}(F_{t_i}^{\mathrm{inc}^{n+1}} + 2F_{t_i}^{\mathrm{inc}^n} + F_{t_i}^{\mathrm{inc}^{n-1}}).$$

The expressions for the $[R_{\mathrm{TE}}]$ and $[R_{\mathrm{TM}}]$ are the same as those defined in (16.56) for $t < t_c$. It is assumed that both ports (the input and output) are identical.

16.3. Numerical Results

The 3D finite-element methods derived in this chapter have been implemented using FORTRAN 77 for a general 3D boundary-value problems using the first-order four-node six degree of freedom (DOF) tetrahedral vector elements (Webb and Forghani, 1993; Mias, 1995; Silvester and Ferrari, 1996), for the volume domains and the first-order triangular vector element for the boundary area domains (Jin, 2002). Edge-based vector elements are used in order to handle sharp metal edges and to avoid the spurious solutions. The code is organized to define edge-based elements and embed the edge components of local elements to their corresponding global location in the final system of equations. The matrix equation is solved directly using the multifrontal sparse Gaussian elimination solvers, MA47 (for FETD) and ME47 (for FEFD) of the Harwell Subroutine Library (AEA Technology, Oxfordshire, UK).

In all the following examples the electric field is chosen as the working variable. For the incident wave, $p = 1$ and $q = 0$. In the FEFD code a TE_{10} mode is sent. In the FETD code the time variation of the incident wave is described by a modulated Gaussian pulse of the form:

$$E_t^{\mathrm{inc}}(t) = \exp\left[-\frac{(t - 4\sigma_t)^2}{2\sigma_t^2}\right]\sin(\omega_c t), \tag{16.58}$$

where $\omega_c = 2\pi f_c$. f_c and σ_t denote the carrier frequency and the time-delay constant respectively. The wave is incident from the left-hand side and has the E_y-TE$_{10}$ profile as the transversal spatial distribution. Furthermore, in all examples, M + 1 ($0 \le m \le$ M) modes and $N + 1$ ($0 \le n \le$ N) modes have been employed where M = N = 3. The surface integrals were evaluated numerically using a seven-point Gaussian–Legendre quadrature integration for triangular (Cowper, 1973). For comparison purposes, the FEFD method has a frequency resolution of Δf. To achieve in the FETD method the same frequency resolution, a total time of $t_{total} = 1/\Delta f$ is required. The same mesh structure is employed for both the FETD and FEFD methods. The density of elements obeys the rule of thumb (Mias, 1995) for a mesh of first-order finite elements, i.e., at least ten first-order elements are required per wavelength. For consistency, the CST mesh properties are set to: lines per wavelength = 10, lower mesh limit = 10, and mesh line ratio limit = 10. In both techniques, the FETD and the FEFD, the Poynting theorem is employed for the evaluation of the normalized reflected power ($\mathrm{P}^{\mathrm{ref}}_{\mathrm{norm}}$) and normalized transmitted power ($\mathrm{P}^{\mathrm{tran}}_{\mathrm{norm}}$) in decibel scale where:

$$\mathrm{P}^{\mathrm{ref}}_{\mathrm{norm}} = \sum_{m=0}^{\infty} \sum_{n=0}^{\infty} \mathrm{P}^{\mathrm{ref}}_{\mathrm{norm}\ mn}$$

$$= \frac{\sum_{m=0}^{\infty} \sum_{n=0}^{\infty} \left(M^{\mathrm{TE}}_{mn\,\Omega_e} \left| a_{mn} F^{\mathrm{ref}}_{t\mathrm{TE}mn}(z_1, s) \right|^2 + M^{\mathrm{TM}}_{mn\,\Omega_e} \left| b_{mn} F^{\mathrm{ref}}_{t\mathrm{TM}mn}(z_1, s) \right|^2 \right)}{M^{\mathrm{TE}}_{\mathrm{inc}\,\Omega_e} \left| a_{\mathrm{inc}} F^{\mathrm{inc}}_{t\mathrm{TE}_{\mathrm{inc}}}(z_1, s) \right|^2 + M^{\mathrm{TM}}_{\mathrm{inc}\,\Omega_e} \left| b_{\mathrm{inc}} F^{\mathrm{inc}}_{t\mathrm{TM}_{\mathrm{inc}}}(z_1, s) \right|^2}$$

$$(16.59)$$

$$\mathrm{P}^{\mathrm{tran}}_{\mathrm{norm}} = \sum_{m=0}^{\infty} \sum_{n=0}^{\infty} \mathrm{P}^{\mathrm{tran}}_{\mathrm{norm}\ mn}$$

$$= \frac{\sum_{m=0}^{\infty} \sum_{n=0}^{\infty} \left(M^{\mathrm{TE}}_{mn\,\Omega_t} \left| c_{mn} F^{\mathrm{tran}}_{t\mathrm{TE}mn}(z_2, s) \right|^2 + M^{\mathrm{TM}}_{mn\,\Omega_t} \left| d_{mn} F^{\mathrm{tran}}_{t\mathrm{TM}mn}(z_2, s) \right|^2 \right)}{M^{\mathrm{TE}}_{\mathrm{inc}\,\Omega_e} \left| a_{\mathrm{inc}} F^{\mathrm{inc}}_{t\mathrm{TE}_{\mathrm{inc}}}(z_1, s) \right|^2 + M^{\mathrm{TM}}_{\mathrm{inc}\,\Omega_e} \left| b_{\mathrm{inc}} F^{\mathrm{inc}}_{t\mathrm{TM}_{\mathrm{inc}}}(z_1, s) \right|^2}$$

$$(16.60)$$

and $M^{\mathrm{TE}}_{mn\,\Omega} = 1/Z^{\mathrm{TE}}_{mn} = k_{z_{mn}}/\omega\mu_0\mu_r$, $M^{\mathrm{TM}}_{mn\,\Omega} = 1/Z^{\mathrm{TM}}_{mn} = \omega\varepsilon_0\varepsilon_r/k_{z_{mn}}$ for $\mathbf{F} = \mathbf{E}$; $M^{\mathrm{TE}}_{mn\,\Omega} = Z^{\mathrm{TE}}_{mn} = \omega\mu_0\mu_r/k_{z_{mn}}$, $M^{\mathrm{TM}}_{mn\,\Omega} = Z^{\mathrm{TM}}_{mn} = k_{z_{mn}}/\omega\varepsilon_0\varepsilon_r$, for $\mathbf{F} = \mathbf{H}$. Note that the results obtained from the CST package have a fixed number of 1001 frequency sampling points within the chosen frequency range. For the FETD-MABTC method, the time t_c is chosen corresponding to the time at which the dumped maximum magnitude of the $y^{\mathrm{TM}}_{11}(t)$ (which fluctuates from 0 to a dumped maximum value as it oscillates with a dumped oscillation about zero) reaches 0.075.

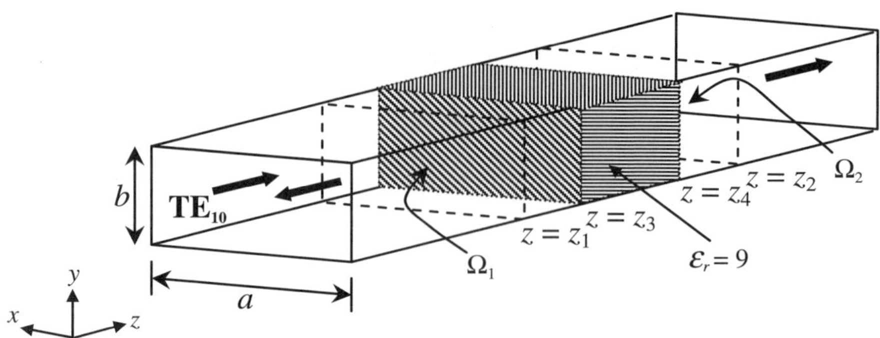

Fig. 16.4.   Sketch of a rectangular waveguide incorporating a dielectric discontinuity.

16.3.1. *Single Dielectric-Layer Situated in a Rectangular Waveguide*

In this example, one considers a WG8 rectangular waveguide incorporating a dielectric discontinuity, which acts as a band-pass filter. As depicted in Fig. 16.4, the waveguide has a width $a = 110\,\text{mm}$, height $b = 55\,\text{mm}$, and the length of the dielectric layer ($\varepsilon_r = 9$) is $\overline{z_3 z_4} = 25\,\text{mm}$. These parameters are chosen such that at around 2 GHz there is a complete transmission. To demonstrate that the MABTC can be positioned very close to the discontinuity, provided a sufficient number of modes is taken into account (here M = N = 3), the length between the port boundary and the discontinuity is chosen to be $\overline{z_1 z_3} = \overline{z_2 z_4} = 10\,\text{mm}$ and $\overline{z_1 z_2} = 45\,\text{mm}$.

Using the FE mesh shown in Fig. 16.5 the FETD results in Fig. 16.6 were obtained. The number of tetrahedra in the finite-element region is 33,977 and the number of the edge unknown to be solved is 38,566. The waveguide is modeled in the frequency range from 1.4 GHz to 2.6 GHz. To obey the sampling theorem, the time step is chosen to be 2.5 ps. The carrier frequency is $f_c = 2\,\text{GHz}$, the time-delay constant is $\sigma_t = 0.5\,\text{ns}$, the approximated convolution cut-off time, $t_c = 6\,\text{ns}$, and the total time is $t_{total} = 16\,\text{ns}$. There is a good agreement between all four results.

16.3.2. *Dielectric Post Situated in a Rectangular Waveguide*

This example consists of a dielectric post discontinuity situated in a WG90 rectangular waveguide shown in Fig. 16.7. The same geometry was analyzed by Wang and Mittra (1994) using the FEFD method. As depicted in Fig. 16.7, the waveguide has a width $a = 22.86\,\text{mm}$ and height $b = 10.16\,\text{mm}$. The dielectric post has height b, width $w = 12\,\text{mm}$,

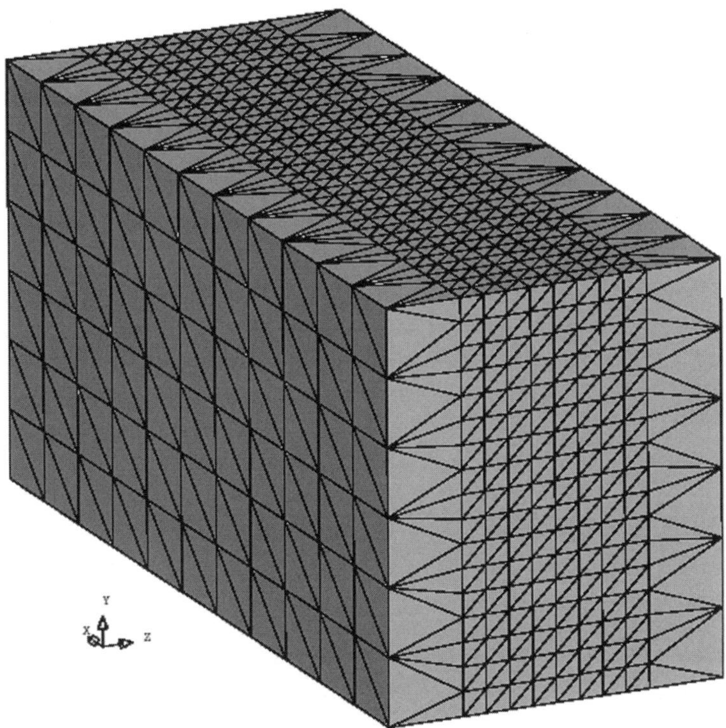

Fig. 16.5. The FE mesh of the rectangular waveguide incorporating a dielectric discontinuity.

and a length of $\overline{z_3 z_4} = 6\,\text{mm}$. The dielectric constant of the post is 8.2. Furthermore, the length between the port boundary and the discontinuity is chosen to be in $\overline{z_1 z_3} = \overline{z_2 z_4} = 7\,\text{mm}$ (instead of $l = a/2$ (i.e., $11.43\,\text{mm}$) as in Wang and Mittra (1994)), and $\overline{z_1 z_2} = 20\,\text{mm}$.

Using the FE mesh shown in Fig. 16.8 the FETD results in Fig. 16.9 were obtained. The number of tetrahedra in the finite-element region is 15,566 and the number of the edge unknowns to be solved in this mesh is 17,485. The waveguide is modeled in the frequency range from $8\,\text{GHz}$ to $12\,\text{GHz}$. To obey the sampling theorem, the time step is chosen to be 2.5 ps. The carrier frequency is $f_c = 10\,\text{GH}z$, the time-delay constant is $\sigma_t = 0.1\,\text{ns}$, the approximated convolution cut-off time is $t_c = 6\,\text{ns}$, and the total time is $t_{total} = 10\,\text{ns}$ (for $\Delta f = 0.1\,\text{GHz}$). Applying the DFT to the time-domain waveform, an excellent agreement is found between the FETD, the FEFD, and the CST results as shown in Fig. 16.9.

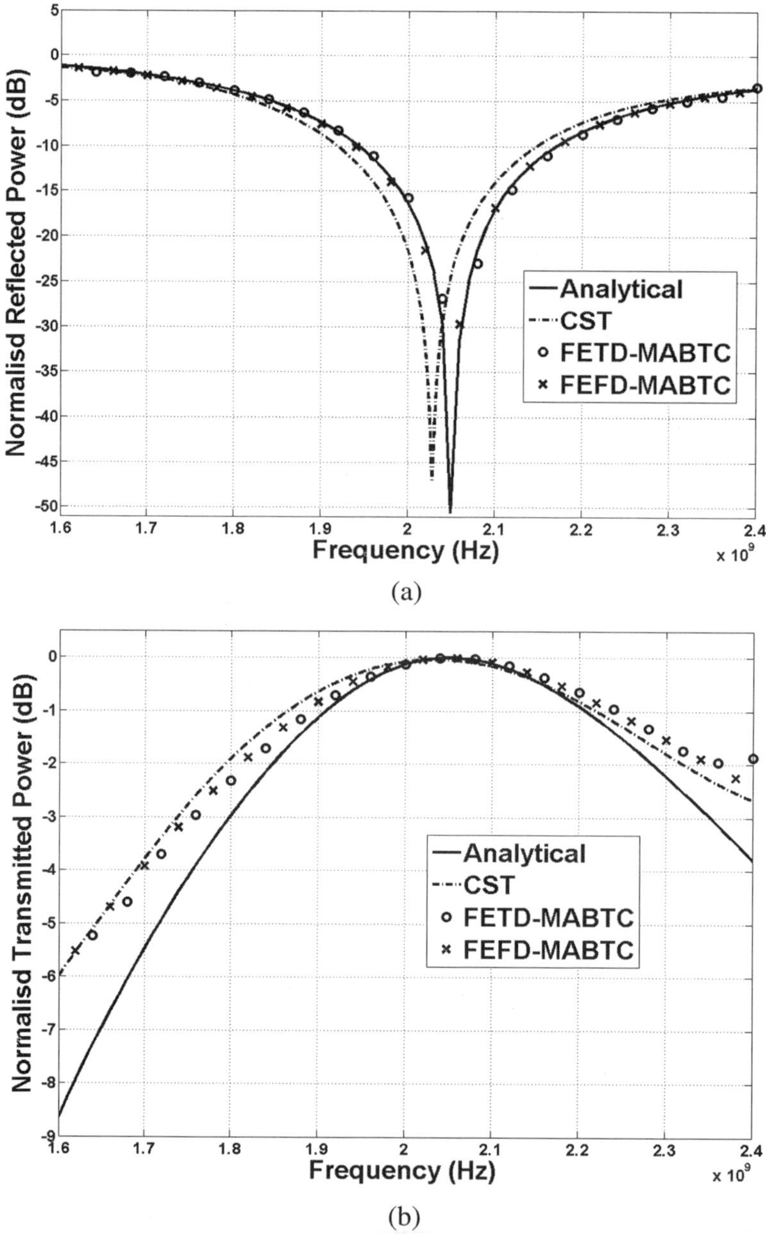

Fig. 16.6. Simulation results of the single dielectric-layer situated in a WG8 rectangular waveguide. Comparison of the normalized reflected/transmitted power obtained by the finite-element methods with analytical results and those obtained by CST Microwave Studio: (a) the normalized reflected power, (b) the normalized transmitted power.

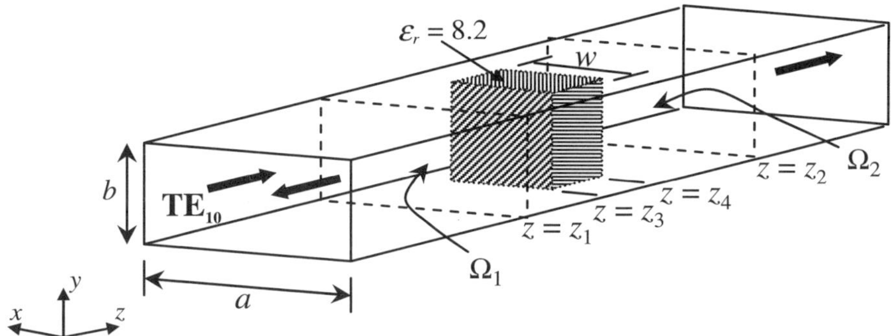

Fig. 16.7.　Sketch of a rectangular waveguide incorporating a dielectric post discontinuity.

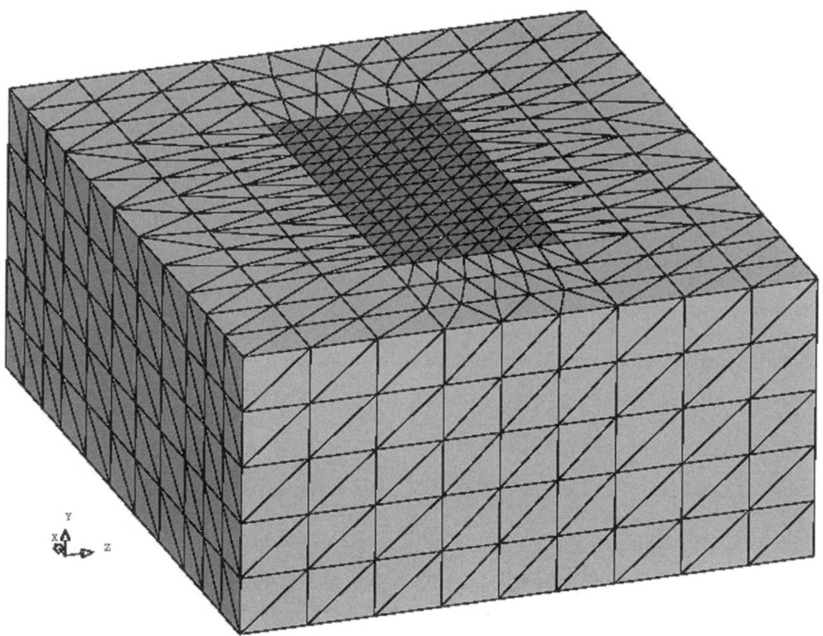

Fig. 16.8.　The FE mesh of the rectangular waveguide incorporating a dielectric discontinuity.

16.3.3. *Inductive Iris Situated in a Rectangular Waveguide*

This example consists of an inductive iris discontinuity situated in a WR90 rectangular waveguide as shown in Fig. 16.10. The same geometry was

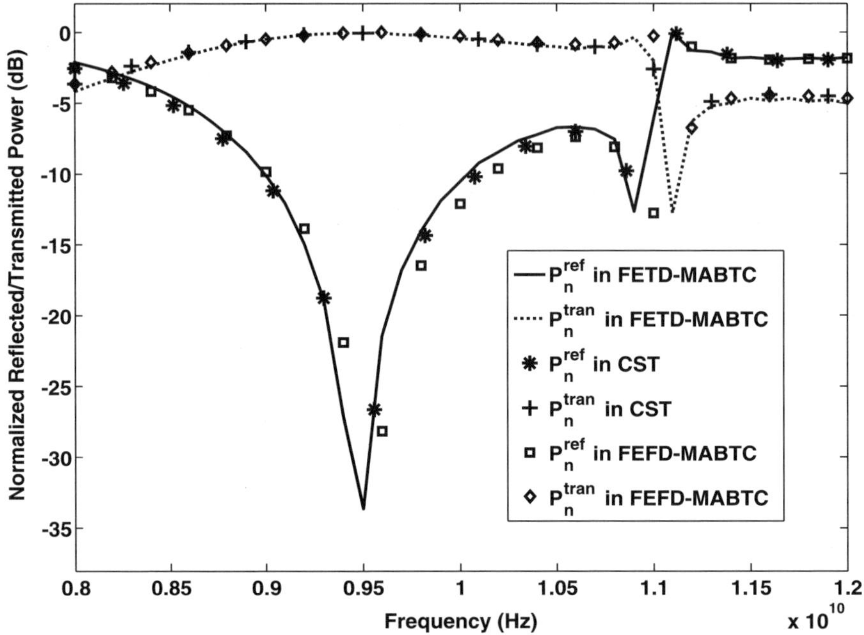

Fig. 16.9. Simulation results of the dielectric post situated in a WG90 rectangular waveguide. Comparison of the normalized reflected/transmitted power obtained by FETD-MABTC and FEFD-MABTC methods with those obtained by the CST Microwave Studio. (P_n^{ref} is the normalized reflected power; P_n^{tran} is the normalized transmitted power.)

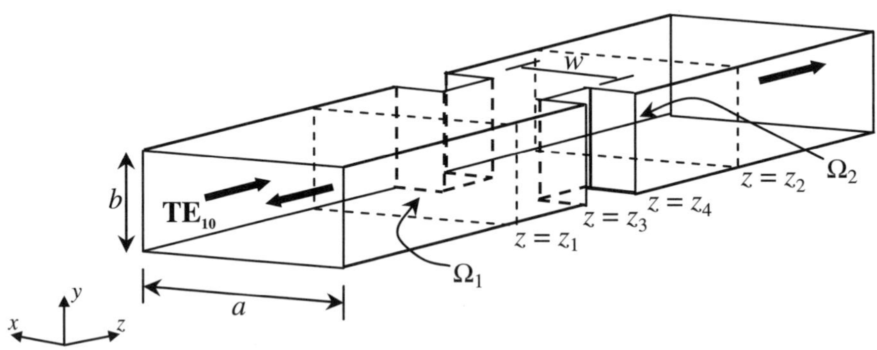

Fig. 16.10. Sketch of a rectangular waveguide incorporating an inductive iris discontinuity.

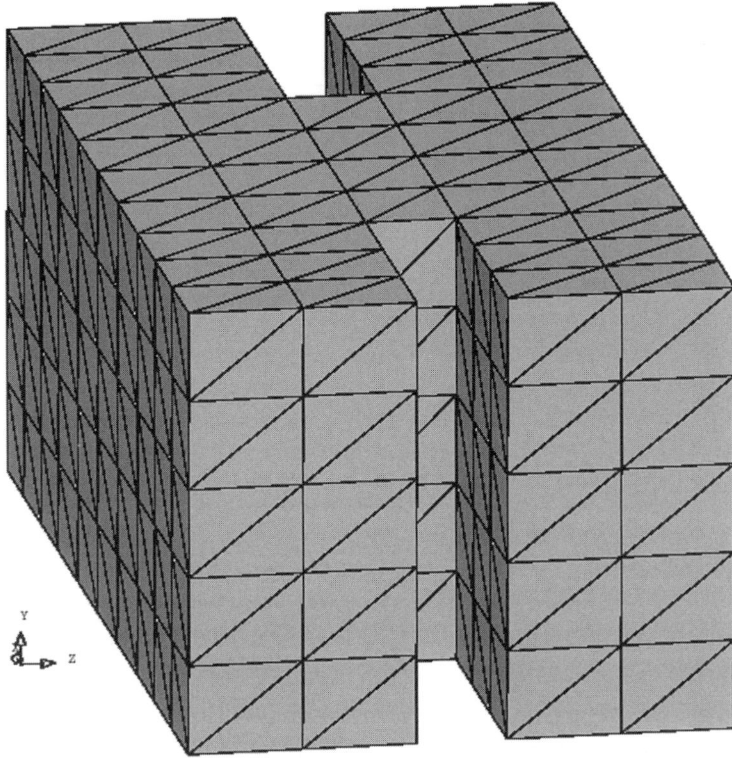

Fig. 16.11.   The FE mesh of the rectangular waveguide incorporating an inductive iris discontinuity.

analyzed by Pierantoni *et al.* using the TLM technique (Pierantoni *et al.*, 2002) and Weisshaar *et al.* (1996) using the mode matching technique. As depicted in Fig. 16.10, the waveguide has a width $a = 22.86\,\text{mm}$ and height $b = 10.16\,\text{mm}$. The inductive iris has a thickness $\overline{z_3 z_4} = 2\,\text{mm}$, with an equal height as the guide, an aperture $w = 10\,\text{mm}$, and a length between the port boundary and the discontinuity of value $l = 5\,\text{mm}$. In addition, $\overline{z_1 z_2} = 12\,\text{mm}$. The results obtained from the FETD-MABTC and FEFD-MABTC methods are then compared.

Using the FE mesh shown in Fig. 16.11 the FETD results in Fig. 16.12 were obtained. The number of tetrahedra in the finite-element region is 15,566 and the number of the edge unknown to be solved in this mesh is 17,485. The waveguide is modeled in the frequency range from $8\,\text{GHz}$ to $12\,\text{GHz}$. To obey the sampling theorem, the time step is chosen to be $2.5\,\text{ps}$. The carrier frequency is $f_c = 10\,\text{GHz}$, the time-delay constant is

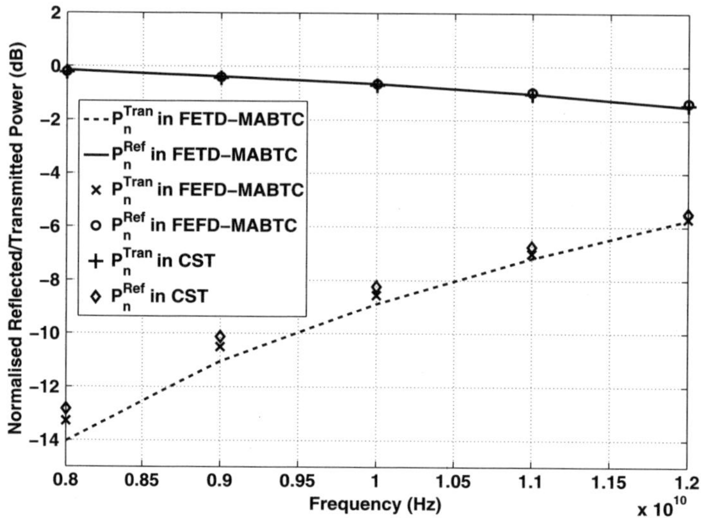

Fig. 16.12. Simulation results of the inductive iris situated in a WR90 rectangular waveguide. Comparison of the normalized reflected/transmitted power obtained by FETD-MABTC and FEFD-MABTC methods with those obtained by CST Microwave Studio, where P_n^{ref} is the normalized reflected power; P_n^{tran} is the normalized transmitted power.

$\sigma_t = 0.1$ ns, the approximated convolution cut-off time is $t_c = 6$ ns, and the total time is $t_{total} = 10$ ns (for $\Delta f = 0.1$ GHz). Figure 16.12 shows that the FETD results agree well with those of the FEFD and the CST.

16.4. Further Investigation on the Criteria for Choosing the Value of t_c in the FETD-MABTC Technique

To investigate further the criteria for choosing the value of t_c, consider first the single dielectric-layer situated in a WG8 rectangular waveguide example, for which the $y_{mn}(t)$ plot has been shown in Fig. 16.3, for $\mathbf{F} = \mathbf{E}$ for different chosen modes (TE_{10}, TE_{01}, TE_{12}, TM_{11}, and TM_{23} modes).

Figures 16.13 and 16.14 show, for $\mathbf{F} = \mathbf{E}$, for $t_c = 6$ ns, and $t_c = 2$ ns, the indicative plots of various time-domain reflected and transmitted TE/TM-modal fields obtained from the FETD-MABTC code, and the normalized reflected/transmitted power (in dB scale), respectively. In Fig. 16.13, the various cut-off times t_c are represented by the dashed lines.

For the second example, a dielectric post situated in a rectangular waveguide, Fig. 16.15 shows how the $y_{mn}(t)$ varies with time for $\mathbf{F} = \mathbf{E}$

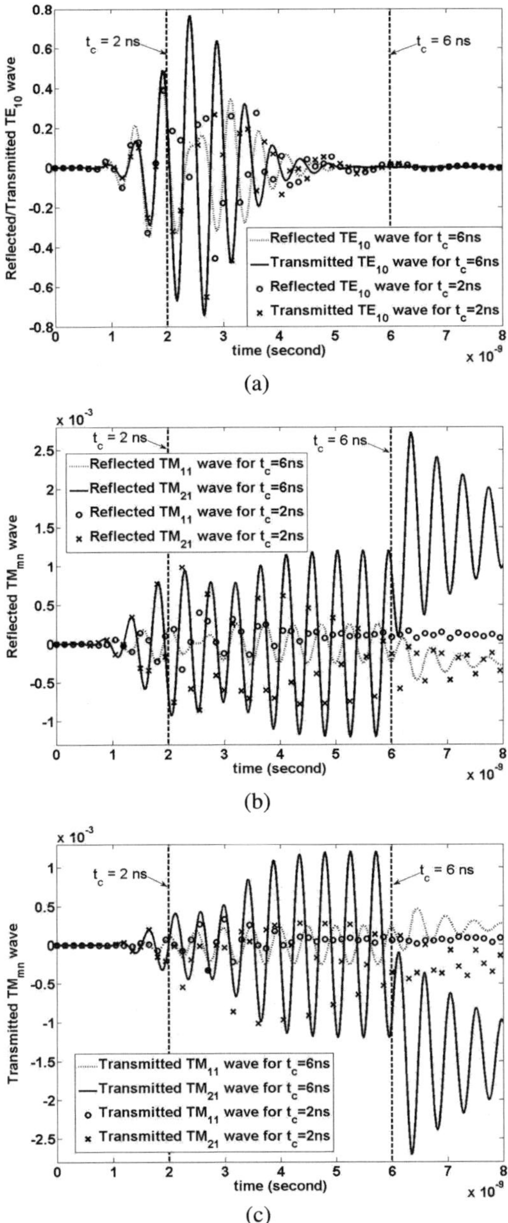

Fig. 16.13.   Time-domain simulation results of the single dielectric-layer situated in rectangular waveguide for various t_c. Comparison of time-domain reflected/transmitted TM/TE wave, obtained by FETD-MABTC, for $t_c = 6$ ns and $t_c = 2$ ns: (a) TE$_{10}$ reflected/transmitted wave, (b) TM$_{mn}$ reflected wave, (c) TM$_{mn}$ transmitted wave.

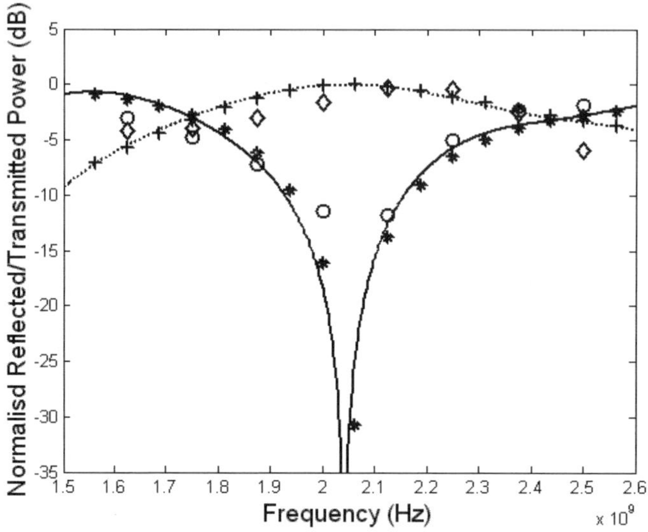

Fig. 16.14. Frequency-domain simulation results of the single dielectric-layer situated in rectangular waveguide for various t_c. Comparison of normalized reflected/transmitted power obtained by FETD-MABTC, for $t_c = 6$ ns and $t_c = 2$ ns, with those obtained by CST package. Solid line: P_n^{ref} in CST package; dotted line: P_n^{tran} in CST package; *: P_n^{ref} in FETD method ($t_c = 6$ ns); +: P_n^{tran} in FETD method ($t_c = 6$ ns); o: P_n^{ref} in FETD method ($t_c = 2$ ns); $\Diamond$: P_n^{tran} in FETD method ($t_c = 2$ ns) where P_n^{ref} is the normalized reflected power; P_n^{tran} is the normalized transmitted power.

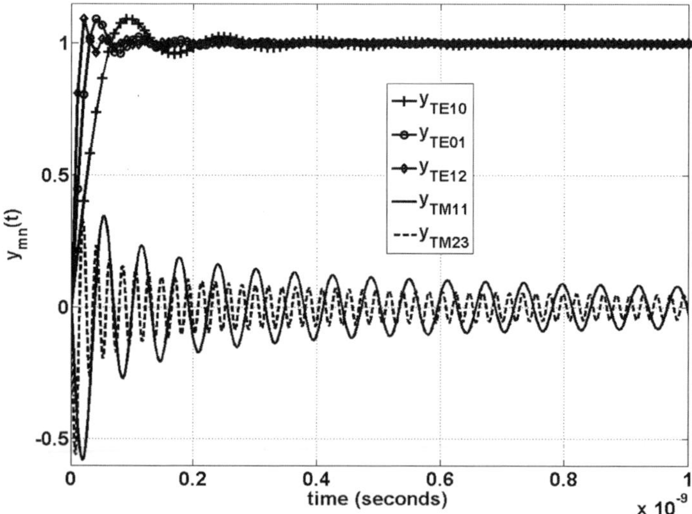

Fig. 16.15. Plot of $y_{mn}(t)$ versus time for different chosen TE/TM modes (TE$_{10}$, TE$_{01}$, TE$_{12}$, TM$_{11}$, and TM$_{23}$ modes), for $\mathbf{F} = \mathbf{E}$, with time step $\Delta t = 2.5$ ps, the height, and the width of a rectangular waveguide, $a = 28.16$ mm and $b = 10.16$ mm, respectively.

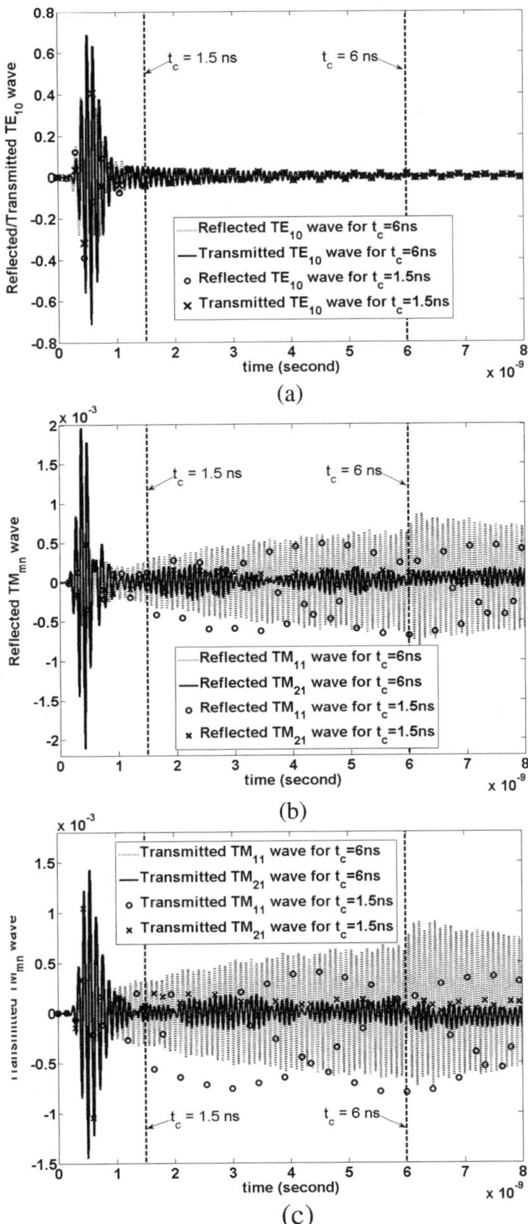

Fig. 16.16.   Simulation results of the dielectric post situated in a rectangular waveguide for various t_c. Comparison of time-domain reflected/transmitted TM/TE wave, obtained by FETD-MABTC, for $t_c = 6\,\mathrm{ns}$ and $t_c = 1.5\,\mathrm{ns}$: (a) TE_{10} reflected/transmitted wave, (b) TM_{mn} reflected wave, (c) TM_{mn} transmitted wave.

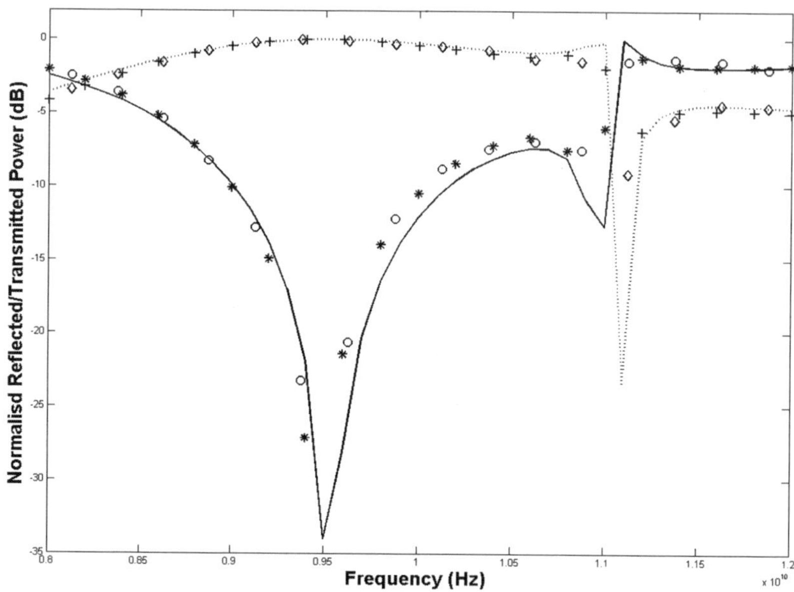

Fig. 16.17. Frequency-domain simulation results of the dielectric post situated in rectangular waveguide for various t_c. Comparison of normalized reflected/transmitted power obtained by FETD-MABTC, for $t_c = 6\,\text{ns}$ and $t_c = 1.5\,\text{ns}$, with those obtained by FEFD-MABTC method. Solid line: P_n^{ref} in FEFD package; dotted line: P_n^{tran} in FEFD package; *: P_n^{ref} in FETD method ($t_c = 6\,\text{ns}$); +: P_n^{tran} in FETD method ($t_c = 6\,\text{ns}$); o: P_n^{ref} in FETD method ($t_c = 1.5\,\text{ns}$); $\Diamond$: P_n^{tran} in FETD method ($t_c = 1.5\,\text{ns}$) where P_n^{ref} is the normalized reflected power; P_n^{tran} is the normalized transmitted power.

for different chosen modes (TE_{10}, TE_{01}, TE_{12}, TM_{11}, and TM_{23} modes). Figures 16.16 and 16.17 show, for $\mathbf{F} = \mathbf{E}$, for $t_c = 6\,\text{ns}$, and $t_c = 1.5\,\text{ns}$, the indicative plots of various time-domain reflected and transmitted TE/TM-modal fields obtained from the FETD-MABTC code, and the normalized reflected/transmitted power (in dB scale), respectively.

From the results, one observes that the TE mode value has settled to a steady state. However this is not the case for the TM modes. In these two examples, the TM modes are evanescent and have a comparatively small value with respect to the TE modes, in which $y_{11}^{\text{TM}}(t)$ has the slowest converge rate among all the $y_{mn}(t)$ considered. However, this might not be the case for other waveguide problems, for example, TM mode propagation for $\mathbf{F} = \mathbf{E}$ or TE mode propagation for $\mathbf{F} = \mathbf{H}$. Hence, one can reasonably conclude that the value of t_c can be chosen as the time over which the magnitude of the slowest converging $y_{mn}(t)$ is less than a reasonable

Table 16.2. Computational performances of the FETD-MABTC for the dielectric post examples considered in Fig. 16.7. The run time column shows the time for each time step taken by the program to solve the problem while the memory column shows the memory needed to run the program at the particular step. The computations were performed on an AMD 2.6-GHz PC Machine. Note that the frequency resolution is 0.1 GHz, $t_c = 1.5$ ns, $\Delta t = 2.5$ ps, and total number of time steps, NT $= 4001$.

Computation Times of the Waveguide in Fig. 16.7 on an AMD 2.6-GHz PC Machine

Problem	Time-Stage	Run Time (HH:MM:SS)	Memory
Fig. 16.7	at $t = 0$ and $t = t_c$	00:40:00*	175.1 MB
17485 DOF	at $t < t_c$	$\approx$00:00:20	175.1 MB
15566 elements	at $t > t_c$	<00:00:20	58.2 MB

NT: Total number of time steps.
DOF: Degrees of freedom (unknowns).
*: This is the time needed for the assembly of the matrix.

threshold value. Furthermore, it is envisaged that the accuracy in FETD-MABTC can further be improved by increasing the number of elements, order of elements, and convolution accuracy. In addition, the computational performance for the dielectric post problem with $t_c = 1.5$ ns is given in Table 16.2.

16.5. Scientific Contributions and Concluding Remarks

Chapters 15 (2D) and 16 (3D), demonstrated that the MABTC boundary condition is a useful tool in the finite-element modeling of the electro-magnetic behavior of microwave devices. Unlike terminations based on PML, it does not require additional finite-element layers. In addition, in contrast to a simple ABC, it can be placed very close to a discontinuity in waveguide applications. Several such waveguide scenarios were simulated using both the frequency- and time-domain methods that confirmed the good accuracy of the methodologies employed. The disadvantages of the MABTC boundary condition are the reduction in the sparsity of the matrix (due to the boundary integral) and, in time domain, the presence of a time-consuming convolution integral. These issues are the subject of an ongoing research work.

References

Berenger, J.P. (1994). A perfectly matched layer for the absorption of electro-magnetic waves, *J. Comput. Phys.*, **114**, 185–200.

Collin, R.E. (1991). *Field Theory of Guided Waves*, (2nd ed.), IEEE Press, New York: NY.

Cowper, E.R. (1973). Gaussian quadrature formulae for triangles, *Int. J. Numer. Meth. Eng.*, **7**, 405–408.

CST (2012). *CST Microwave Studio*. [Online] Available at: www.cst.com. Date accessed: 5 March 2012.

Dibben, D.C. and Metaxas, R. (1996). Time domain finte element analysis of multimode microwave applicators, *IEEE Trans. Magn.*, **32**, 942–945.

Harrington, R.F. (1961). *Time-Harmonic Electromagnetic Fields*, McGraw-Hill, New York: NY.

Jeffrey, A. (2002). *Advanced Engineering Mathematics*, Harcount Academic Press, New York: NY.

Jin, J.M. (2002). *The Finite Element Method in Electromagnetics*, (2nd ed.), Wiley, New York: NY.

Kreyszig, E. (2010). *Advanced Engineering Mathematics*, (10th ed.), Wiley, New York: NY.

Lee, J.F. (1995). Whitney elements time domain (WETD) method, *IEEE T. Magn.*, **31**, 1325–1329.

Lee, J.F., Lee, R. and Cangellaris, A. (1997). Time-domain finite-element methods, *IEEE T. Antenn. Propag.*, **45**, 430–442.

Loh, T.H. (2004). An exact port boundary condition for the finite-element time-domain modelling of microwave devices. Ph.D. Thesis, University of Warwick, Coventry.

Loh, T.H. and Mias, C. (2004). Implementation of an exact modal absorbing boundary termination condition for the application of the finite-element time-domain technique to discontinuity problems in closed homogeneous waveguides, *IEEE T. Microw. Theory*, **52**, 882–888.

Lou, Z. and Jin, J.M. (2005). An accurate waveguide port boundary condition for the time-domain finite-element method, *IEEE T. Microw. Theory*, **53**, 3014–3023.

Maradei, F. (2001). A frequency-dependent WETD formulation for dispersive materials, *IEEE T. Magn.*, **37**, 3303–3306.

Mias, C. (1995). Finite element modelling of the electromagnetic behaviour of spatially periodic structures. Ph.D. Thesis, Cambridge University, Cambridge.

Moglie, F., Rozzi, T., Marcozzi, P. and Schiavoni, A. (1992). A new termination condition for the Application of FDTD techniques to discontinuity problems in close homogeneous waveguide, *IEEE Microw. Guided W.*, **2**, 475–477.

Mur, G. (1981). Absorbing boundary conditions for finite-difference approximation of the time-domain electromagnetic-field equations, *IEEE T. Electromagn. C.*, **EMC-23**, 1073–1077.

Newmark, N.M. (1959). A method of computation for structural dynamics, *J. Eng. Mech. Div-ASCE*, **85**, 67–94.

Olivier, J.C. and McNamara, D.A. (1994). Analysis of multiport discontinuities in waveguide using a pulsed FDTD approach, *IEEE T. Microw. Theory*, **42**, 2229–2238.

Pierantoni, L., Tomassoni, C. and Rozzi, T. (2002). A new termination condition for the application of the TLM method to discontinuity problems in closed homogeneous waveguide, *IEEE T. Microw. Theory*, **50**, 2513–2518.

Pozar, D.M. (2005). *Microwave Engineering*, (3rd ed.), Wiley, New York: NY.

Ramo, S., Whinnery, J.R. and Duzer, T.V. (1994). *Fields and Waves in Communication Electronics*, (3rd ed.), Wiley, New York: NY.

Rao, S.M. (1999). *Time Domain Electromagnetics*, Academic Press, Orlando: FL.

Silvester, P.P. and Ferrari, R.L. (1996). *Finite Elements for Electrical Engineers*, Cambridge University Press, Cambridge.

Taflove, A. and Hugness, S.C. (2005). *Computational Electrodynamics: The Finite-Difference Time-Domain Method*, Artech House, Norwood: MA.

Tsai, H. P., Wang, Y. X. and Itoh, T. (2002). Efficient analysis of microwave passive structures using 3D envelope-finite-element (EVFE), *IEEE T. Microw. Theor.* **50**, 2721–2727.

Wang, J.S. and Mittra, R. (1994). Finite element analysis of MMIC structures and electronic packages using absorbing boundary conditions, *IEEE T. Microw. Theory*, **42**, 441–449.

Wang, Y.X. and Itoh, T. (2001). Envelope-finite-element (EVFE) technique — a more efficient time domain scheme, *IEEE T. Microw. Theory*, **49**, 2241–2247.

Webb, J.P. and Forghani, B. (1993). Hierarchal scalar and vector tetrahedra, *IEEE T. Magn.*, **29**, 1495–1498.

Weisshaar, A., Mongiardo, M., Tripathi, A. and Tripathi, K. (1996). CAD-oriented full-wave equivalent circuit models for waveguide components and circuits, *IEEE T. Microw. Theory*, **44**, 2564–2570.

Zienkiewicz, O.C. (1977). A new look at the Newmark, Houboult and other time stepping formulas, A weighted residual approach, *Earthq. Eng. Struct. D*, **5**, 413–418.

Index